T0291767

CAMBRIDGE LIBRARY COLLECTION

Books of enduring scholarly value

Physical Sciences

From ancient times, humans have tried to understand the workings of the world around them. The roots of modern physical science go back to the very earliest mechanical devices such as levers and rollers, the mixing of paints and dyes, and the importance of the heavenly bodies in early religious observance and navigation. The physical sciences as we know them today began to emerge as independent academic subjects during the early modern period, in the work of Newton and other 'natural philosophers', and numerous sub-disciplines developed during the centuries that followed. This part of the Cambridge Library Collection is devoted to landmark publications in this area which will be of interest to historians of science concerned with individual scientists, particular discoveries, and advances in scientific method, or with the establishment and development of scientific institutions around the world.

Mathematical and Physical Papers

William Thomson, first Baron Kelvin (1824–1907), is best known for devising the Kelvin scale of absolute temperature and for his work on the first and second laws of thermodynamics, though throughout his 53-year career as a mathematical physicist and engineer at the University of Glasgow he investigated a wide range of scientific questions in areas ranging from geology to transatlantic telegraph cables. The extent of his work is revealed in the six volumes of his *Mathematical and Physical Papers*, published from 1882 until 1911, consisting of articles that appeared in scientific periodicals from 1841 onwards. Volume 1, published in 1882, includes articles from the period 1841–1853 and covers issues relating to heat, especially its linear motion and theories about it. Other topics include aspects of electricity, thermodynamics and research relating to magnetism.

Cambridge University Press has long been a pioneer in the reissuing of out-of-print titles from its own backlist, producing digital reprints of books that are still sought after by scholars and students but could not be reprinted economically using traditional technology. The Cambridge Library Collection extends this activity to a wider range of books which are still of importance to researchers and professionals, either for the source material they contain, or as landmarks in the history of their academic discipline.

Drawing from the world-renowned collections in the Cambridge University Library, and guided by the advice of experts in each subject area, Cambridge University Press is using state-of-the-art scanning machines in its own Printing House to capture the content of each book selected for inclusion. The files are processed to give a consistently clear, crisp image, and the books finished to the high quality standard for which the Press is recognised around the world. The latest print-on-demand technology ensures that the books will remain available indefinitely, and that orders for single or multiple copies can quickly be supplied.

The Cambridge Library Collection will bring back to life books of enduring scholarly value (including out-of-copyright works originally issued by other publishers) across a wide range of disciplines in the humanities and social sciences and in science and technology.

Mathematical and
Physical Papers

VOLUME 1

LORD KELVIN

CAMBRIDGE
UNIVERSITY PRESS

CAMBRIDGE UNIVERSITY PRESS

Cambridge, New York, Melbourne, Madrid, Cape Town,
Singapore, São Paolo, Delhi, Tokyo, Mexico City

Published in the United States of America by Cambridge University Press, New York

www.cambridge.org
Information on this title: www.cambridge.org/9781108028981

© in this compilation Cambridge University Press 2011

This edition first published 1882
This digitally printed version 2011

ISBN 978-1-108-02898-1 Paperback

MATHEMATICAL

AND

PHYSICAL PAPERS.

London: C. J. CLAY, M.A. & SON.
CAMBRIDGE UNIVERSITY PRESS WAREHOUSE,
17, PATERNOSTER ROW.

Cambridge: DEIGHTON, BELL, AND CO.
Leipzig: F. A. BROCKHAUS.

MATHEMATICAL

AND

PHYSICAL PAPERS

BY

SIR WILLIAM THOMSON, LL.D., D.C.L., F.R.S.,

PROFESSOR OF NATURAL PHILOSOPHY IN THE UNIVERSITY OF GLASGOW,
AND FELLOW OF ST PETER'S COLLEGE, CAMBRIDGE.

*COLLECTED FROM DIFFERENT SCIENTIFIC PERIODICALS FROM
MAY, 1841, TO THE PRESENT TIME.*

VOLUME I.

Cambridge:
AT THE UNIVERSITY PRESS.
1882

PREFACE.

In arranging for publication the present collection of my papers on Mathematical and Physical subjects, I have taken the dates of their first publication as a general rule in fixing the order of the various papers.

In certain cases, however, I have considered it advisable to bring together under one article all that I have written on one particular line of research; the most important instances of this being Articles XLVIII., XLIX., and L. Article XLVIII., "On the Dynamical Theory of Heat," and Article L., "On Thermodynamic Motivity," includes all of my papers on this subject, published between the years 1851 and the present time, except the Article "Heat" of the *Encyclopædia Britannica*, published in 1880. Article XLIX., "On the Thermal Effects of Fluids in Motion," consists of a joint series of papers by Dr Joule and myself published at various intervals between 1853 and 1862.

In the text the papers are given as originally published, without even verbal change, and where additions or annotations seemed necessary these are enclosed in square brackets and dated. Corrections, where errors have been found, are also distinctly marked in every case and in most cases dated.

This first volume includes the titles of all my papers published between the years 1841 and 1853, and the text of all of them except those which appeared ten years ago in my volume of Collected Papers on Electrostatics and Magnetism. It will be followed as speedily as possible by other volumes completing the series to the present date.

W. T.

YACHT "LALLA ROOKH,"
ROSHVEN BAY, LOCH AYLORT,
July 21, 1882.

CONTENTS.

MATHEMATICAL AND PHYSICAL PAPERS.

[From the *Cambridge Mathematical Journal*, Vol. II. May, 1841.]

ART. I. ON FOURIER'S EXPANSIONS OF FUNCTIONS IN TRIGONOMETRICAL SERIES.

THE object of the following paper is to endeavour to shew in what cases a function, arbitrary between certain limits, can be developed in a series of cosines, and in what cases in a series of sines. Though the whole subject has been treated very fully by Fourier, yet, as he nowhere directly demonstrates, that a function can be developed in a series of sines or cosines separately, and as, from the want of this direct demonstration, many of his formulæ have been believed to be erroneous, the following paper may be interesting to some readers.

It has been rigorously demonstrated, first, so far as I know, by Fourier and afterwards by Poisson, that

$$\pi fx = \tfrac{1}{2}\int_0^{2\pi} fa\, da + \cos x \int_0^{2\pi} \cos a fa\, da + \ldots\ldots$$
$$+ \sin x \int_0^{2\pi} \sin a fa\, da + \ldots\ldots,$$

where fx is completely arbitrary, between the limits $x = 0$ and $x = 2\pi$. Now, putting the first part of this series equal to $\pi\phi x$, and the second to $\pi\psi x$, we have

$$fx = \phi x + \psi x \ldots\ldots\ldots\ldots(a).$$

But, since

$$\sin nx = - \sin n\,(2\pi - x), \text{ and } \cos nx = \cos n\,(2\pi - x),$$

it follows that

$$\psi x = - \psi\,(2\pi - x), \text{ and } \phi x = \phi\,(2\pi - x),$$

and hence $\quad\quad f\,(2\pi - x) = \phi x - \psi x \ldots\ldots\ldots (b).$

Hence, by this equation, and (a), we have

$$\tfrac{1}{2}\,\{fx + f\,(2\pi - x)\} = \phi x \ldots\ldots\ldots (c)$$

and $\quad\quad\quad \tfrac{1}{2}\,\{fx - f\,(2\pi - x)\} = \psi x \ldots\ldots\ldots (d).$

Now, when x is between 0 and π, fx and $f\,(2\pi - x)$ are perfectly arbitrary, and independent of one another; and therefore it follows that ϕx and ψx are likewise perfectly arbitrary between the same limits. Now

$$\pi\phi x = \tfrac{1}{2}\int_0^{2\pi} fa\,da + \cos x \int_0^{2\pi} \cos \alpha fa\,da$$
$$+ \cos 2x \int_0^{2\pi} \cos 2\alpha fa\,d\alpha + \&c.$$

$$= \tfrac{1}{2}\int_0^{\pi} fa\,da + \cos x \int_0^{\pi} \cos \alpha fa\,da$$
$$+ \cos 2x \int_0^{\pi} \cos 2\alpha fa\,d\alpha + \&c.$$

$$+ \tfrac{1}{2}\int_0^{\pi} f\,(2\pi - \alpha)\,da + \cos x \int_0^{\pi} \cos \alpha f\,(2\pi - \alpha)\,d\alpha$$
$$+ \cos 2x \int_0^{\pi} \cos 2\alpha f\,(2\pi - \alpha)\,d\alpha + \&c.$$

$$= \tfrac{1}{2}\int_0^{\pi} \{f\alpha + f\,(2\pi - \alpha)\}\,d\alpha + \cos x \int_0^{\pi} \cos \alpha\,\{f\alpha + f\,(2\pi - \alpha)\}\,d\alpha$$
$$+ \cos 2x \int_0^{\pi} \cos 2\alpha\,\{f\alpha + f\,(2\pi - \alpha)\}\,d\alpha + \&c.$$

or by (c),

$$\frac{\pi}{2}\,\phi x = \tfrac{1}{2}\int_0^{\pi} \phi a\,da + \cos x \int_0^{\pi} \cos \alpha \phi a\,da$$
$$+ \cos 2x \int_0^{\pi} \cos 2\alpha \phi a\,da + \&c.$$

In a similar manner it may be shewn, that

$$\frac{\pi}{2}\psi x = \sin x \int_0^\pi \sin a\psi a\, da + \sin 2x \int_0^\pi \sin 2a\psi a\, da + \&\text{c.}$$

Hence we see, that any function whatever of x, may be represented between the limits 0 and π, by a series either of cosines or sines of multiples of x. If it be represented by a series of cosines, then the value of the function corresponding to any [other] value of x, will be found from its values between the limits $x = 0$ and $x = \pi$, by the condition,

$$fx = f(2\pi - x):$$

while if it be represented by a series of sines, the condition will be

$$fx = -f(2\pi - x).$$

The same results might have been obtained by taking, in the original equation, fx subject to either of these conditions. If it be subject to the former, $\int_0^{2\pi} fa \sin na\, da$ will vanish, and fx will be expanded in a series of cosines. If it be subject to the second condition $\int_0^{2\pi} fa \cos na\, da$ will vanish, and fx will be expanded in a series of sines. The two series thus obtained will be equal to one another, when x is between 0 and π; but when x is between π and 2π, the value of one will be the negative of the value of the other.

Fourier gives many expansions of functions in series of sines or cosines alone, obtained from the formulæ given above; which, however, he demonstrates merely by assuming fx to be developed in such a series, and then determining the coefficients. Now with regard to these series, Mr Kelland, in his excellent *Treatise on Heat*, remarks, that they are "nearly all erroneous." This remark has probably arisen from finding that they differ from the developments obtained from the general formula for functions which follow the given assumption through their whole periods from 0 to 2π, instead of following it merely between the limits 0 and π, as is the case if sines or cosines alone be used.

Thus, Mr Kelland gives the following expansion of a function which is equal to c, when x is between 0 and α, and to zero, when x is between α and 2π :

$$\pi \frac{\phi x}{c} = \tfrac{1}{2}\alpha + \sin\alpha\cos x + \tfrac{1}{2}\sin 2\alpha\cos 2x + \ldots\ldots$$

$$+ \text{vers }\alpha\sin x + \tfrac{1}{2}\text{ vers }2\alpha\sin 2x + \ldots\ldots ;$$

an expression differing, as he remarks, from Fourier's, "which embraces only the second line of this." Now, as long as x lies between 0 and π, the two series of which this expression is composed, are equal to one another, if we suppose, for the present, $\alpha < \pi$, and their sum is the required function between these limits. When, however, x is between π and 2π, the value of the one series is the negative of the value of the other, as is readily seen from what has been said above. Hence, when x is between π and 2π the value of the expression is nothing. Now Fourier proposes to find an expansion of a function which is equal to c, when x is between 0 and α, and to zero when x is between α and π, without any supposition regarding the values of the function, when x is between π and 2π. Now it is clear, that the double of either of the series in the expansion given by Mr Kelland, will be sufficient for this purpose, and Fourier, *Théorie de la chaleur*, page 244, [Art. 226] chooses to make use of the second; that is, he develops the required function in a series of sines of multiples of x.

The series given by Fourier may be verified in the following manner :

Let $\qquad u = \text{vers }\alpha\sin x + \tfrac{1}{2}\text{ vers }2\alpha\sin 2x + \&\text{c.}$

$$= \sin x + \tfrac{1}{2}\sin 2x + \&\text{c.}$$

$$- \tfrac{1}{2}\left\{\sin(\alpha + x) + \tfrac{1}{2}\sin 2(\alpha + x) + \&\text{c.}\right\}$$

$$+ \tfrac{1}{2}\left\{\sin(\alpha - x) + \tfrac{1}{2}\sin 2(\alpha - x) + \&\text{c.}\right\}.$$

Now it is well known, and it is demonstrated by Mr Kelland, p. 59, that

$$\sin\theta + \tfrac{1}{2}\sin 2\theta + \&\text{c.} = \tfrac{1}{2}(\pi - \theta),$$

which obviously holds for any value of θ between 0 and 2π, and for no others. Hence, x and α being, of course, each less than π, we have, when $x < \alpha$,

$$u = \tfrac{1}{2}\pi.$$

When $x > a$, the last series may be put under the form

$$- \tfrac{1}{2} \{\sin (x - a) + \tfrac{1}{2} \sin 2 (x - a) + \&c.\},$$

and consequently we have

$$u = 0.$$

The values of u are thus found for all values of a and x between 0 and π; and, from them, those for any values of a and x are readily found.

In exactly a similar manner it may be shewn, that the series given by Mr Kelland, p. 64, is really the expansion of a very different function from the one of which the series given by Fourier, p. 246, [Art. 228] is the expansion. Thus, the trapezium represented by Mr Kelland's series occupies the whole space along the axis of x, from 0 to 2π; while, in that space, two trapeziums are represented by Fourier's, one extending from 0 to π, and the other from π to 2π, and inverted, with regard to the axes both of x and y. It may also be remarked, that the form of the trapezium represented by Mr Kelland's series is much more general than that of the trapezium represented by Fourier's.

Fourier's series may be verified in the following manner :

Let $\qquad u = \dfrac{4}{\pi} \left(\sin a \sin x + \dfrac{1}{3^2} \sin 3a \sin 3x + \ldots \right).$

Then $\qquad \dfrac{du}{dx} = \dfrac{4}{\pi} \left(\sin a \cos x + \dfrac{1}{3} \sin 3a \cos 3x + \ldots \right),$

$$= \dfrac{2}{\pi} \left\{ \begin{array}{l} \sin (a + x) + \tfrac{1}{3} \sin 3 (a + x) + \ldots \ldots \\ + \sin (a - x) + \tfrac{1}{3} \sin 3 (a - x) + \ldots \end{array} \right\}$$

Now, if a and x be each less than $\tfrac{1}{2}\pi$, we have (Fourier, p. 237, [Art. 222] or Kelland, p. 65),

$$\dfrac{du}{dx} = 1, \quad \text{when } x < a,$$

and $\qquad \dfrac{du}{dx} = 0, \quad \text{when } x > a.$

[And obviously $u = 0$, and du/dx is finite, when $x = 0$.] Hence we have,

$$u = x, \quad \text{from } x = 0 \text{ to } x = a$$

and $\qquad u = a, \quad \text{from } x = a \text{ to } x = \tfrac{1}{2}\pi.$

Now, since $\sin (2n + 1)\, x = \sin (2n + 1)\, (\pi - x)$, the values of u are the same for x as for $\pi - x$. Hence

$$u = x, \quad \text{from } x = 0 \text{ to } x = a,$$

$$u = a, \quad \text{from } x = a \text{ to } x = \pi - a,$$

and $\qquad\quad u = \pi - x, \quad \text{from } x = \pi - a \text{ to } x = \pi.$

I have examined the other series given by Fourier, on this subject, and they seem to be all correct, with the exception of misprints
and mistakes in transcription, which, unfortunately, are very numerous.   The only one of these mistakes which can cause any serious
embarrassment, is with regard to the series at the top of p. 245 [end
of Art. 226]. The value of this series, between the limits $x = 0$ and
$x = a$, is $\sin \dfrac{\pi x}{a}$, and not $\sin x$, as is there stated.   The same error
occurs in p. 444, where an application of this series is made to
determine the value of the definite integral,

$$\int_0^\infty \frac{dq \sin qx \sin q\pi}{1 - q^2} \, ;$$

but Fourier must have been in possession of the correct value, as it
is that value which he employs in this application.

FRANKFORT, *July*, 1840, AND GLASGOW, *April*, 1841.

[From the *Cambridge Mathematical Journal*, Vol. III. Nov. 1841.]

ART. II.—NOTE ON A PASSAGE IN FOURIER'S HEAT.

In finding the motion of heat in a sphere, Fourier expands a function Fx, arbitrary between the limits $x = 0$ and $x = X$, in a series of the form

$$a_1 \sin n_1 x + a_2 \sin n_2 x + \&c.$$

where n_1, n_2, &c. are the successive roots of the equation

$$\frac{\tan nX}{nX} = 1 - hX.$$

Now Fourier gives no demonstration of the possibility of this expansion, but he merely determines what the coefficients a_1, a_2, &c. would be, if the function were represented by a series of this form. Poisson arrives, by another method, at the same conclusion as Fourier, and then states this objection to Fourier's solution; but, as is remarked by Mr Kelland (*Theory of Heat*, p. 81, note), he "does not appear, as far as I can see, to get over the difficulty." The writer of the following article hopes that the demonstration in it will be considered as satisfactory, and consequently as removing the difficulty.

Let $\qquad n_i X = \epsilon_i, \quad \dfrac{\pi x}{X} = x', \quad$ and $Fx = fx'.$

Then the preceding series will take the form

$$a_1 \sin \frac{\epsilon_1 x}{\pi} + a_2 \sin \frac{\epsilon_2 x}{\pi} + \&c.,$$

the accents being omitted above x.

Now it is shewn by Fourier, that

$$\epsilon_i = \left(\frac{2i-1}{2} - c_i\right)\pi,$$

where c_i is always less than $\frac{1}{2}$, and is equal to 0, when i is infinitely great. Hence the series becomes

$$a_1 \sin\left(\frac{1}{2} - c_1\right)x + a_2 \sin\left(\frac{3}{2} - c_2\right)x + \&c.....(a).$$

Now it is easily shewn, from the fact that any function of x can be represented, between the limits 0 and π, by a series of either sines or cosines of multiples of x, that it may be represented, between the same limits, by a series of the form

$$A \sin\frac{1}{2}x + B \sin\frac{3}{2}x + \&c.$$

Hence each of the quantities

$$\sin\left(\frac{1}{2} - c_1\right)x, \quad \sin\left(\frac{3}{2} - c_2\right)x, \&c.,$$

can be developed in a series of this form. We may consequently assume $\sin\left(\frac{1}{2} - c_1\right)x$ equal i terms of a series of sines of odd multiples of $\frac{1}{2}x$, together with a quantity, $'e_1$; $\sin\left(\frac{3}{2} - c_2\right)x$ equal to i terms of a similar series, together with a quantity $'e_2$; and so with all the terms of the series (a), up to the term

$$\sin\left(\frac{2i-1}{2} - c_i\right)x,$$

which may be assumed equal to i terms, together with a quantity $'e_i$; and it is readily seen, that each of the quantities $'e_1, 'e_2,......'e_i,$ is infinitely small when i is infinitely great. Hence we shall have

$$a_1 \sin\left(\frac{1}{2} - c_1\right)x + a_2 \sin\left(\frac{3}{2} - c_2\right)x + ...+ a_i \sin\left(\frac{2i-1}{2} - c_i\right)x$$

$$= A_1 \sin\frac{1}{2}x + A_2 \sin\frac{3}{2}x + ...+ A_i \sin\frac{2i-1}{2}x$$

$$+ a_1{'e_1} + a_2{'e_2} + ...+ a^i{'e_i},$$

$A_1, A_2,......A_i,$ being known, in terms of $a_1, a_2, ... a_i$. Hence, conversely, any series,

$$A_1 \sin\frac{1}{2}x + A_2 \sin\frac{3}{2}x + ...+ A_i \sin\frac{2i-1}{2}x,$$

where $A_1, A_2, \ldots\ldots A_i$, are arbitrary, may be represented by another series of the form

$$a_1\left\{\sin\left(\frac{1}{2}-c_1\right)x - {}^ie_1\right\} + a_2\left\{\sin\left(\frac{3}{2}-c_2\right)x - {}^ie_2\right\} + \ldots\ldots$$
$$+ a_i\left\{\sin\left(\frac{2i-1}{2}x - c_i\right) - {}^ie_i\right\},$$

where $a_1, a_2, \ldots\ldots a_i$ are determined, in terms of $A_1, A_2, \ldots\ldots A_i$, by i equations, giving the latter quantities in terms of the former.

Let now $i = \infty$; then each of the quantities ${}^ie_1, {}^ie_2, \ldots\ldots {}^ie_i$, will vanish, and it will follow that any series,

$$A_1\sin\frac{1}{2}x + A_2\sin\frac{3}{2}x + \&c.,$$

may be represented by a series of the form

$$a_1\sin\left(\frac{1}{2}-c_1\right)x + a_2\sin\left(\frac{3}{2}-c_2\right)x + \&c.$$

Now any function, fx, can be represented, between the limits $x = 0$ and $x = \pi$, by the former series, and consequently by the latter also, between the same limits. But the latter series is equal to

$$a_1\sin\frac{\epsilon_1 x}{\pi} + a_2\sin\frac{\epsilon_2 x}{\pi} + \&c.;$$

and hence fx can be represented, between the limits 0 and π, by this series; and therefore it follows, that any function, Fx, can be represented, between the limits 0 and X, by the series

$$a_1\sin n_1 x + a_2\sin n_2 x + \&c.$$

GLASGOW, *April*, 1841.

ART. III. ON THE UNIFORM MOTION OF HEAT IN HOMOGENEOUS SOLID BODIES, AND ITS CONNECTION WITH THE MATHEMATICAL THEORY OF ELECTRICITY.

LAMLASH, *Aug.* 1841.

[*Camb. Math. Jour.* Vol. III. Feb. 1842. ELECTROSTATICS AND MAGNETISM, Article I.]

[From the *Cambridge Mathematical Journal*, Vol. III. Nov. 1842.]

ART. IV. ON THE LINEAR MOTION OF HEAT. PART I.

THE differential equation which expresses the linear motion of heat in an infinite solid, is

$$\frac{dv}{dt} = \frac{d^2v}{dx^2},$$

where v is the temperature at the time t, of a point at the distance x from a fixed plane, which, for brevity, may be called the *zero plane*, and the conducting power is taken as unity. Its integral may be put under two forms, one containing an arbitrary function of x, and the other containing two* arbitrary functions of t. I propose to deduce the latter of these solutions from the former, and to shew, so far as possible, the relation which they bear to one another, with regard to the physical problem.

The first of the solutions referred to is

$$\pi^{\frac{1}{2}} v = \int_{-\infty}^{\infty} d\alpha\, \epsilon^{-\alpha^2} f\left(x + 2\alpha t^{\frac{1}{2}}\right) \dots\dots\dots\dots\dots(1).$$

Let $_0v$, v_0, represent the values of v corresponding to $t = 0$, and $x = 0$, respectively. Hence, when $t = 0$, we have

$$\pi^{\frac{1}{2}}\,_0v = \int_{-\infty}^{\infty} d\alpha\, \epsilon^{-\alpha^2} fx,$$

or, since $\quad \displaystyle\int_{-\infty}^{\infty} d\alpha\, \epsilon^{-\alpha^2} = \pi^{\frac{1}{2}}, \quad _0v = fx.$

[* See (14) of Part II.]

Hence fx is the function expressing the initial distribution of heat, which therefore is, as it should be, quite arbitrary, and sufficient for determining all the succeeding distributions of the temperature. If, however, the varying temperature of any plane, as for instance the zero plane, be subject to any condition, it is obvious that the initial distribution will cease to be altogether arbitrary, as it alone is sufficient to determine the temperatures at all future times. If, however, the initial distribution be given on the positive side of the zero plane, it is clear that a certain initial distribution on the negative side will enable us to subject the variation of the temperature of the zero plane to any condition we please. By applying this principle, we can determine, in the following manner, the variable temperature of any point in the cases; first, when, the initial distribution on the positive side being given, the temperature of the zero plane is a given function of the time; and secondly, when the part of the solid on the negative side is removed, and the given initial distribution of temperature on the remaining part is dissipated by radiation across the zero plane, into a medium of constant or varying temperature.

No. 1.   Let the conditions be

$$_0v = \phi x \text{ when } x > 0 \dots\dots\dots\dots\dots(2),$$
$$v_0 = \xi t \dots\dots\dots\dots\dots\dots\dots\dots(3).$$

[ξt denoting any given function of t.]

Let ψx be the distribution on the negative side, necessary to produce (3), when ϕx is the distribution on the positive side. Hence, when $x > 0$, $fx = \phi x$, and when $x < 0$, $fx = \psi x$, and therefore (1) becomes

$$\pi^{\frac{1}{2}}v = \int_{-\frac{x}{2t^{\frac{1}{2}}}}^{\infty} d\alpha\,\epsilon^{-\alpha^2}\phi\left(x + 2\alpha t^{\frac{1}{2}}\right) + \int_{-\infty}^{-\frac{x}{2t^{\frac{1}{2}}}} d\alpha\,\epsilon^{-\alpha^2}\psi\left(x + 2\alpha t^{\frac{1}{2}}\right)\dots(4),$$

and (3) gives

$$\pi^{\frac{1}{2}}\xi t = \int_0^{\infty} d\alpha\,\epsilon^{-\alpha^2}\phi\left(2\alpha t^{\frac{1}{2}}\right) + \int_{-\infty}^0 d\alpha\,\epsilon^{-\alpha^2}\psi\left(2\alpha t^{\frac{1}{2}}\right)$$

$$= \int_0^{\infty} d\alpha\,\epsilon^{-\alpha^2}\{\phi\left(2\alpha t^{\frac{1}{2}}\right) + \psi\left(-2\alpha t^{\frac{1}{2}}\right)\}.$$

Hence we must find a function F, such that

$$\pi^{\frac{1}{2}}\xi t = \int_0^{\infty} d\alpha\,\epsilon^{-\alpha^2} F\left(2\alpha t^{\frac{1}{2}}\right)\dots\dots\dots\dots(5);$$

and, when this is done, we have, for determining ψ,

$$\psi(-x) = Fx - \phi x \dots\dots\dots\dots\dots\dots(6),$$

[where x is positive.]

To determine F, let, in the first place, ξt be a periodical function of t, and let

$$\xi t = \Sigma \left(A_i \cos \frac{2i\pi t}{p} + B_i \sin \frac{2i\pi t}{p} \right) \dots\dots\dots\dots(7).$$

Then, by taking $p = \infty$, any unperiodical function may, by Fourier's theorem, be represented in this form. Hence the problem is reduced to that of representing terms of the form $\cos\dfrac{2i\pi t}{p}$, or $\sin\dfrac{2i\pi t}{p}$, by the definite integral $\displaystyle\int_0^\infty d\alpha\, \epsilon^{-\alpha^2} F(2\alpha t^{\frac{1}{2}})$. To effect this, let $q = \alpha + \sqrt{\{\pm 2mt \sqrt{(-1)}\}}$, in the first member of the equation

$$\int_{-\infty}^\infty dq\, \epsilon^{-q^2} = \pi^{\frac{1}{2}}.$$

Then, dividing by $\epsilon^{\mp 2mt\sqrt{(-1)}}$, we have

$$\int_{-\infty}^\infty d\alpha\, \epsilon^{-\alpha^2}\, \epsilon^{-2\alpha\sqrt{(mt)}\,[1\pm\sqrt{(-1)}]} = \pi^{\frac{1}{2}}\epsilon^{\pm 2mt\sqrt{(-1)}}.$$

Hence, by addition and subtraction,

$$\mp \int_{-\infty}^\infty d\alpha\, \epsilon^{-\alpha^2} \epsilon^{-2\alpha\sqrt{(mt)}}\, \genfrac{}{}{0pt}{}{\sin}{\cos}\{2\alpha\sqrt{(mt)}\} = \pi^{\frac{1}{2}}\genfrac{}{}{0pt}{}{\sin}{\cos}(2mt)\dots.(a),$$

the upper sign being taken along with the sines, and the lower with the cosines. Changing the sign of $\sqrt{(mt)}$, we have

$$\int_{-\infty}^\infty d\alpha\, \epsilon^{-\alpha^2}\epsilon^{2\alpha\sqrt{(mt)}}\,\genfrac{}{}{0pt}{}{\sin}{\cos}\{2\alpha\sqrt{(mt)}\} = \pi^{\frac{1}{2}}\genfrac{}{}{0pt}{}{\sin}{\cos}(2mt)\dots\dots(b).$$

Hence, by addition,

$$\int_{-\infty}^\infty d\alpha\, \epsilon^{-\alpha^2}\{\epsilon^{2\alpha\sqrt{(mt)}} \mp \epsilon^{-2\alpha\sqrt{(mt)}}\}\genfrac{}{}{0pt}{}{\sin}{\cos}\{2\alpha\sqrt{(mt)}\} = 2\pi^{\frac{1}{2}}\genfrac{}{}{0pt}{}{\sin}{\cos}(2mt);$$

or, since the multiplier of $d\alpha$ remains the same when α is changed into $-\alpha$,

$$\int_0^\infty d\alpha\, \epsilon^{-\alpha^2}\{\epsilon^{2\alpha\sqrt{(mt)}} \mp \epsilon^{-2\alpha\sqrt{(mt)}}\}\genfrac{}{}{0pt}{}{\sin}{\cos}\{2\alpha\sqrt{(mt)}\} = \pi^{\frac{1}{2}}\genfrac{}{}{0pt}{}{\sin}{\cos}(2mt)\dots(c).$$

Hence, if we put $m = \dfrac{i\pi}{p}$, we have

$$\int_0^\infty d\alpha\, \epsilon^{-\alpha^2} \Sigma \left\{ A_i \left(\epsilon^{2a\sqrt{\frac{i\pi t}{p}}} + \epsilon^{-2a\sqrt{\frac{i\pi t}{p}}} \right) \cos\left(2\alpha \sqrt{\frac{i\pi t}{p}} \right) \right.$$

$$\left. + B_i \left(\epsilon^{2a\sqrt{\frac{i\pi t}{p}}} - \epsilon^{-2a\sqrt{\frac{i\pi t}{p}}} \right) \sin\left(2a \sqrt{\frac{i\pi t}{p}} \right) \right\}$$

$$= \pi^{\frac{1}{2}} \Sigma \left(A_i \cos\frac{2i\pi t}{p} + B_i \sin\frac{2i\pi t}{p} \right) ;$$

and therefore we have, for the form of the function F,

$$Fx = \Sigma \left\{ A_i \left(\epsilon^{x\sqrt{\frac{i\pi}{p}}} + \epsilon^{-x\sqrt{\frac{i\pi}{p}}} \right) \cos\left(x \sqrt{\frac{i\pi}{p}} \right) \right.$$

$$\left. + B_i \left(\epsilon^{x\sqrt{\frac{i\pi}{p}}} - \epsilon^{-x\sqrt{\frac{i\pi}{p}}} \right) \sin\left(x \sqrt{\frac{i\pi}{p}} \right) \dots\dots(8).$$

[Remark that $F(x) = F(-x)$. And, (another affair) consider the intensity of the convergence of (7) required to make (8) convergent for all values of x less than any given quantity a. Compare Art. XI. below.]

Now, to satisfy (7),

$$A_i = \frac{2}{p} \int_0^p dt'\, \xi t' \cos\frac{2i\pi t'}{p}, \text{ when } i > 0,$$

$$A_0 = \frac{1}{p} \int_0^p dt'\, \xi t',$$

$$B_i = \frac{2}{p} \int_0^p dt'\, \xi t' \sin\frac{2i\pi t'}{p}$$

Hence, the expression for Fx becomes

$$pFx = \int_0^p dt'\, \xi t' \left(\epsilon^0 + \epsilon^{-0} \right)$$

$$+ 2\sum_1^\infty \int_0^p dt'\, \xi t' \left[\epsilon^{x\sqrt{\frac{i\pi}{p}}} \cos\left\{ \sqrt{\frac{i\pi}{p}} \left(x - 2t' \sqrt{\frac{i\pi}{p}} \right) \right\} \right.$$

$$\left. + \epsilon^{-x\sqrt{\frac{i\pi}{p}}} \cos\left\{ \sqrt{\frac{i\pi}{p}} \left(x + 2t' \sqrt{\frac{i\pi}{p}} \right) \right\} \right] ;$$

or, since $(\epsilon^z + \epsilon^{-z}) \cos z = \left\{ \epsilon^{z\sqrt{(-1)}} + \epsilon^{-z\sqrt{(-1)}} \right\} \cos\left\{ z\sqrt{(-1)} \right\}$,

and $\quad (\epsilon^z - \epsilon^{-z}) \sin z = -\left\{ \epsilon^{z\sqrt{(-1)}} - \epsilon^{-z\sqrt{(-1)}} \right\} \sin\left\{ z\sqrt{(-1)} \right\}$, and therefore each term of the series remains the same when i is changed into $-i$,

$$pFx = \sum_{-\infty}^{\infty} \int_0^p dt' \xi t' \left[\epsilon^{x\sqrt{\frac{i\pi}{p}}} \cos \left\{ \sqrt{\frac{i\pi}{p}} \left(x - 2t' \sqrt{\frac{i\pi}{p}} \right) \right\} \right.$$

$$\left. + \epsilon^{-x\sqrt{\frac{i\pi}{p}}} \cos \left\{ \sqrt{\frac{i\pi}{p}} \left(x + 2t' \sqrt{\frac{i\pi}{p}} \right) \right\} \right] \ldots\ldots(9).$$

If ξt be not periodical, let $p = \infty$. Then, changing the limits of t' to $-\frac{1}{2}p$ and $\frac{1}{2}p$, instead of 0 and p, and putting $\frac{i\pi}{p} = m$, $\frac{\pi}{p} = dm$, we have

$$\pi Fx = \int_{-\infty}^{\infty} dm \int_{-\infty}^{\infty} dt' \xi t' \left[\epsilon^{m^{\frac{1}{2}}x} \cos \{ m^{\frac{1}{2}} (x - 2t'm^{\frac{1}{2}}) \} \right.$$

$$\left. + \epsilon^{-m^{\frac{1}{2}}x} \cos \{ m^{\frac{1}{2}} (x + 2t'm^{\frac{1}{2}}) \} \right] \ldots\ldots(10).$$

Hence, if we determine F from (9) or (10), and ψ from (6), the solution of the problem is found by using this result in (4).

Equation (6) shews that $\psi (x)$, the initial distribution on the negative side of the zero plane, is composed of two parts, $- \phi (-x)$ and $F (-x)$. The first of these, together with ϕx, on the positive side, would obviously have the effect of retaining the temperature of the zero plane at zero. But, in addition to them, there is the distribution $F(-x)$, on the negative side, which is so determined from (9) or (10), that it alone would have the effect of making the subsequent temperature of the zero plane be ξt. Hence, since the result of the two initial distributions coexisting is equal to the sum of the results in the cases in which they exist separately, it follows that, on the whole, the varying temperature of the zero plane is ξt. Hence we see how it is that, without altering the initial distribution on the positive side, the initial temperature on the negative side may be so distributed as to make the temperature of the zero plane be ξt.

From (9) and (5), and from (10) and (5), we have the following theorems :

$$p\pi^{\frac{3}{2}}\xi t = \overset{\infty}{\underset{-\infty}{\Sigma}}\int_0^p dt'\,\xi t'\int_0^\infty d\alpha\,\epsilon^{-\alpha^2}\left[\epsilon^{2\alpha\sqrt{\frac{i\pi t}{p}}}\cos\left\{2\sqrt{\frac{i\pi}{p}}\left(\alpha t^{\frac{1}{2}}-t'\sqrt{\frac{i\pi}{p}}\right)\right\}\right.$$
$$\left. +\,\epsilon^{-2\alpha\sqrt{\frac{i\pi t}{p}}}\cos\left\{2\sqrt{\frac{i\pi}{p}}\left(\alpha t^{\frac{1}{2}}+t'\sqrt{\frac{i\pi}{p}}\right)\right\}\right]$$
$$\pi^{\frac{3}{2}}\xi t = \int_{-\infty}^\infty dm \int_{-\infty}^\infty dt'\,\xi t'\int_0^\infty d\alpha\,\epsilon^{-\alpha^2}\left[\epsilon^{2\alpha\,\sqrt{(mt)}}\cos\left\{2m^{\frac{1}{2}}\left(\alpha t^{\frac{1}{2}}-t'm^{\frac{1}{2}}\right)\right\}\right.$$
$$\left. +\,\epsilon^{-2\alpha\,\sqrt{(mt)}}\cos\left\{2m^{\frac{1}{2}}\left(\alpha t^{\frac{1}{2}}+t'm^{\frac{1}{2}}\right)\right\}\right]$$

$$\tag{11}$$

the first or second being used according as ξt is or is not periodical.

These theorems obviously hold when t is negative as well as when it is positive. Hence we have found the distribution on the negative side of the zero plane, which not only produces in every succeeding time the given temperature of the zero plane, but would also follow if, for negative values of t, the temperature had been the same function of these negative values. In general, however, the temperature of any plane except the zero plane, as given by (4), will be impossible for negative values of t, since, except on a particular assumption with respect to ϕx, or the value of $_0 v$, when x is positive, the initial distribution, represented by ψx and ϕx, is not of such a form as to be any stage, except the first, in a system of varying possible temperatures, or is not producible by any previous possible distribution. Thus, if $_0 v = 0$ when x is positive, and $_0 v = F(-x)$ when x is negative, the state represented cannot be the result of any *possible* distribution of temperature which has previously existed, though if in (4) we put $\phi x = 0$, and give t a negative value, we find a distribution, probably impossible except when $x = 0$, which will produce the distribution $_0 v$, when $t = 0$. [Compare Article XI. below.]

KNOCK, *May*, 1842.

[From the *Cambridge Mathematical Journal*, Vol. III. Feb. 1843.]

Art. V.—On the Linear Motion of Heat. Part II.

[Time-periodic Solution examined.]

Let us now endeavour to find the general form of $_0 v$, for positive and negative values of x, which is producible by any distribution of heat, an infinite time previously, or which is the same, to find the form of the function f, which renders v, or $\pi^{-\frac{1}{2}} \int_{-\infty}^{\infty} dx\, \epsilon^{-a^2} f\,(x + 2at^{\frac{1}{2}})$, possible for all values of x, and for all values of t, back to $-\infty$. [For right solution of this problem see Art. XI. below.]

If v be possible for all values of t, it may be represented by

$$\Sigma \left(P_i \cos \frac{2i\pi t}{p} + Q_i \sin \frac{2i\pi t}{p} \right) \ldots\ldots\ldots\ldots(d),$$

where P_i and Q_i are functions of x, which it is our object to determine. [For explanation and correction of this statement, see note added at the end of the Article. Several passages from the original Article involving imperfect or erroneous statements regarding the problem of which the right solution is given in Article XI. below are now deleted; and the rest of Article V., given here as follows, may be regarded as merely an examination of the time-periodic solution.]

Modifying (a) and (b), so that the multiplier of $d\alpha \epsilon^{-a^2}$, in the first members, may be of the form $f(x + 2at^{\frac{1}{2}})$, and the second members of the forms $P\cos(2mt)$, $Q\sin(2mt)$, and putting $m = \dfrac{i\pi}{p}$, we have

$$\int_{-\infty}^{\infty} d\alpha\, \epsilon^{-a^2}\, \epsilon^{-(x+2at^{\frac{1}{2}})\sqrt{\frac{i\pi}{p}}} \begin{matrix}\sin \\ \cos\end{matrix} \left\{ \sqrt{\frac{i\pi}{p}}\,(x + 2at^{\frac{1}{2}}) \right\}$$

$$= \pi^{\frac{1}{2}} \epsilon^{-x\sqrt{\frac{i\pi}{p}}} \begin{matrix}\sin \\ \cos\end{matrix} \left(x\sqrt{\frac{i\pi}{p}} - \frac{2i\pi t}{p} \right),$$

$$\int_{-\infty}^{\infty} d\alpha\, \epsilon^{-a^2}\, \epsilon^{(x+2at^{\frac{1}{2}})\sqrt{\frac{i\pi}{p}}} \begin{matrix}\sin \\ \cos\end{matrix} \left\{ \sqrt{\frac{i\pi}{p}}\,(x + 2at^{\frac{1}{2}}) \right\}$$

$$= \pi^{\frac{1}{2}} \epsilon^{x\sqrt{\frac{i\pi}{p}}} \begin{matrix}\sin \\ \cos\end{matrix} \left(x\sqrt{\frac{i\pi}{p}} + \frac{2i\pi t}{p} \right).$$

Hence we see that the most general expression for v, when it is of the form (d), is

$$v = \sum_0^\infty \left[\epsilon^{x\sqrt{\frac{i\pi}{p}}} \left\{ A_i \cos\left(x\sqrt{\frac{i\pi}{p}} + \frac{2i\pi t}{p} \right) + B_i \sin\left(x\sqrt{\frac{i\pi}{p}} + \frac{2i\pi t}{p} \right) \right\} \right.$$
$$\left. + \epsilon^{-x\sqrt{\frac{i\pi}{p}}} \left\{ A_i' \cos\left(-x\sqrt{\frac{i\pi}{p}} + \frac{2i\pi t}{p} \right) + B_i' \sin\left(-x\sqrt{\frac{i\pi}{p}} + \frac{2i\pi t}{p} \right) \right\} \right] (12).$$

Putting $t = 0$, we find

$$_0 v = \sum_0^\infty \left\{ \epsilon^{x\sqrt{\frac{i\pi}{p}}} \left(A_i \cos x\sqrt{\frac{i\pi}{p}} + B_i \sin x\sqrt{\frac{i\pi}{p}} \right) \right.$$
$$\left. + \epsilon^{-x\sqrt{\frac{i\pi}{p}}} \left(A_i' \cos x\sqrt{\frac{i\pi}{p}} - B_i' \sin x\sqrt{\frac{i\pi}{p}} \right) \right\} \dots\dots(13).$$

This expression consists of two independent parts, one containing $\epsilon^{x\sqrt{\frac{i\pi}{p}}}$ as a factor in each term, and the other $\epsilon^{-x\sqrt{\frac{i\pi}{p}}}$. By examining (12), we see that the former of these gives rise to a series of *waves* of heat, proceeding in the negative direction; and the latter to a series of waves proceeding in the positive direction; and that while a wave in the former system moves from

$x = \infty$ to $x = -\infty$, and a wave in the latter from $x = -\infty$ to $x = \infty$, its *amplitude* diminishes from ∞ to 0.

If, in (12), $f_1 t$ and $f_2 t$ be the parts of v_0 arising from the series, of which the coefficients are A_i, B_i, and A_i', B_i', the value of v becomes

$$
\left.
\begin{aligned}
v = \frac{1}{p} &\sum_{-\infty}^{\infty} \left[\epsilon^{x\sqrt{\frac{i\pi}{p}}} \int_{-\frac{1}{2}p}^{\frac{1}{2}p} at' f_1 t' \cos\left\{ x\sqrt{\frac{i\pi}{p}} + \frac{2i\pi t}{p}(t - t') \right\} \right. \\
&\left. + \epsilon^{-x\sqrt{\frac{i\pi}{p}}} \int_{-\frac{1}{2}p}^{\frac{1}{2}p} dt' f_2 t' \cos\left\{ x\sqrt{\frac{i\pi}{p}} - \frac{2i\pi t}{p}(t - t') \right\} \right] \\
\text{or, when } & p = \infty, \\
\pi v = \int_{-\infty}^{\infty} d\beta &\int_{-\infty}^{\infty} dt' \left[\epsilon^{\beta^{\frac{1}{2}}x} f_1 t' \cos\{ \beta^{\frac{1}{2}}x + 2\beta (t - t') \} \right. \\
&\left. + \epsilon^{-\beta^{\frac{1}{2}}x} f_2 t' \cos\{ \beta^{\frac{1}{2}}x - 2\beta (t - t') \} \right].
\end{aligned}
\right\} \quad \ldots\ldots(14).
$$

From the latter of these forms, that given by Fourier (*Théorie de la Chaleur*, p. 544) may be readily deduced, and by putting $f_1 t' = 0$ in the former, we have the solution (given by Kelland, in his *Treatise on Heat*, p. 127) which Fourier employed to express the diurnal and annual variations in the temperature of the earth at small depths. It is obviously suited to the case in which the temperature of the body, below the surface, is naturally constant, and all the periodical variations are produced by external causes, and proceed downwards, from the surface.

No. 2. Let the body be supposed to be terminated by the zero plane, and to radiate heat across it, according to Newton's law; and let the external temperature be a given function, ξt, of the time. To find the state of the temperature of the body, after any time has elapsed, the initial distribution in the body, or on the positive side of the zero plane, being ϕx.

Let the medium into which the surface radiates be supposed to be removed, and, instead of it, let the body extend infinitely on the negative side. The first thing to be done is to find the distribution on the negative side which will exactly supply the place of the radiation. The conditions which this must be chosen to satisfy, are

$$\left(\frac{dv}{dx}\right)_0 = h\,(v_0 - \xi t),$$
$$_0 v = \phi x, \text{ when } x \text{ is positive,}$$
$$\left.\right\}\dots\dots\dots\dots(a),$$

and

where h is the radiating power of the surface. If, in addition to the latter of these equations, we assume ψx to be the required initial distribution on the negative side, the variable temperature, v, of any point, will be given by (4).

Differentiating this equation, and putting $x = 0$ in the result, we have

$$\pi^{\frac{1}{2}}\left(\frac{dv}{dx}\right)_0 = \int_0^\infty d\alpha\,\epsilon^{-\alpha^2}\phi'\,(2\alpha t^{\frac{1}{2}}) + \int_{-\infty}^0 d\alpha\,\epsilon^{-\alpha^2}\psi'\,(2\alpha t^{\frac{1}{2}}) + \frac{\phi 0 - \psi 0}{2t^{\frac{1}{2}}}$$

Now $\psi 0$ must be $= \phi 0$, as otherwise, at the commencement of the variation, the radiation would be infinite. Hence we have, from (a),

$$\int_0^\infty d\alpha\,\epsilon^{-\alpha^2}\phi'\,(2\alpha t^{\frac{1}{2}}) + \int_{-\infty}^0 d\alpha\,\epsilon^{-\alpha^2}\psi'\,(2\alpha t^{\frac{1}{2}})$$

$$= h\left\{\int_0^\infty d\alpha\,\epsilon^{-\alpha^2}\phi\,(2\alpha t^{\frac{1}{2}}) + \int_{-\infty}^0 d\alpha\,\epsilon^{-\alpha^2}\psi\,(2\alpha t^{\frac{1}{2}}) - \pi^{\frac{1}{2}}\xi t\right\},$$

or

$$\int_0^\infty d\alpha\,\epsilon^{-\alpha^2}[\phi'\,(2\alpha t^{\frac{1}{2}}) + \psi'\,(-2\alpha t^{\frac{1}{2}}) - h\,\{\phi\,(2\alpha t^{\frac{1}{2}}) + \psi\,(-2\alpha t^{\frac{1}{2}})\}] = -\pi^{\frac{1}{2}}h\xi t.$$

Hence, if

$$\xi t = \Sigma\left(A_i \cos\frac{2i\pi t}{p} + B_i \sin\frac{2i\pi t}{p}\right)\dots\dots\dots\dots(b),$$

and if Fx be determined by (8), then, using x instead of $2\alpha t^{\frac{1}{2}}$, we must have

$$\phi'x + \psi'\,(-x) - h\,\{\phi x + \psi\,(-x)\} = -hFx\dots\dots\dots(c),$$

or

$$\frac{d\psi\,(-x)}{-dx} - h\psi\,(-x) = h\,(\phi x - Fx) - \frac{d\phi x}{dx};$$

therefore $\quad \psi(-x) = -\epsilon^{-hx} \int \epsilon^{hx} \left\{ h(\phi x - Fx) - \dfrac{d\phi x}{dx} \right\} dx$ $\left.\vphantom{\int} \right\}$ .....(15).

or $\qquad \psi(-x) = \phi x - h\epsilon^{-hx} \int \epsilon^{hx} (2\phi x - Fx)\, dx,$

The function ψ being determined from this, the solution of the problem is found by using the result in (4).

After the motion has continued for a long time, the irregularities of the initial distribution disappear, and the variations of the temperature of the body are reduced, by the periodical variations of the external temperature, to a permanently periodical state. Let us suppose that this permanent state has been reached when $t = 0$. That this may be the case, we must choose ϕx of such a form, that ψx when x is negative, and ϕx when x is positive, may make $_0 v$ be of the form shewn in the second line of (13). Let us therefore assume

when x is negative $\psi x =$
when x is positive $\phi x =$ $\left.\vphantom{\begin{matrix}a\\b\end{matrix}}\right\}$ $\Sigma \epsilon^{-x\sqrt{\frac{i\pi}{p}}} \left(a_i \cos x \sqrt{\dfrac{i\pi}{p}} - b_i \sin x \sqrt{\dfrac{i\pi}{p}} \right),$

where a_i and b_i are to be determined so as to satisfy (c).

Using these values of ψx and ϕx in (c), and using for Fx its value (8), we have, by equating the coefficients of

$$\left(\epsilon^{x\sqrt{\frac{i\pi}{p}}} + \epsilon^{-x\sqrt{\frac{i\pi}{p}}}\right) \cos x \sqrt{\dfrac{i\pi}{p}}, \text{ and } \left(\epsilon^{x\sqrt{\frac{i\pi}{p}}} - \epsilon^{-x\sqrt{\frac{i\pi}{p}}}\right) \sin x \sqrt{\dfrac{i\pi}{p}},$$

in the two members of the resulting equation,

$$a_i \left(h + \sqrt{\dfrac{i\pi}{p}} \right) + b_i \sqrt{\dfrac{i\pi}{p}} = hA_i,$$

$$a_i \sqrt{\dfrac{i\pi}{p}} - b_i \left(\sqrt{\dfrac{i\pi}{p}} + h \right) = -hB_i;$$

whence $\qquad a_i = \dfrac{h\left\{ A_i \left(\sqrt{\frac{i\pi}{p}} + h \right) - B_i \sqrt{\frac{i\pi}{p}} \right\}}{h^2 + 2h\sqrt{\frac{i\pi}{p}} + 2\frac{i\pi}{p}}$ $\left.\vphantom{\begin{matrix}a\\a\\b\\b\end{matrix}}\right\}$ ....(16).

$$b_i = \dfrac{h\left\{ A_i \sqrt{\frac{i\pi}{p}} + B_i \left(\sqrt{\frac{i\pi}{p}} + h \right) \right\}}{h^2 + 2h\sqrt{\frac{i\pi}{p}} + 2\frac{i\pi}{p}}$$

If, in (12), we use these values of a_i and b_i instead of A_i' and B_i', and put A_i and B_i each equal to zero, the resulting expression is the variable temperature of any point. Let, for brevity,

$$D_i = \sqrt{\left(h^2 + 2h \sqrt{\frac{i\pi}{p}} + 2\frac{i\pi}{p} \right)},$$

$$\cos \delta_i = \frac{\sqrt{\dfrac{i\pi}{p}} + h}{D_i}, \qquad \sin \delta_i = \frac{\sqrt{\dfrac{i\pi}{p}}}{D_i}.$$

Hence, using for A_i and B_i, their values, which satisfy (b), we have

$$pv = h\sum_{-\infty}^{\infty} \epsilon^{-x\sqrt{\frac{i\pi}{p}}} \int_{-\frac{1}{2}p}^{\frac{1}{2}p} dt' \, \xi t' \, \frac{\cos\left\{ x\sqrt{\frac{i\pi}{p} - \frac{2i\pi}{p}} (t - t') + \delta_i \right\}}{D_i} \dots\dots(17),$$

which agrees with the expression given by Poisson, in p. 431 of his *Théorie de la Chaleur*.

[NOTE ON ART. V. Added Jan. 27, 1881.—To the statement regarding formula (d) above, the condition *that the series must be convergent* ought to have been added. It was not till about a year after this Article was written that I found the true principle (given in Article XI. below) for answering the question as to the age of a thermal distribution with which it opens. It is also to be remarked that the deduction of the non-periodic from periodic functions of t, fundamentally used in Article V., is, though valid, by no means convenient for the special question with which this Article commences; and that the less analytical and more direct treatment of Article XI., which I applied when I was a year older, and published in 1844 (Art. XI. below), is much more suitable and clearer.]

ART. VI. PROPOSITIONS IN THE THEORY OF ATTRACTION.
(ELECTROSTATICS AND MAGNETISM. XII.)

ART. VII. ON THE ATTRACTION OF CONDUCTING AND NON-CONDUCTING ELECTRIFIED BODIES.
(ELECTROSTATICS AND MAGNETISM. VII.)

[From the *Cambridge Mathematical Journal*, Vol. III. May, 1843.]

ART. VIII.   NOTE ON ORTHOGONAL ISOTHERMAL SURFACES.

THE object of this article is to state an important question relative to Orthogonal Isothermal Surfaces, which does not seem to have been yet considered generally.

It is known that if a system of surfaces be given, there are two, and only two, other systems cutting one another and the first system at right angles.   Lamé has proved many general properties of such conjugate systems of surfaces, and has also considered particularly the case in which each of the three systems is a series of isothermal surfaces.   He has nowhere however proved the possibility of the existence of three such systems of surfaces in general, or, in other words, he has not shewn that if one system be isothermal, the two others which are determined by it are isothermal also.   There is however a considerable probability that this proposition is generally true*.   For, in the first place, it is true, as will be shewn below, of any system whatever of isothermal *cylindrical* surfaces, and it is also true of all isothermal surfaces of the second order, whether cylindrical or not.   As these two cases, the only ones which have as yet been discussed, are very distinct in their nature, they afford a considerable presumptive evidence of the truth of the general proposition.

The case of isothermal surfaces of the second order has been fully discussed by Lamé, in a paper on Isothermal Surfaces in Liouville's *Journal* [for 1837] (Vol. II. p. 147), where he has shewn that all *confocal* surfaces of the second order are isothermal.   Now the two systems of orthogonal surfaces conjugate to a given series of confocal surfaces of the second order are also confocal surfaces of the second order, and therefore they are likewise isothermal. Hence the proposition is true for isothermal surfaces of the second order.

---

* [It is proved to be *not* generally true, at the end of Art. IX. below.]

It may be proved in the following manner to be true for a series of isothermal cylindrical surfaces.

Of the two systems of orthogonal surfaces, conjugate to the given series of cylindrical surfaces, one is a series of cylindrical surfaces cutting them at right angles, and having their generating lines parallel to those of the first system, and the other is a series of planes perpendicular to the cylinders. Now a series of parallel planes is obviously a system of isothermal surfaces, and it therefore only remains to be proved that, if one series of cylindrical surfaces be isothermal, the series which cuts them at right angles will be isothermal also.

To prove this, let us suppose any two surfaces of the first system to be retained at uniform temperatures, and let v be the temperature of any point (xy) between them. The equation of any one of the isothermal surfaces will be

$$v = \alpha \quad \dotfill (1),$$

which is therefore the general equation, comprehending all surfaces of the first series.

By the equation of equilibrium of heat, moving in two directions, we must have

$$\frac{d^2v}{dx^2} + \frac{d^2v}{dy^2} = 0 \quad \dotfill (2).$$

Now let
$$v_1 = \alpha_1 \dotfill (3)$$

be the general equation of the system of cylindrical surfaces, which cuts the first system at right angles.

Hence
$$\frac{dv}{dx}\frac{dv_1}{dx} + \frac{dv}{dy}\frac{dv_1}{dy} = 0 \quad \dotfill (4).$$

Let
$$\frac{dv_1}{dx} = k\frac{dv}{dy} \dotfill (a),$$

therefore
$$\frac{dv_1}{dy} = -k\frac{dv}{dx} \dotfill (b).$$

In these equations k is arbitrary, with the exception that it must be so chosen that $\frac{dv_1}{dx}dx + \frac{dv_1}{dy}dy$ shall be a complete differential. Hence we must have

$$\frac{d}{dy}(a) - \frac{d}{dx}(b) = 0,$$

and therefore

$$k\left(\frac{d^2v}{dx^2} + \frac{d^2v}{dy^2}\right) + \frac{dk}{dx}\frac{dv}{dx} + \frac{dk}{dy}\frac{dv}{dy} = 0,$$

or, on account of (2),

$$\frac{dk}{dx}\frac{dv}{dx} + \frac{dk}{dy}\frac{dv}{dy} = 0.$$

This equation is satisfied if $k = 1$, and therefore we may use this value for k in (a) and (b). Hence

$$\frac{dv_1}{dx} = \frac{dv}{dy} \quad\dots\dots\dots\dots\dots\dots\dots\dots(c),$$

$$\frac{dv_1}{dy} = -\frac{dv}{dx} \quad\dots\dots\dots\dots\dots\dots\dots(d).$$

These equations determine a form of v_1 which satisfies equation (4), the only condition to which v_1 is subject, in order that $v_1 = \alpha_1$ may be the equation to the series of cylinders cutting the first series at right angles.

Now

$$\frac{d}{dx}(c) + \frac{d}{dy}(d)$$

gives

$$\frac{d^2v_1}{dx^2} + \frac{d^2v_1}{dy^2} = 0 \dots\dots\dots\dots\dots\dots(5).$$

Hence a function v_1 of x and y, which satisfies (5), may be found such that $v_1 = \alpha_1$ represents the series of cylindrical surfaces cutting the first series at right angles. Hence the second system is isothermal. [For a remarkable and interesting extension of this theorem, and of a proper theorem of integrability of lines of force or of stream lines, to conical isothermal surfaces, and to cases of fluid motion in which the stream lines lie on concentric spherical surfaces, see Art. XII. below.]

The theorem which has just been proved, was first given by Lamé in a digression on Orthogonal Surfaces, contained in a paper entitled, "Mémoire sur les Lois de l'Équilibre du Fluide Ethéré," in the *Journal de l'École Polytechnique* (Vol. III. cahier XXIII). His proof however is deduced from some general properties of orthogonal surfaces which he has been discussing, and in a subsequent paper on the motion of heat in cylindrical bodies (Liouville's *Journal*, Vol. I. [1836]), he has not noticed the proposition at all.

CAMBRIDGE, *April*, 1843.

[From the *Cambridge Mathematical Journal*, Vol. IV. Nov. 1843.]

ART. IX.   ON THE EQUATIONS OF THE MOTION OF HEAT REFERRED TO CURVILINEAR CO-ORDINATES.

LET x, y, z be the rectangular co-ordinates of any point in space, and let λ, λ_1, λ_2 be any functions of x, y, z, such that

$$\lambda = a, \quad \lambda_1 = a_1, \quad \lambda_2 = a_2. \ldots\ldots\ldots\ldots\ldots(1)$$

are the equations of three surfaces cutting one another at right angles for any values of the variable parameters a, a_1, a_2. The three series produced by giving all possible values to these parameters form what is called a system of conjugate orthogonal surfaces.

It has been proved by Dupin that the surfaces of any two of the three series cut each surface of the other series in its lines of curvature. Hence if one series be given, the other two are determinable from it, except in such extreme cases as those in which the lines of curvature of the given series are indeterminate.

In the method of curvilinear co-ordinates, as proposed by Lamé, the position of any point in space is determined by the three conjugate surfaces of the system (1) which intersect in the point. Thus, if any point P be given, there will be at least one of the surfaces of the first series which passes through it, and in general, but especially in such cases as we shall consider, there will be only one. a the value of λ corresponding to this surface is one of the co-ordinates of P. The other two co-ordinates are the values of λ_1 and λ_2 corresponding to the surfaces of the second and third series which contain the two lines of curvature through P of the first surface.

This general method of co-ordinates comprehends the two systems, rectangular and polar, in ordinary use. Thus, if

$$\lambda = x, \quad \lambda_1 = y, \quad \lambda_2 = z,$$

equations (1) will become

$$x = a, \quad y = a_1, \quad z = a_2,$$

the equations to three planes at right angles, by their intersection, determining the point P, whose rectangular co-ordinates are a, a_1, a_2. Similarly, if the first be a series of concentric spheres, the second a series of planes through a diameter of the sphere, and the third a series of cones having this diameter for axis, and the centre for vertex, λ, λ_1, λ_2 will be polar co-ordinates.

The equations of the motion of heat in a solid body may be referred to the general system of curvilinear co-ordinates in the following manner, which is exactly similar to that by which Fourier establishes them for rectangular rectilinear co-ordinates. For simplicity we shall suppose the body homogeneous, though the investigation would be in principle as simple if this were not the case. Let d be its density, h its conducting power, and c its capacity for heat, or the quantity of heat necessary to raise the temperature of a unit of its mass by unity.

Let λ, λ_1, λ_2 be the co-ordinates of any point P in the body, and let $\lambda + d\lambda$, $\lambda_1 + d\lambda_1$, $\lambda_2 + d\lambda_2$ be those of an adjacent point P'. The portions of the six surfaces corresponding to

$$\lambda, \ \lambda + d\lambda, \ \lambda_1, \ \lambda_1 + d\lambda_1, \ \lambda_2, \text{ and } \lambda_2 + d\lambda_2$$

adjacent to the points P and P' will form a rectangular parallelepiped, of which P and P' are opposite angles. Let dp, dp_1, dp_2 be the three edges of this parallelepiped which are respectively perpendicular to the surfaces corresponding to λ, λ_1, and λ_2: dp will be the element of the normal to the surface λ, commencing at this surface and terminated by the surface $\lambda + d\lambda$, and similarly with dp_1 and dp_2. Hence, by a known theorem,

$$dp = \left(\frac{d\lambda^2}{dx^2} + \frac{d\lambda^2}{dy^2} + \frac{d\lambda^2}{dz^2}\right)^{-\frac{1}{2}} d\lambda = H d\lambda, \text{ for brevity } \ldots\ldots(a),$$

$$dp_1 = \left(\frac{d\lambda_1^2}{dx^2} + \frac{d\lambda_1^2}{dy^2} + \frac{d\lambda_1^2}{dz^2}\right)^{-\frac{1}{2}} d\lambda_1 = H_1 d\lambda_1 \ldots\ldots\ldots\ldots(b),$$

$$dp_2 = \left(\frac{d\lambda_2^2}{dx^2} + \frac{d\lambda_2^2}{dy^2} + \frac{d\lambda_2^2}{dz^2}\right)^{-\frac{1}{2}} d\lambda_2 = H_2 d\lambda_2 \ldots\ldots\ldots\ldots(c).$$

Let v be the temperature of the body at P, and let

$$v + d_\lambda v + d_{\lambda_1} v + d_{\lambda_2} v$$

be the temperature at P'; $d_\lambda v$, $d_{\lambda_1} v$, and $d_{\lambda_2} v$ denoting the increments of v which correspond to the increments $d\lambda$, $d\lambda_1$, $d\lambda_2$ of λ, λ_1, λ_2. This notation, for *partial differentials*, we shall find it convenient to use in those cases in which it is not convenient to employ *partial differential coefficients*. The quantity of heat which flows into the rectangular parallelepiped $dp \cdot dp_1 \cdot dp_2$ across the side which coincides with λ, in the time dt, is

$$- k \frac{d_\lambda v}{dp} dp_1 dp_2 \cdot dt,$$

and the quantity which flows out across the opposite side, in the same time,

$$- k \left(\frac{d_\lambda v}{dp} dp_1 dp_2\right) dt - k d_\lambda \left(\frac{d_\lambda v}{dp} dp_1 dp_2\right) dt.$$

The difference of these two expressions,

or
$$k d_\lambda \left(\frac{d_\lambda v}{dp} dp_1 dp_2\right) dt,$$

is the whole quantity of heat which flows into the element during the time dt in a direction perpendicular to λ.

Similarly, the quantities of heat which the element gains in the same time, by motion in directions perpendicular to λ_1 and λ_2, are

$$k d_{\lambda_1} \left(\frac{d_{\lambda_1} v}{dp_1} dp dp_2\right) dt, \text{ and } k d_{\lambda_2} \left(\frac{d_{\lambda_2} v}{dp_2} dp dp_1\right) dt,$$

and therefore the entire quantity of heat gained by the element in the time dt, is

$$k \left\{ d_\lambda \left(\frac{d_\lambda v}{dp} dp_1 dp_2\right) + d_{\lambda_1} \left(\frac{d_{\lambda_1} v}{dp_1} dp dp_2\right) + d_{\lambda_2} \left(\frac{d_{\lambda_2} v}{dp_2} dp dp_1\right) \right\} dt ;$$

or by (a), (b), (c),

$$k \left\{ \frac{d}{d\lambda}\left(\frac{dv}{d\lambda}\frac{H_1 H_2}{H}\right) + \frac{d}{d\lambda_1}\left(\frac{dv}{d\lambda_1}\frac{HH_2}{H_1}\right) + \frac{d}{d\lambda_2}\left(\frac{dv}{d\lambda_2}\frac{HH_1}{H_2}\right) \right\} d\lambda d\lambda_1 d\lambda_2 dt,$$

where, in accordance with the ordinary notation of partial differential coefficients, the suffixes are omitted in the partial differentials.

Now if dv be the alteration of the temperature of the element $dp\,dp_1 dp_2$, in the time dt, the corresponding addition of heat is

$$c.d.dv dp dp_1 dp_2, \quad \text{or} \quad cd.dv.HH_1 H_2 d\lambda d\lambda_1 d\lambda_2.$$

Hence we have the equation

$$HH_1 H_2 \frac{dv}{dt} = \frac{k}{c.d}\left\{\frac{d}{d\lambda}\left(\frac{dv}{d\lambda}\frac{H_1 H_2}{H}\right) + \frac{d}{d\lambda_1}\left(\frac{dv}{d\lambda_1}\frac{HH_2}{H_1}\right)\right.$$
$$\left. + \frac{d}{d\lambda_2}\left(\frac{dv}{d\lambda_2}\frac{HH_1}{H_2}\right)\right\}\dots\dots\dots\dots(2),$$

which expresses fully, by means of curvilinear co-ordinates, the motion of heat in the interior of homogeneous solid bodies.

If the motion has continued so long as to have become uniform, then $\dfrac{dv}{dt} = 0$, and the equation becomes

$$\frac{d}{d\lambda}\left(\frac{dv}{d\lambda}\frac{H_1 H_2}{H}\right) + \frac{d}{d\lambda_1}\left(\frac{dv}{d\lambda_1}\frac{HH_2}{H_1}\right) + \frac{d}{d\lambda_2}\left(\frac{dv}{d\lambda_2}\frac{HH_1}{H_2}\right) = 0\dots\dots\dots(3).$$

This equation was first given by Lamé, in a paper entitled "Mémoire sur les Lois de l'Équilibre du Fluide Ethéré," in the *Journal de l'École Polytechnique* (Vol. III. cahier XXIII.), who deduced it by a very laborious transformation from the equation

$$\frac{d^2 v}{dx^2} + \frac{d^2 v}{dy^2} + \frac{d^2 v}{dz^2} = 0\dots\dots\dots\dots\dots(4),$$

in which the motion is referred to rectilinear co-ordinates. Equation (3) comprehends, as particular forms, equation (4), and the equation in which the motion is referred to polar co-ordinates. The former of these is obtained at once, if we put

$$\lambda = x, \quad \lambda_1 = y, \quad \lambda_2 = z,$$

which gives $\qquad H = H_1 = H_2 = 1.$

To find the latter, let

$$\lambda = (x^2 + y^2 + z^2)^{\frac{1}{2}} = r,$$

$$\lambda_1 = \tan^{-1}\frac{y}{x} = \phi,$$

$$\lambda_2 = \frac{z}{(x^2 + y^2 + z^2)^{\frac{1}{2}}} = \cos\theta.$$

Hence

$$H = \left(\frac{dr^2}{dx^2} + \frac{dr^2}{dy^2} + \frac{dr^2}{dz^2}\right)^{-\frac{1}{2}} = 1,$$

$$H_1 = \left(\frac{d\phi^2}{dx^2} + \frac{d\phi^2}{dy^2} + \frac{d\phi^2}{dz^2}\right)^{-\frac{1}{2}} = r\sin\theta,$$

$$H_2 = \left\{\frac{(d\cos\theta)^2}{dx^2} + \frac{(d\cos\theta)^2}{dy^2} + \frac{(d\cos\theta^2)}{dz^2}\right\}^{-\frac{1}{2}} = \frac{r^2}{(r^2-z^2)^{\frac{1}{2}}} = \frac{r}{\sin\theta}.$$

Therefore (3) becomes

$$\frac{d}{dr}\left(r^2\frac{dv}{dr}\right) + \frac{d}{d\cos\theta}\left(\sin^2\theta\,\frac{dv}{d\cos\theta}\right) + \frac{d}{d\phi}\left(\frac{1}{\sin^2\theta}\frac{dv}{d\phi}\right) = 0$$

or

$$r\frac{d^2(rv)}{dr^2} + \frac{d}{d\cos\theta}\left(\sin^2\theta\,\frac{dv}{d\cos\theta}\right) + \frac{1}{\sin^2\theta}\frac{d^2v}{d\phi^2} = 0,$$

the well-known equation of which so much use is made in the determination of the properties of the expansion of v in a series of powers of r. Equation (3) may also be applied to find the conditions which must be satisfied so that each series of a system of conjugate orthogonal surfaces may be isothermal; and we shall thus be enabled to answer the question proposed in a paper in the last number of this *Journal* (Vol. III. p. 286) [Art. VIII. above].

If the first series of surfaces, or a series represented by the equation $\lambda = a$, be isothermal, (3) must be satisfied by the assumption $v = f(\lambda)$, and therefore

$$\frac{d}{d\lambda}\left(\frac{H_1 H_2}{H}f'\lambda\right) = 0 \dots\dots\dots\dots\dots\dots(5).$$

Similarly, if the second and third series be isothermal,

$$\frac{d}{d\lambda_1}\left(\frac{HH_2}{H_1}f_1'\lambda_1\right) = 0 \dots\dots\dots\dots\dots\dots(6),$$

$$\frac{d}{d\lambda_2}\left(\frac{HH_1}{H_2}f_2'\lambda_2\right) = 0 \dots\dots\dots\dots\dots\dots(7).$$

Integrating these equations, we have

$$\frac{H_1 H_2}{H} f'\lambda = F(\lambda_1,\ \lambda_2) \dots\dots\dots\dots\dots\dots (8),$$

$$\frac{H H_2}{H_1} f_1'\lambda_1 = F_1(\lambda,\ \lambda_2) \dots\dots\dots\dots\dots\dots (9),$$

$$\frac{H H_1}{H_2} f_2'\lambda_2 = F_2(\lambda,\ \lambda_1) \dots\dots\dots\dots\dots\dots (10).$$

These three equations, together with the following,

$$\frac{d\lambda_1}{dx}\frac{d\lambda_2}{dx} + \frac{d\lambda_1}{dy}\frac{d\lambda_2}{dy} + \frac{d\lambda_1}{dz}\frac{d\lambda_2}{dz} = 0 \dots\dots\dots\dots (11),$$

$$\frac{d\lambda_2}{dx}\frac{d\lambda}{dx} + \frac{d\lambda_2}{dy}\frac{d\lambda}{dy} + \frac{d\lambda_2}{dz}\frac{d\lambda}{dz} = 0 \dots\dots\dots\dots (12),$$

$$\frac{d\lambda}{dx}\frac{d\lambda_1}{dx} + \frac{d\lambda}{dy}\frac{d\lambda_1}{dy} + \frac{d\lambda}{dz}\frac{d\lambda_1}{dz} = 0 \dots\dots\dots\dots (13),$$

which make the surfaces cut one another at right angles, are the conditions which must be satisfied if the three series be orthogonal and isothermal. If we could eliminate all the quantities depending on two of the series, the second and third for instance, we should find two equations relative to the first series which make it both be isothermal itself, and possess the property that the two series which cut its surfaces orthogonally shall also be isothermal. This elimination is probably quite impracticable in general. There is however an extensive class of surfaces which we see by inspection satisfies each condition, the class of cylindrical isothermal surfaces. For, if the axis of z be parallel to the generating lines of a series of isothermal cylindrical surfaces represented by the equation $\lambda = a$, λ will be independent of z, and $\lambda_1 = a_1$ being the equation of a series of orthogonal cylindrical surfaces, λ_1 will also be independent of z. The third series of the conjugate system will be a series of planes parallel to xy. Hence H and H_1 are functions of x and y alone, and therefore of λ and λ_1, and H_2 is a function of z and therefore of λ_2. Now since $\lambda = a$ is an isothermal system, (5) must be satisfied, and therefore $\dfrac{H_1 H_2}{H}$ must contain the factor $\dfrac{1}{f'(\lambda)}$, and λ must enter in no other manner. Hence

$\frac{H_1}{H}$ is the product of two functions, one of λ and the other of λ_1, and therefore $\frac{HH_2}{H_1}$ must consist of three factors, functions of $\lambda, \lambda_1, \lambda_2$ separately. Hence if we choose the arbitrary function $f'(\lambda_1)$ properly, (6) will be satisfied. Also, in every case, whether $\lambda = a$ be isothermal or not, we may choose $f'_2(\lambda_2)$ in such a manner that (7) shall be satisfied; in fact a series of parallel planes is necessarily isothermal. We have thus seen that the two conjugate orthogonal series to a series of isothermal cylinders are themselves isothermal, which agrees with what was proved in the paper already alluded to (Vol. III. p. 286) [Art. VIII. above].

In the case in which λ and λ_1 are surfaces of revolution we can obtain from (5), (6), and (7) a simple condition, which being satisfied, λ and λ_1 will each be isothermal, if one of them is so. To effect this, let the axis of revolution be taken for axis of x, and let $\rho = (y^2 + z^2)^{\frac{1}{2}}$. The generating lines of the surfaces of revolution of which the series λ and λ_1 are composed will be plane curves expressed by two equations between x and ρ. If in these equations we write $(y^2 + z^2)^{\frac{1}{2}}$ for ρ, the results will be the equations $\lambda = a$, $\lambda_1 = a_1$ of the surfaces of revolution. Hence λ and λ_1 are functions of x and ρ, and therefore, conversely, x and ρ are functions of λ and λ_1. Hence we see readily that H and H_1 depend only on x and ρ, or on λ and λ_1. Also, the equation $\lambda_2 = a_2$ must represent a series of planes passing through the axis of x, and we may therefore assume $\lambda_2 = \omega = \tan^{-1}\frac{y}{z}$. From this we readily deduce $H_2 = \rho$. Hence (7) is always satisfied, independently of λ and λ_1; that is to say, a series of planes passing through a fixed axis, is isothermal. Hence we have only to consider the conditions relative to λ and λ_1, or equations (5) and (6). If we substitute for H_2 its value, these become

$$\frac{d}{d\lambda}\left\{\frac{\rho H_1}{H}f'(\lambda)\right\} = 0 \dots\dots\dots\dots\dots(14),$$

$$\frac{d}{d\lambda_1}\left\{\frac{\rho H}{H_1}f'(\lambda_1)\right\} = 0 \dots\dots\dots\dots\dots(15).$$

From these we readily deduce

$$\rho = \phi(\lambda)\phi_1(\lambda_1)\dots\dots\dots\dots\dots\dots(16),$$

[which is the analytical expression of a theorem given by Bertrand, in Liouville's *Journal* for April, 1843, p. 130], ϕ and ϕ_1 being entirely arbitrary. Hence, if the equation $\lambda = a$ of a series of surfaces of revolution be given; and $\lambda_1 = a_1$, the equation to the orthogonal system be deduced, and if it be found that the distance of any point from the axis of revolution can be expressed in the form (16), then both systems, or neither, will be isothermal.

If the given series be isothermal this test may be applied very readily, as in that case we are always able at once to find the equation of the conjugate series. To shew this, let the series λ be isothermal. Then (14) will be true, and therefore we have, by integration,

$$\frac{\rho H_1}{H} f'(\lambda) = F(\lambda_1),$$

Therefore

$$H_1 = \frac{H F(\lambda_1)}{\rho f'(\lambda)} \dots\dots\dots\dots\dots\dots(17).$$

Also, since the sections of the two surfaces made by any plane through x are perpendicular to one another, we have

$$\frac{d\lambda}{dx}\frac{d\lambda_1}{dx} + \frac{d\lambda}{d\rho}\frac{d\lambda_1}{d\rho} = 0 \dots\dots\dots\dots\dots(18);$$

this equation gives

$$\frac{\dfrac{d\lambda_1}{dx}}{\dfrac{d\lambda}{d\rho}} = -\frac{\dfrac{d\lambda_1}{d\rho}}{\dfrac{d\lambda}{dx}} = \frac{H}{H_1},$$

and therefore, by (14),

$$F(\lambda_1)\frac{d\lambda_1}{dx} = \frac{dL_1}{dx} = \rho f'(\lambda)\frac{d\lambda}{d\rho},$$

$$F(\lambda_1)\frac{d\lambda_1}{d\rho} = \frac{dL_1}{d\rho} = -\rho f'(\lambda)\frac{d\lambda}{dx},$$

if $L_1 = \int F(\lambda_1)\,d\lambda_1$; therefore

$$L_1 = \int \rho f'(\lambda)\left(\frac{d\lambda}{d\rho}\,dx - \frac{d\lambda}{dx}\,d\rho\right) \dots\dots\dots\dots(19),$$

and $L_1 = \alpha$ is the equation of the series of orthogonal trajectories to the series of curves in which the surfaces of revolution of the given isothermal system are cut by any plane through the axis. We may verify this solution by observing that the criterion of integrability for the expressions given above for $\dfrac{dL_1}{dx}$ and $\dfrac{dL_1}{d\rho}$ is satisfied; since, if we transform the equation

$$\frac{d^2(f\lambda)}{dx^2} + \frac{d^2(f\lambda)}{dy^2} + \frac{d^2(f\lambda)}{dz^2} = 0,$$

to the independent variables x and ρ, $f\lambda$ being independent of $\dfrac{y}{z}$, we have

$$f'(\lambda)\left(\frac{d^2\lambda}{dx^2} + \frac{d^2\lambda}{d\rho^2}\right) + f''(\lambda)\left(\frac{d\lambda^2}{dx^2} + \frac{d\lambda^2}{d\rho^2}\right) + \frac{1}{\rho}f'(\lambda)\frac{d\lambda}{d\rho} = 0.$$

If λ itself satisfy (4), we may take $f(\lambda) = \lambda$, and the transformed equation becomes

$$\frac{d^2\lambda}{dx^2} + \frac{d^2\lambda}{d\rho^2} + \frac{1}{\rho}\frac{d\lambda}{d\rho} = 0 \ldots\ldots\ldots\ldots\ldots(20).$$

Also, if we take $\lambda_1 = L_1$, (19) becomes

$$\lambda_1 = \int\rho\left(\frac{d\lambda}{d\rho}dx - \frac{d\lambda}{dx}d\rho\right) \ldots\ldots\ldots\ldots\ldots(21).$$

As an example of the application of these formulæ, let λ represent the series of *surfaces of equilibrium* in the case of a sphere having matter distributed over it, according to the law of the distribution of electricity on a neutral conducting sphere under the influence of a distant electrified body. It is readily shewn that, if λ be proportional to the *potential* of such a system, on external points, and if the centre of the sphere be origin, and the line joining this point and the influencing body axis of x, we have

$$\lambda = \frac{x}{(x^2 + \rho^2)^{\frac{3}{2}}} = \frac{x}{r^3}\ldots\ldots\ldots\ldots\ldots(a).$$

Hence $\lambda = a$ represents a series of isothermal surfaces of revolution, and λ satisfies (20), as may be readily verified. Hence, if $\lambda_1 = a_1$ be the orthogonal system of surfaces of revolution, we have, by (21),

$$\lambda_1 = \frac{\rho^2}{(x^2 + \rho^2)^{\frac{3}{2}}} = \frac{\rho^2}{r^3} \ldots\ldots\ldots\ldots\ldots(b).$$

T.   3

If between (*a*) and (*b*) we eliminate *x*, we have

$$\rho^{\frac{8}{3}} + \frac{\lambda_1^{\,2}}{\lambda^2}\,\rho^{\frac{2}{3}} - \frac{\lambda_1^{\,\frac{4}{3}}}{\lambda^2} = 0 \quad\ldots\ldots\ldots\ldots\ldots\ldots(c).$$

The value of ρ deduced from this equation is not of the form $F(\lambda)\,.\,F_1(\lambda_1)$; and hence, by (16), the second series is not isothermal. Similarly, if λ represent the potential of two equal material points situated at the distance $2a$ from one another on the point xyz, we have

$$\lambda = \frac{1}{\{(x-a)^2+\rho^2\}^{\frac{1}{2}}} + \frac{1}{\{(x+a)^2+\rho^2\}^{\frac{1}{2}}} \quad\ldots\ldots\ldots\ldots(d).$$

Then, by (18), we have, for the orthogonal system,

$$\lambda_1 = \frac{x-a}{\{(x-a)^2+\rho^2\}^{\frac{1}{2}}} + \frac{x+a}{\{(x+a)^2+\rho^2\}^{\frac{1}{2}}} \quad\ldots\ldots\ldots\ldots(e).$$

The value of ρ deduced from these two equations cannot, I think, be of the form $F(\lambda)\,F(\lambda_1)$, and hence in this case also the second series is not isothermal. Hence the question proposed in the paper already referred to (Vol. III. p. 286) [Art. IX. above] must be answered in the negative, the proposition to which it refers being not generally true, since we have found cases of surfaces of revolution with regard to which it does not hold; but it holds with regard to every system of conjugate orthogonal surfaces, for which equations (8), (9), and (10) are satisfied. Also, since it does not appear that any two of these equations imply the third, it may happen that two series of a system may be isothermal and the third not, as is exemplified in the particular cases above considered, in each of which a series of surfaces of revolution and of orthogonal planes are isothermal, and the third series, consisting of surfaces of revolution, is not isothermal.

[The theorem expressed in symbols by equation (21) above, that the differential equation of the lines of force, symmetrical round an axis, is rendered integrable by the factor ρ, was anticipated by Stokes in the corresponding proposition for stream-lines in irrotational fluid-motion, which he gave in his paper " On the Steady Motion of Incompressible Fluids," read before the Cambridge Philosophical Society, on the 25th of April, 1842, his first scientific paper, published first in the *Transactions* of the Society, and recently republished at the beginning of the first

volume (pp. 13, 14) of his "Mathematical and Physical Papers."
It is interesting to remark that the same theorem of integrability holds of the differential equation of stream-lines for every
case of motion symmetrical round an axis.   To prove this let
$\dot{x}$, $\dot{\rho}$ be components of velocity parallel and perpendicular to the
axis.   The "equation of continuity" becomes, in virtue of the
symmetry round the axis of x,

$$\frac{d\dot{x}}{dx} + \frac{d\dot{\rho}}{d\rho} + \frac{\dot{\rho}}{\rho} = 0 :$$

and the differential equation of the stream-lines is

$$\dot{\rho}\,dx - \dot{x}\,d\rho = 0.$$

The first member of this is rendered a complete differential by
the factor ρ, as we verify that $d/d\rho \,.\, (\rho\dot{\rho}) = d/dx \,.\, (-\rho\dot{x})$, in virtue
of the first equation.   This theorem is interesting not only in
the mathematics of vortex motion : it is applicable also to lines
of magnetic force through a region traversed by electric currents,
symmetrical round an axis.]

[From the *Cambridge Mathematical Journal*, Vol. IV. February, 1844.]

ART. X. ELEMENTARY DEMONSTRATION OF DUPIN'S THEOREM.

IF there be three series of surfaces, such that all the surfaces of each series cut the surfaces of the other two series at right angles, the theorem to be proved is that the lines of intersection of any one of the surfaces of one of the three series, with the surfaces of the two conjugate series, are its lines of curvature.

Let O be any point in which three conjugate surfaces intersect, and let the rectangular axes OX, OY, OZ be perpendicular to the tangent planes of the three surfaces at O. Let

$$f(x,\ y,\ z) = \lambda \quad\dotfill(a)$$

$$f_1(x,\ y,\ z) = \lambda_1 \quad\dotfill(a_1)$$

$$f_2(x,\ y,\ z) = \lambda_2 \quad\dotfill(a_2)$$

be the equations of the three series; and, when proper values are attached to λ, λ_1, λ_2, let (a) be the surface touched by YOZ, (a_1) by ZOX, and (a_2) by XOY.

Hence, when $x = 0$, $y = 0$, $z = 0$, we have

$$\left. \begin{array}{ll} \dfrac{df}{dy} = 0, & \dfrac{df}{dz} = 0 \\[2mm] \dfrac{df_1}{dz} = 0, & \dfrac{df_1}{dx} = 0 \\[2mm] \dfrac{df_2}{dx} = 0, & \dfrac{df_2}{dy} = 0 \end{array} \right\} \quad\dotfill(b).$$

Now, since the system is orthogonal,

$$\left.\begin{array}{c}\dfrac{df_1}{dx}\dfrac{df_2}{dx}+\dfrac{df_1}{dy}\dfrac{df_2}{dy}+\dfrac{df_1}{dz}\dfrac{df_2}{dz}=0\\[2mm]\dfrac{df_2}{dx}\dfrac{df}{dx}+\dfrac{df_2}{dy}\dfrac{df}{dy}+\dfrac{df_2}{dz}\dfrac{df}{dz}=0\\[2mm]\dfrac{df}{dx}\dfrac{df_1}{dx}+\dfrac{df}{dy}\dfrac{df_1}{dy}+\dfrac{df}{dz}\dfrac{df_1}{dz}=0\end{array}\right\}\quad\ldots\ldots\ldots\ldots(c).$$

Differentiating the first of these equations with respect to x, the second with respect to y, and the third with respect to z; putting x, y, and z each $= 0$, in the result, and making use of equations (b), we have

$$\left(\frac{df_1}{dy}\right)\left(\frac{d^2f_2}{dx\,dy}\right)+\left(\frac{df_2}{dz}\right)\left(\frac{d^2f_1}{dz\,dx}\right)=0,$$

$$\left(\frac{df}{dx}\right)\left(\frac{d^2f_2}{dx\,dy}\right)+\left(\frac{df_2}{dz}\right)\left(\frac{d^2f}{dy\,dz}\right)=0,$$

$$\left(\frac{df}{dx}\right)\left(\frac{d^2f_1}{dx\,dz}\right)+\left(\frac{df_1}{dy}\right)\left(\frac{d^2f}{dy\,dz}\right)=0,$$

the brackets denoting that in the quantities enclosed, x, y, z are put $= 0$. From these equations we conclude that

$$\left(\frac{d^2f}{dy\,dz}\right),\quad\left(\frac{d^2f_1}{dz\,dx}\right),\quad\left(\frac{d^2f_2}{dx\,dy}\right)\text{ are each }=0.$$

Now, since YOZ is the tangent plane to the surface (a) at O, we have

$$x=\frac{1}{2}\left\{\left(\frac{d^2x}{dy^2}\right)y^2+2\left(\frac{d^2x}{dy\,dz}\right)yz+\left(\frac{d^2x}{dz^2}\right)z^2\right\}+\&c.\ldots\ldots(d)$$

for points in the surface adjacent to O. To determine $\left(\dfrac{d^2x}{dy\,dz}\right)$, we have, from (a),

$$\frac{dx}{dy}=-\frac{\dfrac{df}{dy}}{\dfrac{df}{dx}}.$$

Hence, differentiating with regard to z, and putting x, y, and z each $= 0$, in the result,

$$\left(\frac{d^2x}{dy\,dz}\right)=-\frac{\left(\dfrac{d^2f}{dy\,dz}\right)}{\left(\dfrac{df}{dx}\right)}=0.$$

Hence (d) becomes

$$x = \frac{1}{2}\left\{\left(\frac{d^2x}{dy^2}\right)y^2 + \left(\frac{d^2x}{dz^2}\right)z^2\right\} + \&c.$$

Hence the planes of xy and xz contain the principal sections of (a) through O, and therefore the lines of intersection of (a_1) and (a_2) with (a), touch the principal sections of (a) at O. Now O may be any point in one of the surfaces (a), and therefore each of these surfaces has its lines of curvature traced upon it by the surfaces of the series (a_1), (a_2); similarly it may be shewn that each surface (a_1) has its lines of curvature traced by (a_2) and (a), and each surface (a_2) by (a) and (a_1): which is the theorem to be proved.

By differentiating the first of equations (c) with regard to y and z respectively, and putting x, y, and z each $= 0$ in the results, and performing corresponding operations on the second and third, we obtain expressions for

$$\left(\frac{d^2f_1}{dx^2}\right), \left(\frac{d^2f_2}{dx^2}\right), \left(\frac{d^2f}{dy^2}\right), \left(\frac{d^2f_2}{dy^2}\right), \left(\frac{d^2f}{dz^2}\right), \left(\frac{d^2f_1}{dz^2}\right),$$

in terms of $$\left(\frac{d^2f}{dx^2}\right), \left(\frac{d^2f_1}{dy^2}\right), \left(\frac{d^2f_2}{dz^2}\right),$$

which lead directly to the expression for the curvatures of the principal sections of the three surfaces at O, given by Lamé, in his *Memoir on Curvilinear Co-ordinates*.

[It is interesting to remark as an obvious consequence of Dupin's theorem that not every series of surfaces can be cut at right angles by two mutually orthogonal series.]

[From the *Cambridge Mathematical Journal*, Vol. IV. February, 1844.]

ART. XI. NOTE ON SOME POINTS IN THE THEORY OF HEAT.

[An application to Terrestrial Temperature, of the principle set forth in the first part of this paper relating to the Age of thermal distributions, was made the subject of the author's Inaugural Dissertation on the occasion of his induction to the professorship of Natural Philosophy in the University of Glasgow, in October, 1846, "De Motu Caloris per Terrae Corpus:" which, more fully developed afterwards, gave a very decisive limitation to the possible age of the earth as a habitation for living creatures; and proved the untenability of the enormous claims for TIME which, uncurbed by physical science, geologists and biologists had begun to make and to regard as unchallengable. See "Secular Cooling of the Earth," " Geological Time," and several other Articles below.]

IN problems relative to the motion of heat in solid bodies, the initial distribution, which is entirely arbitrary, is usually one of the data. When this is the case, and the circumstances in which the body is placed are known, the distribution at any subsequent period is fully determined, and if our analysis had sufficient power, would become known in every case. The solution of the problem would be an expression for the temperature of any point in the body in terms of the co-ordinates of the point, and the times measured from the instant at which the distribution is given.

It is in many cases an interesting investigation to examine what this expression becomes when negative values are assigned to the time. If, for a particular negative value $-\tau$, we find that the expression gives an actual arithmetical value for the temperature of every point in the body, the distribution represented will be such that, if it had existed at a time τ before the initial instant, the given initial distribution would have been produced by the spontaneous motion of the heat.

It is clear, however, that the arbitrary initial distribution may be of such a nature that it cannot be the natural result of any previous possible distribution, or that it cannot be any stage except the first in a system of varying temperatures. This, for instance, will be the case if there be any abrupt transitions in the initial temperatures of adjacent points, or if the curves representing the initial temperatures of points situated along any straight line through the body, have cusps or angular points; for, though we may suppose such a distribution to be arbitrarily made, it could obviously not be produced by the spontaneous motion of heat from any other preceding distribution, and all such abrupt transitions or angles which may exist in an initial distribution, will disappear instantaneously after the motion has commenced.

If, however, in any case, a complete solution of the problem has been obtained, that is, if an expression for the temperature of any point, in terms of its co-ordinates, and the time has been found, this expression must assume some *form,* when negative values are given to the time; and it is my object here to examine the nature of this form, when the initial state is such as not to be deducible from a previous distribution.

The simplest case is that of the linear motion of heat, and as from it we are enabled to understand the nature of the problem in its most general form, I shall, for the present, confine myself to this case. The case of a thin rod, protected from any lateral radiation, or of an infinite solid, in which the temperature is distributed in parallel isothermal planes, is what is usually contemplated in speaking of the linear motion of heat; but the case of a thin rod losing heat by radiation from its sides may also be readily reduced to the same as the two former. For simplicity, however, we may suppose that there is no lateral radiation.

Let a fixed point O in the rod be considered as origin, let the distance to any point P in the rod be x, and let the temperature of P, at the time t from the initial instant, be v. The equation of the motion of the heat will be

$$\frac{dv}{dt} = \frac{d^2v}{dx^2} \dots\dots\dots\dots\dots\dots\dots(1);$$

if, for brevity, we choose the unit of heat such that the conducting power of the body, referred to a unit of its volume, may be unity.

The complete integral of this equation is

$$v = \Sigma A_i \epsilon^{-m_i^2 t} \cos(m_i x + n_i) \dots\dots\dots\dots\dots\dots(2).$$

Let $_0v$ be the initial temperature of P. Then, putting $t = 0$, in (2), we have

$$_0v = \Sigma A_i \cos(m_i x + n_i) \dots\dots\dots\dots\dots\dots(3).$$

Now, whatever be the arbitrary nature of the function $_0v$, whether continuous or discontinuous, it has been shewn by Fourier that it may be represented by a series such as the second member of (3), and he has shewn how, when the value of $_0v$ corresponding to any value of x, is given, the constants A_i, m_i, n_i may be determined. The series thus found is an actual quantitative representation of $_0v$, and is necessarily convergent. In fact, the proof of its convergence in every case must be included in the demonstration of the possibility of representing any function $_0v$ by the series (3).

Now the expression for v at time t, is found by multiplying the respective terms of the second members of (3), by the quantities

$$\epsilon^{-m_1^2 t}, \quad \epsilon^{-m_2^2 t}, \quad \&c. \dots\dots\dots\dots\dots(a);$$

a series which converges when t is positive (m_1, m_2, &c. being arranged in ascending order of magnitude). Hence, when t is positive, the series for v is still more convergent than the series for $_0v$, and therefore (2) gives a convergent series for v for every positive value of t. If, however, the time considered be τ *before* the initial instant, the series (a) will become

$$\epsilon^{m_1^2 \tau}, \quad \epsilon^{m_2^2 \tau}, \quad \&c. \dots\dots\dots\dots\dots\dots(b),$$

which is divergent. Hence it will depend on the degree of the convergence of the series for $_0v$, whether the expression of v in this case be convergent or divergent. In the latter case the distribution represented will be impossible, and therefore there will be no distribution from which the initial distribution can be derived, by spontaneous motion, in the time τ. There are three cases of the initial distribution, which we must specially consider.

1. If the convergence of the expression for $_0v$, or of the series of coefficients

$$A_1, \; A_2, \; \&c.$$

be of a *lower order* than that of the series (a), for every positive value of t, then for any finite value, however small, of τ, the series

$$A_1 \epsilon^{m_1^2 \tau}, \quad A_2 \epsilon^{m_2^2 \tau}, \; \&c.$$

will be ultimately divergent. Hence, in this case, for any time, however small, before the initial instant, the distribution will be impossible, and therefore the given distribution cannot be any stage but the first in a system of varying temperatures.

2. If the convergence of the series for $_0v$ be ultimately of the *same order* as that of the series (a); that is, if the coefficients A_1, A_2, &c. can be put under the forms

$$a_1 \epsilon^{-m_1^2 \tau'}, \quad a_2 \epsilon^{-m_2^2 \tau'}, \; \&c.,$$

where a_1, a_2, &c. form a series which ultimately has a convergence of a *lower order* than that of the series (a); then, at a time τ' before the initial instant, the distribution is represented by

$$v' = \Sigma a_i \cos (m_i x + n_i),$$

which belongs to the first class considered, and is therefore an essentially primitive distribution. Hence, in this case, a finite *age*, τ' may be assigned to the given initial distribution.

3. If the convergence of the series for $_0v$ be of a *higher order* than that of the series (a); or if, for any finite value of τ, however great, the series

$$A_1 \epsilon^{m_1^2 \tau}, \quad A_2 \epsilon^{m_2^2 \tau}, \; \&c.$$

is ultimately convergent; then, for every finite negative value of t, the series for v will converge, and therefore v will have a possible expression. Hence, in this case, no limit can be assigned to the *age* of the initial distribution.

The simplest case of distributions which belong to the third class, is that in which the expression for $_0v$ is composed of a finite number of terms of the form

$$A \cos{(mx + n)};$$

but we may also readily form series of an infinite number of terms, which will be sufficiently convergent to satisfy the condition stated for the third class, and therefore any initial distribution represented by such a series may be deduced from distributions existing previously for any length of time, however great.

These remarks are sufficient to indicate the *nature* of the question proposed. The details of the convergence or divergence of the series employed will depend on the values which must be assigned to m_1, m_2, &c. for satisfying the [boundary] conditions of any particular problem.

If, instead of the initial distribution of heat in the rod, from $-\infty$ to $+\infty$ being given, we have only a part, that for instance on the positive portion of the rod, we must, to make the problem determinate, have some other condition given. In such cases, the part of the rod, over which the initial distribution is not given may be conceived to be removed, and the second condition will then generally relate to the extremity of the part of the rod considered. In a previous paper (*On the Linear Motion of Heat*, Vol. III. p. 170), [Art. IV. above], I have shewn how a solution of the problem may be obtained by determining the initial distribution which must be made on the negative part of the rod, in order that the condition relative to the zero point may be fulfilled. Thus, if the temperature at the zero point, or the extremity of the part of the rod originally given, be constrained to be a given arbitrary function of the time, this constraint may be effected by producing the rod indefinitely in the negative direction, and impressing on the part produced a certain initial distribution, determined by equations (6) and (10) of the Article already referred to, in terms of the arbitrary function of the time, and the given initial distribution on the positive side. In many cases, however, the distribution so determined will be impossible, either over the whole extent of the negative part of the rod, or from some point at a finite distance on the negative side, to an infinite distance. Notwithstanding this impossibility, the solution of the problem obtained by the method described above will still give a finite possible expression for the temperature of any point on the positive side, at any time subsequent to the initial instant, which will be the temperature that would be actually assumed by the point, if the temperature of the zero point had, by some external application, been constrained to satisfy the given condition; and the only thing indicated by the impossibility of the distribution on the negative side will be that this constraint cannot be effected actually, by adding the negative part of the rod, with a certain initial distribution of heat, to the given positive part. It is unnecessary to enter into this question separately here, as the details [founded on consideration of the intensity of the convergence of the series (7) of Art. IV. above] are very similar to those given above. It should be remarked however that the possibility or impossibility of the distribution on the negative part will depend entirely on the function of the time, which expresses the variable temperature of the zero point, and not at all on the given initial distribution on the positive part; and on this account, if the initial distribution on the negative part be impossible, all its subsequent forms will generally be impossible also.

Before leaving this subject, it may be well to notice a point relative to the theory of isothermal surfaces, or surfaces of equilibrium, which involves considerations analogous to those with which we have been occupied above.

It is a known theorem that an attraction every where perpendicular to any given closed surface S, may be produced by the distribution of a given quantity of matter m over the surface, according to a law which is in every instance determinate; but in general it will be impossible to produce the same effect by the distribution of matter over any surface in the interior of S, not coinciding with it. If, for instance, S were the surface of a cube, or any surface containing points or edges, this would ob-

viously be impossible*, and it would probably also be impossible†
in the case considered by Poisson, in which S is composed of two
spherical surfaces. In every case there will be an infinite series of
surfaces of equilibrium without S, becoming ultimately a series
of spheres having the centre of gravity of m for their common
centre, each of which is such that m exerts an attraction on any
point in it in the direction of the normal. Hence, if one of these
surfaces, s, be given, we may not only produce an attraction every
where perpendicular to it, by matter distributed over itself, but
by matter distributed over S, or over any surface between S and s,
enclosing the former. Hence s is analogous to a distribution of
heat which may be produced by a previously existing distribution.
But, unless S itself have the particular property we are considering
in s, we cannot produce an attraction perpendicular to s by any
distribution on a surface within S. Thus s is analogous to a distribution of heat which cannot be produced by any previous distribution existing at a time before it greater than a certain limit τ.

Again, in some cases the series of surfaces of equilibrium,
of which s is one, may be continued indefinitely inwards, till
we arrive at a surface enclosing no space, as for instance when s is
an ellipsoid, in which case the surfaces of equilibrium are confocal
ellipsoids, and the series within s may be continued till we arrive
at an elliptical disc. Such a case is analogous to that of a distribution of heat, which may be produced from a distribution existing
an unlimited time before.

[*Note added June* 26, 1881. About a year after this Article
was written, the method of Electric Images (Electrostatics and
Magnetism XIV., and v. §§ 113—127) shewed me that I had been
wrong in the surmise regarding two electrified spherical conductors
which it contains: and now I find that I was wrong also in the
positive statement regarding a surface having "points or edges."
Considering the infinite series of images within each of two
detached spheres, and the finite series of $2i-1$ images within a
conductor bounded by two spherical surfaces cutting one another
at an angle π/i, we see that the series of equipotentials may be
continued from the surface inwards, becoming ultimately a group

* [Not so. See Note at end of this Article.]
† [*Ibid.*]

of infinitely small spherical surfaces, or mere points, containing among them the whole of the given quantity of electricity; and thus is disproved not only my original surmise regarding the case considered by Poisson. but the corresponding supposition which I should no doubt have entertained for the case of two spherical surfaces cutting at π/i.

Look for example at the annexed diagram representing an insulated electric conductor bounded by portions of spherical surfaces cutting one another at right angles; and indicating by fine circular arcs the internal completions of these spherical surfaces. The electricity on the bounding surface may be annulled, and instead two positive electric points may be placed at A and B, and a negative electric point at C, without altering the potential or force at the bounding surface and through

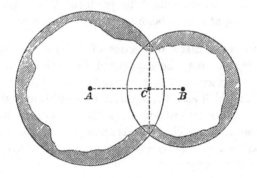

external space: provided the whole quantity, reckoned algebraically, is unchanged, and the quantities at A, B, C are in the proportions of $+a$, $+b$, $-\dfrac{ab}{\sqrt{(a^2+b^2)}}$; where a and b denote the radii, and therefore (in virtue of the orthogonality of the spherical surfaces) the distance AB between their centres is $\sqrt{(a^2+b^2)}$. The values of the three quantities are clearly Va, Vb, $-V\dfrac{ab}{\sqrt{(a^2+b^2)}}$, if V denote the potential of the boundary of the given electrified conductor; and the potential due to these three if placed at A, B, C is clearly equal to V not only over the portions of the spherical surfaces constituting the boundary of the given body, but also over the internal completions of these surfaces. The series of equipotentials, for potential greater than V, lies in the two lateral lens-shaped spaces; for positive equipotentials less than V they lie in external space, and in the middle lenticular space; and for negative potentials they lie wholly within the middle lenticular space. It is interesting to trace the whole series, beginning with an infinitely large positive potential (for which the equipotential is a pair of infinitely small spherical surfaces having their centres at A and B, and their radii in the proportion of a to b), going on outwards to the V equipotential, and thence outwards and inwards to the equipotential for an infinitely small positive value, which consists of an infinitely large spherical surface, having for its centre the centre of gravity of the electricity on the given conductor (or of the three replacing electric points).

Another case also instructive in respect to the wrongness of my old idea is the distribution of electricity on a conductor bounded by a surface of which a part or parts lie on a spherical surface, and the rest (which may be quite arbitrarily discontinuous, with projecting or re-entrant angles,) lies all outside this spherical surface. It is easily proved that the actual distribution of positive electricity on the protuberance, and its image (negative) inside the spherical surface, together with an electric point at the centre of the sphere, of the right quantity to make up the whole given amount of electricity, produces the same potential and force as the given conductor at and without its surface. The reader may find it interesting to draw an illustrative diagram.]

[From the *Cambridge Mathematical Journal*, Vol. IV. November, 1844.]

ART. XII. NOTE ON ORTHOGONAL ISOTHERMAL SURFACES.

IN a previous paper in this *Journal* (" On the Equations of the Motion of Heat referred to Curvilinear Co-ordinates," Vol. IV. p. 33), [Art. IX. above], I expressed the conditions which must be satisfied by a system of conjugate orthogonal surfaces which are all isothermal, and considered some particular cases of such systems. In addition to those, however, there is another class of surfaces which satisfy the conditions. For it is readily seen that all that is necessary in the demonstration of the theorem relative to cylindrical surfaces in Art. IX. above, is that H_2 shall be independent of λ and λ_1, and that $\dfrac{H_1}{H}$ shall be independent of λ_2. Now a series of concentric spheres (including the case of a series of parallel planes) is such that the value of H_2 at any point of one of them is independent of the position of the point on that surface. Hence we may consider $\lambda_2 = a_2$ as representing a series of concentric spheres, and consequently $\lambda = a$, and $\lambda_1 = a_1$, a system of conjugate orthogonal cones having the centre of the spheres for their common vertex. Any cone of one series will intersect any one of the conjugate series along a generating line, and the values of H and H_1 at points along this line will vary as the distances from the centre. Hence $\dfrac{H_1}{H}$ is independent of λ_2. We therefore see that the demonstration in Art. IX. above is applicable to the general class of orthogonal cones, as well as to the particular included case of orthogonal cylinders. Thus we have the general theorem that, if a series of cones having a common vertex be isothermal, the series of orthogonal cones will also be isothermal.

[Added June 27, 1881; with reference to the promise inserted at the conclusion of Art. VIII. above. The *lines of thermal flux* in this case, and the lines of force in the corresponding cases of "fields of force," and the stream-lines in the corresponding case of motion of an incompressible fluid (or liquid as we may call it for brevity), lie on spherical surfaces concentric with the origin of co-ordinates: and their differential equation is integrable by a factor, supposing the *temperature*, or the *potential*, or the *velocity potential* to be a known function of proper co-ordinates. And not only for these mutually equivalent cases, but for the less restricted case of motion of a liquid, whether rotational or irrotational, and the corresponding cases of electro-magnetic force, the differential equation of the stream-lines is integrable by a factor provided they lie on concentric spherical surfaces. To prove this proposition let $\dot{r}$, $r\dot{\theta}$, $r\sin\theta\dot{\phi}$ be components, referred to polar co-ordinates, of the velocity of the liquid at any point (r, θ, ϕ). The equation of continuity is

$$\frac{1}{r^2}\frac{d}{dr}(r^2\dot{r}) + \frac{d}{\sin\theta\,d\theta}(\sin\theta\dot{\theta}) + \frac{d\dot{\phi}}{d\phi} = 0 \quad\ldots\ldots\ldots\ldots(a),$$

and the differential equations of the stream-lines are

$$\frac{dr}{\dot{r}} = \frac{d\theta}{\dot{\theta}} = \frac{d\phi}{\dot{\phi}} \quad\ldots\ldots\ldots\ldots\ldots\ldots\ldots(b).$$

Now suppose every stream-line to lie on a spherical surface concentric with the origin of co-ordinates: that is to say, let $\dot{r} = 0$. The equation of continuity becomes

$$\frac{d}{\sin\theta\,d\theta}(\sin\theta\dot{\theta}) + \frac{d\dot{\phi}}{d\phi} = 0 \quad\ldots\ldots\ldots\ldots\ldots(c),$$

which may be written

$$\frac{d(-\sin\theta\dot{\theta})}{d\theta} = \frac{d(\sin\theta\dot{\phi})}{d\phi} \quad\ldots\ldots\ldots\ldots\ldots(d);$$

and the residual equation of the stream-lines may be written

$$\dot{\phi}d\theta - \dot{\theta}d\phi = 0 \quad\ldots\ldots\ldots\ldots\ldots\ldots(e).$$

Equation (d) shews that the first member of (e) is made a complete differential by the factor $\sin\theta$.]

This class, and the class of confocal surfaces of the second degree, are all the triple systems which as yet have been found to be isothermal; and, in the paper already referred to, some particular cases were considered in which it was shewn that one, or two of the partial series of an orthogonal system are isothermal, and the remaining series not. Since the publication of that paper I have received a Mémoire by M. Lamé, which was published about the same time (*Journal de Mathématiques*, Vol. VIII. p. 397, Oct. and Nov. 1843), in which he shews that no other triple isothermal system can exist. This interesting result is the complete answer of a question proposed in this *Journal*, in May 1843 (Vol. III. p. 286) [Art. VIII. above], and to which a partial answer was given in the paper already referred to (Nov. 1843) [Art. IX. above]. The same question has been proposed by M. Bertrand, in the April number of the *Journal de Mathématiques* of the present year (Vol. IX. p. 117) [Art. IX. above]; and he answers it to a similar extent, by shewing that an isothermal series of surfaces has not in every case its* two conjugate orthogonal series isothermal also. The reason, however, which he assigns for coming to this conclusion does not seem to be quite satisfactory: for it is founded on the assumption that "it is always possible to take two consecutive isothermal surfaces arbitrarily, and that the law of the temperature of the rest of the body is then determined." Now, by considerations analogous to those brought forward in a paper "On some Points in the Theory of Heat" in this *Journal* (Vol. IV. p. 71) [Art. XI. above], it is readily seen that if two consecutive isothermal surfaces be arbitrarily assumed, it will in general be only for *points between them* that a possible system of isothermal surfaces can be determined according to a continuous law. The temperatures of points not lying between them will follow a different law depending on the sources of heat or cold which we must suppose to be distributed over the two assumed surfaces, to retain them at their constant temperatures. Thus, if we assume arbitrarily two consecutive isothermal surfaces indefinitely near one another, the system of isothermal surfaces through the whole body, to which these two belong, will in general be impossible. In fact, it will generally be impossible to find any two surfaces, containing the two assumed ones between them, which will be such that if they be retained at different constant temperatures,

* [For correction of the idea implied here by "its," and by the "the" twice annotated below, see addition to Art. X. above.]

the two assumed surfaces will each be isothermal. But M. Bertrand's conclusion, though correct, is drawn from the assumption that this is generally possible. It may be remarked, however, that though some restriction is necessary in assuming two consecutive surfaces of a possible isothermal system, it will probably be found to be not so narrow as a restriction which M. Bertrand shews to be necessary in choosing two consecutive surfaces of an isothermal series of which the* conjugate orthogonal series are isothermal also. If this were previously shewn, M. Bertrand's inference would be correct.

M. Bertrand also specially considers the case of isothermal orthogonal surfaces of revolution, and arrives at the interesting theorem that, if each of the† conjugate series be isothermal, the traces on the meridian planes will form a system of conjugate isothermal plane curves, or the traces of a system of conjugate isothermal cylinders, on their orthogonal planes.

This follows at once from the equations (14) and (15) (Vol. IV. p. 39 of this *Journal*) [Art. IX. above], though it did not occur to me till I saw M. Bertrand's paper. For, from them we deduce

$$\frac{H_1}{H} = \phi(\lambda)\,\phi(\lambda_1),$$

which is the sole condition that each of the series of orthogonal plane curves represented by $\lambda = a$, $\lambda_1 = a_1$, shall be isothermal.

Hence we see that if this condition be satisfied, and at the same time the condition expressed by equation (16), the two conjugate orthogonal series of surfaces of revolution will also be isothermal. M. Bertrand states the latter condition in geometrical language as follows—

"Two systems of isothermal orthogonal lines being given, in order that their rotation round an axis may generate isothermal surfaces of revolution, it is necessary that the distances from the axis, of the four corners of a curvilinear rectangle formed by the given lines, shall be the four terms of an analogy."

We may also add, that if a single series of surfaces of revolution be isothermal, and if the traces on a meridian plane be

* [See addition to Art. x. above.]
† [*Ibid.*]

isothermal lines, then the conjugate orthogonal series of surfaces will also be isothermal.

Also it follows, from the result of M. Lamé's investigations mentioned above, that confocal surfaces of revolution of the second order form the only isothermal system which trace a series of isothermal lines on a meridian plane.

[From the *Cambridge Mathematical Journal*, Vol. IV. November, 1844.]

ART. XIII.   NOTE ON THE LAW OF GRAVITY AT THE SURFACE
OF A REVOLVING HOMOGENEOUS FLUID.

IT has been shewn by Maclaurin that a homogeneous fluid,
revolving uniformly round a fixed axis, and acted upon only by
the attractive force of its own particles, may, with the same
angular velocity, have two different figures of equilibrium, each a
spheroid of revolution round the shorter axis : and Jacobi has
shewn that it may be in equilibrium in the form of an ellipsoid
with three unequal axes, the shortest coinciding with the axis
of rotation.   The following simple consideration determines the
law of gravity at the surface in each case.

Let any surface concentric with the free surface of the fluid,
and similar to it, be described in the interior of the fluid.   If all
the fluid exterior to the surface were removed, the fluid would
still be in equilibrium, since the *proportions* of the free surface
depend only on the density and angular velocity.   Hence the
accelerating* force at this surface, as far as it is due to the
centrifugal force, and the attraction of the interior mass [that
is the apparent resultant force], must be everywhere perpendicular to the surface.   But the mass without it, being con-

* [The senseless adjective "accelerating" here is happily now a mere monument
of the difficulties created for students of dynamics by the curiously illogical usage
of the terms "moving force" and "accelerating force" still in vogue less than
thirty-six years ago, which have been annulled by the introduction of Gauss'
kinetic unit of force, and the associated reformation of elementary dynamical
language.   See Thomson and Tait's *Treatise on Natural Philosophy*, §§ 219—223,
or *Elements of Natural Philosophy*, §§ 185—191.]

tained between two concentric similar ellipsoids exerts no attraction on any point in the surface, and therefore the direction
of the force* on any point of this surface in the interior of the
fluid is in the normal. Hence the surface must be of equal
pressure. Now, let it be supposed to approach the free surface
so as to be indefinitely near it. In order that the pressure
on every point of it may be the same, the force† on any
point of the indefinitely thin shell between it and the free
surface, or the force of gravity at any point of the free surface,
must be inversely proportional to the thickness of the shell at the
point, or inversely as the perpendicular from the centre to the
tangent plane at the point. A very simple analytical proof of this
result was given by Liouville (*Journal de Mathématiques*, Vol. VIII.
p. 360). We may also state it, that the force of gravity at any
point of the free surface is inversely proportional to the electrical
tension [electric density] at the point, supposing the surface an
electrified conductor.

* [For the sake of clearness the adjective "accelerating," meaninglessly prefixed
to force here as above in the original, is now omitted.]

† [*Ibid.*]

ART. XIV. DEMONSTRATION OF A FUNDAMENTAL THEOREM IN
THE MECHANICAL THEORY OF ELECTRICITY.

[ELECTROSTATICS AND MAGNETISM, VIII.]

[From the *Cambridge Mathematical Journal*, Vol. IV. February, 1845.]

ART. XV. ON THE REDUCTION OF THE GENERAL EQUATION OF SURFACES OF THE SECOND ORDER.

In the following paper, by a simple assumption with reference to the coefficients in the general equation of a surface of the second order, the cubic, by means of which the three principal axes are determined is made to assume a very simple form, which enables us to prove the reality of the roots, and to find the limits between which they lie, with great ease. It also leads to a very simple analytical proof, that the three principal axes are at right angles to one another.

Let
$$Ax^2 + By^2 + Cz^2 + 2A'yz + 2B'zx + 2C'xy$$
$$+ 2A''x + 2B''y + 2C''z = H,$$

or, for brevity, $H_2(x, y, z) + H_1(x, y, z) = H,$

be the equation to a surface of the second order.

Let
$$\left.\begin{array}{lll} A' = (gh)^{\frac{1}{2}}, & B' = (hf)^{\frac{1}{2}}, & C' = (fg)^{\frac{1}{2}}, \\ A = f+\alpha, & B = g+\beta, & C = h+\gamma, \end{array}\right\} \ldots\ldots(1),$$

from which we deduce

$$\left.\begin{array}{lll} f = \dfrac{B'C'}{A'}, & g = \dfrac{C'A'}{B'}, & h = \dfrac{A'B'}{C'}, \\[2mm] \alpha = A - \dfrac{B'C'}{A'}, & \beta = B - \dfrac{C'A'}{B'}, & \gamma = C - \dfrac{A'B'}{C'}, \end{array}\right\} \ldots\ldots(2);$$

which express real determinate values for f, g, &c. in terms of the given coefficients. Making the substitutions (1), we find

$$H_2 = \alpha x^2 + \beta y^2 + \gamma z^2 + (f^{\frac{1}{2}}x + g^{\frac{1}{2}}y + h^{\frac{1}{2}}z)^2 \ldots\ldots(a).$$

Now we may define a principal axis to be a line such that, if it be taken for the axis of x', and the axes of y' and z' be in the plane perpendicular to it, the products $x'y'$ and $x'z'$ shall disappear in the transformed expression H_2. This will be ensured if the product $x'y'$ vanishes for every point in the plane $x'y'$, and for every position of this plane passing through the principal axis OX'; a definition equivalent to the one in which a principal axis is defined as an axis which is perpendicular to its diametral plane.

Let l, m, n be the direction-cosines of a principal axis OX'; l', m', n' those of any line OY' perpendicular to it; x', y' the co-ordinates of any point in the plane $X'OY'$; x, y, z the co-ordinates of the same point referred to the original axes: we have

$$x = lx' + l'y',$$
$$y = mx' + m'y',$$
$$z = nx' + n'y'.$$

Therefore
$$\alpha x^2 + \beta y^2 + \gamma z^2 + (f^{\frac{1}{2}}x + g^{\frac{1}{2}}y + h^{\frac{1}{2}}z)^2$$
$$= (\alpha l^2 + \beta m^2 + \gamma n^2) x'^2 + (\alpha l'^2 + \beta m'^2 + \gamma n'^2) y'^2$$
$$+ 2 (\alpha l l' + \beta m m' + \gamma n n') x'y'$$
$$+ \{(f^{\frac{1}{2}}l + g^{\frac{1}{2}}m + h^{\frac{1}{2}}n) x' + (f^{\frac{1}{2}}l' + g^{\frac{1}{2}}m' + h^{\frac{1}{2}}n') y'\}^2 ;$$

which becomes
$$P x'^2 + P' y'^2,$$

if we put for brevity

$$S = f^{\frac{1}{2}}l + g^{\frac{1}{2}}m + h^{\frac{1}{2}}n \dots\dots\dots\dots\dots\dots(3),$$
$$P = S^2 + \alpha l^2 + \beta m^2 + \gamma n^2 \dots\dots\dots\dots\dots(4),$$

and similarly for $l'm'n'$, and if we assume the coefficient of $x'y' = 0$, which gives

$$(Sf^{\frac{1}{2}} + \alpha l) l' + (Sg^{\frac{1}{2}} + \beta m) m' + (Sh^{\frac{1}{2}} + \gamma n) n' = 0.$$

If OX' be a principal axis, this must hold for all values of l', m', n' consistent with

$$l l' + mm' + nn' = 0 ;$$

we must therefore have

$$\frac{Sf^{\frac{1}{2}} + \alpha l}{l} = \frac{Sg^{\frac{1}{2}} + \beta m}{m} = \frac{Sh^{\frac{1}{2}} + \gamma n}{n} :$$

therefore each member is

$$= l\,(Sf^{\frac{1}{2}} + al) + m\,(Sf^{\frac{1}{2}} + \beta m) + n\,(Sf^{\frac{1}{2}} + \gamma n),$$
$$= S^2 + al^2 + \beta m^2 + \gamma n^2,$$
$$= P.$$

Therefore

$$\left.\begin{aligned}
l &= \frac{Sf^{\frac{1}{2}}}{P-\alpha}, \\[1ex]
m &= \frac{Sg^{\frac{1}{2}}}{P-\beta}, \\[1ex]
n &= \frac{Sh^{\frac{1}{2}}}{P-\gamma},
\end{aligned}\right\} \quad \dots\dots\dots\dots\dots(5).$$

Hence, by (3),

$$S\left(\frac{f}{P-\alpha} + \frac{g}{P-\beta} + \frac{h}{P-\gamma} - 1\right) = 0;$$

and therefore, unless $S = 0$, which, on account of (5), would require $P = \alpha = \beta = \gamma$, a case that will be considered below, we must have

$$\frac{f}{P-\alpha} + \frac{g}{P-\beta} + \frac{h}{P-\gamma} - 1 = 0 \dots\dots\dots\dots(6),$$

which determines P. This equation, being a cubic, gives three values for P. Now, from equations (2) it follows that f, g, h must either be all positive or all negative; the former being the case when two or none of A', B', C', and the latter when one or three are negative*. Hence, if α, β, γ be in descending order of magnitude, and e be an indefinitely small quantity, and if we substitute

$$\alpha - e, \ \beta + e, \ \text{and} \ \beta - e, \ \gamma + e,$$

for P in the first member of (6), the first and second values will have contrary signs, and so will the third and fourth. Hence the cubic has a real root between α, β, and another between β, γ, and its remaining root must therefore also be real, and between ∞ and α, or between γ and $-\infty$. The first is obviously the case when f, g, and h are positive, and the second when they are negative.

* Hence (6) is a particular case of a certain equation of any order, which has been shewn by M. M. Plana and Liouville to have all its roots real. See Moigno, *Calc. Int.* p. 296.

Let P_1, P_2 be any two roots of (6), and let l_1, m_1, n_1, l_2, m_2, n_2 be the corresponding values of l, m, n deduced from equations (5). Writing down (6) for each value, and subtracting, we have

$$(P_1 - P_2) \left\{ \frac{f}{(P_1 - \alpha)(P_2 - \alpha)} + \frac{g}{(P_2 - \beta)(P_2 - \beta)} + \frac{h}{(P_1 - \gamma)(P_2 - \gamma)} \right\} = 0.$$

If P_1 be different from P_2, the second factor must vanish, or, by (5),

$$l_1 l_2 + m_1 m_2 + n_1 n_2 = 0 \dots \dots \dots \dots \dots (b).$$

Hence any two of the axes determined by equations (6) and (5) are at right angles, and therefore the three must form a rectangular system. If we take it for axes of co-ordinates, and if P, Q, R be the three roots of (6), the equation to the surface becomes

$$Px^2 + Qy^2 + Rz^2 + H_1 = H,$$

accents being omitted. If none of the quantities P, Q, R vanishes, we may obviously, by changing the origin, make H_1 disappear, and the equation will be reduced to the form

$$Px^2 + Qy^2 + Rz^2 = H',$$

which is the equation of surfaces of the second order referred to principal axes through the centre.

If in this equation H' be $= 0$, the surface represented will be either a point or a cone, according as P, Q, and R have the same or different signs. Excluding this case, we may, without losing generality, consider H' as positive.

The equation will then represent an ellipsoid if P, Q, R be all positive; a hyperboloid of one sheet if one of them only be negative, and a hyperboloid of two sheets if two of them be negative. If all three be negative, the surface will be imaginary.

Hence, from equation (6) we infer that if f, g, h, α, β, γ, be all positive the surface will be an ellipsoid, and if they be all negative it will be imaginary.

In addition to these we have the following cases.

I. $\dfrac{f}{-\alpha}+\dfrac{g}{-\beta}+\dfrac{h}{-\gamma}-1>0.$

(1) $fgh>0.$

$\alpha>0,\quad \beta>0,\quad \gamma<0,\quad$ Ellipsoid,

$\alpha>0,\quad \beta<0,\quad \gamma<0,\quad$ Hyperboloid of one sheet,

$\alpha<0,\quad \beta<0,\quad \gamma<0,\quad$ Hyperboloid of two sheets.

(2) $fgh<0.$

$\alpha>0,\quad \beta>0,\quad \gamma>0,\quad$ Hyperboloid of one sheet,

$\alpha>0,\quad \beta>0,\quad \gamma<0,\quad$ Hyperboloid of two sheets,

$\alpha>0,\quad \beta<0,\quad \gamma<0,\quad$ Imaginary.

II. $\dfrac{f}{-\alpha}+\dfrac{g}{-\beta}+\dfrac{h}{-\gamma}-1<0.$

(1) $fgh=0.$

$\alpha>0,\quad \beta>0,\quad \gamma<0,\quad$ Hyperboloid of one sheet,

$\alpha>0,\quad \beta<0,\quad \gamma<0,\quad$ Hyperboloid of two sheets,

$\alpha<0,\quad \beta<0,\quad \gamma<0,\quad$ Imaginary.

(2) $fgh<0.$

$\alpha>0,\quad \beta>0,\quad \gamma>0,\quad$ Ellipsoid,

$\alpha>0,\quad \beta>0,\quad \gamma<0,\quad$ Hyperboloid of one sheet,

$\alpha>0,\quad \beta<0,\quad \gamma<0,\quad$ Hyperboloid of two sheets.

These tests enable us, when A, B, C, A', B', C' are given numerically, to find the nature of the surface represented, provided it has a centre, by calculating α, β, γ, f, g, h from equations (2).

If the surface have not a centre, it can belong to neither of the four cases considered above, and we must therefore have

$$\frac{f}{\alpha}+\frac{g}{\beta}+\frac{h}{\gamma}+1=0,$$

which is the condition that one root of (6) may be $= 0$. If we substitute for α, β, γ, their values, by (1), and clear of fractions, this becomes

$$ABC + 2fgh - Agh - Bhf - Cfg = 0,$$

or $\qquad ABC + 2A'B'C' - AA'^2 - BB'^2 - CC'^2 = 0,$

which agrees with the condition given in Gregory's *Solid Geometry*, p. 64.

This is the condition that must be satisfied in the case in which it is impossible to make H_1 vanish, from the general equation, by any finite change in the position of the origin, as may be readily verified.

If the surface be of revolution, two of the roots of (6) must be equal. Hence each of the two must be equal to one of the quantities α, β, γ, on account of the limits found above for the roots. Hence, clearing of fractions, we find for the conditions

$$\alpha = \beta = \gamma,$$

and the remaining root will be given by

$$f + g + h - (P - \alpha) = 0.$$

Hence, restoring the original constants, we have, in the case of surfaces of revolution,

$$A - \frac{B'C'}{A'} = B - \frac{C'A'}{B'} = C - \frac{A'B'}{C'} = Q = R,$$

which agrees with the condition given by *Gregory*, p. 109,

and $\qquad P = Q + \dfrac{B'C'}{A'} + \dfrac{C'A'}{B'} + \dfrac{A'B'}{C'},$

or $\qquad = A + B + C - 2Q.$

The formulæ which have been proved above furnish us with a very simple proof of the following theorem of Chasles. [Compare Thomson and Tait's *Natural Philosophy*, Second ed. § 523.]

The principal axes of a cone touching a surface of the second order, are perpendicular to the three confocal surfaces of the second order which intersect in the vertex.

Let
$$\frac{x^2}{a} + \frac{y^2}{b} + \frac{z^2}{c} = 1,$$

be the equation to the surface. That of the tangent cone through ξ, η, ζ is

$$\left(\frac{\xi^2}{a} + \frac{\eta^2}{b} + \frac{\zeta^2}{c} - 1\right)\left(\frac{x^2}{a} + \frac{y^2}{b} + \frac{z^2}{c} - 1\right) = \left(\frac{\xi x}{a} + \frac{\eta y}{b} + \frac{\zeta z}{c} - 1\right)^2.$$

Here
$$H_2 = \left(\frac{\xi x}{a} + \frac{\eta y}{b} + \frac{\zeta z}{c}\right)^2 - K\left(\frac{x^2}{a} + \frac{y^2}{b} + \frac{z^2}{c}\right),$$

where
$$K = \frac{\xi^2}{a} + \frac{\eta^2}{b} + \frac{\zeta^2}{c} - 1.$$

Hence, comparing with (a), we have

$$f = \frac{\xi^2}{a^2}, \quad g = \frac{\eta^2}{b^2}, \quad h = \frac{\zeta^2}{c^2},$$

$$\alpha = -\frac{K}{a}, \quad \beta = -\frac{K}{b}, \quad \gamma = -\frac{K}{c}.$$

Therefore (6) becomes

$$\frac{\xi^2}{a(aP+K)} + \frac{\eta^2}{b(bP+K)} + \frac{\zeta^2}{c(cP+K)} = 1;$$

but
$$\frac{\xi^2}{aK} + \frac{\eta^2}{bK} + \frac{\zeta^2}{cK} = 1 + \frac{1}{K}.$$

Hence, by subtraction,

$$\frac{\xi^2}{a + \dfrac{K}{P}} + \frac{\eta^2}{b + \dfrac{K}{P}} + \frac{\zeta^2}{c + \dfrac{K}{P}} = 1 \ldots\ldots\ldots\ldots\ldots\ldots(c).$$

Also, equations (5) give

$$\frac{l\left(a + \dfrac{K}{P}\right)}{\xi} = \frac{m\left(b + \dfrac{K}{P}\right)}{\eta} = \frac{n\left(c + \dfrac{K}{P}\right)}{\zeta}.$$

Hence l, m, n, are the direction-cosines of a normal at ξ, η, ζ, to the surface represented by (c) when either of the three values of $\dfrac{K}{P}$ is used in that equation. If the value $> a$ be used, the surface represented will be the ellipsoid confocal with the given surface which passes through the vertex of the cone; the value between *a* and *b* corresponds to the confocal hyperboloid of one sheet; and the value between *b* and *c* to the confocal hyperboloid of two sheets. Thus we infer that these three surfaces intersect at right angles in the vertex of the cone, and that the principal axes of the cone touch their lines of intersection.

The same theorem might be proved separately for the different cases of surfaces without centres; but, as these are only extreme cases of central surfaces; we may infer the truth of the theorem for them, as whatever is true in general, is true in limiting cases, provided the result remain definite.

St Peter's College, Cambridge,
Jan. 11, 1845.

[From the *Cambridge Mathematical Journal*, Vol. IV. May, 1845.]

Art. XVI. On the Lines of Curvature of Surfaces of the Second Order.

The method usually followed in works on Geometry of Three Dimensions, in treating the differential equation to the lines of curvature of an ellipsoid, leads to an unsymmetrical integral, involving only two of the co-ordinates, and therefore representing the projection of the lines of curvature on one of the co-ordinate planes. In the following paper, by making use of an equivalent process, but preserving the symmetry with respect to the two variables which are involved, an integral is obtained which enables us from the symmetry to infer the equations to the projections on the other two co-ordinate planes. By combining the three forms of the integral thus obtained, we arrive at the integral given by Mr Ellis, and at other symmetrical formulæ.

Let the equation to the surface be

$$\frac{x^2}{a^2} + \frac{y^2}{b^2} + \frac{z^2}{c^2} = 1.$$

The differential equation to the lines of curvature is consequently

$$\frac{(b^2 - c^2)x}{dx} + \frac{(c^2 - a^2)y}{dy} + \frac{(a^2 - b^2)z}{dz} = 0.$$

Let $\frac{x^2}{a^2} = u$, $\frac{y^2}{b^2} = v$, $\frac{z^2}{c^2} = w$. The preceding equations become

$$u + v + w = 1 \dots\dots\dots\dots\dots\dots\dots(1),$$

$$\frac{(b^2 - c^2)u}{du} + \frac{(c^2 - a^2)v}{dv} + \frac{(a^2 - b^2)w}{dw} = 0 \dots\dots\dots\dots(2).$$

Eliminating u from the latter equation, by means of the former, we have

$$\frac{b^2 - c^2}{du} + v\left(\frac{c^2 - a^2}{dv} - \frac{b^2 - c^2}{du}\right) + w\left(\frac{a^2 - b^2}{dw} - \frac{b^2 - c^2}{du}\right) = 0,$$

or $\quad b^2 - c^2 = \dfrac{v\left(-c^2 du - c^2 dv + a^2 du + b^2 dv\right)}{dv}$

$$- \frac{w\left(-b^2 du - b^2 dw + a^2 du + c^2 dw\right)}{dw}.$$

But $du + dv + dw = 0$, by (1); and hence the equation to the lines of curvature may be put under the form

$$\frac{v}{dv} - \frac{w}{dw} = \frac{b^2 - c^2}{a^2 du + b^2 dv + c^2 dw}.$$

If from this equation we eliminate du, we obtain an equation of Clairaut's form, of which the integral is found by substituting for $\dfrac{dv}{dw}$ an arbitrary constant. For the sake of symmetry we shall denote this constant by $\dfrac{g}{h}$; and we must consequently substitute g and h for dv and dw in the differential equation, and therefore also for du, $-(g + h)$, which we shall denote by f. Thus we have

$$\left.\begin{array}{c} \dfrac{du}{f} = \dfrac{dv}{g} = \dfrac{dw}{h} \\ f + g + h = 0, \end{array}\right\} \dots\dots\dots\dots\dots(3),$$

where

and the complete integral is

$$\frac{v}{g} - \frac{w}{h} = \frac{b^2 - c^2}{a^2 f + b^2 g + c^2 h} \dots\dots\dots\dots (4).$$

Also, by the symmetry, we have for the integrals involving the variables wu and uv,

$$\left.\begin{array}{c} \dfrac{w}{h} - \dfrac{u}{f} = \dfrac{c^2 - a^2}{a^2 f + b^2 g + c^2 h} \\ \dfrac{u}{f} - \dfrac{v}{g} = \dfrac{a^2 - b^2}{a^2 f + b^2 g + c^2 h} \end{array}\right\} \dots\dots\dots\dots (4).$$

The manner in which the quantities f, g, h have been introduced, shews clearly how they represent only one arbitrary constant. Any one of the equations (4) may be written in such a form as to contain only one arbitrary constant explicitly; and it will be shewn below how f, g, h may be expressed symmetrically by two arbitrary constants, one of which is irrelevant, as it enters as a factor in the integral.

From the equations (4), as from the ordinary forms, the properties of the projections of the lines of curvature may be readily deduced. Thus, taking the second, and substituting for w, u, and g their values $\dfrac{z^2}{c^2}$, $\dfrac{x^2}{a^2}$, and $-(f+h)$, we have

$$\frac{z^2}{c^2 h} - \frac{x^2}{a^2 f} = \frac{a^2 - c^2}{(b^2 - c^2) h - (a^2 - b^2) f}.$$

Let a^2, b^2, c^2 be positive quantities in descending order of magnitude. Then, unless f and h have opposite signs, this equation cannot be satisfied by any values of z, x which satisfy the inequality

$$\frac{z^2}{c^2} + \frac{x^2}{a^2} < 1;$$

that is, by values which correspond to any point of the ellipsoid. Hence we may write the equation as

$$\frac{z^2}{\gamma^2} + \frac{x^2}{\alpha^2} = 1 \dots\dots\dots\dots\dots\dots(5),$$

where

$$\gamma^2 = \frac{c^2(a^2 - c^2) h}{(b^2 - c^2) h - (a^2 - b^2) f},$$

$$\alpha^2 = -\frac{a^2(a^2 - c^2) f}{(b^2 - c^2) h - (a^2 - b^2) f}.$$

Eliminating $f : h$ between these equations, we have

$$\frac{\gamma^2(b^2 - c^2)}{c^2} + \frac{\alpha^2(a^2 - b^2)}{a^2} = a^2 - c^2 \dots\dots\dots\dots(6).$$

Hence we conclude that the projections of the lines of curvature on the plane of the greatest and least axes of the ellipsoid, are the ellipses whose semiaxes, γ, α, are connected by the equation (6). Thus the construction for describing them is as follows. Draw an ellipse, concentric with the ellipsoid, in the plane ca, with the lines

$$c\left(\frac{a^2 - c^2}{b^2 - c^2}\right)^{\frac{1}{2}}, \quad a\left(\frac{a^2 - c^2}{a^2 - b^2}\right)^{\frac{1}{2}},$$

as semiaxes. Take any point in this ellipse, draw perpendiculars to the axes, and with the intersections as vertices, describe a concentric ellipse. This will be the projection of a line of curvature. Also, by giving the point assumed in the auxiliary ellipse every possible position in its circumference, we obtain the projections of all the lines of curvature. Similar constructions are applicable to the projections of the lines of curvature on the other two principal planes; and, by taking one or two of the quantities a^2, b^2, c^2 negative, we may extend the rules to hyperboloids of one or of two sheets.

In the case we have taken, of an ellipsoid, and the plane of the greatest and least axes for the plane of projection, each curve intersects the consecutive one, and the locus of these intersections may be found from (5) and (6) by the ordinary process. Thus, by differentiation,

$$\frac{z^2}{\gamma^3}\,d\gamma + \frac{x^2}{a^3}\,d\alpha = 0,$$

$$\frac{a^2-b^2}{a^2}\,a\,d\alpha + \frac{b^2-c^2}{c^2}\,\gamma\,d\gamma = 0.$$

Hence
$$\frac{c^2}{b^2-c^2}\frac{z^2}{\gamma^4} = \frac{a^2}{a^2-b^2}\frac{x^2}{a^4},$$

which gives
$$\frac{\gamma^2}{cz\,(b^2-c^2)^{-\frac{1}{2}}} = \frac{a^2}{ax\,(a^2-b^2)^{-\frac{1}{2}}}.$$

By combining this equation, first with (6) and then with (5), we find each member

$$= \frac{a^2-c^2}{\dfrac{z}{c}\,(b^2-c^2)^{\frac{1}{2}} + \dfrac{x}{a}\,(a^2-b^2)^{\frac{1}{2}}} = \frac{z}{c}\,(b^2-c^2)^{\frac{1}{2}} + \frac{x}{a}\,(a^2-b^2)^{\frac{1}{2}}.$$

Hence $$\frac{z}{c}\,(b^2-c^2)^{\frac{1}{2}} + \frac{x}{a}\,(a^2-b^2)^{\frac{1}{2}} = (a^2-c^2)^{\frac{1}{2}} \ \ldots\ldots\ldots\ldots\ldots (7),$$

is the equation to the required locus, which is therefore a group of four straight lines (on account of the double signs of the radicals) forming a rhombus, of which the diagonals coincide with the axes of c and a.

Thus we see that the projections of the lines of curvature on the plane of the greatest and least axes, are ellipses inscribed in a rhombus, with their axes coincident with those of the ellipsoid. If we consider b^2 as not of intermediate magnitude between a^2 and c^2, the equation (7) represents an imaginary group of straight lines, which shews that the projection of any line of curvature on the plane of the greatest and mean, or of the mean and least axes, does not meet its consecutive.

It may be remarked with respect to equations (4), that any one of them may be deduced from any other, by combining it with the equation $u + v + w = 1$, as is easily verified. Also, by multiplying the first by a^2, the second by b^2, and the third by c^2, and adding, we have

$$\frac{u(b^2 - c^2)}{f} + \frac{v(c^2 - a^2)}{g} + \frac{w(a^2 - b^2)}{h} = 0 \ \ldots\ldots\ldots (8),$$

which is the symmetrical integral given by Mr Ellis (Vol. II. p. 133). This equation might also have been found directly, when it was proved that

$$\frac{du}{f} = \frac{dv}{g} = \frac{dw}{h} \, ;$$

by eliminating by means of these relations, du, dv, dw from (2), the differential equation to the lines of curvature, which is the method followed by Mr Ellis.

Without losing generality, we may substitute for f, g, h, any expressions in terms of two distinct constants which satisfy the condition $f + g + h = 0$. Thus, if we take k and v for the constants, we may assume

$$f = ka^2(b^2 - c^2) - kv(b^2 - c^2),$$

or

$$\left. \begin{aligned} f &= k(b^2 - c^2)(a^2 - v), \\ g &= k(c^2 - a^2)(b^2 - v), \\ h &= k(a^2 - b^2)(c^2 - v), \end{aligned} \right\} \ \ldots\ldots\ldots\ldots\ldots\ldots (9).$$

Making these substitutions in (8), we have

$$\left. \begin{aligned} \frac{u}{a^2 - v} + \frac{v}{b^2 - v} + \frac{w}{c^2 - v} &= 0, \\ \frac{x^2}{a^2(a^2 - v)} + \frac{y^2}{b^2(b^2 - v)} + \frac{z^2}{c^2(c^2 - v)} &= 0, \end{aligned} \right\} \ \ldots\ldots\ldots (10).$$

or

Adding this equation, multiplied by v, to

$$\frac{x^2}{a^2} + \frac{y^2}{b^2} + \frac{z^2}{c^2} = 1 \dots\dots\dots\dots\dots (a),$$

we have

$$\frac{x^2}{a^2 - v} + \frac{y^2}{b^2 - v} + \frac{z^2}{c^2 - v} = 1 \dots\dots\dots\dots (11).$$

This equation shews that the lines of curvature of any surface of the second order with a centre, are its intersections with confocal surfaces of the second order. Since this property is independent of the centre, it follows that it must also be true for the case of a surface without a centre.

If the co-ordinates x, y, z, of any point in the line of curvature be given, by substituting these values for x, y, z in (10), which will be a quadratic in v, we may determine two values of this parameter, which, substituted in (11), will give the equations to the hyperboloids of one sheet and of two sheets confocal with the ellipsoid, which cut it in the two lines of curvature passing through the given point x_1, y_1, z_1. In general, equation (11) may be considered as a cubic for determining v, when x, y, z have any given real values whatever, x_1, y_1, z_1. The three roots, which may readily be shewn to be real, correspond to the three species of surfaces confocal with the ellipsoid (a, b, c) which intersect in the point x_1, y_1, z_1. In the present case, when this point is on the surface of the ellipsoid (a, b, c), and therefore x_1, y_1, z_1 satisfy the equation (a), one root of the cubic is zero, and the other two are the roots of the quadratic (10), which is the *reduced equation*. Thus, whether we take the form (10) or (11) of the integral, the two arbitrary constants may be determined by the solution of a quadratic equation, from the condition that the curve passes through a given point.

If we wish to determine the direction of the tangent at any point of a line of curvature, we may follow the usual process of differentiation for curves of double curvature.

Thus, taking any two of the equations (4), we find

$$\frac{du}{f} = \frac{dv}{g} = \frac{dw}{h}.$$

But if l, m, n be the direction-cosines of the tangent, we have

$$\frac{l}{\frac{a^2}{x}\,du} = \frac{m}{\frac{b^2}{y}\,dv} = \frac{n}{\frac{c^2}{z}\,dw};$$

and hence

$$\frac{l}{\frac{a^2}{x}f} = \frac{m}{\frac{b^2}{y}g} = \frac{n}{\frac{c^2}{z}h} \quad\dots\dots\dots\dots (12).$$

If we substitute for f, g, h their values by (9), these equations become

$$\frac{l}{\frac{a^2}{x}(b^2-c^2)(a^2-v)} = \frac{m}{\frac{b^2}{y}(c^2-a^2)(b^2-v)} = \frac{n}{\frac{c^2}{z}(a^2-b^2)(c^2-v)} \quad\dots(13).$$

We may also determine l, m, n by the ordinary formulæ for the principal directions of curvature. Thus, if ρ be the radius of curvature of the normal section touching the line of curvature at the point considered, we have

$$l = \frac{\mu x}{a^2 - p\rho}, \quad m = \frac{\mu y}{b^2 - p\rho}, \quad n = \frac{\mu z}{c^2 - p\rho} \quad\dots\dots(14);$$

and $p\rho$ is one of the values of Q deduced from the quadratic equation

$$\frac{x^2}{a^2(a^2 - Q)} + \frac{y^2}{b^2(b^2 - Q)} + \frac{z^2}{c^2(c^2 - Q)} = 0 \quad\dots\dots(15).$$

Also v is a root of this equation, since, when it is substituted for Q, the equation is identical with (10), one of the equations of the line of curvature considered. Thus a line of curvature is the locus of points on the surface, for which one root of (15) is constant. Now, by combining equations (13) and (14), we have

$$\left.\begin{aligned} &\frac{a^2(b^2 - c^2)(a^2 - v)(a^2 - p\rho)}{x^2} \\ =\ &\frac{b^2(c^2 - a^2)(b^2 - v)(b^2 - p\rho)}{y^2} \\ =\ &\frac{c^2(a^2 - b^2)(c^2 - v)(c^2 - p\rho)}{z^2} \end{aligned}\right\} \quad\dots\dots\dots\dots(16).$$

These equations shew that v and $p\rho$ cannot both be constant, and therefore they must be different roots of the equation (15). Hence, if ρ' be the radius of curvature of a normal section perpendicular to the line of curvature at P, we have

$$v = p\rho';$$

which shews that the radii of curvature of sections perpendicular to a line of curvature at different points, are inversely proportional to the perpendiculars from the centre upon the tangent planes at those points.

The equations (16) may be verified directly, since, $p\rho$ and v being the two roots of (15), we have

$$v \cdot p\rho = a^2 b^2 c^2 \left(\frac{x^2}{a^4} + \frac{y^2}{b^4} + \frac{z^2}{c^4} \right),$$

$$v + p\rho = \frac{(b^2 + c^2) x^2}{a^2} + \frac{(c^2 + a^2) y^2}{b^2} + \frac{(a^2 + b^2) z^2}{c^2}.$$

Hence $(a^2 - v)(a^2 - p\rho)$

$$= a^4 - \{a^2(b^2 + c^2) - b^2 c^2\} \frac{x^2}{a^2} - \{a^2(c^2 + a^2) - c^2 a^2\} \frac{y^2}{b^2} - \{a^2(a^2 + b^2) - a^2 b^2\} \frac{z^2}{c^2}$$

$$= a^4 \left(1 - \frac{y^2}{b^2} - \frac{z^2}{c^2} \right) - \{a^2(b^2 + c^2) - b^2 c^2\} \frac{x^2}{a^2}$$

$$= (a^2 - b^2)(a^2 - c^2) \frac{x^2}{a^2}.$$

Hence $\dfrac{a^2 (b^2 - c^2)(a^2 - v)(a^2 - p\rho)}{x^2} = -(b^2 - c^2)(c^2 - a^2)(a^2 - b^2);$

and therefore we infer that each member of (16) is equal to this expression, on account of the symmetry.

Let l', m', n' be the direction-cosines of the principal section corresponding to the root v of the quadratic equation (15). We shall have, by the formulæ which correspond to (14),

$$l' = \frac{\mu x}{a^2 - v}, \quad m' = \frac{\mu y}{b^2 - v}, \quad n' = \frac{\mu z}{c^2 - v} \quad \ldots\ldots\ldots\ldots(17).$$

Thus by means of the same root of the quadratic equation, we have, in (13) and (17), expressed the direction-cosines of each of the two principal sections. If we put λ for each member of (13), these equations give

$$ll' + mm' + nn' = \lambda\mu \left\{ a^2(b^2 - c^2) + b^2(c^2 - a^2) + c^2(a^2 - b^2) \right\} = 0,$$

which proves that the principal directions of curvature are at right angles to one another; a theorem, of which many other different proofs have been given.

December, 1844.

ART. XVII. DEMONSTRATION D'UN THÉORÈME D'ANALYSE.
[*Liouville Mathematical Journal*, Vol. x. 1845. Incorporated in other papers in "ELECTROSTATICS AND MAGNETISM."]

ART. XVIII. NOTE SUR LES LOIS ÉLÉMENTAIRES D'ÉLECTRICITÉ STATIQUE.
[*Liouville Mathematical Journal*, Vol. x. 1845. "ELECTROSTATICS AND MAGNETISM," Article II.]

ART. XIX. EXTRAIT D'UNE LETTRE SUR L'APPLICATION DU PRINCIPE DES IMAGES À LA SOLUTION DE QUELQUES PROBLÈMES RELATIFS À LA DISTRIBUTION D'ÉLECTRICITÉ.
[*Liouville Mathematical Journal*, Vol. x. 1845. "ELECTROSTATICS AND MAGNETISM," Article XIV.]

ART. XX. NOTE ON INDUCED MAGNETISM IN A PLATE.
[*Cambridge and Dublin Mathematical Journal*, Vol. I. 1846. "ELECTROSTATICS AND MAGNETISM," Article IX.]

ART. XXI. ON THE MATHEMATICAL THEORY OF ELECTRICITY IN EQUILIBRIUM.
[*Cambridge and Dublin Mathematical Journal*, Vol. I. 1846. "ELECTROSTATICS AND MAGNETISM," Article V.]

[From the *Cambridge and Dublin Mathematical Journal*, Vol. I. 1846.]

ART. XXII.　Note on the Rings and Brushes in the Spectra produced by Biaxal Crystals.

It has been shewn in this Journal (Vol. III. p. 286) that if any system of isothermal plane curves be given, the orthogonal system, which is proved to be necessarily isothermal also, may in every case be determined. Thus if $v = a$ be the equation to the first system, v being a function of x and y which satisfies the equation

$$\frac{d^2v}{dx^2} + \frac{d^2v}{dy^2} = 0,$$

we shall have for the equation to the orthogonal system

$$u = \int \left(\frac{dv}{dy}\, dx - \frac{dv}{dx}\, dy \right) = \beta,$$

the expression under the sign of integration being in this case a complete differential, and the equation

$$\frac{d^2u}{dx^2} + \frac{d^2u}{dy^2} = 0$$

will be satisfied; results which may be readily verified.

To take an example, let Q, Q', &c. be any number of fixed points determined by the coordinates (a, b), (a', b'), &c., and let r, r', &c. be the distances of the point $P(xy)$ from those points. We may take for the equation of an isothermal system of curves

$$v = m \log r + m' \log r' + \&c. = a \quad\dots\dots\dots\dots(1),$$

where $r^2 = (x - a)^2 + (y - b)^2$, &c., and m, m', &c. are constants.

In this case we have

$$u = m \tan^{-1} \frac{x-a}{y-b} + m' \tan^{-1} \frac{x-a'}{y-b'} + \&c. = \beta \ \text{.......} \ (2)$$

for the orthogonal system, which, as may be readily verified, is also isothermal.

To take a simple case, let there be only two fixed points, Q, Q', and let $m = m' = 1$. The equation of the first system becomes

$$rr' = \epsilon^a = c \ \text{..........................} \ (3);$$

and, if we take the origin as the point of bisection of QQ', and make this line the axis of x, the equation of the second system becomes

$$\tan^{-1} \frac{x-a}{y} + \tan^{-1} \frac{x+a}{y} = \beta \ \text{.................} \ (4),$$

or

$$\frac{2xy}{y^2 - x^2 + a^2} = \tan \beta,$$

which may be put under the form

$$x^2 + 2kxy - y^2 = a^2 \ \text{.....................} \ (5).$$

The equation (3) represents the series of *lemniscates* which Herschel has shewn to be the forms of the rings in a biaxal crystal. Also (4) is the equation of a brush, since, if we draw PD bisecting the angle QPQ' and meeting QQ' in D, we have

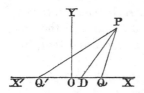

$$PDQ = PQX - DPQ = PQ'X + Q'PD$$

$$= \tfrac{1}{2}(PQX + PQ'X) = \tfrac{1}{2}\left(\tan^{-1}\frac{x-a}{y} + \tan^{-1}\frac{x+a}{y}\right).$$

Hence (4) represents the locus of the points P, when the angle PDQ is constant, which is the characteristic property of a brush.

Thus we see that the rings in a biaxal crystal form a system of isothermal plane curves, and the brushes the conjugate orthogonal system.

Some curious properties of the second system, which is a series of hyperbolas, may be deduced from equation (5). Let $\dfrac{a}{h_1^{\frac{1}{2}}}$ and $\dfrac{a}{h_2^{\frac{1}{2}}}$ be the semiaxes, real and imaginary, and θ the angle which the former makes with OX. To determine h_1, h_2, and θ, we have

$$(h-1)(h+1) = k^2,$$

$$\tan^2 \theta = \frac{h_1 - 1}{h_1 + 1},$$

from which we deduce $h_1 = (k^2 + 1)^{\frac{1}{2}},$

$$h_2 = -(k^2 + 1)^{\frac{1}{2}},$$

$$\tan^2 \theta = \frac{(k^2 + 1)^{\frac{1}{2}} - 1}{(k^2 + 1)^{\frac{1}{2}} + 1}.$$

Hence $$(k^2 + 1)^{\frac{1}{2}} = \frac{1 + \tan^2 \theta}{1 - \tan^2 \theta} = \frac{1}{\cos 2\theta},$$

and $$\left(\frac{a}{h_1^{\frac{1}{2}}}\right)^2 = a^2 \cos 2\theta.$$

Thus the second system is a series of rectangular hyperbolas whose vertices lie on the lemniscate of which the equation is

$$\rho^2 = a^2 \cos 2\theta.$$

By putting $y = 0$ in (5), we have, for the two values of x, $\pm a$, and therefore each hyperbola passes through the points Q and Q'. The series is determined by this and the preceding property.

In addition it may be remarked that, by putting $\epsilon^a = a^2$ in (3), we find

$$(x^2 + y^2)^2 = 2a^2 (x^2 - y^2),$$

or, in polar coordinates,

$$\rho^2 = 2a^2 \cos 2\theta,$$

for the equation of one of the curves of the first system. This curve is a lemniscate similar to that which is the locus of the vertices of the second system, and similarly situated, but of different magnitude.

ART. XXIII.   ON THE PRINCIPAL AXES OF A RIGID BODY.

[*Cambridge and Dublin Mathematical Journal*, Vol. I. 1846.   The substance
of this is given in Thomson and Tait's *Natural Philosophy*, §§ 282, 284.]

ART. XXIV.   NOTE ON A PAPER "SUR UNE PROPRIÉTÉ DE LA
COUCHE ELECTRIQUE EN EQUILIBRE A LA SURFACE D'UN
CORPS CONDUCTEUR."

[*Cambridge and Dublin Mathematical Journal*, Vol. I. 1846.   "ELECTROSTATICS
AND MAGNETISM," Article x.]

ART. XXV.   ON ELECTRICAL IMAGES.

[*British Association Report*, 1847.   "ELECTROSTATICS AND MAGNETISM,"
Article xiv.]

ART. XXVI.   ON THE ELECTRIC CURRENTS BY WHICH THE
PHENOMENA OF TERRESTRIAL MAGNETISM MAY BE PRODUCED.

[*British Association Report*, 1847.   "ELECTROSTATICS AND MAGNETISM,"
Article xxix.]

[From the *Cambridge and Dublin Mathematical Journal*, Vol. II. 1847.]

ART. XXVII. ON A MECHANICAL REPRESENTATION OF ELECTRIC, MAGNETIC, AND GALVANIC FORCES.

Mr FARADAY, in the eleventh series of his *Experimental Researches in Electricity*, has set forth a theory of Electrostatical Induction, which suggests the idea that there may be a problem in the theory of elastic solids corresponding to every problem connected with the distribution of electricity on conductors, or with the forces of attraction and repulsion exercised by electrified bodies. The clue to a similar representation of magnetic and galvanic forces is afforded by Mr Faraday's recent discovery of the affection with reference to polarized light, of transparent solids subjected to magnetic or electromagnetic forces. I have thus been led to find three distinct particular solutions of the equations of equilibrium of an elastic solid, of which one expresses a state of distortion such that the absolute displacement of a particle, in any part of the solid, represents the resultant attraction at this point, produced by an electrified body; another gives a state of the solid in which each element has a certain resultant angular displacement, representing in magnitude and direction the force at this point, produced by a magnetic body; and the third represents in a similar manner the force produced by any portion of a galvanic wire; the directions of the forces in the latter cases being given by the axes of the resultant rotations impressed upon the elements of the solid.

The general equations of equilibrium of an elastic solid have been investigated by Mr Stokes*, without the assumption of any relation between the "cubical compressibility" and the elasticity, with reference to variations of form which are not accompanied by change of volume. If we denote by α, β, γ the projections on three rectangular axes of coordinates, of the infinitely small displacement of a point (x, y, z) of the solid, it follows from Mr Stokes' results that the equations of equilibrium, when the body is acted on by no forces except at its bounding surfaces, may be written as follows:

$$\left. \begin{aligned} -\frac{dp}{dx} + \frac{d^2\alpha}{dx^2} + \frac{d^2\alpha}{dy^2} + \frac{d^2\alpha}{dz^2} &= 0 \\ -\frac{dp}{dy} + \frac{d^2\beta}{dx^2} + \frac{d^2\beta}{dy^2} + \frac{d^2\beta}{dz^2} &= 0 \\ -\frac{dp}{dz} + \frac{d^2\gamma}{dx^2} + \frac{d^2\gamma}{dy^2} + \frac{d^2\gamma}{dz^2} &= 0 \\ p = -k\left(\frac{d\alpha}{dx} + \frac{d\beta}{dy} + \frac{d\gamma}{dz}\right) & \end{aligned} \right\} \dots\dots\dots (1).$$

In the ideal limiting case in which the solid is incompressible, k will have an infinite value, and we shall have the relation

$$\frac{d\alpha}{dx} + \frac{d\beta}{dy} + \frac{d\gamma}{dz} = 0 \dots\dots\dots\dots (2).$$

Hence equations (1) and (2) express the conditions of the interior equilibrium of an incompressible elastic solid. These equations are to be employed for the representation of the forces in the several physical problems considered in this paper.

Now equations (1) merely shew that the expression

$$\nabla^2\alpha . dx + \nabla^2\beta . dy + \nabla^2\gamma . dz \dots\dots\dots\dots (a),$$

$\left(\text{in which } \nabla^2 \text{ denotes the operation } \dfrac{d^2}{dx^2} + \dfrac{d^2}{dy^2} + \dfrac{d^2}{dz^2}\right)$, must be a

complete differential, and therefore any expressions for α, β, γ subject to this condition, which satisfy (2), will represent an interior state of the body which can be produced by the action of forces at its bounding surface or surfaces.

---

* In a paper "On the Friction of Fluids in Motion, and the Equilibrium and Motion of Elastic Solids," read at the Cambridge Philosophical Society, April 14, 1845. See *Trans.*, Vol. VIII. Part 3.

We may obtain a particular solution by assuming

$$\alpha dx + \beta dy + \gamma dz$$

to be a complete differential. Again, if we suppose this expression not to be a complete differential, we may assume

$$\left(\frac{d\beta}{dz} - \frac{d\gamma}{dy}\right) dx + \left(\frac{d\gamma}{dx} - \frac{d\alpha}{dz}\right) dy + \left(\frac{d\alpha}{dy} - \frac{d\beta}{dx}\right) dz \ \dots\dots (c)$$

to be a complete differential and find another solution; or lastly we may obtain a particular solution by means of a third supposition, according to which neither of these expressions is a complete differential. These three solutions I shall now proceed to consider, with reference to the representation of Electrical, Magnetic, and Galvanic forces.

I. *Electrical Forces.*

Let $$r^2 = x^2 + y^2 + z^2,$$

and assume $$\alpha dx + \beta dy + \gamma dz = - d\left(\frac{1}{r}\right).$$

Then, since $$\frac{d^2}{dx^2}\frac{1}{r} + \frac{d^2}{dy^2}\frac{1}{r} + \frac{d^2}{dz^2}\frac{1}{r} = 0,$$

equation (2) is satisfied, and the coefficients of the differentials in (a) vanish; so that all the conditions of equilibrium are satisfied. Now $\frac{1}{r}$ is the potential at (x, y, z), due to a unit of electricity at the origin, and

$$\alpha = \frac{x}{r^3}, \quad \beta = \frac{y}{r^3}, \quad \gamma = \frac{z}{r^3} \dots\dots\dots\dots\dots(I.)$$

are the components of the force exerted at the point (xyz).

II. *Magnetic Forces.*

Let

$$\left(\frac{d\beta}{dz} - \frac{d\gamma}{dy}\right) dx + \left(\frac{d\gamma}{dx} - \frac{d\alpha}{dz}\right) dy + \left(\frac{d\alpha}{dy} - \frac{d\beta}{dx}\right) dz = d\frac{lx + my + nz}{r^3}.$$

This equation is satisfied by

$$\alpha = \frac{mz - ny}{r^3}, \quad \beta = \frac{nx - lz}{r^3}, \quad \gamma = \frac{ly - mx}{r^3} \ \dots\dots(3),$$

which also satisfy equation (2), and make the coefficients of the differentials in (*a*) vanish. Hence, displacements expressed in this way may be produced by externally applied forces. Now

$$\frac{lx + my + nz}{r^3}$$

is the potential due to a small magnet, of which the 'moment' is unity, placed at the origin, with its axis of polarization in the direction $l : m : n$. The components X, Y, Z of the force which this magnet exerts upon an ideal unit of magnetism (one end of a thin uniformly magnetized bar) at the point x, y, z being the differential coefficients of this expression, we have

$$X = \frac{d\beta}{dz} - \frac{d\gamma}{dy}, \quad Y = \frac{d\gamma}{dx} - \frac{d\alpha}{dz}, \quad Z = \frac{d\alpha}{dy} - \frac{d\beta}{dx} \dots\dots(\text{II.}).$$

The halves of the expressions $\frac{d\beta}{dz} - \frac{d\gamma}{dy}$, &c. indicate the components round lines parallel to the axes, of the infinitely small rotation which an element of the solid receives, besides its change of form, when $\alpha dx + \beta dy + \gamma dz$ is not a complete differential. This rotation therefore represents the resultant magnetic force, in direction and magnitude.

III. *Galvanic Forces.*

Let $\nabla^2\alpha . dx + \nabla^2\beta . dy + \nabla^2\gamma . dz = - d \dfrac{lx + my + nz}{r^3},$

which is true if

$$\left. \begin{aligned} \alpha &= \tfrac{1}{2} \frac{d}{dx} \frac{lx + my + nz}{r} - \frac{l}{r} \\ \beta &= \tfrac{1}{2} \frac{d}{dy} \frac{lx + my + nz}{r} - \frac{m}{r} \\ \gamma &= \tfrac{1}{2} \frac{d}{dz} \frac{lx + my + nz}{r} - \frac{n}{r} \end{aligned} \right\} \dots\dots\dots (4).$$

It is readily verified that these expressions also satisfy equation (2), and hence they represent an interior state of the body which may be produced by externally applied forces.   Now, we find by means of these equations

$$\left. \begin{aligned} \frac{d\beta}{dz} - \frac{d\gamma}{dy} &= \frac{mz - ny}{r^3} \\[1.5ex] \frac{d\gamma}{dx} - \frac{d\alpha}{dz} &= \frac{nx - lz}{r^3} \\[1.5ex] \frac{d\alpha}{dy} - \frac{d\beta}{dx} &= \frac{ly - mx}{r^3} \end{aligned} \right\} \quad\dots\dots\dots\dots (\text{III.})$$

which are the expressions for the components of the force an infinitely small element of a galvanic current, in the direction l, m, n, at the origin, produces on a unit of magnetism at the point (x, y, z); the intensity of the current, multiplied by the length of the element, being unity. Thus we conclude that the rotation of any element of the solid, in the state expressed by (4), represents in direction and magnitude, the force of an element of a galvanic wire.

I should exceed my present limits were I to enter into a special examination of the states of a solid body representing various problems in electricity, magnetism, and galvanism, which must therefore be reserved for a future paper.

GLASGOW COLLEGE, *Nov.* 28, 1846.

ART. XXVIII. ON CERTAIN DEFINITE INTEGRALS SUGGESTED BY PROBLEMS IN THE THEORY OF ELECTRICITY.

[*Cambridge and Dublin Mathematical Journal*, Vol. II. 1847. "ELECTROSTATICS AND MAGNETISM," Article XI.]

ART. XXIX. ON THE FORCES EXPERIENCED BY SMALL SPHERES UNDER MAGNETIC INFLUENCE; AND ON SOME OF THE PHENOMENA PRESENTED BY DIAMAGNETIC SUBSTANCES.

[*Cambridge and Dublin Mathematical Journal*, Vol. II. 1847. "ELECTROSTATICS AND MAGNETISM," Article XXXIII.]

[From the *Cambridge and Dublin Mathematical Journal*, May, 1847.]

ART. XXX.   ON A SYSTEM OF MAGNETIC CURVES.

LET λ be the potential produced by a magnet symmetrical about an axis OX at a point $P(x, y)$. The magnetic curves, or lines of force, being the orthogonal trajectories of the surfaces for which the potential is constant, will lie in planes through OX; and the system in the plane YOX will be the orthogonal trajectory of the system of curves, $\lambda = $ const. Their equation, as was shewn in a paper "On the Equations of Motion of Heat referred to Curvilinear Coordinates" (Vol. IV. p. 40), is

$$\int y \left(\frac{d\lambda}{dy} dx - \frac{d\lambda}{dx} dy \right) = C \dots\dots\dots\dots (1).$$

As an example, let λ be due to two small needles placed in the line OX, at points M, M'; so that we may take

$$\lambda = \frac{\mu (x-f)}{\{(x-f)^2 + y^2\}^{\frac{3}{2}}} + \frac{\mu' (x-f')}{\{(x-f')^2 + y^2\}^{\frac{3}{2}}} = \frac{\mu (x-f)}{\Delta^3} + \frac{\mu' (x-f')}{\Delta'^3}.$$

By integration we find, from (1),

$$\frac{\mu y^2}{\Delta^3} + \frac{\mu' y^2}{\Delta'^3} = C,$$

for the equation of the system of magnetic curves.

If we take as a particular case, $C = 0$, we find $y^2 = 0$, which shews that the axis is a line of force; we have also, for another branch, corresponding to the same value of C,

$$\frac{\mu}{\Delta^3} + \frac{\mu'}{\Delta'^3} = 0.$$

As Δ and Δ' are essentially positive in the physical problem, this can only be satisfied if μ and μ' have different signs.   For instance, if $\mu = 1$, $\mu' = -m$, we have

$$\Delta' = m^{\frac{1}{3}}\Delta.$$

The locus of this equation is, as is well known, a circle, which may be described thus.   Divide MM' in A, and produce it to A_1, so that

$$M'A = m^{\frac{1}{3}} . MA \text{ and } M'A_1 = m^{\frac{1}{3}} . MA_1;$$

on AA_1 as diameter describe a circle.

This result was suggested to me by the solution of a corresponding problem (of much greater interest however) in fluid motion, verbally communicated to me by Mr Stokes.

St Peter's College, *May* 18, 1847.

ART. XXXI. NOTES ON HYDRODYNAMICS.

(1) *On the Equation of Continuity.*

[Not reprinted because its substance is repeated in Thomson and Tait's *Nat. Phil.* § 194, and is also to be found in most text-books on Hydrodynamics.]

[From the *Cambridge and Dublin Mathematical Journal*, Feb., 1848.]

(2) *On the Equation of the Bounding Surface.*

IF a fluid mass be in motion, under any conceivable circumstances, its bounding surface will always be such that there will be no motion of the fluid across it. By expressing this circumstance analytically, we obtain the differential equation of the bounding surface*.

In most problems of Hydrodynamics certain conditions are given with reference to the surface within which the fluid mass

---

* The condition to be expressed is entirely equivalent to this : that the surface must always contain the same fluid matter within it, and it therefore appears that the differential equation expressing this condition, according to the investigation in the text, is satisfied, without exception, at the extreme boundary of the fluid, as well as at the varying surface bounding any portion of the entire mass. Mr Stokes is, I believe, the only writer who has given this view of the subject, all other authors having taken for the condition to be expressed, that a particle which is once in the bounding surface always remains in the bounding surface. Poisson has justly remarked that cases may actually occur in which this condition is violated; but we cannot infer, as he and subsequent authors have done, that the differential equation is liable to exception in its applications, although we may conclude that the demonstration they have given fails in certain cases. In the demonstration given in the text, founded on a *necessary* condition, there are no such cases of failure. See Poisson, *Traité de Mécanique*, Vol. II. No. 652; Duhamel, *Cours de Mécanique, Partie deuxième*, p. 266; Stokes, *Cambridge Philosophical Transactions*, Vol. VIII. p. 299.

considered is contained : for instance, that this surface is given as
fixed in form and position ; that its form or position varies in a
given arbitrary manner ; or, as may be the case when the mass
considered is a *liquid*, that the bounding surface is free, but subjected to a given constant pressure.   Hence, in the analysis of all
such problems it will be necessary to have an expression of the
fact that the fluid is actually contained within the surface at which
the given conditions are satisfied.   This will be effected by combining the differential equation of the surface with the special
equation, or equations, of condition in each particular problem, all
of which must be satisfied at the bounding surface of the fluid.
Hence the investigation of the equation of the surface ought to
find a place in a complete treatise on the mathematical theory of
hydrodynamics.   It is wanting in many of the elementary works,
and this paper may therefore be considered to be useful as supply-
ing the deficiency.

It will not be necessary for us in the present investigation to
consider at all whether the fluid be compressible or incompressible;
and we may avoid all restriction by supposing the boundary to be
a varying surface S.   This will be expressed by considering t, the
time, as a variable parameter in the equation of S, which we may
therefore assume to be

$$F(x, y, z, t) = 0 \quad \dots\dots\dots\dots\dots\dots\dots (a).$$

To express the fact that every particle of the fluid remains on the
same side of this surface, or that there is no *flux* across it, we must
find the normal motion of the surface, at any part, in an infinitely
small time dt, and equate this to the normal component of the
motion of a neighbouring fluid particle during the same time.
The normal motion of the surface will be the projection upon the
normal of the line PP', joining two points P and P' infinitely near
one another, of which the former is on the surface S, at the time t,
and the latter on the altered surface, at the time $t + dt$, their
coordinates therefore satisfying the equations

$$\left.\begin{array}{l} F(x, y, z, t) = 0 \\ F(x', y', z', t + dt) = 0 \end{array}\right\} \quad \dots\dots\dots\dots\dots (b).$$

Now, with the notation

$$\{F'(x)\}^2 + \{F'(y)\}^2 + \{F'(z)\}^2 = R^2,$$

we have, for the projection of PP' on the normal, the expression

$$\frac{(x'-x)\,F''(x) + (y'-y)\,F''(y) + (z'-z)\,F''(z)}{R}.$$

But, since $x'-x$, $y'-y$, $z'-z$ are infinitely small, we deduce, from equations (b),

$$(x'-x)\,F''(x) + (y'-y)\,F''(y) + (z'-z)\,F''(z) + dt \cdot F''(t) = 0.$$

Consequently we have, for the normal motion of the surface, during the time dt,

$$-\frac{F''(t)}{R}\,dt.$$

Now the normal component of the motion of a fluid particle in the neighbourhood of the point (x, y, z), during the time dt, is equal to

$$u\,dt \cdot \frac{F''(x)}{R} + v\,dt \cdot \frac{F''(y)}{R} + w\,dt \cdot \frac{F''(z)}{R},$$

and we therefore have

$$\frac{u\,dt \cdot F''(x) + v\,dt \cdot F''(y) + w\,dt \cdot F''(z)}{R} = -\frac{F''(t)\,dt}{R}.$$

This may be reduced to the simpler form

$$u\,F''(x) + v\,F''(y) + w\,F''(z) + F''(t) = 0 \ \ldots\ldots\ldots\ (1),$$

which is the required differential equation of the bounding surface.

To illustrate the applicability of this equation to the class of cases considered by Poisson as exceptional*, let us, for simplicity,

---

* The following passage, extracted from Art. 652 of the *Traité de Mécanique*, contains Poisson's statement with reference to the restriction. "Dans les mouvements des fluides que l'on soumet au calcul, on a coutume de supposer que les points qui se trouvent à une époque déterminée sur une paroi fixe ou mobile, ou qui appartiennent à la surface libre d'un liquide, demeureront sur cette paroi, ou appartiendront à cette surface pendant toute la durée du mouvement; en sorte que l'on exclut les mouvements très compliqués dans lesquels des points d'un fluide après avoir appartenu à sa superficie, rentreraient dans l'intérieur de la masse, ou réciproquement; et l'on exclut même les cas où des points d'un liquide passeraient alternativement de la surface libre à la surface en contact avec une paroi fixe ou mobile. Ces conditions particulières auxquelles on assujettit les mouvements que l'on considère, s'expriment par les équations suivantes." This passage is followed by a demonstration of the differential equation of the surface.

suppose the fluid to be contained within a fixed surface S. The condition on which the preceding demonstration depends may, in this case, be satisfied in three different ways by the motion of a fluid particle which at some instant is in contact with the bounding surface.

First, if its line of motion lie entirely on the surface, S;

Secondly, if its line of motion lie within S, but touch it in one or more points;

Thirdly, if its velocity as it approaches the surface diminish continually, and vanish when it reaches the surface, in which case the line of motion may meet the surface at any angle.

The second and third kinds of motion (which are excluded according to the restrictions already referred to) may be illustrated by taking as examples, actual cases of motion such as the following.

Let part of the fixed bounding surface S be plane, and let a certain cylindrical portion of the fluid, touched along a straight line by this plane, revolve continually about its axis, while the rest of the fluid moves within the surface S in any way that would be possible were the cylindrical portion a fixed solid; or, whatever be the form of S, let a spherical portion of the fluid revolve in any way, touching S in a point, the rest of the fluid moving as if this portion were solid. Thus we have two easily imagined cases of motion, in one of which, at a certain straight line, and in the other at a certain point, particles of the fluid are continually *coming to the surface, and then retreating into the interior of the fluid.*

To illustrate the third kind of motion, a certain portion of the fluid may be supposed to move as if it were solid, with a continually decreasing velocity, till it comes to rest, when one point of it reaches the surface; or, we may suppose the portion of the fluid which moves as if it were solid, to be of such a form that a finite portion of its surface may be made to coincide with part of S, and we may suppose it to move with continually decreasing velocity, till, when this coincidence takes place, it comes to rest. In such cases particles of the fluid which were originally in the interior, come to occupy positions on the bounding surface; yet the condition, on which the demonstration given above depends, is not violated; and the equation arrived at does not become nugatory in form, so that we must consider it as literally satisfied.

With reference to the additional restriction made by Poisson, excluding cases in which particles of a liquid *pass alternately from the free surface to the surface in contact with a fixed or moveable vessel containing it*, it may be remarked that, although there will in general be the same kind of difficulty in such cases of motion, as in the others considered above, yet even Poisson's investigation is perfectly applicable, provided that the free and constrained surfaces have a common tangent plane along their line of intersection, a condition which may be actually ensured by arbitrarily constraining, by means of a flexible and extensible envelope, the motion of a band of the bounding surface of the liquid, contiguous to the free surface.

GLASGOW COLLEGE, *Jan.* 10, 1848.

ART. XXXII. SYSTÈME NOUVEAUX DE COORDONNÉS ORTHOGONALES.

[Liouville, *Jour. Math.* Vol. XII. 1847. "ELECTROSTATICS AND MAGNETISM," Article XIV. §§ 211, 220.]

ART. XXXIII. NOTE SUR UNE ÉQUATION AUX DIFFÉRENCES PARTIELLES QUI SE PRÉSENTE DANS PLUSIEURS QUESTIONS DE PHYSIQUE MATHÉMATIQUE.

[Liouville, *Math. Jour.* Vol. XII. 1847. "ELECTROSTATICS AND MAGNETISM," Article XIII.]

[From the *British Association Report*, 1848 (Part II.). Omitted accidentally from Reprint " Electrostatics and Magnetism."]

ART. XXXIV. ON THE EQUILIBRIUM OF MAGNETIC OR DIA-
MAGNETIC BODIES OF ANY FORM, UNDER THE INFLUENCE
OF THE TERRESTRIAL MAGNETIC FORCE.

IF a body composed of a magnetic substance, such as soft
iron, or of a diamagnetic substance, be supported by its centre of
gravity, the effects of the terrestrial magnetic force in producing
magnetism by induction, and in acting on the magnetism so
developed, are in general such as to impress a certain directive
tendency on the mass. The investigation of these circumstances
leads to results according to which the conditions of equilibrium
of a body of any irregular form may be expressed in a very elegant
and simple manner.

In the present communication I shall merely give a brief
general explanation of the conclusions at which I have arrived,
without attempting to state fully the process of reasoning on which
they are founded, as this could not be rendered intelligible without
entering upon mathematical details, which must be reserved for a
paper of greater length.

In the first place, if the body considered be an ellipsoid of
homogeneous matter, supported by its centre of gravity, it is clear
that it will be in equilibrium with any one of its three principal
axes in the direction of the lines of force ; and if it be put into
any other position, the action of the earth upon it will be a couple,
of which the moment may be expressed very simply in terms of
the quantities denoting the position of the axes with reference to

the direction of the terrestrial force, and certain constant magnetic elements depending on the substance and dimensions of the body.

In considering the corresponding problem for a body of any irregular form, we readily obtain for the components of the directive couple, round three rectangular axes chosen arbitrarily in the body, expressions involving nine constant magnetic elements. I have succeeded in proving that six of these elements must be equal, two and two; so that the entire number of independent constants is reduced to six. I have thus arrived at the interesting theorem, that there are, in any irregular body, *three principal magnetic axes at right angles to one another*, such that if the body be supported by its centre of gravity, it will be in equilibrium with any one of these axes in the direction of the terrestrial magnetic force. If the body be held in any other position, there will be a directive couple of which the moment is expressible in precisely the same form as in the case of an ellipsoid, in terms of the magnetic elements of the body, and of the quantities denoting its position.

From this it follows, that, as far as regards the directive action of terrestrial magnetism, *the ellipsoid with three unequal axes may be taken as the general type for a body of any form whatever.*

Besides other special cases of interest, to which it is unnecessary for me at present to call attention, on account of the close analogy which is presented by the well-known theory of principal axes in dynamics, there is one to which I shall allude, on account of its importance with reference to the general principles on which the directive agency depends. If the body considered be a cube, the three principal magnetic elements will be equal, and therefore the corresponding ellipsoid must have its three axes equal; that is, it must be a sphere. Hence a cube supported by its centre of gravity cannot experience any directive tendency, and will therefore be *astatic*.

Now a mass of any form may be divided into an infinite number of small cubes, and the resultant of the actual directive couples on all of these cubes will determine the directive tendency of the whole mass. Hence if each small cube were acted upon only by the terrestrial magnetic force, there would be no directive agency on the body; and it is to the modification of the circumstances introduced by the mutual action of the different parts of the body that we must ascribe the directive tendency which is actually experienced, in general by irregular, but especially by elongated masses. This modification is distinctly alluded to by Mr Faraday in his memoir on the General Magnetic Condition of Matter (Experimental Researches, § 2264), and the directive tendency which he has observed in needles of diamagnetic substances is shewn to depend on essentially different physical circumstances (§§ 2269, 2418) connected with the variation of the total intensity of the resultant magnetic force in the neighbourhood of the poles of a magnet, and quite independent of the actual directions of the lines of force. A mathematical investigation of the circumstances on which these phænomena depend will be found in the *Cambridge and Dublin Mathematical Journal* (May 1847), ELECTRO-STATICS AND MAGNETISM, Art. XXXIII.

From the principles alluded to above, we may draw the following general conclusions with reference to the action experienced by a body subjected to magnetic influence when the intensity of the magnetizing force is constant in its neighbourhood.

1.   The directive tendency on a diamagnetic substance of any form must be extremely small, probably quite insensible in any actual experiment that can be made; depending as it does upon the mutual action of the parts of the body which are primarily influenced to but a very feeble extent in the case of every known diamagnetic.

2.   An elongated body, whether of a magnetic or of a diamagnetic substance, will tend to place itself in the direction of the lines of force; so that, for instance, either a bar of soft iron, or a diamagnetic bar, supported by its centre of gravity, would, if perfectly free, assume the position of the dipping-needle.

[For further information on this subject, see ELECTROSTATICS AND MAGNETISM, §§ 647—732.]

[From the *British Association Report*, 1848 (Part II.).]

Art. XXXV. On the Theory of Electro-magnetic Induction.

THE theory of electro-magnetic induction, founded on the elementary experiments of Faraday and Lenz, has been subjected to mathematical analysis by Neumann, who has recently laid some very valuable researches on this subject before the Berlin Academy of Sciences. The case of a closed linear conductor (a bent metallic wire with its ends joined) under the influence of a magnet in a state of relative motion is considered in Neumann's first memoir*, and a very beautiful theorem is demonstrated, completely expressing the circumstances which determine the intensity of the induced current. It has appeared to me that a very simple *à priori* demonstration of this theorem may be founded on the axiom that the amount of work expended in producing the relative motion on which the electro-magnetic induction depends must be equivalent to the mechanical effect lost by the current induced in the wire.

In the first place, it may be proved that the amount of the mechanical effect continually *lost* or spent in some physical agency (according to Joule the generation of heat) during the existence of a galvanic current in a given closed wire is, for a given time, proportional to the square of the intensity of the current. For, whatever be the actual source of the galvanism, an equivalent current might be produced by the motion of a magnetic body in the neighbourhood of the closed wire. If now, other circumstances remaining the same, the intensity of the magnetism in the influencing body be altered in any ratio, the intensity of the induced current must be proportionately changed; hence the amount of work spent in the motion, as it depends on the mutual influence of the magnet and the induced current, is altered in the

---

* A translation of this memoir into French is published in the last April number of Liouville's *Journal des Mathématiques*.

duplicate ratio of that in which the current is altered; and therefore the amount of mechanical effect lost in the wire, being equivalent to the work spent in the motion, must be proportional to
the square of the intensity of the current.   Hence if i denote the
intensity of a current existing in a closed conductor, the amount
of work lost by its existence for an interval of time dt, so small
that the intensity of the current remains sensibly constant during
it, will be $k \cdot i^2 \cdot dt$; where k is a certain constant depending on
the resistance of the complete wire.

Let us now suppose this current to be actually produced by
induction in the wire, under the influence of a magnetic body in a
state of relative motion.   The entire mutual force between the
magnetic and the galvanic wire may, according to Ampère's theory,
be expressed by means of the differential coefficients of a certain
"force function."   This function, which may be denoted by U, will
be a quantity depending solely on the form and position of the
wire at any instant, and on the magnetism of the influencing body.
During the very small time dt, let U change from U to $U + dU$,
by the relative motion which takes place during that interval.
Then idU will be the amount of work spent in sustaining the
motion; but the mechanical effect lost in the wire during the
same interval is equal to $k i^2 dt$ [if we neglect "self-induction"];
and therefore we must have

$$i\,dU = k i^2\,dt.$$

Hence, dividing both members by $kidt$, we deduce

$$i = \frac{1}{k} \cdot \frac{dU}{dt},$$

which expresses the theorem of Neumann, the subject of the present communication.   We may enunciate the result in general
language thus:—

When a current is induced in a closed wire by a magnet in
relative motion [and when "self-induction" is negligible], the intensity of the current produced is proportional to the actual rate of
variation of the "force function" by the differential coefficients of
which the mutual action between the magnet and the wire would be
represented if the intensity of the current in the wire were unity.

[For more on this subject, and particularly for the correction to take "self-
induction" into the account, see Articles on "Electromagnetic Induction"
below.]

[From the *Cambridge and Dublin Mathematical Journal*, Feb., 1848.
ELECTROSTATICS AND MAGNETISM, Article XIII.]

ART. XXXVI. THEOREMS WITH REFERENCE TO THE SOLUTION
OF CERTAIN PARTIAL DIFFERENTIAL EQUATIONS.

THEOREM 1. It is possible to find a function V, of x, y, z*,
which shall satisfy, for all real values of these variables, the differential equation

$$\frac{d\left(\alpha^2 \frac{dV}{dx}\right)}{dx} + \frac{d\left(\alpha^2 \frac{dV}{dy}\right)}{dy} + \frac{d\left(\alpha^2 \frac{dV}{dz}\right)}{dz} = 4\pi\rho \quad\ldots\ldots\ldots\ldots(A),$$

α being any real continuous or discontinuous function of x, y, z,
and ρ a function which vanishes for all values of x, y, z, exceeding
certain finite limits (such as may be represented geometrically by
a finite closed surface), within which its value is finite, but entirely
arbitrary.

THEOREM 2. There cannot be two different solutions of equation (A) for all real values of the variables.

1. (*Demonstration.*) Let U be a function of x, y, z, given by
the equation

$$U = \iiint \frac{\rho' dx' dy' dz'}{\{(x - x')^2 + (y - y')^2 + (z - z')^2\}^{\frac{1}{2}}} \quad\ldots\ldots\ldots\ldots(a),$$

the integrations in the second member including all the space for
which ρ' is finite; so that, if we please, we may conceive the
limits of each integration to be $-\infty$ and $+\infty$, as thus all the

---

* The case of three variables, which includes the applications to physical
problems, is alone considered here; although the analysis is equally applicable
whatever be the number of variables.

values of the variables for which ρ' is finite will be included, and the amount of the integral will not be affected by those values of the variables for which ρ' vanishes being included. Again, V being any real function of x, y, z, let

$$Q = \int_{-\infty}^{\infty}\int_{-\infty}^{\infty}\int_{-\infty}^{\infty}\left\{\left(\alpha\frac{dV}{dx}-\frac{1}{\alpha}\frac{dU}{dx}\right)^2+\left(\alpha\frac{dV}{dy}-\frac{1}{\alpha}\frac{dU}{dy}\right)^2\right.$$
$$\left.+\left(\alpha\frac{dV}{dz}-\frac{1}{\alpha}\frac{dU}{dz}\right)^2\right\}dx\,dy\,dz \ldots\ldots\ldots(b).$$

It is obvious that, although V may be assigned so as to make Q as great as we please, it is impossible to make the value of Q less than a certain limit, since we see at once that it cannot be negative. Hence Q, considered as depending on the arbitrary function V, is susceptible of a minimum value; and the calculus of variations will lead us to the assigning of V according to this condition.

Thus we have

$$-\tfrac{1}{2}\delta Q = \iiint\left\{\left(\alpha\frac{dV}{dx}-\frac{1}{\alpha}\frac{dU}{dx}\right).\alpha\frac{d\delta V}{dx}+\left(\alpha\frac{dV}{dy}-\frac{1}{\alpha}\frac{dU}{dy}\right).\alpha\frac{d\delta V}{dy}\right.$$
$$\left.+\left(\alpha\frac{dV}{dz}-\frac{1}{\alpha}\frac{dU}{dz}\right).\alpha\frac{d\delta V}{dz}\right\}dx\,dy\,dz.$$

Hence, by the ordinary process of integration by parts, the integrated terms vanishing at each limit*, we deduce

$$-\tfrac{1}{2}\delta Q = \iiint\delta V\left\{\frac{d}{dx}\left(\alpha^2\frac{dV}{dx}-\frac{dU}{dx}\right)+\frac{d}{dy}\left(\alpha^2\frac{dV}{dy}-\frac{dU}{dy}\right)\right.$$
$$\left.+\frac{d}{dz}\left(\alpha^2\frac{dV}{dz}-\frac{dU}{dz}\right)\right\}dx\,dy\,dz.$$

But by a well-known theorem (proved in Pratt's *Mechanics*, and in the treatise on Attraction in Earnshaw's *Dynamics*), we have

$$\frac{d^2U}{dx^2}+\frac{d^2U}{dy^2}+\frac{d^2U}{dz^2}=4\pi\rho.$$

Hence the preceding expression becomes

$$-\tfrac{1}{2}\delta Q = \iiint\delta V.\left\{\frac{d}{dx}\left(\alpha^2\frac{dV}{dx}\right)+\frac{d}{dy}\left(\alpha^2\frac{dV}{dy}\right)\right.$$
$$\left.+\frac{d}{dz}\left(\alpha^2\frac{dV}{dz}\right)-4\pi\rho\right\}dx\,dy\,dz.$$

---

* All the functions of x, y, z contemplated in this paper are supposed to vanish for infinite values of the variables.

We have therefore, for the condition that Q may be a maximum or minimum, the equation,

$$\frac{d}{dx}\left(\alpha^2 \frac{dV}{dx}\right) + \frac{d}{dy}\left(\alpha^2 \frac{dV}{dy}\right) + \frac{d}{dz}\left(\alpha^2 \frac{dV}{dz}\right) = 4\pi\rho,$$

to be satisfied for all values of the variables.

Now it is possible to assign V so that Q may be a minimum, and therefore there exists a function, V, which satisfies equation (A).

2. (*Demonstration.*) Let V be a solution of (A), and let V_1 be any different function of x, y, z, that is to say, any function such that $V_1 - V$, which we may denote by ϕ, does not vanish for all values of x, y, z. Let us consider the integral Q_1, obtained by substituting V_1 for V, in the expression for Q. Since

$$\left(\alpha \frac{dV_1}{dx} - \frac{1}{\alpha} \frac{dU}{dx}\right)^2$$

$$= \left(\alpha \frac{dV}{dx} - \frac{1}{\alpha} \frac{dU}{dx}\right)^2 + 2\left(\alpha \frac{dV}{dx} - \frac{1}{\alpha} \frac{dU}{dx}\right)\alpha \frac{d\phi}{dx} + \alpha^2 \frac{d\phi^2}{dx^2},$$

we have

$$Q_1 = Q + 2 \iiint \left\{ \left(\alpha \frac{dV}{dx} - \frac{1}{\alpha} \frac{dU}{dx}\right)\alpha \frac{d\phi}{dx} + \left(\alpha \frac{dV}{dy} - \frac{1}{\alpha} \frac{dU}{dy}\right)\alpha \frac{d\phi}{dy} \right.$$

$$\left. + \left(\alpha \frac{dV}{dz} - \frac{1}{\alpha} \frac{d\phi}{dz}\right)\alpha \frac{d\phi}{dz} \right\} dx\,dy\,dz$$

$$+ \iiint \alpha^2 \left(\frac{d\phi^2}{dx^2} + \frac{d\phi^2}{dy^2} + \frac{d\phi^2}{dz^2}\right) dx\,dy\,dz.$$

Now, by integration by parts, we find

$$\int_{-\infty}^{\infty}\int_{-\infty}^{\infty}\int_{-\infty}^{\infty} \left(\alpha \frac{dV}{dx} - \frac{1}{\alpha} \frac{dU}{dx}\right)\alpha \frac{d\phi}{dx} . dx\,dy\,dz$$

$$= -\iiint \phi . \frac{d}{dx}\left(\alpha^2 \frac{dV}{dx} - \frac{dU}{dx}\right) dx\,dy\,dz,$$

the integrated term vanishing at each limit. Applying this and similar processes with reference to y and z, we find an expression for the second term of Q_1 which, on account of equation (A), vanishes. Hence

$$Q_1 = Q + \iiint \alpha^2 \left(\frac{d\phi^2}{dx^2} + \frac{d\phi^2}{dy^2} + \frac{d\phi^2}{dz^2}\right) dx\,dy\,dz \ldots\ldots\ldots\ldots(c),$$

which shews that Q_1 is greater than Q. Now the only peculiarity of Q is, that V, from which it is obtained, satisfies the equation (A), and therefore V_1 cannot be a solution of (A). Hence no function different from V can be a solution of (A).

The analysis given above, especially when interpreted in various cases of abrupt variations in the value of α, and of infinite or evanescent values, through finite spaces, possesses very important applications in the theories of heat, electricity, magnetism, and hydrodynamics, which may form the subject of future communications.

EDENBARNET, DUMBARTONSHIRE, *Oct.* 9, 1847.

[From the *Cambridge and Dublin Mathematical Journal*, Feb. 1848.]

ART. XXXVII. NOTE ON THE INTEGRATION OF THE EQUATIONS
OF EQUILIBRIUM OF AN ELASTIC SOLID.

IN a short paper published at the beginning of this year (1847;
see *Mathematical Journal*, Vol. II. p. 61) [Art. XXVII. above] certain
particular integrals of the equations of equilibrium of an incompressible elastic solid were given.   It may readily be shewn that
any integral whatever may be expressed by the sum of such particular integrals as those which are given in equations (4) of the
paper referred to ; or that any possible state of distortion of an
incompressible solid may be considered as the resultant of coexistent elementary states of distortion, each of which is represented
by a particular solution of the form there expressed.   Hence the
consideration of the particular integrals alluded to may be of great
importance in the mathematical theory of incompressible solids;
and it would be very desirable to have a similar method for treating the theory of elastic solids in general.   I have recently found
that a slight modification may be introduced in the expression
already referred to, by which their application will be extended
from the case of incompressible solids to that of solids in general.
Thus, l, m, n, and f, g being arbitrary constants, if we take the
expressions written below, for α, β, γ, the three components of the
displacement of a point (x, y, z) from its position of equilibrium,
we have, by means of the following assumptions, a solution of the
general equations of equilibrium, for a solid of any kind, which
will become the same as equations (4) of the former paper, corresponding to the case of an incompressible solid, if we put $f = g$.

T.   7

Let
$$\alpha = \tfrac{1}{2} f \frac{d}{dx} \frac{lx + my + nz}{r} - g \frac{l}{r},$$

$$\beta = \tfrac{1}{2} f \frac{d}{dy} \frac{lx + my + nz}{r} - g \frac{m}{r},$$

$$\gamma = \tfrac{1}{2} f \frac{d}{dz} \frac{lx + my + nz}{r} - g \frac{n}{r},$$

where $r = (x^2 + y^2 + z^2)^{\frac{1}{2}}$.

To shew that these equations satisfy the three general equations of equilibrium, of which the following is one, and the other two complete a symmetrical system*,

$$k \frac{d}{dx} \left(\frac{d\alpha}{dx} + \frac{d\beta}{dy} + \frac{d\gamma}{dz} \right) + \frac{d^2\alpha}{dx^2} + \frac{d^2\alpha}{dy^2} + \frac{d^2\alpha}{dz^2} = 0,$$

we have merely to perform the necessary differentiations and substitutions. Thus we find

$$\frac{d^2\alpha}{dx^2} + \frac{d^2\alpha}{dy^2} + \frac{d^2\alpha}{dz^2} = -f \frac{d}{dx} \frac{lx + my + nz}{r^3},$$

$$\frac{d\alpha}{dx} + \frac{d\beta}{dy} + \frac{d\gamma}{dz} = (-f + g) \frac{lx + my + nz}{r^3}$$

Hence, by substitution in the first member of the preceding equation, we find that it vanishes if

$$k (g - f) - f = 0.$$

This equation may be put under the form

$$\frac{f}{g} = \frac{k}{k + 1}.$$

If now we suppose the solid to be incompressible, we must take $k = \infty$. This gives $f = g$, and we have the case of the former paper.

The application of the solution expressed above, in the mathematical theory of elastic solids, is analogous in some degree to a method of treating certain questions in the theory of heat, which was indicated in a paper "On the Uniform Motion of Heat in Solid Bodies, and its Connexion with the Mathematical Theory of

---

* Already referred to in the *Math. Journal*, 1847 [Article XXVII. above]. See Stokes on Elastic Solids, *Camb. Trans.* April, 1845, the only work in which the true formulæ are given.

Electricity." [*Electrostatics and Magnetism*, Art. I.]   Instead of the three components of the flux of heat at any point, we have to consider the three components of the rotation of an element of a distorted solid; instead of a source of heat, we have a *source of strain* round the point of application of a force.   If the solid were incompressible, there would be as close a connexion with the mathematical theory of electro-magnetism, as was shewn to subsist between the theories of heat and electricity : this follows at once from the theorem given in the former paper, with reference to the "mechanical representation" of the electro-magnetic force due to an element of a galvanic arc.

I cannot at present enter upon the general mathematical theory of elastic solids; but the method here indicated may be readily imagined by examining the meaning of the particular solution given above.   It is easily shewn that those forms for α, β, γ express the three components of the displacement produced at a point (x, y, z) in an infinite homogeneous elastic solid, where a force is applied to it at the origin of coordinates, in the direction (l, m, n).

GLASGOW COLLEGE, *December* 10, 1847.

ART. XXXVIII.   ON THE MATHEMATICAL THEORY OF ELECTRICITY IN EQUILIBRIUM.

[*Cambridge and Dublin Mathematical Journal, March, May, and Nov.* 1848, *Nov.* 1849, *and Feb.* 1850.
ELECTROSTATICS AND MAGNETISM, Articles IV. and V.]

[*Cambridge Philosophical Society Proceedings for June* 5, 1848; *and Phil. Mag., Oct.* 1848.]

ART. XXXIX. ON AN ABSOLUTE THERMOMETRIC SCALE FOUNDED
ON CARNOT'S THEORY OF THE MOTIVE POWER OF HEAT*, AND
CALCULATED FROM REGNAULT'S OBSERVATIONS†.

THE determination of temperature has long been recognized as
a problem of the greatest importance in physical science. It has
accordingly been made a subject of most careful attention, and,
especially in late years, of very elaborate and refined experimental
researches‡; and we are thus at present in possession of as complete
a practical solution of the problem as can be desired, even for the
most accurate investigations. The theory of thermometry is however as yet far from being in so satisfactory a state. The principle
to be followed in constructing a thermometric scale might at first
sight seem to be obvious, as it might appear that a perfect thermometer would indicate equal additions of heat, as corresponding to
equal elevations of temperature, estimated by the numbered divisions of its scale. It is however now recognized (from the variations in the specific heats of bodies) as an experimentally demonstrated fact that thermometry under this condition is impossible,

* Published in 1824 in a work entitled *Réflexions sur la Puissance Motrice du
Feu*, by M. S. Carnot. Having never met with the original work, it is only through
a paper by M. Clapeyron, on the same subject, published in the *Journal de l'École
Polytechnique*, Vol. XIV. 1834, and translated in the first volume of Taylor's *Scientific
Memoirs*, that the Author has become acquainted with Carnot's Theory.—W. T.
[Note of Nov. 5th, 1881. A few months later through the kindness of my late
colleague Prof. Lewis Gordon, I received a copy of Carnot's original work and was
thus enabled to give to the Royal Society of Edinburgh my "Account of Carnot's
theory" which is reprinted as Art. XLI. below. The original work has since been
republished, with a biographical notice, Paris, 1878.]
† An account of the first part of a series of researches undertaken by M.
Regnault by order of the French Government, for ascertaining the various physical
data of importance in the Theory of the Steam Engine, is just published in the
Mémoires de l'Institut, of which it constitutes the twenty-first volume (1847). The
second part of the researches has not yet been published. [Note of Nov. 5, 1881.
The continuation of these researches has now been published: thus we have for the
whole series, Vol. I. in 1847; Vol. II. in 1862; and Vol. III. in 1870.]
‡ A very important section of Regnault's work is devoted to this object.

and we are left without any principle on which to found an absolute thermometric scale.

Next in importance to the primary establishment of an absolute scale, independently of the properties of any particular kind of matter, is the fixing upon an arbitrary system of thermometry, according to which results of observations made by different experimenters, in various positions and circumstances, may be exactly compared. This object is very fully attained by means of thermometers constructed and graduated according to the clearly defined methods adopted by the best instrument-makers of the present day, when the rigorous experimental processes which have been indicated, especially by Regnault, for interpreting their indications in a comparable way, are followed. The particular kind of thermometer which is least liable to uncertain variations of any kind is that founded on the expansion of air, and this is therefore generally adopted as the standard for the comparison of thermometers of all constructions. Hence the scale which is at present employed for estimating temperature is that of the air-thermometer; and in accurate researches care is always taken to reduce to this scale the indications of the instrument actually used, whatever may be its specific construction and graduation.

The principle according to which the scale of the air-thermometer is graduated, is simply that equal absolute expansions of the mass of air or gas in the instrument, under a constant pressure, shall indicate equal differences of the numbers on the scale; the length of a "degree" being determined by allowing a given number for the interval between the freezing- and the boiling-points. Now it is found by Regnault that various thermometers, constructed with air under different pressures, or with different gases, give indications which coincide so closely, that, unless when certain gases, such as sulphurous acid, which approach the physical condition of vapours at saturation, are made use of, the variations are inappreciable*. This remarkable circumstance enhances very much the practical value of the air-thermometer; but still a

---

* Regnault, *Relation des Expériences*, &c., Fourth Memoir, First Part. The differences, it is remarked by Regnault, would be much more sensible if the graduation were effected on the supposition that the coefficients of expansion of the different gases are equal, instead of being founded on the principle laid down in the text, according to which the freezing- and boiling-points are experimentally determined for each thermometer.

rigorous standard can only be defined by fixing upon a certain
gas at a determinate pressure, as the thermometric substance.
Although we have thus a strict principle for constructing a *definite*
system for the estimation of temperature, yet as reference is
essentially made to a specific body as the standard thermometric
substance, we cannot consider that we have arrived at an *absolute*
scale, and we can only regard, in strictness, the scale actually
adopted as *an arbitrary series of numbered points of reference sufficiently close for the requirements of practical thermometry.*

In the present state of physical science, therefore, a question
of extreme interest arises: *Is there any principle on which an absolute thermometric scale can be founded?* It appears to me that
Carnot's theory of the motive power of heat enables us to give an
affirmative answer.

The relation between motive power and heat, as established
by Carnot, is such that *quantities of heat,* and *intervals of temperature,* are involved as the sole elements in the expression for the
amount of mechanical effect to be obtained through the agency of
heat; and since we have, independently, a definite system for the
measurement of quantities of heat, we are thus furnished with a
measure for intervals according to which absolute differences of
temperature may be estimated. To make this intelligible, a few
words in explanation of Carnot's theory must be given; but for a full
account of this most valuable contribution to physical science, the
reader is referred to either of the works mentioned above (the original
treatise by Carnot, and Clapeyron's paper on the same subject.

In the present state of science no operation is known by which
heat can be absorbed, without either elevating the temperature of
matter, or becoming latent and producing some alteration in the
physical condition of the body into which it is absorbed; and the
conversion of heat (or *caloric*) into mechanical effect is probably
impossible*, certainly undiscovered. In actual engines for ob-

---

* This opinion seems to be nearly universally held by those who have written on
the subject. A contrary opinion however has been advocated by Mr Joule of
Manchester; some very remarkable discoveries which he has made with reference to
the *generation* of heat by the friction of fluids in motion, and some known experiments with magneto-electric machines, seeming to indicate an actual conversion of
mechanical effect into caloric. No experiment however is adduced in which the
converse operation is exhibited; but it must be confessed that as yet much is
involved in mystery with reference to these fundamental questions of natural
philosophy.

taining mechanical effect through the agency of heat, we must consequently look for the source of power, not in any absorption and conversion, but merely in a transmission of heat. Now Carnot, starting from universally acknowledged physical principles, demonstrates that it is by the *letting down* of heat from a hot body to a cold body, through the medium of an engine (a steam-engine, or an air-engine for instance), that mechanical effect is to be obtained; and conversely, he proves that the same amount of heat may, by the expenditure of an equal amount of labouring force, be *raised* from the cold to the hot body (the engine being in this case *worked backwards*); just as mechanical effect may be obtained by the descent of water let down by a water-wheel, and by spending labouring force in turning the wheel backwards, or in working a pump, water may be elevated to a higher level. The amount of mechanical effect to be obtained by the transmission of a given quantity of heat, through the medium of any kind of engine in which the economy is perfect, will depend, as Carnot demonstrates, not on the specific nature of the substance employed as the medium of transmission of heat in the engine, but solely on the interval between the temperature of the two bodies between which the heat is transferred.

Carnot examines in detail the ideal construction of an air-engine and of a steam-engine, in which, besides the condition of perfect economy being satisfied, the machine is so arranged, that at the close of a complete operation the substance (air in one case and water in the other) employed is restored to precisely the same physical condition as at the commencement. He thus shews on what elements, capable of experimental determination, either with reference to air, or with reference to a liquid and its vapour, the absolute amount of mechanical effect due to the transmission of a unit of heat from a hot body to a cold body, through any given interval of the thermometric scale, may be ascertained. In M. Clapeyron's paper various experimental data, confessedly very imperfect, are brought forward, and the amounts of mechanical effect due to a unit of heat descending a degree of the air-thermometer, in various parts of the scale, are calculated from them, according to Carnot's expressions. The results so obtained indicate very decidedly, that what we may with much propriety call *the value of a degree* (estimated by the mechanical effect to be obtained from the descent of a unit of heat through it) of the air-thermometer depends on the part of the scale in which it is taken, being less for high than for low temperatures*

The characteristic property of the scale which I now propose is, that all degrees have the same value; that is, that a unit of heat descending from a body A at the temperature T^0 of this scale, to a body B at the temperature $(T-1)^0$, would give out the same mechanical effect, whatever be the number T. This may justly be termed an absolute scale, since its characteristic is quite independent of the physical properties of any specific substance.

To compare this scale with that of the air-thermometer, the *values* (according to the principle of estimation stated above) of degrees of the air-thermometer must be known. Now an expression, obtained by Carnot from the consideration of his ideal steam-engine, enables us to calculate these values, when the latent heat of a given volume and the pressure of saturated vapour at any temperature are experimentally determined. The determination of these elements is the principal object of Regnault's great work, already referred to, but at present his researches are not complete. In the first part, which alone has been as yet published, the latent heats of a given *weight*, and the pressures of saturated vapour, at all temperatures between 0° and 230° (Cent. of the air-thermometer), have been ascertained; but it would be necessary in addition to know the densities of saturated vapour at different temperatures, to enable us to determine the latent heat of a given volume at any temperature. M. Regnault announces his intention of instituting researches for this object; but till the results are made known, we have no way of completing the data necessary for the present problem, except by estimating the density of saturated vapour at any temperature (the corresponding pressure being known by Regnault's researches already published) according to the approximate laws of compressibility and expansion (the laws

---

* This is what we might anticipate, when we reflect that infinite cold must correspond to a finite number of degrees of the air-thermometer below zero; since, if we push the strict principle of graduation, stated above, sufficiently far, we should arrive at a point corresponding to the volume of air being reduced to nothing, which would be marked as − 273° of the scale (− 100/·366, if ·366 be the coefficient of expansion); and therefore − 273° of the air-thermometer is a point which cannot be reached at any finite temperature, however low.

of Mariotte and Gay-Lussac, or Boyle and Dalton). Within the limits of natural temperature in ordinary climates, the density of saturated vapour is actually found by Regnault (*Études Hygro-métriques* in the *Annales de Chimie*) to verify very closely these laws ; and we have reason to believe from experiments which have been made by Gay-Lussac and others, that as high as the tem-perature 100° there can be no considerable deviation ; but our estimate of the density of saturated vapour, founded on these laws, may be very erroneous at such high temperatures as 230°. Hence a completely satisfactory calculation of the proposed scale cannot be made till after the additional experimental data shall have been obtained ; but with the data which we actually possess, we may make an approximate comparison of the new scale with that of the air-thermometer, which at least between 0° and 100° will be tolerably satisfactory.

The labour of performing the necessary calculations for effecting a comparison of the proposed scale with that of the air-thermo-meter, between the limits 0° and 230° of the latter, has been kindly undertaken by Mr William Steele, lately of Glasgow College, now of St Peter's College, Cambridge. His results in tabulated forms were laid before the Society, with a diagram, in which the com-parison between the two scales is represented graphically. In the first table*, the amounts of mechanical effect due to the descent of a unit of heat through the successive degrees of the air-thermo-meter are exhibited. The unit of heat adopted is the quantity necessary to elevate the temperature of a kilogramme of water from 0° to 1° of the air-thermometer ; and the unit of mechanical effect is a metre-kilogramme ; that is, a kilogramme raised a metre high.

In the second table, the temperatures according to the pro-posed scale, which correspond to the different degrees of the air-thermometer from 0° to 230°, are exhibited. [The arbitrary points which coincide on the two scales are 0° and 100°].

Note.—If we add together the first hundred numbers given in the first table, we find 135·7 for the amount of work due to a unit of heat descending from a body A at 100° to B at 0°. Now 79 such units of heat would, according to Dr Black (his result being

---

* [Note of Nov. 4, 1881. This table (reduced from metres to feet) was repeated in my " Account of Carnot's Theory of the Motive power of Heat," republished as Article XLI. below, in § 38 of which it will be found.]

very slightly corrected by Regnault), melt a kilogramme of ice. Hence if the heat necessary to melt a pound of ice be now taken as unity, and if a *metre-pound* be taken as the unit of mechanical effect, the amount of work to be obtained by the descent of a unit of heat from $100°$ to $0°$ is $79 \times 135\cdot7$, or $10,700$ nearly. This is the same as $35,100$ foot pounds, which is a little more than the work of a one-horse-power engine ($33,000$ foot pounds) in a minute; and consequently, if we had a steam-engine working with perfect economy at one-horse-power, the boiler being at the temperature $100°$, and the condenser kept at $0°$ by a constant supply of ice, rather less than a pound of ice would be melted in a minute.

[Note of Nov. 4, 1881. This paper was wholly founded on Carnot's uncorrected theory, according to which the quantity of heat taken in in the hot part of the engine, (the boiler of the steam engine for instance), was supposed to be equal to that abstracted from the cold part (the condenser of the steam engine), in a complete period of the regular action of the engine, when every varying temperature, in every part of the apparatus, has become strictly periodic. The reconciliation of Carnot's theory with what is now known to be the true nature of heat is fully discussed in Article XLVIII. below; and in §§ 24—41 of that article, are shewn in detail the consequently required corrections of the thermodynamic estimates of the present article. These corrections however do not in any way affect the absolute scale for thermometry which forms the subject of the present article. Its relation to the practically more convenient scale (agreeing with air thermometers nearly enough for most purposes, throughout the range from the lowest temperatures hitherto measured, to the highest that can exist so far as we know) which I gave subsequently, Dynamical Theory of Heat (Art. XLVIII. below), Part VI., §§ 99, 100; *Trans. R. S. E.*, May, 1854: and Article 'Heat,' §§ 35—38, 47—67, *Encyclopædia Britannica*, is shewn in the following formula:

$$\theta = 100 \, \frac{\log t - \log 273}{\log 373 - \log 273},$$

where θ and t are the reckonings of one and the same temperature, according to my first and according to my second thermodynamic absolute scale.]

[From the *Cambridge and Dublin Mathematical Journal*, Feb. 1849.]

ART. XL.　NOTES ON HYDRODYNAMICS*　ON THE VIS-VIVA OF
A LIQUID IN MOTION.

1.　IF a liquid of finite dimensions be set in motion, it will
go on moving in a manner determined in general by the circumstances affecting its bounding surface, and the forces which operate
through its interior.　Now all the forces which are observed in
nature to act upon the mass of a liquid at rest, whatever may be
the agencies to which it is subjected, are such that if the liquid
be enclosed in a fixed envelope they cannot disturb its equilibrium,
but are in all cases balanced by the resistance which the fluid
pressure experiences from the bounding solid.　Hence, if a liquid
in motion were acted on by such forces only as it might experience when at rest, its motion within the bounding surface would,
by d'Alembert's principle, be entirely independent of the forces
operating on its mass.　The pressure through the interior and at
the bounding surface would, however, in general depend, partly
upon these forces, and partly upon the state of motion of the
liquid; and therefore, in all cases in which the form of the bounding surface is susceptible of alteration by the pressure of the fluid,
the forces through the mass will, by the effect they may thus
produce on the form of the bounding surface, exercise an indirect

---

* [Note of Nov. 5, 1881.　The series of "Notes on Hydrodynamics" which are
printed in Vols. II., III., and IV. of the *Cambridge and Dublin Mathematical Journal*,
were written by agreement between Prof. Stokes and myself.　It may be convenient
to give here the references to the whole series.

I.　On the Equation of Continuity (Thomson), Nov., 1847.
II.　On the Equation of the Bounding Surface (Thomson), Jan., 1848.
III.　On the Dynamical Equations (Stokes), March, 1848.
IV.　Demonstration of a Fundamental Theorem (Stokes), Nov., 1848.
V.　On the *Vis Viva* of a Liquid in motion (Thomson), Jan., 1849.
VI.　On Waves (Stokes), Nov., 1849.

Of these No. I. is referred to, and No. II. is reprinted, in Art. XXXI. above:
Nos. III., IV., and VI. will be found in Stokes' *Mathematical and Physical Papers*.]

influence on the motion which takes place within it. Thus it is
that gravity, which could not affect the motion of a liquid entirely
filling a rigid closed vessel, will exercise a most important influence
on the motion of a liquid contained in an open vessel, and exposing a free surface to the atmospheric pressure. The same
remark is applicable to the forces which Faraday has discovered
to be exerted by electrified bodies upon liquid non-conductors,
and by magnets or galvanic wires upon ferro-magnetic or diamagnetic liquids ; but the internal forces of friction which are found
to operate in actual liquids (see Note III., by Mr Stokes), and the
forces experienced by a liquid traversed by electric currents, under
the action of a magnet, belong to a different class, and none of
the preceding remarks are applicable to them*. In this paper it
will generally be understood that the forces considered belong to
the former class.

2. The bounding surface of a liquid mass may be given as
varying in any arbitrary manner with the time, under the sole
condition of containing a constant volume ; since, if the liquid be
enclosed in a perfectly flexible and extensible envelope, we may
clearly, by external agency, mould it arbitrarily at each instant,
altering it gradually from one form to another. Hence, among
the *data* of a hydrodynamical problem, we may have

$$F (x, y, z, t) = 0 \dots\dots\dots\dots\dots\dots\dots (1),$$

as the equation of the bounding surface, where F may denote an
arbitrary function of the four variables subject to the single condition that the volume of the surface so represented may be constant, but with besides the practical limitation, that there can be
no finite variations of the surface in infinitely small times†. From
this we may deduce, as was shewn in Note II., the following
equation, which must be satisfied by the components u, v, w, of
the fluid velocity at any point (x, y, z) infinitely near the boundary

$$F'(x) . u + F'(y) . v + F'(z) . w + F'(t) = 0 \dots\dots\dots\dots (a).$$

* The analytical characteristic of this class of forces is, that
$$Xdx + Ydy + Zdz,$$
is *not* a complete differential, if Xdm, Ydm, Zdm be the components of the force
experienced by an element, dm, of the liquid mass.

† If it be considered that perfectly impulsive action is practically impossible,
the farther limitation ought to be introduced that there can be no finite variations
in the "normal velocity" of the surface at any point in infinitely small times.

If we denote by l, m, n, the direction cosines of the normal, and by $H\,dt$ the normal motion of the surface at the point (x, y, z) in the time dt; we shall have (Vol. II. p. 91),

$$\left. \begin{aligned} l &= \frac{F'(x)}{R} \\[1ex] m &= \frac{F'(y)}{R} \\[1ex] n &= \frac{F'(z)}{R} \end{aligned} \right\} \quad \dots\dots\dots\dots\dots\dots\dots\dots\dots(2),$$

and
$$H = -\frac{F'(t)}{R} \quad \dots\dots\dots\dots\dots\dots\dots (3) ;$$

so that, when the bounding surface is given at each instant, we may regard these four quantities as known. By introducing them in (a), we may put it under the form,

$$lu + mv + nw = H \dots\dots\dots\dots\dots\dots\dots(4),$$

which will be convenient in the investigations to follow.

3.   Since, according to the article referred to in § 2, the distance in the neighbourhood of a point (x, y, z), between the surface at the time t and the surface at the time $t + dt$, is $H\,dt$, it follows that at any instant, the form of the bounding surface being known, the value of H may be arbitrarily given at each point of it, subject to the sole condition, that the volume contained within the surface must not be changed during the time dt, a condition expressed by the equation

$$\iint H\,ds = 0 \dots\dots\dots\dots\dots\dots\dots\dots(5),$$

where the integration is to include all the elements, ds, of the surface.

4.   The components u, v, w, of the fluid velocity at any point (x, y, z), which, when this point is infinitely near the boundary, must satisfy the preceding equation (4), are, through the whole liquid, subject to the equation

$$\frac{du}{dx} + \frac{dv}{dy} + \frac{dw}{dz} = 0 \dots\dots\dots\dots\dots\dots(6),$$

as was proved in Note I. (Vol. II. p. 286). These two equations, (4) and (5), express all the cinematical relations of the problem.

5.   THEOREM.   *If the bounding surface of a liquid, primitively at rest, be made to vary in a given arbitrary manner, the vis-viva*

*of the entire liquid at each instant will be less than it would be if
the liquid had any other motion consistent with the given motion of
the bounding surface.*

Let u_1, v_1, w_1, be the components of the fluid velocity at (x, y, z)
in any other state of motion cinematically possible at the time dt,
which must consequently satisfy the following conditions:

at the surface, $lu_1 + mv_1 + nw_1 = H$ ....(b),

through the interior, $\dfrac{du_1}{dx} + \dfrac{dv_1}{dy} + \dfrac{dw_1}{dz} = 0$ ....(c):

u_1, v_1, w_1, may be taken as any three quantities whatever for which
these equations hold.

Let Q be the actual vis-viva of the liquid, and Q_1 the vis-viva
it would have if u_1, v_1, w_1 represented its motion. We shall have,
calling ρ the density,

$$Q = \rho \iiint (u^2 + v^2 + w^2)\,dx\,dy\,dz \ldots\ldots\ldots(7),$$
$$Q_1 = \rho \iiint (u_1{}^2 + v_1{}^2 + w_1{}^2)\,dx\,dy\,dz.$$

From these we deduce, by subtraction, and by an algebraical
modification,

$$Q - Q_1 = \rho\iiint \{2u(u_1 - u) + 2v(v_1 - v) + 2w(w_1 - w) + (u_1 - u)^2 + (v_1 - v)^2 + (w_1 - w)^2\}\,dx\,dy\,dz.$$

Now, according to the proposition proved by Mr Stokes in
Note V., since the fluid was primitively at rest,

$$u\,dx + v\,dy + w\,dz$$

must be the differential of some function of x, y, z (involving also,
in general, t as another independent variable), which we may
denote by ϕ; so that we have

$$u = \frac{d\phi}{dx}, \quad v = \frac{d\phi}{dy}, \quad w = \frac{d\phi}{dz}$$

Hence

$$\iiint \{u(u_1 - u) + v(v_1 - v) + w(w_1 - w)\}\,dx\,dy\,dz$$
$$= \iiint \left\{\frac{d\phi}{dx}(u_1 - u) + \frac{d\phi}{dy}(v_1 - v) + \frac{d\phi}{dz}(w_1 + w)\right\}\,dx\,dy\,dz.$$

Integrating the first term by parts with reference to x, the second
with reference to y, and the third with reference to z, we reduce
the second member to the form

$$\iint \phi \cdot \{(u_1 - u)\,dy\,dz + (v_1 - v)\,dz\,dx + (w_1 - w)\,dx\,dy\}$$
$$- \iiint \phi \cdot \left\{\frac{d(u_1 - u)}{dx} + \frac{d(v_1 - v)}{dy} + \frac{d(w_1 - w)}{dz}\right\}\,dx\,dy\,dz.$$

The triple integral here vanishes in virtue of equations (b) and (c); and the double integral, which is to be extended over the entire bounding surface, may (see Note I. Vol. II. p. 285) be put under the form

$$\iint \phi \cdot \{(u_1 - u)\, l + (v_1 - v)\, m + (w_1 - w)\, n\}\, ds.$$

This also is equal to nothing in virtue of equations (4) and (b); and hence the definite integral under consideration vanishes. Thus we see that the expression for $Q_1 - Q$ becomes reduced to

$$Q_1 - Q = \rho \iiint \{(u_1 - u)^2 + (v_1 - v)^2 + (w_1 - w)^2\}\, dx\, dy\, dz \ldots \ldots (8).$$

In this expression the factor of $dx\, dy\, dz$, and consequently the entire integral, is essentially positive, unless $u_1 = u$, $v_1 = v$, $w_1 = w$. Hence, of all the states of motion of the fluid cinematically possible at each instant, any one which differs from the actual state of motion possesses a greater vis-viva. Q.E.D.[*]

6. COR. 1. *The condition that* $u\,dx + v\,dy + w\,dz$ *must be a complete differential is, in addition to the cinematical relations, sufficient to determine the motion.* For in the preceding demonstration no other condition was introduced to characterise u, v, w[†].

COR. 2. *The motion of the fluid at any time is independent of the preceding motion, and depends solely on the given form and normal motion of the bounding surface at the instant.*

COR. 3. *If the bounding surface, after having been in motion, be brought to rest in any position, the liquid will, at the same instant, be reduced to rest.*

7. The expression for the vis-viva of the liquid may be put into a very remarkable form, by making use of the differential coefficients of ϕ in place of u, v, and w; and then integrating by parts, in the following manner. Thus we have

$$Q = \rho \iiint \left(u\,\frac{d\phi}{dx} + v\,\frac{d\phi}{dy} + w\,\frac{d\phi}{dz} \right) dx\, dy\, dz,$$

$$= \rho \iint \phi \cdot (u\,dy\,dz + v\,dz\,dx + w\,dx\,dy)$$

$$\qquad - \rho \iiint \phi \cdot \left(\frac{du}{dx} + \frac{dv}{dy} + \frac{dw}{dz} \right) dx\, dy\, dz.$$

[*] [Note of Nov. 5, 1881. This is a particular case of a general theorem of minimum energy announced in the *Proc. of Royal Society of Edinburgh*, for April, 1863, and fully treated in Thomson and Tait's *Natural Philosophy*, §§ 312, and 317—319.]

[†] See Vol. III. p. 84 [Art. XXXVI. above], where similar reasoning was applied to prove a theorem, of which the corollary in the text is a particular case.

The triple integral vanishes in virtue of equation (5), and the double integral, extended over the bounding surface, may be modified by the transformation employed in § 4, so that we have the following expression for the vis-viva,

$$Q = \rho \iint \phi \cdot H \cdot ds^* \dots\dots\dots\dots\dots\dots(9).$$

The variation of the function ϕ within the bounding surface will not affect the value of this integral, in which ϕ may be considered as merely a function of the co-ordinates of a point in the surface itself. Hence, while the factor H expresses the given normal velocities at the different points of the containing surface, the other factor, ϕ, under the integral sign, is such as to express by its differential components, with reference to superficial co-ordinates, the tangential component of the velocity of a particle of the fluid in contact with this surface.

GLASGOW COLLEGE, *Jan.* 11, 1849.

---

* This is a particular case of a general theorem proved in an article entitled "Propositions in the Theory of Attraction," Part II. [*Electrostatics and Magnetism*, Art. XII.], being obtained by taking $R = R_1 = \sqrt{(u^2 + v^2 + w^2)}$ and $\theta = 0$, in equations (3) and (4) of that article (*Camb. Math. Journ.*, Feb. 1843).

[From *Transactions of the Edinburgh Royal Society*, XVI. 1849; *Annal. de Chimie*, XXXV. 1852.]

XLI. AN ACCOUNT OF CARNOT'S THEORY OF THE MOTIVE POWER OF HEAT*; WITH NUMERICAL RESULTS DEDUCED FROM REGNAULT'S EXPERIMENTS ON STEAM†.

(Read January 2, 1849.)

1. THE presence of heat may be recognised in every natural object; and there is scarcely an operation in nature which is not more or less affected by its all-pervading influence. An evolution and subsequent absorption of heat generally give rise to a variety of effects; among which may be enumerated, chemical combinations or decompositions; the fusion of solid substances; the vaporisation of solids or liquids; alterations in the dimensions of bodies, or in the statical pressure by which their dimensions may be modified; mechanical resistance overcome; electrical currents generated. In many of the actual phenomena of nature, several or all of these effects are produced together; and their complication will, if we attempt to trace the agency of heat in producing any individual effect, give rise to much perplexity. It will, therefore, be desirable, in laying the foundation of a physical

---

* Published in 1824, in a work entitled, "Réflexions sur la Puissance Motrice du Feu, et sur les Machines Propres à Développer cette Puissance. Par S. Carnot." An account of Carnot's Theory is also published in the *Journal de l'Ecole Polytechnique*, Vol. XIV., 1834, in a paper by Mons. Clapeyron. [Note of Nov. 5, 1881. The original work has now been republished, with a biographical notice, Paris, 1878.]

† An account of the first part of a series of researches undertaken by Mons. Regnault, by order of the late French Government, for ascertaining the various physical data of importance in the theory of the steam-engine, has been recently published (under the title, "Relation des Expériences," &c.) in the *Mémoires de l'Institut*, of which it constitutes the twenty-first volume (1847). The second part of these researches has not yet been published. [Note of Nov. 5, 1881. The continuation of these researches has now been published; thus we have for the whole series, Vol. I. in 1847; Vol. II. in 1862; and Vol. III. in 1870.]

theory of any of the effects of heat, to discover or to imagine phenomena free from all such complication, and depending on a definite thermal agency ; in which the relation between the cause and effect, traced through the medium of certain simple operations, may be clearly appreciated. Thus it is that Carnot, in accordance with the strictest principles of philosophy, enters upon the investigation of the theory of the motive power of heat.

2.   The sole effect to be contemplated in investigating the motive power of heat is *resistance overcome,* or, as it is frequently called, *"work performed,"* or *"mechanical effect."* The questions to be resolved by a complete theory of the subject are the following :

(1)   What is the precise nature of the thermal agency by means of which *mechanical effect* is to be produced, without effects of any other kind ?

(2)   How may the amount of this thermal agency necessary for performing a given quantity of work be estimated ?

3.   In the following paper I shall commence by giving a short abstract of the reasoning by which Carnot is led to an answer to the first of these questions ; I shall then explain the investigation by which, in accordance with his theory, the experimental elements necessary for answering the second question are indicated ; and, in conclusion, I shall state the *data* supplied by Regnault's recent observations on steam, and apply them to obtain, as approximately as the present state of experimental science enables us to do, a complete solution of the question.

I.   On the nature of Thermal agency, considered as a motive power.

4.   There are [at present known] two, and only two, distinct ways in which mechanical effect can be obtained from heat. One of these is by means of the alterations of volume which bodies may experience through the action of heat; the other is through the medium of electric agency. Seebeck's discovery of thermo-electric currents enables us at present to conceive of an electro-magnetic engine supplied from a thermal origin, being used as a motive power : but this discovery was not made until 1821, and the subject of thermo-electricity can only have been

generally known in a few isolated facts, with reference to the electrical effects of heat upon certain crystals, at the time when Carnot wrote. He makes no allusion to it, but confines himself to the method for rendering thermal agency available as a source of mechanical effect, by means of the expansions and contractions of bodies.

5. A body expanding or contracting under the action of force, may, in general, either produce mechanical effect by overcoming resistance, or receive mechanical effect by yielding to the action of force. The amount of mechanical effect thus developed will depend not only on the calorific agency concerned, but also on the alteration in the physical condition of the body. Hence, after allowing the volume and temperature of the body to change, we must restore it to its original temperature and volume; and then we may estimate the aggregate amount of mechanical effect developed as due solely to the thermal origin.

6. Now the ordinarily-received, and almost universally-acknowledged, principles with reference to " quantities of caloric" and "latent heat," lead us to conceive that, at the end of a cycle of operations, when a body is left in precisely its primitive physical condition, if it has absorbed any heat during one part of the operations, it must have given out again exactly the same amount during the remainder of the cycle. The truth of this principle is considered as axiomatic by Carnot, who admits it as the foundation of his theory; and expresses himself in the following terms regarding it, in a note on one of the passages of his treatise*.

"In our demonstrations we tacitly assume that after a body has experienced a certain number of transformations, if it be brought identically to its primitive physical state as to density, temperature, and molecular constitution, it must contain the same quantity of heat as that which it initially possessed ; or, in other words, we suppose that the quantities of heat lost by the body under one set of operations are precisely compensated by those which are absorbed in the others. This fact has never been doubted; it has at first been admitted without reflection, and afterwards verified, in many cases, by calorimetrical experiments. To deny it would be to overturn the whole theory

* Carnot, p. 37.

of heat, in which it is the fundamental principle. It must be admitted, however, that the chief foundations on which the theory of heat rests, would require a most attentive examination. Several experimental facts appear nearly inexplicable in the actual state of this theory."

7. Since the time when Carnot thus expressed himself, the necessity of a most careful examination of the entire experimental basis of the theory of heat has become more and more urgent. Especially all those assumptions depending on the idea that heat is a *substance,* invariable in quantity; not convertible into any other element, and incapable of being *generated* by any physical agency; in fact the acknowledged principles of latent heat; would require to be tested by a most searching investigation before they ought to be admitted, as they usually have been, by almost every one who has been engaged on the subject, whether in combining the results of experimental research, or in general theoretical investigations.

8. The extremely important discoveries recently made by Mr Joule of Manchester, that heat is evolved in every part of a closed electric conductor, moving in the neighbourhood of a magnet*, and that heat is *generated* by the friction of fluids in

---

* The *evolution* of heat in a fixed conductor, through which a galvanic current is sent from any source whatever, has long been known to the scientific world; but it was pointed out by Mr Joule that we cannot infer from any previously-published experimental researches, the actual *generation* of heat when the current originates in electro-magnetic induction; since the question occurs, *is the heat which is evolved in one part of the closed conductor merely transferred from those parts which are subject to the inducing influence?* Mr Joule, after a most careful experimental investigation with reference to this question, finds that it must be answered in the negative.—(See a paper *"On the Calorific Effects of Magneto-Electricity, and on the Mechanical Value of Heat;* by J. P. Joule, Esq." Read before the British Association at Cork in 1843, and subsequently communicated by the Author to the *Philosophical Magazine,* Vol. xxiii., pp. 263, 347, 435.)

Before we can finally conclude that heat is absolutely generated in such operations, it would be necessary to prove that the inducing magnet does not become lower in temperature, and thus compensate for the heat evolved in the conductor. I am not aware that any examination with reference to the truth of this conjecture has been instituted; but, in the case where the inducing body is a pure electro-magnet (without any iron), the experiments actually performed by Mr Joule render the conclusion probable that the heat evolved in the wire of the electro-magnet is not affected by the inductive action, otherwise than through the reflected influence which increases the strength of its own current.

motion, seem to overturn the opinion commonly held that heat cannot be *generated*, but only produced from a source, where it has previously existed either in a sensible or in a latent condition.

In the present state of science, however, no operation is known by which heat can be absorbed into a body without either elevating its temperature, or becoming latent, and producing some alteration in its physical condition; and the fundamental axiom adopted by Carnot may be considered as still the most probable basis for an investigation of the motive power of heat; although this, and with it every other branch of the theory of heat, may ultimately require to be reconstructed upon another foundation, when our experimental data are more complete. On this understanding, and to avoid a repetition of doubts, I shall refer to Carnot's fundamental principle, in all that follows, as if its truth were thoroughly established.

9.   We are now led to the conclusion that the origin of motive power, developed by the alternate expansions and contractions of a body, must be found in the agency of heat entering the body and leaving it; since there cannot, at the end of a complete cycle, when the body is restored to its primitive physical condition, have been any absolute absorption of heat, and consequently no conversion of heat, or caloric, into mechanical effect; and it remains for us to trace the precise nature of the circumstances under which heat must enter the body, and afterwards leave it, so that mechanical effect may be produced. As an example, we may consider that machine for obtaining motive power from heat with which we are most familiar—the steam-engine.

10.   Here, we observe, that heat enters the machine from the furnace, through the sides of the boiler, and that heat is continually abstracted by the water employed for keeping the condenser cool. According to Carnot's fundamental principle, the quantity of heat thus discharged, during a complete revolution (or double stroke) of the engine must be precisely equal to that which enters the water of the boiler*; provided the total mass

---

* So generally is Carnot's principle tacitly admitted as an axiom, that its application in this case has never, so far as I am aware, been questioned by practical engineers.

of water and steam be invariable, and be restored to its primitive physical condition (which will be the case rigorously, if the condenser be kept cool by the external application of cold water, instead of by injection, as is more usual in practice), and if the condensed water be restored to the boiler at the end of each complete revolution. Thus, we perceive, that a certain quantity of heat is *let down* from a hot body, the metal of the boiler, to another body at a lower temperature, the metal of the condenser; and that there results from this transference of heat a certain development of mechanical effect.

11. If we examine any other case in which mechanical effect is obtained from a thermal origin, by means of the alternate expansions and contractions of any substance whatever, instead of the water of a steam-engine, we find that a similar transference of heat is effected, and we may therefore answer the first question proposed, in the following manner:—

The thermal agency by which mechanical effect may be obtained, is the transference of heat from one body to another at a lower temperature.

II. On the measurement of Thermal Agency, considered with reference to its equivalent of mechanical effect.

12. A *perfect* thermo-dynamic engine of any kind, is a machine by means of which the greatest possible amount of mechanical effect can be obtained from a given thermal agency; and, therefore, if in any manner we can construct or imagine a perfect engine which may be applied for the transference of a given quantity of heat from a body at any given temperature, to another body, at a lower given temperature, and if we can evaluate the mechanical effect thus obtained, we shall be able to answer the question at present under consideration, and so to complete the theory of the motive power of heat. But whatever kind of engine we may consider with this view, it will be necessary for us to prove that it is a perfect engine; since the transference of the heat from one body to the other may be wholly, or partially, effected by conduction through a solid*, without the

* When "thermal agency" is thus spent in conducting heat through a solid, what becomes of the mechanical effect which it might produce? Nothing can be lost in the operations of nature—no energy can be destroyed. What effect then is

development of mechanical effect; and, consequently, engines may be constructed in which the whole, or any portion of the thermal agency is wasted. Hence it is of primary importance to discover the criterion of a perfect engine. This has been done by Carnot, who proves the following proposition :—

13. *A perfect thermo-dynamic engine is such that, whatever amount of mechanical effect it can derive from a certain thermal agency ; if an equal amount be spent in working it backwards, an equal reverse thermal effect will be produced* *.

14. This proposition will be made clearer by the applications of it which are given below (§ 29), in the cases of the air-engine and the steam-engine, than it could be by any general explanation ; and it will also appear, from the nature of the operations described in those cases, and the principles of Carnot's reasoning, that a perfect engine may be constructed with any substance of an indestructible texture as the alternately expanding and contracting medium. Thus we might conceive thermo-dynamic engines founded upon the expansions and contractions of a perfectly elastic solid, or of a liquid; or upon the alterations of volume experienced by substances, in passing from the liquid to

produced in place of the mechanical effect which is lost? A perfect theory of heat imperatively demands an answer to this question; yet no answer can be given in the present state of science. A few years ago, a similar confession must have been made with reference to the mechanical effect lost in a fluid set in motion in the interior of a rigid closed vessel, and allowed to come to rest by its own internal friction ; but in this case, the foundation of a solution of the difficulty has been actually found, in Mr Joule's discovery of the generation of heat, by the internal friction of a fluid in motion. Encouraged by this example, we may hope that the very perplexing question in the theory of heat, by which we are at present arrested, will, before long, be cleared up. [Note of Sep. 1881. The Theory of the Dissipation of Energy (Article LVIII.; below) completely answers this question and removes the difficulty.]

It might appear, that the difficulty would be entirely avoided, by abandoning Carnot's fundamental axiom; a view which is strongly urged by Mr Joule (at the conclusion of his paper "On the Changes of Temperature produced by the Rarefaction and Condensation of Air." *Phil. Mag.*, May 1845, Vol. XXVI.) If we do so, however, we meet with innumerable other difficulties—insuperable without farther experimental investigation, and an entire reconstruction of the theory of heat from its foundation. It is in reality to experiment that we must look—either for a verification of Carnot's axiom, and an explanation of the difficulty we have been considering; or for an entirely new basis of the Theory of Heat.

* For a demonstration, see § 29, below.

the solid state*, each of which being perfect, would produce the same amount of mechanical effect from a given thermal agency; but there are two cases which Carnot has selected as most worthy of minute attention, because of their peculiar appropriateness for illustrating the general principles of his theory, no less than on account of their very great practical importance; the steam-engine, in which the substance employed as the transferring medium is water, alternately in the liquid state, and in the state of vapour; and the air-engine, in which the transference is effected by means of the alternate expansions and contractions of a medium, always in the gaseous state. The details of an actually practicable engine of either kind are not contemplated by Carnot, in his general theoretical reasonings, but he confines himself to the ideal construction, in the simplest possible way in each case, of an engine in which the economy is perfect. He thus determines the degree of perfectibility which cannot be surpassed; and, by describing a conceivable method of attaining to this perfection by an air-engine or a steam-engine, he points out the proper objects to be kept in view in the practical construction and working of such machines. I now proceed to give an outline of these investigations.

CARNOT'S *Theory of the Steam-Engine.*

15.   Let CDF_2E_2 be a cylinder, of which the curved surface is perfectly impermeable to heat, with a piston also impermeable to heat, fitted in it; while the fixed bottom CD, itself with no capacity for heat, is possessed of perfect conducting power. Let K be an impermeable stand, such that when the cylinder is placed upon it, the contents below the piston can neither gain nor lose heat. Let A and B be two .bodies permanently retained at constant temperatures, S^0 and T^0, respectively, of which the former is higher than the latter. Let the cylinder, placed on the impermeable stand, K, be partially filled with water, at the temperature S, of the body A, and (there being no air below it) let the piston be placed in a position EF, near the surface of

---

* A case minutely examined in another paper, to be laid before the Society at the present meeting. " Theoretical considerations on the Effect of Pressure in lowering the Freezing Point of Water," by Prof. James Thomson. [Appended at the end of the present Article by his permission Nov. 5, 1881.]

the water. The pressure of the vapour above the water will tend to push up the piston, and must be resisted by a force applied to the piston*, till the commencement of the operations, which are conducted in the following manner.

(1) The cylinder being placed on the body A, so that the water and vapour may be retained at the temperature S, *let the*

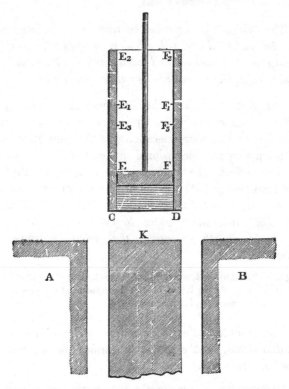

piston rise any convenient height EE_1, to a position E_1F_1, performing work by the pressure of the vapour below it during its ascent.

[During this operation a certain quantity, H, of heat, the amount of latent heat in the fresh vapour which is formed, is abstracted from the body A.]

* In all that follows, the pressure of the atmosphere on the upper side of the piston will be included in the applied forces, which, in the successive operations described, are sometimes overcome by the upward motion, and sometimes yielded to in the motion downwards. It will be unnecessary, in reckoning at the end of a cycle of operations, to take into account the work thus spent upon the atmosphere, and the restitution which has been made, since these precisely compensate for one another.

(2) The cylinder being removed, and placed on the impermeable stand K, *let the piston rise gradually, till, when it reaches a position E_2F_2, the temperature of the water and vapour is T, the same as that of the body B.*

[During this operation the fresh vapour continually formed requires heat to become latent; and, therefore, as the contents of the cylinder are protected from any accession of heat, their temperature sinks.]

(3) The cylinder being removed from K, and placed on B, *let the piston be pushed down, till, when it reaches the position E_3F_3, the quantity of heat evolved and abstracted by B amounts to that which, during the first operation, was taken from A.*

[Note of Nov. 5, 1881. The specification of this operation, with a view to the return to the primitive condition, intended as· the conclusion to the four operations, is the only item in which Carnot's temporary and provisional assumption of the materiality of heat has effect. To exclude this hypothesis, Prof. James Thomson gave (see p. 161) the following corrected specification for the third operation ·—*Let the piston be pushed down, till it reaches a position E_3F_3, determined so as to fulfil the condition, that at the end of the fourth operation, the primitive temperature S shall be reached**:]

[During this operation the temperature of the contents of the cylinder is retained constantly at T^0, and all the latent heat of the vapour which is condensed into water at the same temperature, is given out to B.]

(4) The cylinder being removed from B, and placed on the impermeable stand, *let the piston be pushed down from E_3F_3 to its original position EF.*

[During this operation, the impermeable stand preventing any loss of heat, the temperature of the water and air must rise continually, till (since the quantity of heat evolved during the third operation was precisely equal to that which was

---

* [Note of Nov. 5, 1881. Maxwell has simplified the correction by beginning the cycle with Carnot's second operation, and completing it through his third, fourth, and first operations, with his third operation merely as follows :—

let the piston be pushed down to any position E_3F_3;

then Carnot's fourth operation altered to the following :—

let the piston be pushed down from E_3F_3 until the temperature reaches its primitive value S ;

and lastly Carnot's first operation altered to the following : —

let the piston rise to its primitive position.]

previously absorbed), at the conclusion it reaches its primitive value, S, in virtue of Carnot's fundamental axiom.]

[Note of Nov. 5, 1881. With Prof. James Thomson's correction of operation (3), the words in virtue of "Carnot's Fundamental Axiom" must be replaced by "the condition fulfilled by operation (3)," in the description of the results of operation (4).]

16. At the conclusion of this cycle of operations* the total thermal agency has been the *letting down* of H units of heat from the body A, at the temperature S, to B, at the lower temperature T; and the aggregate of the mechanical effect has been a certain amount of *work produced*, since during the ascent of the piston in the first and second operations, the temperature of the water and vapour, and therefore the pressure of the vapour on the piston, was on the whole higher than during the descent, in the third and fourth operations. It remains for us actually to evaluate this aggregate amount of work performed; and for this purpose the following graphical method of representing the mechanical effect developed in the several operations, taken from Mons. Clapeyron's paper, is extremely convenient.

17. Let OX and OY be two lines at right angles to one another. Along OX measure off distances ON_1, N_1N_2, N_2N_3, N_3O, respectively proportional to the spaces described by the piston during the four successive operations described above; and, with reference to these four operations respectively, let the following constructions be made:—

(1) Along OY measure a length OA, to represent the pressure of the saturated vapour at the temperature S; and draw AA_1 parallel to OX, and let it meet an ordinate through N_1, in A_1.

(2) Draw a curve A_1PA such that, if ON represent, at any instant during the second operation, the distance of the piston from its primitive position, NP shall represent the pressure of the vapour at the same instant.

---

* In Carnot's work some perplexity is introduced with reference to the temperature of the water, which, in the operations he describes, is not brought back exactly to what it was at the commencement; but the difficulty which arises is explained by the author. No such difficulty occurs with reference to the cycle of operations described in the text, for which I am indebted to Mons. Clapeyron.

(3) Through A_2 draw A_2A_3 parallel to OX, and let it meet an ordinate through N_3 in A_3.

(4) Draw the curve A_3A such that the abscissa and ordinate of any point in it may represent respectively the distances of the piston from its primitive position, and the pressure of the

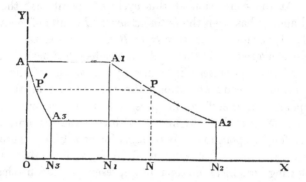

vapour, at each instant during the fourth operation. The last point of this curve must, according to Carnot's fundamental principle, coincide with A, since the piston is, at the end of the cycle of operations, again in its primitive position, and the pressure of the vapour is the same as it was at the beginning.

18. Let us now suppose that the lengths, ON_1, N_1N_2, N_2N_3, and N_3O, *represent numerically* the volumes of the spaces moved through by the piston during the successive operations. It follows that the mechanical effect obtained during the first operation will be *numerically represented* by the area AA_1N_1O; that is, the number of superficial units in this area will be equal to the number of "foot-pounds" of work performed by the ascending piston during the first operation. The work performed by the piston during the second operation will be similarly represented by the area $A_1A_2N_2N_1$. Again, during the third operation a certain amount of work is spent on the piston, which will be represented by the area $A_2A_3N_3N_2$; and lastly, during the fourth operation, work is spent in pushing the piston to an amount represented by the area A_3AON_3.

19. Hence the mechanical effect (represented by the area $OAA_1A_2N_2$) which was obtained during the first and second

operations, exceeds the work (represented by $N_2A_2A_3AO$) spent during the third and fourth, by an amount represented by the area of the quadrilateral figure $AA_1A_2A_3$; and, consequently, it only remains for us to evaluate this area, that we may determine the total mechanical effect gained in a complete cycle of operations. Now, from experimental data, at present nearly complete, as will be explained below, we may determine the length of the line AA_1 for the given temperature S, and a given absorption H, of heat, during the first operation; and the length of A_2A_3 for the given lower temperature T, and the evolution of the same quantity of heat during the fourth operation: and the curves A_1PA_2, $A_3P'A$ may be drawn as graphical representations of actual observations*. The figure being thus constructed, its area may be measured, and we are, therefore, in possession of a graphical method of determining the amount of mechanical effect to be obtained from any given thermal agency. As, however, it is merely the area of the figure which it is required to determine, it will not be necessary to be able to describe each of the curves A_1PA_2, $A_3P'A$, but it will be sufficient to know the difference of the abscissas corresponding to any equal ordinates in the two; and the following analytical method of completing the problem is the most convenient for leading to the actual numerical results.

20.   Draw any line PP' parallel to OX, meeting the curvilineal sides of the quadrilateral in P and P'. Let ξ denote the length of this line, and p its distance from OX. The area of the figure, according to the integral calculus, will be denoted by the expression

$$\int_{p_3}^{p_1} \xi dp,$$

where p_1, and p_3 (the limits of integration indicated according to Fourier's notation) denote the lines OA, and N_3A_3, which represent respectively the pressures during the first and third operations. Now, by referring to the construction described above, we see that ξ is the difference of the volumes below the piston at corresponding instants of the second and fourth operations, or instants at which the saturated steam and the water in the cylinder have the same pressure p, and, consequently, the same temperature

* See Note at the end of this Paper.

which we may denote by t.   Again, throughout the second operation the entire contents of the cylinder possess a greater amount
of heat by H units than during the fourth; and, therefore, at
any instant of the second operation there is as much more steam
as contains H units of latent heat, than at the corresponding
instant of the fourth operation.   Hence, if k denote the latent
heat in a unit of saturated steam at the temperature t, the volume
of the steam at the two corresponding instants must differ by
$\dfrac{H}{k}$.   Now, if σ denote the ratio of the density of the steam to

that of the water, the volume $\dfrac{H}{k}$ of steam will be formed from the

volume $\sigma \dfrac{H}{k}$ of water; and, consequently, we have for the differ-

ence of volumes of the entire contents at the corresponding
instants,

$$\xi = (1 - \sigma)\frac{H}{k}.$$

Hence the expression for the area of the quadrilateral figure
becomes

$$\int_{p_3}^{p_1} (1 - \sigma)\frac{H}{k}\,dp.$$

Now, σ, k, and p, being quantities which depend upon the temperature, may be considered as functions of t; and it will be convenient to modify the integral so as to make t the independent
variable.   The limits will be from $t = T$ to $t = S$, and, if we denote
by M the value of the integral, we have the expression

$$M = H \int_{T}^{S} (1 - \sigma)\frac{dp}{kdt}\,dt \dots\dots\dots\dots(1),$$

for the total amount of mechanical effect gained by the operations
described above.

21.   If the interval of temperatures be extremely small; so
small that $(1 - \sigma)\,dp/kdt$ will not sensibly vary for values of t
between T and S, the preceding expression becomes simply

$$M = (1 - \sigma)\frac{dp}{kdt} . H(S - T) \dots\dots\dots\dots\dots(2).$$

This might, of course, have been obtained at once, by supposing the breadth of the quadrilateral figure AA_1A_2A to be extremely small compared with its length, and then taking for its area, as an approximate value, the product of the breadth into the line AA_1, or the line A_3A_2, or any line of intermediate magnitude.

The expression (2) is rigorously correct for any interval $S - T$, if the mean value of $(1 - \sigma)\, dp/kdt$ for that interval be employed as the coefficient of $H(S - T)$.

CARNOT'S *Theory of the Air-Engine.*

22.   In the ideal air-engine imagined by Carnot four operations performed upon a mass of air or gas enclosed in a closed vessel of variable volume, constitute a complete cycle, at the end of which the medium is left in its primitive physical condition; the construction being the same as that which was described above for the steam-engine, a body A, permanently retained at the temperature S, and B at the temperature T; an impermeable stand K; and a cylinder and piston, which, in this case, contains a mass of air at the temperature S, instead of water in the liquid state, at the beginning and end of a cycle of operations. The four successive operations are conducted in the following manner:—

(1)   The cylinder is laid on the body A, so that the air in it is kept at the temperature S; and the piston is allowed to rise, performing work.

(2)   The cylinder is placed on the impermeable stand K, so that its contents can neither gain nor lose heat, and the piston is allowed to rise farther, still performing work, till the temperature of the air sinks to T.

(3)   The cylinder is placed on B, so that the air is retained at the temperature T, and the piston is pushed down till the air gives out to the body B as much heat as it had taken in from A, during the first operation.

[Note of Nov. 5, 1881. To eliminate the assumption of the materiality of heat,

make Professor James Thomson's correction here also; as above in § 15; or take
Maxwell's re-arrangement of the cycle described in the foot-note to § 15.]

(4) The cylinder is placed on K, so that no more heat can
be taken in or given out, and the piston is pushed down to its
primitive position.

23. *At the end of the fourth operation the temperature must
have reached its primitive value S, in virtue of* CARNOT'S *axiom.*

24. Here, again, as in the former case, we observe that work
is performed by the piston during the first two operations; and,
during the third and fourth, work is spent upon it, but to a
less amount, since the pressure is on the whole less during the
third and fourth operations than during the first and second,
on account of the temperature being lower. Thus, at the end
of a complete cycle of operations, mechanical effect has been
obtained; and the thermal agency from which it is drawn is
the taking of a certain quantity of heat from A, and *letting it
down,* through the medium of the engine, to the body B at a
lower temperature.

25. To estimate the actual amount of effect thus obtained,
it will be convenient to consider the alterations of volume of
the mass of air in the several operations as extremely small.
We may afterwards pass by the integral calculus, or, practically,
by summation, to determine the mechanical effect whatever be the
amplitudes of the different motions of the piston.

26. Let dq be the quantity of heat absorbed during the
first operation, which is evolved again during the third; and
let dv be the corresponding augmentation of volume which takes
place while the temperature remains constant, as it does during
the first operation*. The diminution of volume in the third

---

* Thus, $\dfrac{dq}{dv}$ will be the partial differential coefficient, with respect to v of that
function of v and t, which expresses the quantity of heat that must be added to
a mass of air when in a "standard" state (such as at the temperature zero, and
under the atmospheric pressure), to bring it to the temperature t, and the volume v.
That there is such a function, of two independent variables v and t, is merely an
analytical expression of Carnot's fundamental axiom, as applied to a mass of air.
The general principle may be analytically stated in the following terms:—If Mdv
denote the accession of heat received by a mass of any kind, not possessing a
destructible texture, when the volume is increased by dv, the temperature being

operation must be also equal to dv, or only differ from it by an infinitely small quantity of the second order. During the second operation we may suppose the volume to be increased by an infinitely small quantity ϕ; which will occasion a diminution of pressure, and a diminution of temperature, denoted respectively by ω and τ. During the fourth operation there will be a diminution of volume, and an increase of pressure and temperature, which can only differ, by infinitely small quantities of the second order, from the changes in the other direction, which took place in the second operation, and they also may, therefore, be denoted by ϕ, ω, and τ, respectively. The alteration of pressure, during the first and third operations, may at once be determined by means of Mariotte's law, since, in them, the temperature remains constant. Thus, if, at the commencement of the cycle, the volume and pressure be v and p, they will have become $v + dv$ and $pv/(v + dv)$ at the end of the first operation. Hence the diminution of pressure, during the first operation, is $p - pv/(v + dv)$ or $pdv/(v + dv)$; and, therefore, if we neglect infinitely small quantities of the second order, we have pdv/v for the diminution of pressure during the first operation; which, to the same degree of approximation, will be equal to the increase of pressure during the third. If $t + \tau$ and t be taken to denote the superior and inferior limits of temperature, we shall thus have for the volume, the temperature, and the pressure at the commencements of the four successive operations, and at the end of the cycle, the following values respectively :—

(1) v, $t + \tau$, p;

(2) $v + dv$, $t + \tau$, $p\left(1 - \dfrac{dv}{v}\right)$;

(3) $v + dv + \phi$, t, $p\left(1 - \dfrac{dv}{v}\right) - \omega$;

(4) $v + \phi$, t, $p - \omega$;

(5) v, $t + \tau$, p.

kept constant, and if Ndt denote the amount of heat which must be supplied to raise the temperature by dt, without any alteration of volume; then $Mdv + Ndt$ must be the differential of a function of v and t. [Note of Nov. 5, 1881. In the corrected theory it is $(M - Jp)\,dv + Ndt$, that is a complete differential, not $Mdv + Ndt$. See *Dynamical Theory of Heat* (Art. XLVIII., below), § 20.]

Taking the mean of the pressures at the beginning and end of each operation, we find

$$(1) \quad p\left(1 - \tfrac{1}{2}\frac{dv}{v}\right), \qquad (2) \quad p\left(1 - \frac{dv}{v}\right) - \tfrac{1}{2}\omega,$$

$$(3) \quad p\left(1 - \tfrac{1}{2}\frac{dv}{v}\right) - \omega, \qquad (4) \quad p - \tfrac{1}{2}\omega,$$

which, as we are neglecting infinitely small quantities of the second order, will be the expressions for the mean pressures during the four successive operations. Now, the mechanical effect gained or spent, during any of the operations, will be found by multiplying the mean pressure by the increase or diminution of volume which takes place; and we thus find

$$(1) \quad p\left(1 - \tfrac{1}{2}\frac{dv}{v}\right)dv \qquad (2) \quad \left\{p\left(1 - \frac{dv}{v}\right) - \tfrac{1}{2}\omega\right\}\phi$$

$$(3) \quad \left\{p\left(1 - \tfrac{1}{2}\frac{dv}{v}\right) - \omega\right\}dv \qquad (4) \quad (p - \tfrac{1}{2}\omega)\phi$$

for the amounts gained during the first and second, and spent during the third and fourth operations; and hence, by addition and subtraction, we find

$$\omega dv - p\phi\,\frac{dv}{v}, \quad \text{or} \quad (v\omega - p\phi)\frac{dv}{v},$$

for the aggregate amount of mechanical effect gained during the cycle of operations. It only remains for us to express this result in terms of dq and τ, on which the given thermal agency depends. For this purpose, we remark that ϕ and ω are alterations of volume and pressure which take place along with a change of temperature τ, and hence, by the laws of compressibility and expansion, we may establish a relation* between them in the following manner.

Let p_0 be the pressure of the mass of air when reduced to the temperature zero, and confined in a volume v_0; then, what-

---

* We might also nvestigate another relation, to express the fact that there is no accession or removal of heat during either the second or the fourth operation; but it will be seen that this will not affect the result in the text; although it would enable us to determine both ϕ and ω in terms of τ.

ever be v_0, the product p_0v_0 will, by the law of compressibility, remain constant; and, if the temperature be elevated from 0 to $t + \tau$, and the gas be allowed to expand freely without any change of pressure, its volume will be increased in the ratio of 1 to $1 + E(t + \tau)$, where E is very nearly equal to ·00366 (the centigrade scale of the air-thermometer being referred to), whatever be the gas employed, according to the researches of Regnault and of Magnus on the expansion of gases by heat. If, now, the volume be altered arbitrarily with the temperature continually at $t + \tau$, the product of the pressure and volume will remain constant; and, therefore, we have

$$pv = p_0v_0 \{1 + E(t + \tau)\}.$$

Similarly $\qquad (p - \omega)(v + \phi) = p_0v_0 \{1 + Et\}.$

Hence, by subtraction, we have

$$v\omega - p\phi + \omega\phi = p_0v_0E\tau,$$

or, neglecting the product $\omega\phi$,

$$v\omega - p\phi = p_0v_0E\tau.$$

Hence, the preceding expression for mechanical effect, gained in the cycle of operations, becomes $p_0v_0 \cdot E\tau \cdot dv/v$.

Or, as we may otherwise express it,

$$\frac{Ep_0v_0}{vdq/dv} \cdot dq \cdot \tau.$$

Hence, if we denote by M the mechanical effect due to H units of heat descending through the same interval τ, which might be obtained by repeating the cycle of operations described above, $\frac{H}{dq}$ times, we have

$$M = \frac{Ep_0v_0}{vdq/dv} \cdot H\tau \ldots\ldots\ldots\ldots\ldots\ldots\ldots(3).$$

27. If the *amplitudes* of the operations had been finite, so as to give rise to an absorption of H units of heat during the first operation, and a lowering of temperature from S to T during the second, the amount of work obtained would have been found to be expressed by means of a double definite integral, thus* :—

---

* This result might have been obtained by applying the usual notation of the integral calculus to express the area of the curvilinear quadrilateral, which, ac-

$$M = \int_0^H dq \int_T^S dt \cdot \frac{Ep_0 v_0}{v dq/dv} \quad \text{or} \quad M = Ep_0 v_0 \int_0^H \int_T^S \frac{1}{v} \frac{dv}{dq} \cdot dt \, dq \, ; \ldots (4),$$

this second form being sometimes more convenient.

28. The preceding investigations, being founded on the approximate laws of compressibility and expansion (known as the law of Mariotte and Boyle, and the law of Dalton and Gay-Lussac), would require some slight modifications, to adapt them to cases in which the gaseous medium employed is such as to present sensible deviations from those laws. Regnault's very accurate experiments shew that the deviations are insensible, or very nearly so, for the ordinary gases at ordinary pressures; although they may be considerable for a medium, such as sulphurous acid, or carbonic acid under high pressure, which approaches the physical condition of a vapour at saturation; and therefore, in general, and especially in practical applications to real air-engines, it will be unnecessary to make any modification in the expressions. In cases where it may be necessary, there is no difficulty in making the modifications, when the requisite data are supplied by experiment.

29* Either the steam-engine or the air-engine, according to the arrangements described above, gives all the mechanical effect that can possibly be obtained from the thermal agency employed. For it is clear, that, in either case, the operations may be performed in the reverse order, with every thermal and mechanical effect reversed. Thus, in the steam-engine, we may commence by placing the cylinder on the impermeable stand, allow the piston to rise, performing work, to the position $E_3 F_3$; we may then place it on the body B, and allow it to rise, performing work, till it reaches $E_2 F_2$; after that the cylinder may be placed again on

cording to Clapeyron's graphical construction, would be found to represent the entire mechanical effect gained in the cycle of operations of the air-engine. It is not necessary, however, to enter into the details of this investigation, as the formula (3), and the consequences derived from it, include the whole theory of the air-engine, in the best practical form; and the investigation of it which I have given in the text, will probably give as clear a view of the reasoning on which it is founded, as could be obtained by the graphical method, which, in this case, is not so valuable as it is from its simplicity in the case of the steam-engine.

* This paragraph is the demonstration referred to above, of the proposition stated in § 13; as it is readily seen that it is applicable to any conceivable kind of thermo-dynamic engine.

the impermeable stand, and the piston may be pushed down to E_1F_1; and, lastly, the cylinder being removed to the body A, the piston may be pushed down to its primitive position. In this inverse cycle of operations, a certain amount of work has been spent, precisely equal, as we readily see, to the amount of mechanical effect gained in the direct cycle described above; and heat has been abstracted from B, and deposited in the body A, at a higher temperature, to an amount precisely equal to that which, in the direct cycle, was *let down* from A to B. Hence it is impossible to have an engine which will derive more mechanical effect from the same thermal agency, than is obtained by the arrangement described above; since, if there could be such an engine, it might be employed to perform, as a part of its whole work, the inverse cycle of operations, upon an engine of the kind we have considered, and thus to continually restore the heat from B to A, which has descended from A to B for working itself; so that we should have a complex engine, giving a residual amount of mechanical effect without any thermal agency, or alteration of materials, which is an impossibility in nature. The same reasoning is applicable to the air-engine; and we conclude, generally, that any two engines, constructed on the principles laid down above, whether steam-engines with different liquids, an air-engine and a steam-engine, or two air-engines with different gases, must derive the same amount of mechanical effect from the same thermal agency.

30. Hence, by comparing the amounts of mechanical effect obtained by the steam-engine and the air-engine from the letting down of the H units of heat from A at the temperature $(t + \tau)$ to B at t, according to the expressions (2) and (3), we have

$$M = (1 - \sigma) \frac{dp}{kdt} . 'H\tau = \frac{Ep_0v_0}{vdq/dv} . H\tau \dots\dots\dots\dots(5).$$

If we denote the coefficient of $H\tau$ in these equal expressions by μ, which may be called "Carnot's coefficient," we have

$$\mu = (1 - \sigma) \frac{dp}{kdt} = \frac{Ep_0v_0}{vdq/dv} \dots\dots\dots\dots\dots(6),$$

and we deduce the following very remarkable conclusions :—

(1) For the saturated vapours of all different liquids, at the same temperature, the value of $(1-\sigma)\dfrac{dp}{kdt}$ must be the same.

(2) For any different gaseous masses, at the same temperature, the value of $\dfrac{Ep_0v_0}{vdq/dv}$ must be the same.

(3) The values of these expressions for saturated vapours and for gases, at the same temperature, must be the same.

31. No conclusion can be drawn *a priori* regarding the values of this coefficient μ for different temperatures, which can only be determined, or compared, by experiment. The results of a great variety of experiments, in different branches of physical science (Pneumatics and Acoustics), cited by Carnot and by Clapeyron, indicate that the values of μ for low temperatures exceed the values for higher temperatures; a result amply verified by the continuous series of experiments performed by Regnault on the saturated vapour of water for all temperatures from 0° to 230°, which, as we shall see below, give values for μ gradually diminishing from the inferior limit to the superior limit of temperature. When, by observation, μ has been determined as a function of the temperature, the amount of mechanical effect, M, deducible from H units of heat descending from a body at the temperature S to a body at the temperature T, may be calculated from the expression,

$$M = H\int_{T}^{S}\mu dt \dots\dots\dots\dots\dots\dots\dots\dots(7),$$

which is, in fact, what either of the equations (1) for the steam-engine, or (4) for the air-engine, becomes, when the notation μ, for Carnot's multiplier, is introduced.

The values of this integral may be practically obtained, in the most convenient manner, by first determining, from observation, the mean values of μ for the successive degrees of the thermometric scale, and then adding the values for all the degrees within the limits of the extreme temperatures S and $T*$.

---

* The results of these investigations are exhibited in Tables I. and II. below.

32. The complete theoretical investigation of the motive power of heat is thus reduced to the experimental determination of the coefficient μ; and may be considered as perfect, when, by any series of experimental researches whatever, we can find a value of μ for every temperature within practical limits. The special character of the experimental researches, whether with reference to gases, or with reference to vapours, necessary and sufficient for this object, is defined and restricted in the most precise manner, by the expressions (6) for μ, given above.

33. The object of Regnault's great work, referred to in the title of this paper, is the experimental determination of the various physical elements of the steam-engine ; and when it is complete, it will furnish all the *data* necessary for the calculation of μ. The valuable researches already published in a first part of that work, make known the latent heat of a given weight, and the pressure, of saturated steam for all temperatures between 0° and 230° cent. of the air-thermometer. Besides these data, however, the density of saturated vapour must be known, in order that k, the latent heat of a unit of volume, may be calculated from Regnault's determination of the latent heat of a given weight*. Between the limits of 0° and 100°, it is probable, from various experiments which have been made, that the density of vapour follows very closely the simple laws which are so accurately verified by the ordinary gases†; and thus it may be calculated from Regnault's table giving the pressure at any temperature within those limits. Nothing as yet is known with accuracy as to the density of saturated steam between 100° and 230°, and we must be contented at present to estimate it by calculation from Regnault's table of pressures; although, when accurate experimental researches on the subject shall have been

---

* It is, comparatively speaking, of little consequence to know accurately the value of σ, for the factor $(1-\sigma)$ of the expression for μ, since it is so small (being less than $\frac{1}{1700}$ for all temperatures between 0° and 100°) that, unless all the data are known with more accuracy than we can count upon at present, we might neglect it altogether, and take dp/kdt simply, as the expression for μ, without committing any error of important magnitude.

† This is well established, within the ordinary atmospheric limits, in Regnault's Études Météorologiques, in the *Annales de Chimie*, Vol. xv., 1846.

made, considerable deviations from the laws of Boyle and Dalton,
on which this calculation is founded, may be discovered.

34. Such are the experimental data on which the mean
values of μ for the successive degrees of the air-thermometer,
from 0^0 to 230^0, at present laid before the Royal Society, is
founded. The unit of length adopted is the English foot; the
unit of weight, the pound; the unit of work, a "foot-pound;"
and the unit of heat that quantity which, when added to a pound
of water at 0^0, will produce an elevation of 1^0 in temperature.
The mean value of μ for any degree is found to a sufficient
degree of approximation, by taking, in place of σ, dp/dt, and k;
in the expression

$$(1 - \sigma)\frac{dp}{kdt};$$

the mean values of those elements; or, what is equivalent to
the corresponding accuracy of approximation, by taking, in place
of σ and k respectively, the mean of the values of those elements
for the limits of temperature, and in place of dp/dt, the difference
of the values of p, at the same limits.

35. In Regnault's work (at the end of the eighth Mémoire),
a table of the pressures of saturated steam for the successive
temperatures 0^0, 1^0, 2^0,... 230^0, expressed in millimetres of mercury,
is given. On account of the units adopted in this paper, these
pressures must be estimated in pounds on the square foot, which
we may do by multiplying each number of millimetres by $2\cdot7896$,
the weight in pounds of a sheet of mercury, one millimetre thick,
and a square foot in area.

36. The value of k, the latent heat of a cubic foot, for any
temperature t, is found from λ, the latent heat of a pound of
saturated steam, by the equation

$$k = \frac{p}{760} \cdot \frac{1 + \cdot00366 \times 100}{1 + \cdot00366 \times t} \cdot \times \cdot036869^* . \lambda,$$

---

* It appears that the vol. of 1 kilog. must be $1\cdot69076$ according to the data here
assumed.

The density of saturated steam at 100^0 is taken as $\frac{1}{1608\cdot5}$ of that of water at
its maximum. Rankine takes it as $\frac{1}{1696}$.

where p denotes the pressure in millimetres, and λ the latent heat of a pound of saturated steam; the values of λ being calculated by the empirical formula*

$$\lambda = (606\cdot5 + 0\cdot305t) - (t + \cdot00002t^2 + 0\cdot0000003t^3),$$

given by Regnault as representing, between the extreme limits of his observations, the latent heat of a unit weight of saturated steam.

Explanation of Table I.

37. The mean values of μ for the first, for the eleventh, for the twenty-first, and so on, up to the 231st† degree of the air-thermometer, have been calculated in the manner explained in the preceding paragraphs. These, and interpolated results, which must agree with what would have been obtained, by direct calculation from Regnault's data, to three significant places of figures (and even for the temperatures between 0° and 100°, the experimental data do not justify us in relying on any of the results to a greater degree of accuracy), are exhibited in Table I.

To find the amount of mechanical effect due to a unit of heat, descending from a body at a temperature S to a body at T, if these numbers be integers, we have merely to add the values of μ in Table I. corresponding to the successive numbers.

$$T + 1, \; T + 2, \ldots\ldots S - 2, \; S - 1.$$

Explanation of Table II.

38. The calculation of the mechanical effect, in any case, which might always be effected in the manner described in § 37

---

* The part of this expression in the first vinculum (see Regnault, end of ninth Mémoire) is what is known as "the total heat" of a pound of steam, or the amount of heat necessary to convert a pound of water at 0° into a pound of saturated steam at t^{0}; which, according to "Watt's law," thus approximately verified, would be constant. The second part, which would consist of the single term t, if the specific heat of water were constant for all temperatures, is the number of thermic units necessary to raise the temperature of a pound of water from 0° to t^{0}, and expresses empirically the results of Regnault's experiments on the specific heat of water (see end of the tenth Mémoire), described in the work already referred to.

† In strictness, the 230th is the last degree for which the experimental data are complete; but the data for the 231st may readily be assumed in a sufficiently satisfactory manner.

(with the proper modification for fractions of degrees, when necessary), is much simplified by the use of Table II., where the first number of Table I., the sum of the first and second, the sum of the first three, the sum of the first four, and so on, are successively exhibited. The sums thus tabulated are the values of the integrals

$$\int_0^1 \mu dt, \quad \int_0^2 \mu dt, \quad \int_0^3 \mu dt, \ldots \ldots \int_0^{231} \mu dt;$$

and, if we denote $\int_0^t \mu dt$ by the letter M, Table II. may be regarded as a table of the value of M.

To find the amount of mechanical effect due to a unit of heat descending from a body at a temperature S to a body at T, if these numbers be integers, we have merely to subtract the value of M, for the number T, from the value for the number S, given in Table II.

TABLE I.* *Mean Values of μ for the successive Degrees of the Air-Thermometer from 0° to 230°.*

	μ		μ		μ		μ		μ
1°	4·960	48°	4·366	94°	3·889	140°	3·549	186°	3·309
2	4·946	49	4·355	95	3·880	141	3·543	187	3·304
3	4·932	50	4·343	96	3·871	142	3·537	188	3·300
4	4·918	51	4·331	97	3·863	143	3·531	189	3·295
5	4·905	52	4·319	98	3·854	144	3·525	190	3·291
6	4·892	53	4·308	99	3·845	145	3·519	191	3·287
7	4·878	54	4·296	100	3·837	146	3·513	192	3·282
8	4·865	55	4·285	101	3·829	147	3·507	193	3·278
9	4·852	56	4·273	102	3·820	148	3·501	194	3·274
10	4·839	57	4·262	103	3·812	149	3·495	195	3·269
11	4·826	58	4·250	104	3·804	150	3·490	196	3·265
12	4·812	59	4·239	105	3·796	151	3·484	197	3·261
13	4·799	60	4·227	106	3·788	152	3·479	198	3·257
14	4·786	61	4·216	107	3·780	153	3·473	199	3·253
15	4·773	62	4·205	108	3·772	154	3·468	200	3·249
16	4·760	63	4·194	109	3·764	155	3·462	201	3·245
17	4·747	64	4·183	110	3·757	156	3·457	202	3·241
18	4·735	65	4·172	111	3·749	157	3·451	203	3·237
19	4·722	66	4·161	112	3·741	158	3·446	204	3·233
20	4·709	67	4·150	113	3·734	159	3·440	205	3·229
21	4·697	68	4·140	114	3·726	160	3·435	206	3·225
22	4·684	69	4·129	115	3·719	161	3·430	207	3·221
23	4·672	70	4·119	116	3·712	162	3·424	208	3·217
24	4·659	71	4·109	117	3·704	163	3·419	209	3·213
25	4·646	72	4·098	118	3·697	164	3·414	210	3·210
26	4·634	73	4·088	119	3·689	165	3·409	211	3·206
27	4·621	74	4·078	120	3·682	166	3·404	212	3·202
28	4·609	75	4·067	121	3·675	167	3·399	213	3·198
29	4·596	76	4·057	122	3·668	168	3·394	214	3·195
30	4·584	77	4·047	123	3·661	169	3·389	215	3·191
31	4·572	78	4·037	124	3·654	170	3·384	216	3·188
32	4·559	79	4·028	125	3·647	171	3·380	217	3·184
33	4·547	80	4·018	126	3·640	172	3·375	218	3·180
34	4·535	81	4·009	127	3·633	173	3·370	219	3·177
35	4·522	82	3·999	128	3·627	174	3·365	220	3·173
36	4·510	83	3·990	129	3·620	175	3·361	221	3·169
37	4·498	84	3·980	130	3·614	176	3·356	222	3·165
38	4·486	85	3·971	131	3·607	177	3·351	223	3·162
39	4·474	86	3·961	132	3·601	178	3·346	224	3·158
40	4·462	87	3·952	133	3·594	179	3·342	225	3·155
41	4·450	88	3·943	134	3·586	180	3·337	226	3·151
42	4·438	89	3·934	135	3·579	181	3·332	227	3·148
43	4·426	90	3·925	136	3·573	182	3·328	228	3·144
44	4·414	91	3·916	137	3·567	183	3·323	229	3·141
45	4·402	92	3·907	138	3·561	184	3·318	230	3·137
46	4·390	93	3·898	139	3·555	185	3·314	231	3·134
47	4·378								

* The numbers here tabulated may also be regarded as *the actual values of* μ *for* $t = \frac{1}{2}$, $t = 1\frac{1}{2}$, $t = 2\frac{1}{2}$, $t = 3\frac{1}{2}$, &c.

TABLE II. *Mechanical Effect in Foot-Pounds due to a Thermic Unit Centigrade, passing from a body, at any Temperature less than 230° to a body at 0°.*

Superior Limit of Temperature.	Mechanical Effect.	Superior Limit of Temperature.	Mechanical Effect.	Superior Limit of Temperature.	Mechanical Effect.	Superior Limit of Temperature.	Mechanical Effect.	Superior Limit of Temperature.	Mechanical Effect.
	Foot-Pounds.		Foot-Pounds.		Foot-Pounds.		Foot-Pounds.		Foot-Pounds.
1°	4·960	48°	223·487	94°	412·545	140°	582·981	186°	740·310
2	9·906	49	227·842	95	416·425	141	586·524	187	743·614
3	14·838	50	232·185	96	420·296	142	590·061	188	746·914
4	19·756	51	236·516	97	424·159	143	593·592	189	750·209
5	24·661	52	240·835	98	428·013	144	597·117	190	753·500
6	29·553	53	245·143	99	431·858	145	600·636	191	756·787
7	34·431	54	249·439	100	435·695	146	604·099	192	760·069
8	39·296	55	253·724	101	439·524	147	607·656	193	763·347
9	44·148	56	257·997	102	443·344	148	611·157	194	766·621
10	48·987	57	262·259	103	447·156	149	614·652	195	769·890
11	53·813	58	266·509	104	450·960	150	618·142	196	773·155
12	58·625	59	270·748	105	454·756	151	621·626	197	776·416
13	63·424	60	274·975	106	458·544	152	625·105	198	779·673
14	68·210	61	279·191	107	462·324	153	628·578	199	782·926
15	72·983	62	283·396	108	466·096	154	632·046	200	786·175
16	77·743	63	287·590	109	469·860	155	635·508	201	789·420
17	82·490	64	291·773	110	473·617	156	638·965	202	792·661
18	87·225	65	295·945	111	477·366	157	642·416	203	795·898
19	91·947	66	300·106	112	481·107	158	645·862	204	799·131
20	96·656	67	304·256	113	484·841	159	649·302	205	802·360
21	101·353	68	308·396	114	488·567	160	652·737	206	805·585
22	106·037	69	312·525	115	492·286	161	656·167	207	808·806
23	110·709	70	316·644	116	495·998	162	659·591	208	812·023
24	115·368	71	320·752	117	499·702	163	663·010	209	815·236
25	120·014	72	324·851	118	503·399	164	666·424	210	818·446
26	124·648	73	328·939	119	507·088	165	669·833	211	821·652
27	129·269	74	333·017	120	510·770	166	673·237	212	824·854
28	133·878	75	337·084	121	514·445	167	676·636	213	828·052
29	138·474	76	341·141	122	518·113	168	680·030	214	831·247
30	143·058	77	345·188	123	521·174	169	683·419	215	834·438
31	147·630	78	349·225	124	525·428	170	686·803	216	837·626
32	152·189	79	353·253	125	529·075	171	690·183	217	840·810
33	156·736	80	357·271	126	532·715	172	693·558	218	843·990
34	161·271	81	361·280	127	536·348	173	696·928	219	847·167
35	165·793	82	365·279	128	539·975	174	700·293	220	850·340
36	170·303	83	369·269	129	543·595	175	703·654	221	853·509
37	174·801	84	373·249	130	547·209	176	707·010	222	856·674
38	179·287	85	377·220	131	550·816	177	710·361	223	859·836
39	183·761	86	381·181	132	554·417	178	713·707	224	862·994
40	188·223	87	385·133	133	558·051	179	717·049	225	866·149
41	192·673	88	389·076	134	561·597	180	720·386	226	869·300
42	197·111	89	393·010	135	565·176	181	723·718	227	872·448
43	201·537	90	396·935	136	568·749	182	727·046	228	875·592
44	205·951	91	400·851	137	572·316	183	730·369	229	878·733
45	210·353	92	404·758	138	575·877	184	733·687	230	881·870
46	214·743	93	408·656	139	579·432	185	737·001	231	885·004
47	219·121								

Note.—On the curves described in Clapeyron's graphical method of exhibiting Carnot's Theory of the Steam-Engine.

39. At any instant when the temperature of the water and vapour is t, during the fourth operation (see above, § 16, and suppose, for the sake of simplicity that, at the beginning of the first, and· at the end of the fourth operation, the piston is absolutely in contact with the surface of the water), the latent heat of the vapour must be precisely equal to the amount of heat that would be necessary to raise the temperature of the whole mass, if in the liquid state, from t to S*. Hence, if v' denote the volume of the vapour, c the mean capacity for heat of a pound of water between the temperatures S and t, and W the weight of the entire mass, in pounds, we have

$$kv' = c\,(S - t)\;W.$$

Again, the circumstances during the second operation are such that the mass of liquid and vapour possesses H units of heat more than during the fourth; and consequently, at the instant of the second operation, when the temperature is t, the volume v of the vapour will exceed v' by an amount of which the latent heat is H, so that we have

$$v = v' + \frac{H}{k}\,.$$

40. Now, at any instant, the volume between the piston and its primitive position is less than the actual volume of vapour by the volume of the water evaporated. Hence, if x and x' denote the abscissæ of the curve at the instants of the second and fourth operations respectively, when the temperature is t, we have

$$x = v - \sigma v, \quad x' = v' - \sigma v',$$

and, therefore, by the preceding equations,

$$x = \frac{1-\sigma}{k}\,\{H + c\,(S - t)\;W\}\ldots\ldots\ldots\ldots\ldots(a),$$

$$x' = \frac{1-\sigma}{k}\,c\,(S - t)\;W\ldots\ldots\ldots\ldots(b).$$

These equations, along with $y = y' = p$ $\ldots\ldots\ldots\ldots\ldots\ldots\ldots(c)$

* For, at the end of the fourth operation, the whole mass is liquid, and at the temperature S. Now, this state might be arrived at by first compressing the vapour into water at the temperature t, and then raising the temperature of the liquid to S; and however this state may be arrived at, there cannot, on the whole, be any heat added to or subtracted from the contents of the cylinder, since, during

enable us to calculate, from the data supplied by Regnault, the abscissa and ordinate for each of the curves described above (§ 17), corresponding to any assumed temperature t. After the explanations of §§ 33, 34, 35, 36, it is only necessary to add that c is a quantity of which the value is very nearly unity, and would be exactly so were the capacity of water for heat the same at every temperature as it is between $0°$ and $1°$; and that the value of $c(S-t)$, for any assigned values of S and t, is found, by subtracting the number corresponding to t from the number corresponding to s, in the column headed "*Nombre des unités de chaleur abandonnées par un kilogramme d'eau en descendant de $T°$ à $0°$*", of the last table (at the end of the Tenth Mémoire) of Regnault's work. By giving S the value $230°$, and by substituting successively 220, 210, 200, &c., for t, values for x, y, x', y', have been found, which are exhibited in the following Table :—

Temperatures.	Volumes to be described by the piston, to complete the fourth operation.	Volumes from the primitive position of the piston to those occupied at instants of the second operation.	Pressures of saturated steam, in pounds on the square foot.
t	x'	x	$y=y'=p$
$0°$	1269.W	$x'+5·409$.H	12·832
10	639·6.W	$x'+2·847$.H	25·567
20	337·3.W	$x'+1·571$.H	48·514
30	185·5.W	$x'+·9062$.H	88·007
40	105·9.W	$x'+·5442$.H	153·167
50	62·62.W	$x'+·3392$.H	256·595
60	38·19.W	$x'+·2188$.H	415·070
70	21·94.W	$x'+·1456$.H	650·240
80	15·38.W	$x'+·09962$.H	989·318
90	10·09.W	$x'+·06994$.H	1465·80
100	6·744.W	$x'+·05026$.H	2120·11
110	4·578.W	$x'+·03688$.H	2999·87
120	3·141.W	$x'+·02758$.H	4160·10
130	2·176.W	$x'+·02098$.H	5663·70
140	1·519.W	$x'+·01625$.H	7581·15
150	1·058.W	$x'+·01271$.H	9990·26
160	0·7369.W	$x'+·01010$.H	12976·2
170	0·5085.W	$x'+·008116$.H	16630·7
180	0·3454.W	$x'+·006592$.H	21051·5
190	0·2267.W	$x'+·005406$.H	26341·5
200	0·1409.W	$x'+·004472$.H	32607·7
210	0·0784.W	$x'+·003729$.H	39960·7
220	0·3310.W	$x'+·003130$.H	48512·4
230	0	$x'+·002643$.H	58376·6

the fourth operation, there is neither gain nor loss of heat. This reasoning is, of course, founded on Carnot's fundamental principle, which is tacitly assumed in the commonly-received ideas connected with "Watt's law," the "latent heat of steam," and "the total heat of steam."

Appendix.

(Read April 30, 1849.)

41.   In p. 30, some conclusions drawn by Carnot from his general reasoning were noticed; according to which it appears, that if the value of μ for any temperature is known, certain information may be derived with reference to the saturated vapour of any liquid whatever, and, with reference to any gaseous mass, without the necessity of experimenting upon the specific medium considered.   Nothing in the whole range of Natural Philosophy is more remarkable than the establishment of general laws by such a process of reasoning.   We have seen, however, that doubt may exist with reference to the truth of the axiom on which the entire theory is founded, and it therefore becomes more than a matter of mere curiosity to put the inferences deduced from it to the test of experience.   The importance of doing so was clearly appreciated by Carnot; and, with such data as he had from the researches of various experimenters, he tried his conclusions.   Some very remarkable propositions which he derives from his Theory, coincide with Dulong and Petit's subsequently-discovered experimental laws with reference to the heat developed by the compression of a gas; and the experimental verification is therefore in this case (so far as its accuracy could be depended upon) decisive.   In other respects, the data from experiment were insufficient, although, so far as they were available as tests, they were confirmatory of the theory.

42.   The recent researches of Regnault add immensely to the experimental data available for this object, by giving us the means of determining with considerable accuracy the values of μ within a very wide range of temperature, and so affording a trustworthy standard for the comparison of isolated results at different temperatures, derived from observations in various branches of physical science.

In the first section of this Appendix the Theory is tested, and shewn to be confirmed by the comparison of the values of μ found above, with those obtained by Carnot and Clapeyron from the observations of various experimenters on air, and the vapours of different liquids.   In the second and third sections some striking confirmations of the theory arising from observations by Dulong, on the specific heat of gases, and from Mr Joule's experiments on the heat developed by the compression of air, are pointed out; and in conclusion, the actual methods of obtaining mechanical effect from heat are briefly examined with reference to their economy.

I. *On the values of μ derived by Carnot and Clapeyron from observations on Air, and on the Vapours of various liquids.*

43.   In Carnot's work, p. 80—82, the mean value of μ between $0°$ and $1°$ is derived from the experiments of Delaroche and Berard on the specific heat of gases, by a process approximately equivalent to the calculation of the value of $\dfrac{Ep_0v_0}{vdq/dv}$ for the temperature $\frac{1}{2}°$. There are also, in the same work, determinations of the values of μ from observations on the vapours of alcohol and water; but a table given in M. Clapeyron's paper, of the values of μ derived from the data supplied by various experiments with reference to the vapours of ether, alcohol, water, and oil of turpentine, at the respective boiling-points of these liquids, afford us the means of comparison through a more extensive range of temperature. In the cases of alcohol and water, these results ought of course to agree with those of Carnot. There are, however, slight discrepancies which must be owing to the uncertainty of the experimental data*. In the following table, Carnot's results with reference to air, and Clapeyron's results with reference to the four different liquids, are exhibited, and compared with the

Names of the Media.	Temperatures.	Values of μ.	Values of μ deduced from Regnault's Observations.	Differences.
Air	$0°5$	(CARNOT) 4·377	4·960	·383
Sulphuric Ether . .	(Boiling point) 35·5	(CLAPEYRON) 4·478	4·510	·032
Alcohol . . .	.... 78·8	.... 3·963	4·030	·071
Water . . .	100	.... 3·658	3·837	·179
Essence of Turpentine	...............156·8	.... 3·530	3·449	− ·081

* Thus, from Carnot's calculations, we find, in the case of alcohol, 4·035; and in the case of water, 3·648, instead of, 3·963, and 3·658, which are Clapeyron's results in the same cases.

values of μ which have been given above (Table I.) for the same temperatures, as derived from Regnault's observations on the vapour of water.

44. It may be observed that the discrepancies between the results founded on the experimental data supplied by the different observers with reference to water at the boiling-point, are greater than those which are presented between the results deduced from any of the other liquids, and water at the other temperatures; and we may therefore feel perfectly confident that the verification is complete to the extent of accuracy of the observations*. The considerable discrepancy presented by Carnot's result, deduced from experiments on air, is not to be wondered at when we consider the very uncertain nature of his data.

45. The fact of the gradual decrease of μ through a very extensive range of temperature, being indicated both by Regnault's continuous series of experiments, and by the very varied experiments on different media, and in different branches of Physical Science, must be considered as a striking verification of the theory.

II. *On the Heat developed by the compression of Air.*

46. Let a mass of air, occupying initially a given volume V, under a pressure P, at a temperature t, be compressed to a less volume V', and allowed to part with heat until it sinks to its primitive temperature t. The quantity of heat which is evolved may be determined, according to Carnot's theory, when the particular value of μ, corresponding to the temperature t, is known. For, by § 30, equation (6), we have $vdq/dv = Ep_0v_0/\mu$, where dq is the quantity of heat absorbed, when the volume is allowed to increase from v to $v+dv$; or the quantity evolved by the reverse operation. Hence we deduce

$$dq = \frac{Ep_0v_0}{\mu}\frac{dv}{v} \dots\dots\dots\dots\dots\dots\dots(8).$$

---

* A still closer agreement must be expected, when more accurate experimental data are afforded with reference to the other media. Mons. Regnault informs me that he is engaged in completing some researches, from which we may expect, possibly before the end of the present year, to be furnished with all the data for five or six different liquids which we possess at present for water. It is therefore to be hoped that, before long, a most important test of the validity of Carnot's theory will be afforded.

T. 10

Now, Ep_0v_0/μ is constant, since the temperature remains unchanged; and therefore, we may at once integrate the second number. By taking it between the limits V' and V, we thus find

$$Q = \frac{Ep_0v_0}{\mu} \log \frac{V}{V'}^{*} \dots\dots\dots\dots\dots(9),$$

where Q denotes the required amount of heat, evolved by the compression from V to V'. This expression may be modified by employing the equations $PV = P'V' = p_0v_0(1 + Et)$; and we thus obtain

$$Q = \frac{EPV}{\mu(1 + Et)} \log \frac{V}{V'} = \frac{EP'V'}{\mu(1 + Et)} \log \frac{V}{V'}\dots\dots(10).$$

From this result we draw the following conclusion :—

47. *Equal volumes of all elastic fluids, taken at the same temperature and pressure, when compressed to smaller equal volumes, disengage equal quantities of heat.*

This extremely remarkable theorem of Carnot's was independently laid down as a probable experimental law by Dulong, in his "Recherches sur la Chaleur Spécifique des Fluides Élastiques," and it therefore affords a most powerful confirmation of the theory †.

48. In some very remarkable researches made by Mr Joule upon the heat developed by the compression of air, the quantity of heat produced in different experiments has been ascertained with reference to the amount of work spent in the operation. To compare the results which he has obtained with the indications of theory, let us determine the amount of work necessary actually to produce the compression considered above.

---

* The *Napierian* logarithm of V/V' is here understood.

† Carnot varies the statement of his theorem, and illustrates it in a passage, pp. 52, 53, of which the following is a translation:—

"*When a gas varies in volume without any change of temperature, the quantities of heat absorbed or evolved by this gas are in arithmetical progression, if the augmentation or diminution of volume are in geometrical progression.*

"When we compress a litre of air maintained at the temperature 10^0, and reduce it to half a litre, it disengages a certain quantity of heat. If, again, the volume be reduced from half a litre to a quarter of a litre, from a quarter to an eighth, and so on, the quantities of heat successively evolved will be the same.

"If, in place of compressing the air, we allow it to expand to two litres, four litres, eight litres, &c., it will be necessary to supply equal quantities of heat to maintain the temperature always at the same degree."

49. In the first place, to compress the gas from the volume $v + dv$ to v, the work required is pdv, or, since $pv = p_0 v_0 (1 + Et)$,

$$p_0 v_0 (1 + Et) \frac{dv}{v}.$$

Hence, if we denote by W the total amount of work necessary to produce the compression from V to V', we obtain, by integration, $W = p_0 v_0 (1 + Et) \log (V/V')$. Comparing this with the expression above, we find

$$\frac{W}{Q} = \frac{\mu (1 + Et)}{E} \quad \dots\dots\dots\dots\dots\dots(11).$$

50. Hence we infer that

(1) The amount of work necessary to produce a unit of heat by the compression of a gas, is the same for all gases at the same temperature.

(2) And that the quantity of heat evolved in all circumstances, when the temperature of the gas is given, is proportional to the amount of work spent in the compression.

51. The expression for the amount of work necessary to produce a unit of heat is $\mu (1 + Et)/E$, and therefore Regnault's experiments on steam are available to enable us to calculate its value for any temperature. By finding the values of μ at 0°, 10°, 20°, &c., from Table I., and by substituting successively the values 0, 10, 20, &c., for t, the following results have been obtained.

Table of the Values of $\mu (1 + Et)/E$.

Work requisite to produce a unit of Heat by the compression of a Gas.	Temperature of the Gas.	Work requisite to produce a unit of Heat by the compression of a Gas.	Temperature of the Gas.
Ft.-lbs.	°	Ft.-lbs.	°
1357·1	0	1446·4	120
1368·7	10	1455·8	130
1379·0	20	1465·3	140
1388·0	30	1475·8	150
1395·7	40	1489·2	160
1401·8	50	1499·0	170
1406·7	60	1511·3	180
1412·0	70	1523·5	190
1417·6	80	1536·5	200
1424·0	90	1550·2	210
1430·6	100	1564·0	220
1438·2	110	1577·8	230

Mr Joule's experiments were all conducted at temperatures from 50° to about 60° Fahr., or from 10° to 16° cent.; and, consequently, although some irregular differences in the results, attributable to errors of observation inseparable from experiments of such a very difficult nature are presented, no regular dependence on the temperature is observable. From three separate series of experiments, Mr Joule deduces the following numbers for the work, in foot-pounds, necessary to produce a thermic unit Fahrenheit by the compression of a gas, namely, 820, 814, 760. Multiplying these by 1·8, to get the corresponding number for a thermic unit centigrade, we find 1476, 1465, and 1368.

The largest of these numbers is most nearly conformable with Mr Joule's views of the relation between such experimental "equivalents," and others which he obtained in his electro-magnetic researches; but the smallest agrees almost perfectly with the indications of Carnot's theory; from which, as exhibited in the preceding Table, we should expect, from the temperature in Mr Joule's experiments, to find a number between 1369 and 1379 as the result*.

III.   *On the Specific Heats of Gases.*

52.   The following proposition is proved by Carnot as a deduction from his general theorem regarding the specific heats of gases.

The excess of specific heat† under a constant pressure above the specific heat at a constant volume, is the same for all gases at the same temperature and pressure.

53.   To prove this proposition, and to determine an expression for the "excess" mentioned in its enunciation, let us suppose a unit of volume of a gas to be elevated in temperature by a small amount, τ. The quantity of heat required to do this will be $A\tau$, if A denote the specific heat at a constant volume. Let us next allow the gas to expand without going down in temperature, until its pressure becomes reduced to its primitive value. The

---

* [Note added Mar. 14, 1851; 772 is now the most probable, 1390 foot-pounds for 1° Cent.]

† Or the capacity of a unit of volume for heat.

expansion which will take place will be $E\tau/(1 + Et)$, if the temperature be denoted by t; and hence, by (8), the quantity of heat that must be supplied, to prevent any lowering of temperature, will be

$$\frac{E p_0 v_0}{\mu} \cdot \frac{E\tau}{1 + Et}, \quad \text{or} \quad \frac{E^2 p}{\mu (1 + Et)^2} \tau.$$

Hence, the total quantity added is equal to $A\tau + \dfrac{E^2 p}{\mu (1 + Et)^2} \tau.$

But, since B denotes the specific heat under constant pressure, the quantity of heat requisite to bring the gas into this state, from its primitive condition, is equal to $B\tau$; and hence we have

$$B = A + \frac{E^2 p}{\mu (1 + Et)^2} \quad \dots\dots\dots\dots\dots\dots\dots(12).$$

IV.   *Comparison of the Relative advantages of the Air-Engine and Steam-Engine.*

54.   In the use of water-wheels for motive power, the economy of the engine depends not only upon the excellence of its adaptation for actually transmitting any given quantity of water through it, and producing the equivalent of work, but upon turning to account the entire available fall; so, as we are taught by Carnot, the object of a thermodynamic engine is to economize in the best possible way the transference of all the heat evolved, from bodies at the temperature of the source, to bodies at the lowest temperature at which the heat can be discharged.   With reference then to any engine of the kind, there will be two points to be considered.

(1)   The extent of the *fall* utilized.

(2)   The economy of the engine, with the fall which it actually uses.

55.   In the first respect, the air-engine, as Carnot himself points out, has a vast advantage over the steam-engine; since the temperature of the hot part of the machine may be made very much higher in the air-engine than would be possible in the steam-engine, on account of the very high pressure produced

in the boiler, by elevating the temperature of the water which it contains to any considerable extent above the atmospheric boiling point. On this account, a "perfect air-engine" would be a much more valuable instrument than a "perfect steam-engine."*

Neither steam-engines nor air-engines, however, are nearly perfect ; and we do not know in which of the two kinds of machine the nearest approach to perfection may be actually attained. The beautiful engine invented by Mr Stirling of Galston, may be considered as an excellent beginning for the air-engine†; and it is only necessary to compare this with Newcomen's steam-engine, and consider what Watt has effected, to give rise to the most sanguine anticipations of improvement.

V. *On the Economy of actual Steam-Engines.*

56. The steam-engine being universally employed at present as the means for deriving motive power from heat, it is extremely interesting to examine, according to Carnot's theory, the economy actually attained in its use. In the first place, we remark that, out of the entire "fall" from the temperature of the coals to that of the atmosphere, it is only part—that from the temperature of the boiler to the temperature of the condenser—that is made available; while the very great fall from the temperature of the burning coals to that of the boiler, and the comparatively small fall from the temperature of the condenser to that of the atmosphere, are entirely lost as far as regards the mechanical effect

---

* Carnot suggests a combination of the two principles, with air as the medium for receiving the heat at a very high temperature from the furnace; and a second medium, alternately in the state of saturated vapour and liquid water, to receive the heat, discharged at an intermediate temperature from the air, and transmit it to the coldest part of the apparatus. It is possible that a complex arrangement of this kind might be invented, which would enable us to take the heat at a higher temperature, and discharge it a lower temperature than would be practicable in any simple air-engine or simple steam-engine. If so, it would no doubt be equally possible, and perhaps more convenient, to employ steam alone, but to use it at a very high temperature not in contact with water in the hottest part of the apparatus, instead of, as in the steam-engine, always in a saturated state.

† It is probably this invention to which Carnot alludes in the following passage (p. 112):—"Il a été fait, dit-on, tout récemment en Angleterre des essais heureux sur le développement de la puissance motrice par l'action de la chaleur sur l'air atmosphérique. Nous ignorons entièrement en quoi ces essais ont consisté, si toutefois ils sont réels."

which it is desired to obtain. We infer from this, that the temperature of the boiler ought to be kept as high as, according to the strength, is consistent with safety, while that of the condenser ought to be kept as nearly down at the atmospheric temperature as possible. To take the entire benefit of the actual fall, Carnot shewed that the "principle of expansion" must be pushed to the utmost *.

57. To obtain some notion of the economy which has actually been obtained, we may take the alleged performances of the best Cornish engines, and some other interesting practical cases as examples †.

(1) The engine of *the Fowey Consols mine* was reported, in 1845, to have given 125,089,000 foot-pounds of effect, for the consumption of one bushel or 94 lbs. of coals. Now, the average amount evaporated from Cornish boilers, by one pound of coal, is 8½ lbs. of steam ; and hence, for each pound of steam evaporated 156,556 foot-pounds of work are produced.

The pressure of the saturated steam in the boiler may be taken as 3½ atmospheres ‡; and, consequently, the temperature of the water will be 140°. Now (Regnault, end of Mémoire X.), the latent heat of a pound of saturated steam at 140° is 508, and since, to compensate for each pound of steam removed from the boiler in

---

* From this point of view, we see very clearly how imperfect is the steam-engine, even after all Watt's improvements. For to " push the principle of expansion to the utmost," we must allow the steam, before leaving the cylinder, to expand until its pressure is the same as that of the vapour in the condenser. According to "Watt's law," its temperature would then be the same as (actually a little above, as Regnault has shewn) that of the condenser, and hence the steam-engine worked in this most advantageous way, has in reality the very fault that Watt found in Newcomen's engine. This defect is partially remedied by Hornblower's system of using a separate expansion cylinder, an arrangement, the advantages of which did not escape Carnot's notice, although they have not been recognized extensively among practical engineers, until within the last few years.

† I am indebted to the kindness of Professor Gordon of Glasgow, for the information regarding the various cases given in the text.

‡ In different Cornish engines, the pressure in the boiler is from 2½ to 5 atmospheres; and, therefore, as we find from Regnault's table of the pressure of saturated steam, the temperature of the water in the boiler must, in all of them, lie between 128° and 152°. For the better class of engines, the average temperature of the water in the boiler may be estimated at 140°, the corresponding pressure of steam being 3½ atmospheres.

the working of the engine, a pound of water, at the temperature of the condenser, which may be estimated at 30°, is introduced from the hot well; it follows that 618 units of heat are introduced to the boiler for each pound of water evaporated. But the work produced, for each pound of water evaporated, was found above to be 156,556 foot-pounds. Hence, $\frac{156556}{618}$, or 253 foot-pounds is the amount of work produced for each unit of heat transmitted through the Fowey Consols engine. Now, in Table II., we find 583·0 as the theoretical effect due to a unit descending from 140° to 0°, and 143 as the effect due to a unit descending from 30° to 0°. The difference of these numbers, or 440 *, is the number of foot-pounds of work that a *perfect* engine with its boiler at 140°, and its condenser at 30°, would produce for each unit of heat transmitted. Hence, the Fowey Consols engine, during the experiments reported on, performed $\frac{253}{440}$ of its theoretical duty, or 57½ per cent.

(2) The best duty on record, as performed by an engine at work (not for merely experimental purposes), is that of Taylor's engine, at the United Mines, which, in 1840, worked regularly, for several months, at the rate of 98,000,000 foot-pounds for each bushel of coals burned. This is $\frac{98}{125}$, or ·784 of the experimental duty reported in the case of the Fowey Consols engine. Hence, the best useful work on record, is at the rate of 198·3 foot-pounds for each unit of heat transmitted, and is $\frac{198\cdot3}{440}$, or 45 per cent. of the theoretical duty, on the supposition that the boiler is at 140°, and the condenser at 30°.

(3) French engineers contract (in Lille, in 1847, for example) to make engines for mill power which will produce 30,000 metre-lbs., or 98,427 foot-lbs. of work for each pound of steam used. If we divide this by 618, we find 159 foot-pounds for the work produced by each unit of heat. This is 36·1 per cent. of 440, the theoretical duty†.

* This number agrees very closely with the number corresponding to the fall from 100° to 0°, given in Table II. Hence, the fall from 140° to 30° of the scale of the air-thermometer is equivalent, with reference to motive power, to the fall from 100° to 0°.

† It being assumed that the temperatures of the boiler and condenser are the same as those of the Cornish engines. If, however, the pressure be lower, two atmospheres, for instance, the numbers would stand thus: The temperature in the boiler would be only 121. Consequently, for each pound of steam evaporated,

(4) English engineers have contracted to make engines and boilers which will require only $3\frac{1}{2}$ lbs. of the best coal per horse-power per hour. Hence, in such engines, each pound of coal ought to produce 565,700 foot-pounds of work, and if 7 lbs. of water be evaporated by each pound of coal, there would result 83,814 foot-pounds of work for each pound of water evaporated. If the pressure in the boiler be $3\frac{1}{2}$ atmospheres (temperature 140°) the amount of work for each unit of heat will be found, by dividing this by 618, to be 130·7 foot-pounds, which is $\frac{130·7}{440}$ or 29·7 per cent. of the theoretical duty*.

(5) The actual average of work performed by good Cornish engines and boilers is 55,000,000 foot-pounds for each bushel of coal, or less than half the experimental performance of the Fowey Consols engine, more than half the actual duty performed by the United Mines engine in 1840; in fact about 25 per cent. of the theoretical duty.

(6) The average performances of a number of Lancashire engines and boilers have been recently found to be such as to require 12 lbs. of Lancashire coal per horse-power per hour (i.e., for performing 60 × 33,000 foot-pounds) and of a number of Glasgow engines, such as to require 15 lbs. (of a somewhat inferior coal) for the same effect. There are, however, more than twenty large engines in Glasgow at present†, which work with a consumption of only $6\frac{1}{4}$ lbs. of dross, equivalent to 5 lbs. of the best Scotch, or 4 lbs. of the best Welsh coal, per horse-power per hour. The economy may be estimated from these data, as in the other cases, on the assumption which, with reference to these, is the

---

only 614 units of heat would be required; and, therefore, the work performed for each unit of heat transmitted would be 160·3 foot-pounds, which is *more* than according to the estimate in the text. On the other hand, the range of temperatures, or the fall utilised, is only from 131 to 30, instead of from 140 to 30°, and, consequently (Table II.), the theoretical duty for each unit of heat is only 371 foot-pounds. Hence, if the engine, to work according to the specification, requires a pressure of only 15 lbs. on the square inch (*i.e.* a total steam pressure of two atmospheres), its performance is $\frac{160·3}{371}$, or 43·2 per cent. of its theoretical duty.

* If, in this case again, the pressure required in the boiler to make the engine work according to the contract were only 15 lbs. on the square inch, we should have a different estimate of the economy, for which, see Table B, at the end of this paper.

† These engines are provided with separate expansive cylinders, which have been recently added to them by Mr M'Naught of Glasgow.

most probable we can make, that the evaporation produced by a pound of best coal is 7 lbs. of steam.

58.   The following Tables afford a synoptic view of the performances and theoretical duties in the various cases discussed above.

In Table A the numbers in the second column are found by dividing the numbers in the first by 8½ in cases (1), (2), and (5), and by 7 in cases (4), (6), and (7), the estimated numbers of pounds of steam actually produced in the different boilers by the burning of 1 lb. of coal.

The numbers in the third column are found from those in the second, by dividing by 618, in Table A, and 614 in Table B, which are respectively the quantities of heat required to convert a pound of water taken from the hot well at 30°, into saturated steam, in the boiler, at 140° or at 121°.

With reference to the cases (3), (4), (6), (7), the hypothesis of Table B is probably in general nearer the truth than that of Table A.   In (4), (6), and (7), especially upon hypothesis B, there is much uncertainty as to the amount of evaporation that will be actually produced by 1 lb. of fuel.   The assumption on which the numbers in the second column in Table B are calculated, is, that each pound of coal will send the same number of units of heat into the boiler whether hypothesis A or hypothesis B be followed.   Hence, except in the case of the French contract, in which the *evaporation*, not the fuel, is specified, the numbers in the third column are the same as those in the third column of Table A.

TABLE A. *Various Engines in which the temperature of the Boiler is 140°, and that of the Condenser 30°.*

Theoretical Duty for each Unit of Heat transmitted, 440* foot-pounds.

CASES.	Work produced for each pound of coal consumed.	Work produced for each pound of water evaporated.	Work produced for each unit of heat transmitted.	Per centage of theoretical duty.
	Foot-Pounds.	Foot-Pounds.	Foot-Pounds.	
(1) Fowey Consols Experiment, reported in 1845	1,330,734	156,556	253	57·5
(2) Taylor's Engine at the United Mines, working in 1840	1,042,553	122,653	198·4	45·1
(3) French Engines, according to contract	* * * *	98,427	159	36·1
(4) English Engines, according to contract	565,700	80,814	130·8	29·7
(5) Average actual performance of Cornish Engines	585,106	68,836	111·3	25·3
(6) Common Engines, consuming 12 lbs. of best coal per hour per horse-power	165,000	23,571	38·1	8·6
(7) Improved Engines with expansion cylinders, consuming an equivalent to 4 lbs. of best coal per horse-power per hour . .	495,000	70,710	114·4	26

* [Note added March 15, 1851. Total work for thermal unit, 1390 (Joule), 377·1 corrected by the dynamical theory, March 15, 1851,

$$377·1 = ·2713 \times 1390,$$
$$253 = ·1820 \times 1390 = \frac{1}{5·49} \times 1390.]$$

TABLE B. *Various Engines in which the Temperature of the Boilers is 121*, and that of the Condenser 30°.*

Theoretical Duty for each Unit of Heat transmitted, 371 foot-pounds.

CASES.	Work produced for each pound of coal consumed.	Work produced for each pound of water evaporated.	Work produced for each unit of heat transmitted.	Per centage of theoretical duty.
	Foot-Pounds.	Foot-Pounds.	Foot-Pounds.	
(3) French Engines, according to contract	* * *	98,427	160·3	43·2
(4) English Engines, according to contract	565,700	$\frac{614}{613} \times 80,814$	130·8	35
(6) Common Engines, consuming 12 lbs. of coal per horse-power per hour	165,000	$\frac{614}{613} \times 23,571$	38·1	10·3
(7) Improved Engines with expansion cylinders, consuming an equivalent to 4 lbs. best coal per horse-power per hour	495,000	$\frac{614}{613} \times 70,710$	114·4	30·7

* Pressure 15 lbs. on the square inch.

[Appended to the preceding (Art. XLI.), by permission of my brother
Professor James Thomson.   February 10, 1881.]

THEORETICAL CONSIDERATIONS ON THE EFFECT OF PRESSURE IN
LOWERING THE FREEZING POINT OF WATER.   BY JAMES
THOMSON.

(*Cambridge and Dublin Mathematical Journal*, Nov. 1850; *taken, with some slight
alterations made by the author, from the Transactions of the Royal Society of
Edinburgh*, Jan. 2, 1849.)

SOME time ago my brother, Professor William Thomson,
pointed out to me a curious conclusion to which he had been
led, by reasoning on principles similar to those developed by
Carnot, with reference to the motive power of heat.   It was, that
*water at the freezing point may be converted into ice by a process
solely mechanical, and yet without the final expenditure of any
mechanical work.*   This at first appeared to me to involve an
impossibility, because water expands while freezing; and therefore it seemed to follow, that if a quantity of it were merely
enclosed in a vessel with a moveable piston and frozen, the
motion of the piston, consequent on the expansion, being resisted
by pressure, mechanical work would be given out without any
corresponding expenditure ; or, in other words, a perpetual source
of mechanical work, commonly called a perpetual motion, would
be possible.   After farther consideration, however, the former
conclusion appeared to be incontrovertible ; but then, to avoid
the absurdity of supposing that mechanical work could be got
out of nothing, it occurred to me that it is necessary farther to

conclude, that *the freezing point becomes lower as the pressure to which the water is subjected is increased.*

The following is the reasoning by which these conclusions are proved.

First, to prove that water at the freezing point may be converted into ice by a process solely mechanical, and yet without the final expenditure of any mechanical work :—Let there be supposed to be a cylinder, and a piston fitting water-tight to it, and capable of moving without friction. Let these be supposed to be formed of a substance which is a perfect non-conductor of heat; also, let the bottom of the cylinder be closed by a plate, supposed to be a perfect conductor, and to possess no capacity for heat. Now, to convert a given mass of water into ice without the expenditure of mechanical work, let this imaginary vessel be partly filled with air at 0° C.* and let the bottom of it be placed in contact with an indefinite mass of water, a lake for instance, at the same temperature. Now, let the piston be pushed towards the bottom of the cylinder by pressure from some external reservoir of mechanical work, which, for the sake of fixing our ideas, we may suppose to be the hand of an operator. During this process the air in the cylinder would tend to become heated on account of the compression, but it is constrained to remain at 0° by being in communication with the lake at that temperature. The change, then, which takes place is, that a certain amount of work is given from the hand to the air, and a certain amount of heat is given from the air to the water of the lake. In the next place, let the bottom of the cylinder be placed in contact with the mass of water at 0°, which is proposed to be converted into ice, and let the piston be allowed to move back to the position it had at the commencement of the first process. During this second process, the temperature of the air would tend to sink on account of the expansion, but it is constrained to remain constant at 0° by the air being in communication with the freezing water, which cannot change its temperature so long as any of it remains unfrozen. Hence, so far as the air and the hand are concerned, this process has been exactly the converse of the former one. Thus the air has expanded through the same distance through which it was formerly compressed; and since it has been

* The centigrade thermometric scale is adopted throughout this paper.

constantly at the same temperature during both processes, the law of the variation of its pressure with its volume must have been the same in both. From this it follows, that the hand has received back exactly the same amount of mechanical work in the second process as it gave out in the first. By an analogous reason it is easily shewn that the air also has received again exactly the same amount of heat as it gave out during its compression; and, hence, it is now left in a condition the same as that in which it was at the commencement of the first process. *The only change which has been produced then, is that a certain quantity of heat has been abstracted from a small mass of water at 0°, and dispersed through an indefinite mass at the same temperature, the small mass having thus been converted into ice.* This conclusion, it may be remarked, might be deduced at once by the application, to the freezing of water, of the general principle developed by Carnot, that no work is given out when heat passes from one body to another without a fall of temperature; or rather by the application of the converse of this, which of course equally holds good, namely, that no work requires to be expended to make heat pass from one body to another at the same temperature.

Next, to prove that the freezing point of water is lowered by an increase of the pressure to which the water is subjected:— Let the imaginary cylinder and piston employed in the foregoing demonstration, be again supposed to contain some air at 0°. Let the bottom of the cylinder be placed in contact with the water of an indefinitely large lake at 0°; and let the air be subjected to compression by pressure applied by the hand to the piston. A certain amount of work is thus given from the hand to the air, and a certain amount of heat is given out from the air to the lake. Next, let the bottom of the cylinder be placed in communication with a small quantity of water at 0°, enclosed in a second imaginary cylinder similar in character to the first, and which we may call the water cylinder the first being called the air cylinder; and let this water be, at the commencement, subject merely to the atmospheric pressure. Let, however, resistance be offered by the hand to any motion of the piston of the water cylinder which may take place. Things being in this state, let the piston of the air cylinder move back to its original position. During this process, heat becomes latent in the air on account of the increase of volume, and therefore the air abstracts heat from the water, because the air and water, being in communication with one another, must remain each at the same temperature as the other, whether that temperature changes or not. The first effect of the abstraction of heat from the water must be the conversion of a part of the water into ice, an effect which must be accompanied with an increase of volume of the mass enclosed in the water cylinder. Hence, on account of the resistance offered by the hand to the motion of the piston of this cylinder, the internal pressure is increased, and work is received by the hand from the piston. Towards the end of this process, let the resistance offered by the hand gradually decrease, till, just at the end it becomes nothing, and the pressure within the water cylinder thus becomes again equal to that of the atmosphere. The temperature of the mass of partly frozen water must now be 0°, and the air in the other cylinder, being in communication with this, must have the same temperature. The air is therefore at its original temperature, and it has its original volume, or, in other words, it is in its original state. Farther, let the ice be converted, under atmospheric pressure, into water; the requisite heat being transferred to it from the lake by the mechanical process already pointed out, which involves no loss of mechanical work. Thus, now at the conclusion of the operation, the whole mass of water is left in its original state; and likewise, as has already been shewn, the air is left in its original state. Hence no work can have been developed by any change on the air and water, which have been used. But work has been given out by the piston of the water cylinder to the hand; and therefore an equal quantity* of work must have been given from the hand to the air piston, as there is no other way in which the work developed could have been introduced into the apparatus. Now, the only way in which this can have taken place is by the air having been colder, while it was expanding in the second process, than it was while it was undergoing compression during the first. Hence it was colder than 0° during the course of the second process; or, in other words, *while the water was freezing, under a pressure greater than that of the atmosphere, its temperature was lower than 0°.*

The fact of the lowering of the freezing point being thus

---

* In saying "an equal quantity" I, of course, neglect infinitely small quantities in comparison to quantities not infinitely small.

demonstrated, it becomes desirable, in the next place, to find what is the freezing point of water for any given pressure. The most obvious way to determine this would be by direct experiment with freezing water. I have not, however, made any attempt to do so in this way. The variation to be appreciated is extremely small, so small in fact as to afford sufficient reason for its existence never having been observed by any experimenter. Even to detect its existence, much more to arrive at its exact amount by direct experiment, would require very delicate apparatus which would not be easily planned out or procured. Another, and a better mode of proceeding has, however, occurred to me : and by it we can deduce, from the known expansion of water in freezing, and the known quantity of heat which becomes latent in the melting of ice, together with data founded on the experiments of Regnault on steam at the freezing point, a formula which gives the freezing point in terms of the pressure ; and which may be applied for any pressure, from nothing up to many atmospheres. The following is the investigation of this formula.

Let us suppose that we have a cylinder of the imaginary con- struction described at the commencement of this paper ; and let us use it as an ice-engine analogous to the imaginary steam-engine conceived by Carnot, and employed in his investigations. For this purpose, let the entire space enclosed within the cylinder by the piston be filled at first with as much ice at 0° as would, if melted, form rather more than a cubic foot of water, and let the ice be subject merely to one atmosphere of pressure, no force being applied to the piston. Now, let the following four processes, forming one complete stroke of the ice-engine, be performed.

Process 1. Place the bottom of the cylinder in contact with an indefinite lake of water at 0°, and push down the piston. The effect of the motion of the piston is to convert ice at 0° into water at 0°, and to abstract from the lake at 0° the heat which becomes latent during this change. Continue the compression till one cubic foot of water is melted from ice.

Process 2. Remove the cylinder from the lake, and place it with its bottom on a stand which is a perfect non-conductor of heat. Push the piston a very little farther down, till the pressure inside is increased by any desired quantity which may be denoted, in pounds on the square foot, by p. During this motion of the piston, since the cylinder contains ice and water, the temperature of the mixture must vary with the pressure, being at any instant the freezing point which corresponds to the pressure at that instant. Let the temperature at the end of this process be denoted by $-t^0$ C.

Process 3. Place the bottom of the cylinder in contact with a second indefinitely large lake at $-t^0$, and move the piston upwards. During this motion the pressure must remain constant at p above that of the atmosphere, the water in the cylinder increasing its volume by freezing, since if it did not freeze, its pressure would diminish, and therefore its temperature would increase, which is impossible, since the whole mass of water and ice is constrained by the lake to remain at $-t^0$. Continue the motion till so much heat has been given out to the second lake at $-t^0$, as that if the whole mass contained in the cylinder were allowed to return to its original volume without any introduction or abstraction of heat, it would assume its original temperature and pressure. This, if Carnot's principles be admitted, as they are supposed to be throughout the present investigation, is the same as to say,—Continue the motion till all the heat has been given out to the second lake at $-t^0$, which was taken in during Process 2, from the first lake at 0^0.*

Process 4. Remove the cylinder from the lake at $-t^0$, and place its bottom again on the non-conducting stand. Move the piston back to the position it occupied at the commencement of Process 1. At the end of this fourth process the mass contained in the cylinder must, according to the condition by which the termination of Process 3 was fixed, have its original temperature and pressure, and therefore it must be in every respect in its original physical state.

By representing graphically in a diagram the various volumes

---

* This step, as well as the corresponding one in Carnot's investigation, it must be observed, involves difficult questions, which cannot as yet be satisfactorily answered, regarding the possibility of the absolute formation or destruction of heat as an equivalent for the destruction or formation of other agencies, such as mechanical work; but, in taking it, I go on the almost universally adopted supposition of the perfect conservation of heat.

and corresponding pressures, at all the stages of the four processes
which have just been described, we shall arrive, in a simple and
easy manner, at the quantity of work which is developed in one
complete stroke by the heat which is transferred during that
stroke from the lake at 0^0 to the lake at $-t^0$. For this purpose,
let E be the position of the piston at the beginning of Process 1;

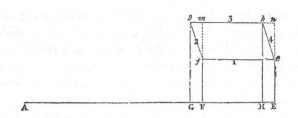

and let some distance, such as EG, represent its stroke in feet,
its area being made a square foot, so that the numbers expressing,
in feet, distances along EG may also express, in cubic feet, the
changes in the contents of the cylinder produced by the motion of
the piston.   Now, when 1·087 cubic feet of ice are melted, one
cubic foot of water is formed.   Hence, if EF be taken equal to
·087 feet, F will be the position of the piston when one cubic
foot of water has been melted from ice, that is, the position at the
end of Process 1, the bottom of the cylinder being at a point A
distant from F by rather more than a foot.   Let FG be the
compression during Process 2, and HE the expansion during
Process 4.   Let ef be parallel to EF, and let Ee represent one
atmosphere of pressure ; that is, let the units of length for the
vertical ordinates be taken such that the number of them in Ee
may be equal to the number which expresses an atmosphere of
pressure.   Also let gh be parallel to EF, and let fm represent
the increase of pressure produced during Process 2.   Then the
straight lines ef and gh will be the lines of pressure for Processes
1 and 3 ; and for the other two processes, the lines of pressure
will be some curves which would extremely nearly coincide with
the straight lines fg and he.   For want of experimental data,
the natures of these two curves cannot be precisely determined ;
but, for our present purpose, it is not necessary that they should
be so, as we merely require to find the area of the figure $efgh$,
which represents the work developed by the engine during one

complete stroke, and this can readily be obtained with sufficient accuracy. For, even though we should adopt a very large value for *fm*, the change of pressure during Process 2, still the changes of volume *gm* and *hn* in Process 2 and Process 4 would be extremely small compared to the expansion during the freezing of the water; and from this it follows evidently that the area of the figure *efgh* is extremely nearly equal to that of the rectangle *efmn*, but *fe* is equal to *FE*, which is 087 feet. Hence the work developed during an entire stroke is ·087 × *p* foot-pounds. Now this is developed by the descent from 0° to −*t*° of the quantity of heat necessary to melt a cubic foot of ice; that is, by 4925 thermic units, the unit being the quantity of heat required to raise a pound of water from 0° to 1° centigrade. Next we can obtain another expression for the same quantity of work; for, by the tables deduced in the preceding paper from the experiments of Regnault, we find that the quantity of work developed by one of the same thermic units descending through one degree about the freezing point, is 4·97 foot-pounds. Hence, the work due to 4925 thermic units descending from 0° to −*t*° is 4925 × 4·97 × *t* foot-pounds. Putting this equal to the expression which was formerly obtained for the work due to the same quantity of heat falling through the same number of degrees, we obtain

$$4925 \times 4\!\cdot\!97 \times t = \cdot087 \times p.$$

Hence $\qquad\qquad t = \cdot00000355p$ ....(1).

This, then, is the desired formula for giving the freezing point −*t*° centigrade, which corresponds to a pressure exceeding that of the atmosphere by a quantity *p*, estimated in pounds on a square foot.

To put this result in another form, let us suppose water to to be subjected to one additional atmosphere, and let it be required to find the freezing point. Here *p* = one atmosphere = 2120 pounds on a square foot; and therefore, by (1),

$$t = \cdot00000355 \times 2120, \text{ or } t = \cdot0075.$$

That is, the freezing point of water, under the pressure of one additional atmosphere, is − ·0075° centigrade; and hence, if the pressure above one atmosphere be now denoted in atmospheres*, as units, by n, we obtain t, the lowering of the freezing point in degrees centigrade, by the following formula,

$$t = \cdot 0075n \dots \dots \dots \dots \dots (2).$$

[The phenomena predicted by the author of the preceding paper, in anticipation of any direct observations on the freezing point of water, have been fully confirmed by experiment. See a short paper published in the *Proceedings of the Royal Society of Edinburgh* (Feb. 1850), and republished in the *Philosophical Magazine* for August, 1850, under the title "The Effect of Pressure in Lowering the Freezing Point of Water experimentally demonstrated. By Prof. William Thomson."]

* The atmosphere is here taken as being the pressure of a column of mercury of 760 millimetres; that is 29·92, or very nearly 30 English inches.

ART. XLII. ON THE THEORY OF MAGNETIC INDUCTION IN CRYSTALLINE SUBSTANCES.

[*Brit. Assoc. Rep.* 1850 (Part II). ELECTROSTATICS AND MAGNETISM, Art. XXX.]

ART. XLIII. NOTES ON A PAPER "PROBLEMS RESPECTING POLYGONS IN A PLANE." BY ROBERT MOON.

[*Cambridge and Dublin Mathematical Journal*, Vol. V. 1850.]

ART. XLIV. ON THE POTENTIAL OF A CLOSED GALVANIC CIRCUIT OF ANY FORM.

[*Cambridge and Dublin Mathematical Journal*, Vol. V. 1850. ELECTROSTATICS AND MAGNETISM, Art. XXV.]

[From the *Proc. R. S. E.* Jan. 1850 ; *Phil. Mag.* XXXVII. 1850; *Annal. de Chimie*, XXXV. 1852; *Journ. de Pharm.* XVIII. 1850; *Poggend. Annal.* LXXXI. 1850.]

ART. XLV. THE EFFECT OF PRESSURE IN LOWERING THE FREEZING POINT OF WATER EXPERIMENTALLY DEMONSTRATED.

ON the 2nd of January 1849, a communication entitled "Theoretical Considerations on the Effect of Pressure in Lowering the Freezing Point of Water, by James Thomson, Esq., of Glasgow," was laid before the Royal Society, and it has since been published in the *Transactions*, Vol. XVI. part 5*. In that paper it was demonstrated that, if the fundamental axiom of Carnot's Theory of the Motive Power of Heat be admitted, it follows, as a rigorous consequence, that the temperature at which ice melts will be lowered by the application of pressure ; and the extent of this effect due to a given amount of pressure was deduced by a reasoning analogous to that of Carnot from Regnault's experimental determination of the latent heat, and the pressure of saturated aqueous vapour at various temperatures differing very little from the ordinary freezing point of water. Reducing to Fahrenheit's scale the final result of the paper, we find

$$t = n \times 0\ 0135 ;$$

where t denotes the depression in the temperature of melting ice produced by the addition of n "atmospheres" (or n times the pressure due to 29·922 inches of mercury), to the ordinary pressure experienced from the atmosphere.

In this very remarkable speculation, an entirely novel physical phænomenon was *predicted* in anticipation of any direct experi-

---

* It will appear also, with some slight alterations made by the author, in the *Cambridge and Dublin Mathematical Journal*, Nov. 1850. [Reprinted as Appendix to Art. XLI. above.]

ments on the subject; and the actual observation of the phænomenon was pointed out as a highly interesting object for experimental research.

To test the phænomenon by experiment without applying
excessively great pressure, a very sensitive thermometer would be
required, since for ten atmospheres the effect expected is little
more than the tenth part of a Fahrenheit degree; and the thermometer employed, if founded on the expansion of a liquid in a
glass bulb and tube, must be protected from the pressure of the
liquid, which, if acting on it, would produce a deformation, or at
least a compression of the glass that would materially affect the
indications. For a thermometer of extreme sensibility, mercury
does not appear to be a convenient liquid; since, if a very fine
tube be employed, there is some uncertainty in the indications on
account of the irregularity of capillary action, due probably to
superficial impurities, and observable even when the best mercury
that can be prepared is made use of; and again, if a very large
bulb be employed, the weight of the mercury causes a deformation
which will produce a very marked difference in the position of the
head of the column in the tube according to the manner in which
the glass is supported, and may therefore affect with uncertainty
the indications of the instrument. The former objection does not
apply to the use of any fluid which perfectly wets the glass; and
the last-mentioned source of uncertainty will be much less for
any lighter liquid than mercury, of equal or greater expansibility
by heat. Now the coefficient of expansion of sulphuric æther
at $0°$ C. being, according to Mr I. Pierre*, ·00151, is eight or nine
times that of mercury (which is ·000179, according to Regnault),
and its density is about the twentieth part of the density of mercury. Hence a thermometer of much higher sensibility may be
constructed with æther than with mercury, without experiencing
inconvenience from the circumstances which have been alluded to.
An æther thermometer was accordingly constructed by Mr Robert
Mansell of Glasgow, for the experiment which I proposed to make.
The bulb of this instrument is nearly cylindrical, and is about
$3\frac{1}{2}$ inches long and $\frac{3}{8}$ths of an inch in diameter. The tube has a
cylindrical bore about $6\frac{1}{2}$ inches long: about $5\frac{1}{2}$ inches of the tube
are divided into 220 equal parts. The thermometer is entirely

---

* See Dixon *On Heat*, p. 72.

inclosed, and hermetically sealed in a glass tube, which is just large enough to admit it freely*. On comparing the indications of this instrument with those of a thermometer of Crichton's with an ivory scale, which has divisions corresponding to degrees Fahrenheit of about $\frac{1}{25}$th of an inch each, I found that the range of the æther thermometer is about 3° Fahrenheit; and that there are about 212 divisions on the tube corresponding to the interval of temperature from 31° to 34°, as nearly as I could discover from such an unsatisfactory standard of reference. This gives $\frac{1}{71}$ of a degree for the mean value of a division. From a rough calibration of the tube which was made, I am convinced that the values of the divisions at no part of the tube differ by more than $\frac{1}{30}$th of this amount from the true mean value; and, taking into account all the sources of uncertainty, I think it probable that each of the divisions on the tube of the æther thermometer corresponds to something between $\frac{1}{68}$ and $\frac{1}{75}$ of a degree Fahrenheit.

With this thermometer in its glass envelope, and with a strong glass cylinder (Œrsted's apparatus for the compression of water), an experiment was made in the following manner:—

The compression vessel was partly filled with pieces of clean ice and water: a glass tube about a foot long and $\frac{1}{10}$th of an inch internal diameter, closed at one end, was inserted with its open end downwards, to indicate the fluid pressure by the compression of the air which it contained; and the æther thermometer was let down and allowed to rest with the lower end of its glass envelope pressing on the bottom of the vessel. A lead ring was let down so as to keep free from ice the water in the compression cylinder round that part of the thermometer tube where readings were expected. More ice was added above; so that both above and below the clear space, which was only about two inches deep, the compression cylinder was full of pieces of ice. Water was then poured in by a tube with a stopcock fitted in the neck of the vessel, till the vessel was full up to the piston, after which the stopcock was shut.

* Following a suggestion made to me by Professor Forbes of Edinburgh, I have in subsequent experiments with this thermometer, used it with enough of mercury introduced into the tube in which it is hermetically sealed to entirely cover its bulb; as I found that, without this, if the experiment was conducted in a warm room, the indications of the thermometer were frequently deranged by the portion of the water which was left free from ice becoming slightly elevated in temperature.

After it was observed that the column of æther in the thermometer stood at about 67°, with reference to the divisions on the tube, a pressure of from 12 to 15 atmospheres was applied, by forcing the piston down with the screw. Immediately the column of æther descended very rapidly, and in a very few minutes it was below 61°. The pressure was then suddenly removed, and immediately the column in the thermometer began to rise rapidly. Several times pressure was again suddenly applied, and again suddenly removed, and the effects upon the thermometer were most marked.

The fact that the freezing point of water is sensibly lowered by a few atmospheres of pressure was thus established beyond all doubt. After that I attempted, in a more deliberate experiment, to determine as accurately as my means of observation allowed me to do, the actual extent to which the temperature of freezing is affected by determinate applications of pressure.

In the present communication I shall merely mention the results obtained, without entering at all upon the details of the experiment.

I found that a pressure of, as nearly as I have been able to estimate it, 8·1 atmospheres produced a depression measured by $7\frac{1}{2}$ divisions of the tube on the column of æther in the thermometer; and again, a pressure of 16·8 atmospheres produced a thermometric depression of $16\frac{1}{2}$ divisions. Hence the observed lowering of temperature was $7\frac{1}{2}/71$, or ·106° F. in the former case, and $16\frac{1}{2}/71$, or ·232° F. in the latter.

Let us compare these results with theory. According to the conclusions arrived at by my brother in the paper referred to above, the lowering of the freezing point of water by 8·1 atmospheres of pressure would be 8·1 × ·0135, or ·109° F.; and the lowering of the freezing point by 16·8 atmospheres would be 16·8 × ·0135, or ·227° F. Hence we have the following highly satisfactory comparison, for the two cases, between the experiment and theory:—

Observed pressures.	Observed depressions of temperatures.	Depressions according to theory, on the hypothesis that the pressures were truly observed.	Differences.
8·1 atmospheres...	·106° F.	·109° F.	− ·003° F.
16·8 atmospheres...	·232° F.	·227° F.	+ ·005° F.

It was, I confess, with some surprise, that, after having completed the observations under an impression that they presented
great discrepancies from the theoretical expectations, I found the
numbers I had noted down indicated in reality an agreement so
remarkably close, that I could not but attribute it in some degree
to chance, when I reflected on the very rude manner in which the
quantitative parts of the experiment (especially the measurement
of the pressure, and the evaluation of the division of the æther
thermometer) had been conducted.

I hope before long to have a thermometer constructed, which
shall be at least three times as sensitive as the æther thermometer
I have used hitherto; and I expect with it to be able to perceive
the effect of increasing or diminishing the pressure by less than
an atmosphere, in lowering or elevating the freezing point of water.

If a convenient *minimum* thermometer could be constructed,
the effects of very great pressures might easily be tested by hermetically sealing the thermometer in a strong glass, or in a metal
tube, and putting it into a mixture of ice and water, in a strong
metal vessel, in which an enormous pressure might be produced
by the forcing-pump of a Bramah's press.

In conclusion, it may be remarked, that the same theory which
pointed out the remarkable effect of pressure on the freezing point
of water, now established by experiment, indicates that a corresponding effect may be expected for all liquids which expand
in freezing ; that a reverse effect, or an elevation of the freezing-
point by an increase of pressure, may be expected for all liquids
which contract in freezing ; and that the extent of the effect to
be expected may in every case be deduced from Regnault's observations on vapour (provided that the freezing point is within the
temperature-limits of his observations), if the latent heat of a
cubic foot of the liquid, and the alteration of its volume in
freezing be known.

ART. XLVI. ON THE FORCES EXPERIENCED BY INDUCTIVELY
MAGNETIZED FERRO-MAGNETIC OR DIA-MAGNETIC NON-CRYS-
TALLINE SUBSTANCES.

[*Phil. Mag.* XXXVII. 1850; *Poggend. Annal.* LXXXII. 1851. ELECTROSTATICS
AND MAGNETISM, Art. XXXIV.]

ART. XLVII. ON A REMARKABLE PROPERTY OF STEAM CON-
NECTED WITH THE THEORY OF THE STEAM-ENGINE.

[*Phil. Mag.* Nov. 1850. Pogg. *Ann.* LXXXI. 1850.]

To the Editors of the Philosophical Magazine and Journal.

GENTLEMEN,

I am permitted by my friend Professor Thomson to communicate the following letter to the *Philosophical Magazine*, containing
an explanation of the true cause of the non-scalding property
of steam issuing from a high-pressure boiler. The proposition
announced by Mr Rankine is certainly one of very great importance; as it would appear from it that when saturated steam is
allowed to expand so as to evolve work, a part of it is condensed,
and that this condensation affords heat for the expansion of the
remainder of the steam. This fact, which is analogous to that
of the production of a cloud when air saturated with vapour is
rarefied in the receiver of an air-pump, explains the approach
of the œconomical duty of the steam-engine to that of the air-
engine, on which I propose to make a few observations shortly.

I have the honour to remain, Gentlemen,

Yours very respectfully,

JAMES P. JOULE.

PARIS, *October* 15, 1850.

MY DEAR SIR,

In Mr Rankine's paper on the Mechanical Action of Heat*,
the following very remarkable result is announced :—" If vapour

* *Transactions of the Royal Society of Edinburgh*, Vol. XX. Part 1. (Read
Feb. 4, 1850.)

at saturation is allowed to expand, and at the same time is maintained at the temperature of saturation, the heat which disappears in producing the expansion is greater than that set free
by the fall of temperature, and the deficiency of heat must be
supplied from without, *otherwise a portion of the vapour will
be liquefied in order to supply the heat necessary for the expansion
of the rest.*" This conclusion can, I think, be reconciled with
known facts only by means of your discovery, that heat is evolved
by the friction of fluids in motion.   For it is well known that the
hand may be held with impunity in a current of steam issuing
from the safety-valve of a high-pressure boiler; and again, it
is known that "Watt's law" does not rigorously express the actual
decrease in the latent heat of saturated steam with an elevation of
temperature; but, on the contrary, Regnault shows that the "total
heat" of saturated steam increases slowly with the temperature,
at an approximately uniform rate.   These two facts are consistent
and connected with one another; for, according to the latter, steam
issuing from a high-pressure boiler ought, in the immediate
neighbourhood and on the outside of the orifice, where, of course,
its pressure scarcely exceeds that of the atmosphere, to be at a
temperature sensibly above 212°, and consequently supersaturated,
and quite dry; and it is well known that the hand experiences
no pain from being exposed to a hot current of a dry gas, even if
the temperature considerably exceeds 212°.   But, according to
Mr Rankine's proposition, steam allowed to expand from saturation will, *if no heat be supplied to it,* remain saturated, except a
small portion which becomes liquefied.   Either then Mr Rankine's
conclusion is opposed to the facts, or *some heat must be acquired by
the steam as it issues from the boiler.*   The pretended explanation
of a corresponding circumstance connected with the rushing of
air from one vessel to another in Gay-Lussac's experiment, on
which you have commented, is certainly not applicable in this
case, since instead of receiving heat from without, the steam must
lose a little in passing through the stop-cock or steam-pipe by
external radiation and convection.   There is no possible way in
which the heat can be acquired except by the friction of the steam
as it rushes through the orifice.   Hence I think I am justified in
saying that your discovery alone can reconcile Mr Rankine's
discovery with known facts.

In connexion with this subject it is to be remarked, that if your fundamental principle regarding the convertibility of heat and mechanical effect, adopted also by Mr Rankine, be true, a quantity of water raised from the freezing-point to any higher temperature, converted into saturated vapour at that temperature, and then allowed to expand through a small orifice wasting all its "work" in friction, will, in its expanded state, possess the "total heat" which has been given to it; but, on the contrary, if it be allowed to expand, pushing out a piston against a resisting force, it will in the expanded state possess less than that total heat by the amount corresponding to the mechanical effect developed. If the proposition quoted above of Mr Rankine's be true, this amount must exceed the amount of deviation from Watt's law measured by Regnault; and must consequently bear a very considerable ratio to the total heat, instead of being, as I believe all experimenters except yourself have hitherto considered it to be, quite inappreciable.

In the paragraph following that from which I have quoted, Mr Rankine remarks,—"There is as yet no experimental proof" of the preceding proposition. "It is true that in the working of non-condensing engines it has been found that the steam which escapes is always at the temperature of saturation corresponding to its pressure, and carries along with it a portion of water in the liquid state; but it is impossible to distinguish between the water which has been liquefied by the expansion of the steam, and that which has been carried over mechanically from the boiler." The circumstances of the passage of steam through the various parts of a non-condensing engine, are certainly very complicated. Even if there were no water "carried over mechanically from the boiler," we could not conclude the truth of Mr Rankine's proposition from the fact of the steam issuing moist and at 212° from the waste steam-pipe, since this might be accounted for by the external loss of heat from the cylinder, steam-pipes, &c.; nor could we conclude that Mr Rankine's proposition is false, if the steam were observed in any case to issue dry from the steam-pipe, and at a temperature above 212°, unless the expansive principle were known to be pushed to the utmost in the actual working of the engine. It is however certain that if Mr Rankine's proposition be true, steam, after having passed through a high-pressure engine in which the expansive principle is pushed to the utmost, whether there be any "priming" or not, and whether there be any heat lost externally from the different parts of the engine or not, will issue at the temperature of 212°, and moist (and consequently scalding to the hand), from the waste steam-pipe; and, Regnault's modification of Watt's law being considered as established, it is certain that steam issuing immediately from a high-pressure boiler into the open air will be above 212°, and dry.

The demonstration which Mr Rankine gives of his proposition is partially founded on certain hypotheses regarding the specific heats of gases and vapours. But, besides this proposition, he derives another conclusion from the same investigation which is experimentally verified by Regnault's modification of Watt's law: and hence, as it is easy to show, if we are contented to take Regnault's result as an experimental fact, and if we adopt your mechanical equivalent for a thermal unit (or Rankine's value, which is about $\frac{7}{8}$ths of yours), we may demonstrate Mr Rankine's remarkable theorem without any other hypothesis than the convertibility of heat and mechanical effect.

In a paper by Clausius, published in Poggendorff's *Annalen* for last April and May, a similar conclusion to that which I have quoted of Mr Rankine's (whose paper was read before the Royal Society of Edinburgh on the 4th of February), is announced. I have not yet been able to make myself fully acquainted with this paper; but, from the principles and methods of reasoning explained at the commencement, which differ from those of Carnot only in the adoption of your axiom instead of Carnot's, I have no doubt but that the demonstration of the proposition in question is the same in substance as Mr Rankine's modified in the manner I have suggested.

I remain, dear Sir, yours most truly,

WILLIAM THOMSON.

J. P. Joule, Esq.

ART. XLVIII. ON THE DYNAMICAL THEORY OF HEAT, WITH
NUMERICAL RESULTS DEDUCED FROM MR JOULE'S EQUIVALENT
OF A THERMAL UNIT, AND M. REGNAULT'S OBSERVATIONS ON
STEAM.

[*Transactions of the Royal Society of Edinburgh*, March, 1851, *and Phil.
Mag.* IV. 1852.]

Introductory Notice.

1. SIR HUMPHRY DAVY, by his experiment of melting two
pieces of ice by rubbing them together, established the following proposition:—"The phenomena of repulsion are not dependent on a
peculiar elastic fluid for their existence, or caloric does not exist."
And he concludes that heat consists of a motion excited among
the particles of bodies. "To distinguish this motion from others,
and to signify the cause of our sensation of heat," and of the
expansion or expansive pressure produced in matter by heat, "the
name *repulsive* motion has been adopted *."

2. The dynamical theory of heat, thus established by Sir
Humphry Davy, is extended to radiant heat by the discovery of
phenomena, especially those of the polarization of radiant heat,
which render it excessively probable that heat propagated through
"vacant space," or through diathermanic substances, consists of
waves of transverse vibrations in an all-pervading medium.

---

* From Davy's first work, entitled *An Essay on Heat, Light, and the Combina-
tions of Light*, published in 1799, in "Contributions to Physical and Medical
Knowledge, principally from the West of England, collected by Thomas Beddoes,
M.D.," and republished in Dr Davy's edition of his brother's collected works, Vol. II.
Lond. 1836.

3. The recent discoveries made by Mayer and Joule*, of the generation of heat through the friction of fluids in motion, and by the magneto-electric excitation of galvanic currents, would either of them be sufficient to demonstrate the immateriality of heat; and would so afford, if required, a perfect confirmation of Sir Humphry Davy's views.

4. Considering it as thus established, that heat is not a substance, but a dynamical form of mechanical effect, we perceive that there must be an equivalence between mechanical work and heat, as between cause and effect. The first published statement of this principle appears to be in Mayer's *Bemerkungen über die Kräfte der unbelebten Natur*†, which contains some correct views regarding the mutual convertibility of heat and mechanical effect, along with a false analogy between the approach of a weight to the earth and a diminution of the volume of a continuous substance, on which an attempt is founded to find numerically the mechanical equivalent of a given quantity of heat. In a paper published about fourteen months later, "On the Calorific Effects of Magneto-Electricity and the Mechanical Value of Heat‡," Mr Joule of Manchester expresses very distinctly the consequences regarding the mutual convertibility of heat and mechanical effect which follow from the fact, that heat is not a substance but a state of motion; and investigates on unquestionable principles the "absolute numerical relations," according to which heat is connected with mechanical power; verifying experimentally, that whenever heat is generated from purely mechanical action, and no other effect produced, whether it be by means of the friction of fluids or by the magneto-electric excitation of galvanic currents, the same quantity is generated by the same amount of work spent; and determining the actual amount of work, in foot-pounds,

---

* In May, 1842, Mayer announced in the *Annalen* of Wöhler and Liebig, that he had raised the temperature of water from 12^0 to 13^0 Cent. by agitating it. In August, 1843, Joule announced to the British Association "That heat is evolved by the passage of water through narrow tubes;" and that he had "obtained one degree of heat per lb. of water from a mechanical force capable of raising 770 lbs. to the height of one foot;" and that heat is generated when work is spent in turning a magneto-electric machine, or an electro-magnetic engine. (See his paper "On the Calorific Effects of Magneto-Electricity, and on the Mechanical Value of Heat."— *Phil. Mag.*, Vol. XXIII., 1843.)

† *Annalen* of Wöhler and Liebig, May, 1842.

‡ British Association, August, 1843; and *Phil. Mag.*, Sept., 1843.

required to generate a unit of heat, which he calls "the mechanical equivalent of heat." Since the publication of that paper,
Mr Joule has made numerous series of experiments for determining with as much accuracy as possible the mechanical equivalent of heat so defined, and has given accounts of them in various
communications to the British Association, to the *Philosophical
Magazine*, to the Royal Society, and to the French Institute.

5. Important contributions to the dynamical theory of heat
have recently been made by Rankine and Clausius; who, by
mathematical reasoning analogous to Carnot's on the motive
power of heat, but founded on an axiom contrary to his fundamental axiom, have arrived at some remarkable conclusions.
The researches of these authors have been published in the
Transactions of this Society, and in Poggendorff's *Annalen*, during
the past year; and they are more particularly referred to below in
connexion with corresponding parts of the investigations at present
laid before the Royal Society.

[Various statements regarding animal heat, and the heat of
combustion and chemical combination, are made in the writings of
Liebig (as, for instance, the statement quoted in the foot-note
added to § 18 below), which virtually imply the convertibility of
heat into mechanical effect, and which are inconsistent with any
other than the dynamical theory of heat.]

6. The object of the present paper is threefold:—

(1) To show what modifications of the conclusions arrived at
by Carnot, and by others who have followed his peculiar mode of
reasoning regarding the motive power of heat, must be made when
the hypothesis of the dynamical theory, contrary as it is to
Carnot's fundamental hypothesis, is adopted.

(2) To point out the significance in the dynamical theory, of
the numerical results deduced from Regnault's observations on
steam, and communicated about two years ago to the Society,
with an account of Carnot's theory, by the author of the present
paper; and to show that by taking these numbers (subject to
correction when accurate experimental data regarding the density
of saturated steam shall have been afforded), in connexion with
Joule's mechanical equivalent of a thermal unit, a complete theory

of the motive power of heat, within the temperature limits of the experimental data, is obtained.

(3)   To point out some remarkable relations connecting the physical properties of all substances, established by reasoning analogous to that of Carnot, but founded in part on the contrary principle of the dynamical theory.

PART I.

Fundamental Principles in the Theory of the Motive Power of Heat.

7.   According to an obvious principle, first introduced, however, into the theory of the motive power of heat by Carnot, mechanical effect produced in any process cannot be said to have been derived from a purely thermal source, unless at the end of the process all the materials used are in precisely the same physical and mechanical circumstances as they were at the beginning. In some conceivable " thermo-dynamic engines," as for instance Faraday's floating magnet, or Barlow's "wheel and axle," made to rotate and perform work uniformly by means of a current continuously excited by heat communicated to two metals in contact, or the thermo-electric rotatory apparatus devised by Marsh, which has been actually constructed; this condition is fulfilled at every instant.   On the other hand, in all thermo-dynamic engines, founded on electrical agency, in which discontinuous galvanic currents, or pieces of soft iron in a variable state of magnetization, are used, and in all engines founded on the alternate expansions and contractions of media, there are really alterations in the condition of materials; but, in accordance with the principle stated above, these alterations must be strictly periodical.   In any such engine, the series of motions performed during a period, at the end of which the materials are restored to precisely the same condition as that in which they existed at the beginning, constitutes what will be called a complete cycle of its operations. Whenever in what follows, *the work done* or *the mechanical effect produced* by a thermo-dynamic engine is mentioned without qualification, it must be understood that the mechanical effect produced, either in a non-varying engine, or in a complete cycle, or any number of complete cycles of a periodical engine, is meant.

8. The *source of heat* will always be supposed to be a hot body at a given constant temperature, put in contact with some part of the engine; and when any part of the engine is to be kept from rising in temperature (which can only be done by drawing off whatever heat is deposited in it), this will be supposed to be done by putting a cold body, which will be called the refrigerator, at a given constant temperature in contact with it.

9. The whole theory of the motive power of heat is founded on the two following propositions, due respectively to Joule, and to Carnot and Clausius.

PROP. I. (Joule).—When equal quantities of mechanical effect are produced by any means whatever from purely thermal sources, or lost in purely thermal effects, equal quantities of heat are put out of existence or are generated.

PROP. II. (Carnot and Clausius).—If an engine be such that, when it is worked backwards, the physical and mechanical agencies in every part of its motions are all reversed, it produces as much mechanical effect as can be produced by any thermodynamic engine, with the same temperatures of source and refrigerator, from a given quantity of heat.

10. The former proposition is shown to be included in the general "principle of mechanical effect," and is so established beyond all doubt by the following demonstration.

11. By whatever direct effect the heat gained or lost by a body in any conceivable circumstances is tested, the measurement of its quantity may always be founded on a determination of the quantity of some standard substance, which it or any equal quantity of heat could raise from one standard temperature to another; the test of equality between two quantities of heat being their capability of raising equal quantities of any substance from any temperature to the same higher temperature. Now, according to the dynamical theory of heat, the temperature of a substance can only be raised by working upon it in some way so as to produce increased thermal motions within it, besides effecting any modifications in the mutual distances or arrangements of its particles which may accompany a change of temperature. The work necessary to produce this total mechanical effect is of course proportional to the quantity of the substance raised from one

standard temperature to another; and therefore when a body, or a group of bodies, or a machine, parts with or receives heat, there is in reality mechanical effect produced from it, or taken into it, to an extent precisely proportional to the quantity of heat which it emits or absorbs. But the work which any external forces do upon it, the work done by its own molecular forces, and the amount by which the half *vis viva* of the thermal motions of all its parts is diminished, must together be equal to the mechanical effect produced from it; and consequently, to the mechanical equivalent of the heat which it emits (which will be positive or negative, according as the sum of those terms is positive or negative). Now let there be either no molecular change or alteration of temperature in any part of the body, or, by a cycle of operations, let the temperature and physical condition be restored exactly to what they were at the beginning; the second and third of the three parts of the work which it has to produce vanish; and we conclude that the heat which it emits or absorbs will be the thermal equivalent of the work done upon it by external forces, or done by it against external forces; which is the proposition to be proved.

12.   The demonstration of the second proposition is founded on the following axiom:—

It is impossible, by means of inanimate material agency, to derive mechanical effect from any portion of matter by cooling it below the temperature of the coldest of the surrounding objects*.

13.   To demonstrate the second proposition, let A and B be two thermo-dynamic engines, of which B satisfies the conditions expressed in the enunciation; and let, if possible, A derive more work from a given quantity of heat than B, when their sources and refrigerators are at the same temperatures, respectively. Then on account of the condition of complete *reversibility* in all its operations which it fulfils, B may be worked backwards, and made to restore any quantity of heat to its source, by the expenditure of the amount of work which, by its forward action, it would derive from the same quantity of heat. If, therefore, B be

---

* If this axiom be denied for all temperatures, it would have to be admitted that a self-acting machine might be set to work and produce mechanical effect by cooling the sea or earth, with no limit but the total loss of heat from the earth and sea, or, in reality, from the whole material world.

worked backwards, and made to restore to the source of A (which we may suppose to be adjustable to the engine B) as much heat as has been drawn from it during a certain period of the working of A, a smaller amount of work will be spent thus than was gained by the working of A. Hence, if such a series of operations of A forwards and of B backwards be continued, either alternately or simultaneously, there will result a continued production of work without any continued abstraction of heat from the source; and, by Prop. I., it follows that there must be more heat abstracted from the refrigerator by the working of B backwards than is deposited in it by A. Now it is obvious that A might be made to spend part of its work in working B backwards, and the whole might be made self-acting. Also, there being no heat either taken from or given to the source on the whole, all the surrounding bodies and space except the refrigerator might, without interfering with any of the conditions which have been assumed, be made of the same temperature as the source, whatever that may be. We should thus have a self-acting machine, capable of drawing heat constantly from a body surrounded by others at a higher temperature, and converting it into mechanical effect. But this is contrary to the axiom, and therefore we conclude that the hypothesis that A derives more mechanical effect from the same quantity of heat drawn from the source than B, is false. Hence no engine whatever, with source and refrigerator at the same temperatures, can get more work from a given quantity of heat introduced than any engine which satisfies the condition of reversibility, which was to be proved.

14. This proposition was first enunciated by Carnot, being the expression of his criterion of a perfect thermo-dynamic engine*. He proved it by demonstrating that a negation of it would require the admission that there might be a self-acting machine constructed which would produce mechanical effect indefinitely, without any source either in heat or the consumption of materials, or any other physical agency; but this demonstration involves, fundamentally, the assumption that, in "a complete cycle of operations," the medium parts with exactly the same quantity of heat as it receives. A very strong expression of doubt regarding the truth of this assumption, as a universal principle, is given by

* Account of Carnot's *Theory*, § 13.

Carnot himself*; and that it is false, where mechanical work is, on he whole, either gained or spent in the operations, may (as I have tried to show above) be considered to be perfectly certain. It must then be admitted that Carnot's original demonstration utterly fails, but we cannot infer that the proposition concluded is false. The truth of the conclusion appeared to me, indeed, so probable, that I took it in connexion with Joule's principle, on account of which Carnot's demonstration of it fails, as the foundation of an investigation of the motive power of heat in air-engines or steam-engines through finite ranges of temperature, and obtained about a year ago results, of which the substance is given in the second part of the paper at present communicated to the Royal Society. It was not until the commencement of the present year that I found the demonstration given above, by which the truth of the proposition is established upon an axiom (§ 12) which I think will be generally admitted. It is with no wish to claim priority that I make these statements, as the merit of first establishing the proposition upon correct principles is entirely due to Clausius, who published his demonstration of it in the month of May last year, in the second part of his paper on the motive power of heat†. I may be allowed to add, that I have given the demonstration exactly as it occurred to me before I knew that Clausius had either enunciated or demonstrated the proposition. The following is the axiom on which Clausius' demonstration is founded :—

It is impossible for a self-acting machine, unaided by any external agency, to convey heat from one body to another at a higher temperature.

It is easily shown, that, although this and the axiom I have used are different in form, either is a consequence of the other. The reasoning in each demonstration is strictly analogous to that which Carnot orginally gave.

15. A complete theory of the motive power of heat would consist of the application of the two propositions demonstrated above, to every possible method of producing mechanical effect from thermal agency‡. As yet this has not been done for the

---

* Account of Carnot's *Theory*, § 6.

† Poggendorff's *Annalen*, referred to above.

‡ "There are at present known two, and only two, distinct ways in which

electrical method, as far as regards the criterion of a perfect engine implied in the second proposition, and probably cannot be done without certain limitations; but the application of the first proposition has been very thoroughly investigated, and verified experimentally by Mr Joule in his researches "On the Calorific Effects of Magneto-Electricity;" and on it is founded one of his ways of determining experimentally the mechanical equivalent of heat. Thus, from his discovery of the laws of generation of heat in the galvanic circuit*, it follows that when mechanical work by means of a magneto-electric machine is the source of the galvanism, the heat generated in any given portion of the fixed part of the circuit is proportional to the whole work spent; and from his experimental demonstration that heat is developed in any moving part of the circuit at exactly the same rate as if it were at rest, and traversed by a current of the same strength, he is enabled to conclude—

(1)   That heat may be created by working a magneto-electric machine.

(2)   That if the current excited be not allowed to produce any other than thermal effects, the total quantity of heat produced is in all circumstances exactly proportional to the quantity of work spent.

16.   Again, the admirable discovery of Peltier, that cold is produced by an electrical current passing from bismuth to antimony, is referred to by Joule†, as showing how it may be proved

mechanical effect can be obtained from heat. One of these is by the alterations of volume which bodies experience through the action of heat; the other is through the medium of electric agency."—"Account of Carnot's Theory," § 4. (*Transactions*, Vol. xvi. part 5.)

* That, in a given fixed part of the circuit, the heat evolved in a given time is proportional to the square of the strength of the current, and for different fixed parts, with the same strength of current, the quantities of heat evolved in equal times are as the resistances. A paper by Mr Joule, containing demonstrations of these laws, and of others on the relations of the chemical and thermal agencies concerned, was communicated to the Royal Society on the 17th of December, 1840, but was not published in the *Transactions*. (See abstract containing a statement of the laws quoted above, in the *Philosophical Magazine*, Vol. xviii. p. 308.) It was published in the *Philosophical Magazine* in October, 1841 (Vol. xix. p. 260).

† [Note of March 20, 1852, added in *Phil. Mag.* reprint. In the introduction to his paper on the "Calorific Effects of Magneto-Electricity," &c., *Phil. Mag.*, 1843. I take this opportunity of mentioning that I have only recently become ac-

that, when an electrical current is continuously produced from a
purely thermal source, the quantities of heat evolved electrically
in the different homogeneous parts of the circuit are only compen-
sations for a loss from the junctions of the different metals, or
that, when the effect of the current is entirely thermal, there must
be just as much heat emitted from the parts not affected by the
source as is taken from the source.

17. Lastly*, when a current produced by thermal agency is
made to work an engine and produce mechanical effect, there will
be less heat emitted from the parts of the circuit not affected by
the source than is taken in from the source, by an amount
precisely equivalent to the mechanical effect produced; since Joule
demonstrates experimentally, that a current from any kind of

quainted with Helmholtz's admirable treatise on the principle of mechanical effect
(*Ueber die Erhaltung der Kraft*, von Dr H. Helmholtz. Berlin. G. Reimer, 1847),
having seen it for the first time on the 20th of January of this year; and that I
should have had occasion to refer to it on this, and on numerous other points of
the dynamical theory of heat, the mechanical theory of electrolysis, the theory of
electro-magnetic induction, and the mechanical theory of thermo-electric currents,
in various papers communicated to the Royal Society of Edinburgh, and to this
Magazine, had I been acquainted with it in time.—W. T., March 20, 1852.]

* This reasoning was suggested to me by the following passage contained in a
letter which I received from Mr Joule on the 8th of July, 1847. "In Peltier's
experiment on cold produced at the bismuth and antimony solder, we have an
instance of the conversion of heat into the mechanical force of the current," which
must have been meant as an answer to a remark I had made, that no evidence could
be adduced to show that heat is ever put out of existence. I now fully admit the
force of that answer; but it would require a proof that there is more heat put out
of existence at the heated soldering [or in this and other parts of the circuit] than is
created at the cold soldering [and the remainder of the circuit, when a machine is
driven by the current] to make the "evidence" be *experimental*. That this is the
case I think is certain, because the statements of § 16 in the text are demonstrated
consequences of the first fundamental proposition; but it is still to be remarked,
that neither in this nor in any other case of the production of mechanical effect
from purely thermal agency, has the ceasing to exist of an equivalent quantity of
heat been demonstrated otherwise than theoretically. It would be a very great
step in the experimental illustration (or *verification*, for those who consider such
to be necessary) of the dynamical theory of heat, to actually show in any one case a
loss of heat; and it might be done by operating through a very considerable range
of temperatures with a good air-engine or steam-engine, not allowed to waste its
work in friction. As will be seen in Part. II. of this paper, no experiment of any
kind could show a considerable loss of heat without employing bodies differing
considerably in temperature; for instance, a loss of as much as ·098, or about one-
tenth of the whole heat used, if the temperature of all the bodies used be between
0° and 30° Cent.

source driving an engine, produces in the engine just as much less heat than it would produce in a fixed wire exercising the same resistance as is equivalent to the mechanical effect produced by the engine.

18. The quality of thermal effects, resulting from equal causes through very different means, is beautifully illustrated by the following statement, drawn from Mr Joule's paper on magneto-electricity*.

Let there be three equal and similar galvanic batteries furnished with equal and similar electrodes; let A_1 and B_1 be the terminations of the electrodes (or wires connected with the two poles) of the first battery, A_2 and B_2 the terminations of the corresponding electrodes of the second, and A_3 and B_3 of the third battery. Let A_1 and B_1 be connected with the extremities of a long fixed wire; let A_2 and B_2 be connected with the "poles" of an electrolytic apparatus for the decomposition of water; and let A_3 and B_3 be connected with the *poles* (or *ports* as they might be called) of an electro-magnetic engine. Then if the length of the wire between A_1 and B_1, and the speed of the engine between A_3 and B_3, be so adjusted that the strength of the current (which for simplicity we may suppose to be continuous and perfectly uniform in each case) may be the same in the three circuits, there will be more heat given out in any time in the wire between A_1 and B_1 than in the electrolytic apparatus between A_2 and B_2, or the working engine between A_3 and B_3. But if the hydrogen were allowed to burn in the oxygen, within the electrolytic vessel, and the engine to waste all its work without producing any other than thermal effects (as it would do, for instance, if all its work were spent in continuously agitating a limited fluid mass), the total heat emitted would be precisely the same in each of these two pieces of apparatus as in the wire between A_1 and B_1. It is worthy of remark that these propositions are *rigorously* true, being demonstrable consequences of the fundamental principle of the dynamical theory of heat, which have been discovered by Joule,

---

* In this paper reference is made to his previous paper "On the Heat of Electrolysis" (published in Vol. VII. part 2, of the second series of the Literary and Philosophical Society of Manchester) for experimental demonstration of those parts of the theory in which chemical action is concerned.

and illustrated and verified most copiously in his experimental
researches*.

19. Both the fundamental propositions may be applied in a
perfectly rigorous manner to the second of the known methods of
producing mechanical effect from thermal agency. This application of the first of the two fundamental propositions has already
been published by Rankine and Clausius; and that of the second,
as Clausius showed in his published paper, is simply Carnot's
unmodified investigation of the relation between the mechanical
effect produced and the thermal circumstances from which it
originates, in the case of an expansive engine working within an
infinitely small range of temperatures. The simplest investigation
of the consequences of the first proposition in this application,
which has occurred to me, is the following, being merely the
modification of an analytical expression of Carnot's axiom regarding the permanence of heat, which was given in my former
paper†, required to make it express, not Carnot's axiom, but
Joule's.

20. Let us suppose a mass‡ of any substance, occupying
a volume v, under a pressure p uniform in all directions, and at a
temperature t, to expand in volume to $v + dv$, and to rise in tem-

---

[* Note of March 20, 1852, added in *Phil. Mag.* reprint. I have recently
met with the following passage in Liebig's *Animal Chemistry* (3rd edit. London,
1846, p. 43), in which the dynamical theory of the heat both of combustion
and of the galvanic battery is indicated, if not fully expressed:—"When we
kindle a fire under a steam-engine, and employ the power obtained to produce
heat by friction, it is impossible that the heat thus obtained can ever be greater
than that which was required to heat the boiler; and if we use the galvanic current
to produce heat, the amount of heat obtained is never in any circumstances
greater than we might have by the combustion of the zinc which has been dissolved
in the acid."

A paper "On the Heat of Chemical Combination," by Dr Thomas Woods, published last October in the *Philosophical Magazine*, contains an independent and direct
experimental demonstration of the proposition stated in the text regarding the
comparative thermal effects in a fixed metallic wire, and an electrolytic vessel
for the decomposition of water, produced by a galvanic current.—W. T., March 20,
1852.]

† "Account of Carnot's Theory," foot-note on § 26.

‡ This may have parts consisting of different substances, or of the same substance in different states, provided the temperature of all be the same. See below
Part III., § 53—56.

perature to $t + dt$. The quantity of work which it will produce will be

$$pdv;$$

and the quantity of heat which must be added to it to make its temperature rise during the expansion to $t + dt$ may be denoted by

$$Mdv + Ndt.$$

The mechanical equivalent of this is

$$J\,(Mdv + Ndt),$$

if J denote the mechanical equivalent of a unit of heat. Hence the mechanical measure of the total external effect produced in the circumstances is

$$(p - JM)\,dv - JNdt.$$

The total external effect, after any finite amount of expansion, accompanied by any continuous change of temperature, has taken place, will consequently be, in mechanical terms,

$$\int \{(p - JM)\,dv - JNdt\};$$

where we must suppose t to vary with v, so as to be the actual temperature of the medium at each instant, and the integration with reference to v must be performed between limits corresponding to the initial and final volumes. Now if, at any subsequent time, the volume and temperature of the medium become what they were at the beginning, however arbitrarily they may have been made to vary in the period, the total external effect must, according to Prop. I., amount to nothing; and hence

$$(p - JM)\,dv - JNdt*$$

must be the differential of a function of two independent variables, or we must have

$$\frac{d\,(p - JM)}{dt} = \frac{d\,(-JN)}{dv} \quad\ldots\ldots\ldots\ldots\ldots\ldots (1),$$

this being merely the analytical expression of the condition, that the preceding integral may vanish in every case in which the

[* The integral function $\int \{(JM - p)\,dv + JNdt\}$ may obviously be called the *mechanical energy* of the fluid mass; as (when the constant of integration is properly assigned) it expresses the whole work the fluid has in it to produce. The consideration of this function is the subject of a short paper communicated to the Royal Society of Edinburgh, Dec. 15, 1851, as an appendix to the paper at present republished; (see below Part v. §§ 81—96).]

initial and final values of v and t are the same, respectively. Observing that J is an absolute constant, we may put the result into the form

$$\frac{dp}{dt} = J\left(\frac{dM}{dt} - \frac{dN}{dv}\right)\dots\dots\dots\dots(2).$$

This equation expresses, in a perfectly comprehensive manner, the application of the first fundamental proposition to the thermal and mechanical circumstances of any substance whatever, under uniform pressure in all directions, when subjected to any possible variations of temperature, volume and pressure.

21. The corresponding application of the second fundamental proposition is completely expressed by the equation

$$\frac{dp}{dt} = \mu M\dots\dots\dots\dots\dots(3),$$

where μ denotes what is called "Carnot's function," a quantity which has an absolute value, the same for all substances for any given temperature, but which may vary with the temperature in a manner that can only be determined by experiment. To prove this proposition, it may be remarked in the first place that Prop. II. could not be true for every case in which the temperature of the refrigerator differs infinitely little from that of the source, without being true universally. Now, if a substance be allowed first to expand from v to $v + dv$, its temperature being kept constantly t; if, secondly, it be allowed to expand further, without either emitting or absorbing heat till its temperature goes down through an infinitely small range, to $t - \tau$; if, thirdly, it be compressed at the constant temperature $t - \tau$, so much (actually by an amount differing from dv by only an infinitely small quantity of the second order), that when, fourthly, the volume is further diminished to v without the medium's being allowed to either emit or absorb heat, its temperature may be exactly t; it may be considered as constituting a thermo-dynamic engine which fulfils Carnot's condition of complete reversibility. Hence, by Prop. II., it must produce the same amount of work for the same quantity of heat absorbed in the first operation, as any other substance similarly operated upon through the same range of temperatures. But $\frac{dp}{dt}\tau \,.\, dv$ is obviously the whole work done in the complete cycle, and (by the definition of M in § 20) $M dv$ is the quantity of heat absorbed in the first operation. Hence the value of

$$\frac{\frac{dp}{dt}\tau . dv}{M dv}, \quad \text{or} \quad \frac{\frac{dp}{dt}}{M}\tau,$$

must be the same for all substances, with the same values of t and τ; or, since τ is not involved except as a factor, we must have

$$\frac{\frac{dp}{dt}}{M} = \mu \dots\dots\dots\dots\dots\dots\dots\dots(4),$$

where μ depends only on t; from which we conclude the proposition which was to be proved.

[Note of Nov. 9, 1881. Elimination of $\frac{dp}{dt}$ by (2) from (4) gives

$$\frac{J\left(\dfrac{dM}{dt} - \dfrac{dN}{dv}\right)}{M} = \mu \dots\dots\dots\dots\dots (4'),$$

a very convenient and important formula.]

22. The very remarkable theorem that $\dfrac{\frac{dp}{dt}}{M}$ must be the same for all substances at the same temperature, was first given (although not in precisely the same terms) by Carnot, and demonstrated by him, according to the principles he adopted. We have now seen that its truth may be satisfactorily established without adopting the false part of his principles. Hence all Carnot's conclusions, and all conclusions derived by others from his theory, which depend merely on equation (3), require no modification when the dynamical theory is adopted. Thus, all the conclusions contained in Sections I., II., and III., of the Appendix to my "Account of Carnot's Theory" [Art. XLI. §§ 43—53 above], and in the paper immediately following it in the *Transactions* [and in the present reprint], entitled "Theoretical Considerations on the Effect of Pressure in Lowering the Freezing Point of Water," by my elder brother, still hold. Also, we see that Carnot's expression for the mechanical effect derivable from a given quantity of heat by means of a perfect engine in which the range of temperatures is infinitely small, expresses truly the greatest effect

which can possibly be obtained in the circumstances; although
it is in reality only an infinitely small fraction of the whole
mechanical equivalent of the heat supplied; the remainder being
irrecoverably lost to man, and therefore "wasted," although not
annihilated.

23.   On the other hand, the expression for the mechanical
effect obtainable from a given quantity of heat entering an engine
from a "source" at a given temperature, when the range down
to the temperature of the cold part of the engine or the "refrigerator" is finite, will differ most materially from that of Carnot;
since, a finite quantity of mechanical effect being now obtained
from a finite quantity of heat entering the engine, a finite fraction
of this quantity must be converted from heat into mechanical
effect.   The investigation of this expression, with numerical determinations founded on the numbers deduced from Regnault's
observations on steam, which are shown in Tables I. and II. of my
former paper, constitutes the second part of the paper at present
communicated.

PART II.

On the Motive Power of Heat through Finite Ranges of Temperature.

24.   It is required to determine the quantity of work which a
perfect engine, supplied from a source at any temperature, S, and
parting with its waste heat to a refrigerator at any lower temperature, T, will produce from a given quantity, H, of heat drawn
from the source.

25.   We may suppose the engine to consist of an infinite
number of perfect engines, each working within an infinitely small
range of temperature, and arranged in a series of which the source
of the first is the given source, the refrigerator of the last the
given refrigerator, and the refrigerator of each intermediate engine
is the source of that which follows it in the series.   Each of these
engines will, in any time, emit just as much less heat to its
refrigerator than is supplied to it from its source, as is the equivalent of the mechanical work which it produces.   Hence if t and
$t + dt$ denote respectively the temperatures of the refrigerator and

source of one of the intermediate engines, and if q denote the quantity of heat which this engine discharges into its refrigerator in any time, and $q + dq$ the quantity which it draws from its source in the same time, the quantity of work which it produces in that time will be Jdq according to Prop. I., and it will also be $q\mu dt$ according to the expression of Prop. II., investigated in § 21; and therefore we must have

$$Jdq = q\mu dt.$$

Hence, supposing that the quantity of heat supplied from the first source, in the time considered is H, we find by integration

$$\log \frac{H}{q} = \frac{1}{J} \int_t^S \mu dt.$$

But the value of q, when $t = T$, is the final remainder discharged into the refrigerator at the temperature T; and therefore, if this be denoted by R, we have

$$\log \frac{H}{R} = \frac{1}{J} \int_T^S \mu dt \dots\dots\dots\dots\dots\dots(5);$$

from which we deduce

$$R = H e^{-\frac{1}{J} \int_T^S \mu dt} \dots\dots\dots\dots\dots\dots(6).$$

Now the whole amount of work produced will be the mechanical equivalent of the quantity of heat lost; and, therefore, if this be denoted by W, we have

$$W = J(H - R)\dots\dots\dots\dots\dots\dots(7),$$

and consequently, by (6),

$$W = JH\{1 - \epsilon^{-\frac{1}{J} \int_T^S \mu dt}\}\dots\dots\dots\dots\dots(8).$$

26. To compare this with the expression $H \int_T^S \mu dt$, for the duty indicated by Carnot's theory*, we may expand the exponential in the preceding equation, by the usual series. We thus

find
$$W = \left(1 - \frac{\theta}{1.2} + \frac{\theta^2}{1.2.3} - \&c.\right) . H \int_T^S \mu dt \left.\vphantom{\int}\right\}$$

where
$$\theta = \frac{1}{J} \int_T^S \mu dt \qquad\qquad \left.\vphantom{\int}\right\} \dots\dots(9),$$

---

* "Account," &c., Equation 7, § 31. [Art. XLI. above.]

This shows that the work really produced, which always falls short of the duty indicated by Carnot's theory, approaches more and more nearly to it as the range is diminished ; and ultimately, when the range is infinitely small, is the same as if Carnot's theory required no modification, which agrees with the conclusion stated above in § 22.

27.   Again, equation (8) shows that the real duty of a given quantity of heat supplied from the source increases with every increase of the range; but that instead of increasing indefinitely in proportion to $\int_{T}^{S} \mu dt$, as Carnot's theory makes it do, it never reaches the value JH, but approximates to this limit, as $\int_{T}^{S} \mu dt$ is increased without limit.   Hence Carnot's remark* regarding the practical advantage that may be anticipated from the use of the air-engine, or from any method by which the range of temperatures may be increased, loses only a part of its importance, while a much more satisfactory view than his of the practical problem is afforded.   Thus we see that, although the full equivalent of mechanical effect cannot be obtained even by means of a perfect engine, yet when the actual source of heat is at a high enough temperature above the surrounding objects, we may get more and more nearly the whole of the admitted heat converted into mechanical effect, by simply increasing the effective range of temperature in the engine.

28.   The preceding investigation (§ 25) shows that the value of Carnot's function, μ, for all temperatures within the range of the engine, and the absolute value of Joule's equivalent, J, are enough of data to calculate the amount of mechanical effect of a perfect engine of any kind, whether a steam-engine, an air-engine, or even a thermo-electric engine; since, according to the axiom stated in § 12, and the demonstration of Prop. II., no inanimate material agency could produce more mechanical effect from a given quantity of heat, with a given available range of temperatures, than an engine satisfying the criterion stated in the enunciation of the proposition.

---

* "Account," &c.   Appendix, Section IV.   [Art. XLI. above.]

29.   The mechanical equivalent of a thermal unit Fahrenheit, or the quantity of heat necessary to raise the temperature of a pound of water from 32° to 33° Fahr., has been determined by Joule in foot-pounds at Manchester, and the value which he gives as his best determination is 772·69.   Mr Rankine takes, as the result of Joule's determination 772, which he estimates must be within $\frac{1}{300}$ of its own amount, of the truth.   If we take $772\frac{2}{3}$ as the number, we find, by multiplying it by $\frac{9}{5}$, 1390 as the equivalent of the thermal unit Centigrade, which is taken as the value of J in the numerical applications contained in the present paper.   [Note of Jan. 12, 1882.   Joule's recent redetermination gives 771·8 Manchester foot-pounds as the work required to warm 1 lb. of water from 32° to 33° Fahr.]

30.   With regard to the determination of the values of μ for different temperatures, it is to be remarked that equation (4) shows that this might be done by experiments upon any substance whatever of indestructible texture, and indicates exactly the experimental data required in each case.   For instance, by first supposing the medium to be air; and again, by supposing it to consist partly of liquid water and partly of saturated vapour, we deduce, as is shown in Part III. of this paper, the two expressions (6), given in § 30 of my former paper ("Account of Carnot's Theory"), for the value of μ at any temperature.   As yet no experiments have been made upon air which afford the required data for calculating the value of μ through any extensive range of temperature; but for temperatures between 50° and 60° Fahr., Joule's experiments* on the heat evolved by the expenditure of a given amount of work on the compression of air kept at a constant temperature, afford the most direct data for this object which have yet been obtained; since, if Q be the quantity of heat evolved by the compression of a fluid subject to "the gaseous laws" of expansion and compressibility, W the amount of mechanical work spent, and t the constant temperature of the fluid, we have by (11) of § 49 of my former paper,

$$\mu = \frac{W . E}{Q (1 + Et)} \quad \text{......................................(10),}$$

* "On the Changes of Temperature produced by the Rarefaction and Condensation of Air," *Phil. Mag.*, Vol. xxvi., May, 1845.

which is in reality a simple consequence of the other expression for μ in terms of data with reference to air. Remarks upon the determination of μ by such experiments, and by another class of experiments on air originated by Joule, are reserved for a separate communication, which I hope to be able to make to the Royal Society on another occasion. [*Dyn. Theory of Heat*, below, Part IV. §§ 61—80.]

31.   The second of the expressions (6), in § 30 of my former paper, or the equivalent expression (32), given below in the present paper, shows that μ may be determined for any temperature from determinations for that temperature of—

(1)   The rate of variation with the temperature, of the pressure of saturated steam.

(2)   The latent heat of a given weight of saturated steam.

(3)   The volume of a given weight of saturated steam.

(4)   The volume of a given weight of water.

The last mentioned of these elements may, on account of the manner in which it enters the formula, be taken as constant, without producing any appreciable effect on the probable accuracy of the result.

32.   Regnault's observations have supplied the first of the data with very great accuracy for all temperatures between -32° Cent. and 230°.

33.   As regards the second of the data, it must be remarked that all experimenters, from Watt, who first made experiments on the subject, to Regnault, whose determinations are the most accurate and extensive that have yet been made, appear to have either explicitly or tacitly assumed the same principle as that of Carnot which is overturned by the dynamical theory of heat; inasmuch as they have defined the "total heat of steam" as the quantity of heat required, to convert a unit of weight of water at 0°, into steam in the particular state considered. Thus Regnault, setting out with this definition for "the total heat of saturated steam," gives experimental determinations of it for the entire range of temperatures from 0° to 230°; and he deduces the

"latent heat of saturated steam" at any temperature, from the "total heat," so determined, by subtracting from it the quantity of heat necessary to raise the liquid to that temperature. Now, according to the dynamical theory, the quantity of heat expressed by the preceding definition depends on the manner (which may be infinitely varied) in which the specified change of state is effected; differing in different cases by the thermal equivalents of the differences of the external mechanical effect produced in the expansion. For instance, the total quantity of heat required to evaporate a quantity of water at 0°, and then, keeping it always in the state of saturated vapour *, bring it to the temperature 100°, cannot be so much as three-fourths of the quantity required, first, to raise the temperature of the liquid to 100°, and then evaporate it at that temperature; and yet either quantity is expressed by what is generally received as a *definition* of the "total heat" of the saturated vapour. To find what it is that is really determined as "total heat" of saturated steam in Regnault's researches, it is only necessary to remark, that the measurement actually made is of the quantity of heat emitted by a certain weight of water in passing through a calorimetrical apparatus, which it enters as saturated steam, and leaves in the liquid state, the result being reduced to what would have been found if the final temperature of the water had been exactly 0°. For there being no external mechanical effect produced (other than that of sound, which it is to be presumed is quite inappreciable), the only external effect is the emission of heat. This must, therefore, according to the fundamental proposition of the dynamical theory, be independent of the intermediate agencies. It follows that, however the steam may rush through the calorimeter, and at whatever reduced pressure it may actually be condensed†, the

---

* See below (Part III. § 58), where the "negative" specific heat of saturated steam is investigated. If the mean value of this quantity between 0° and 100° were − 1·5 (and it cannot differ much from this) there would be 150 units of heat emitted by a pound of saturated vapour in having its temperature raised (by compression) from 0° to 100°. The latent heat of the vapour at 0° being 606·5, the final quantity of heat required to convert a pound of water at 0° into saturated steam at 100°, in the first of the ways mentioned in the text, would consequently be 456·5, which is only about ⁴⁄₅ of the quantity 637 found as "the total heat" of the saturated vapour at 100°, by Regnault.

† If the steam have to rush through a long fine tube, or through a small aperture within the calorimetrical apparatus, its pressure will be diminished before it is

heat emitted externally must be exactly the same as if the condensation took place under the full pressure of the entering saturated steam; and we conclude that *the total heat*, as actually determined from his experiments by Regnault, is the quantity of heat that would be required, first to raise the liquid to the specified temperature, and then to evaporate it at that temperature; and that the principle on which he determines the latent heat is correct. Hence, through the range of his experiments, that is from 0^0 to 230^0, we may consider the second of the data required for the calculation of μ as being supplied in a complete and satisfactory manner.

34. There remains only the third of the data, or the volume of a given weight of saturated steam, for which accurate experiments through an extensive range are wanting; and no experimental researches bearing on the subject having been made since the time when my former paper was written, I see no reason for supposing that the values of μ which I then gave are not the most probable that can be obtained in the present state of science; and, on the understanding stated in § 33 of that paper, that accurate experimental determinations of the densities of saturated steam at different temperatures may indicate considerable errors in the densities which have been assumed according to the "gaseous laws," and may consequently render considerable alterations in my results necessary, I shall still continue to use Table I.

condensed; and there will, therefore, in two parts of the calorimeter be saturated steam at different temperatures (as, for instance, would be the case if steam from a high pressure boiler were distilled into the open air); yet, on account of the heat developed by the fluid friction, which would be precisely the equivalent of the mechanical effect of the expansion wasted in the rushing, the heat measured by the calorimeter would be precisely the same as if the condensation took place at a pressure not appreciably lower than that of the entering steam. The circumstances of such a case have been overlooked by Clausius (Poggendorff's *Annalen*, 1850, No. 4, p. 510), when he expresses with some doubt the opinion that the latent heat of saturated steam will be truly found from Regnault's "total heat," by deducting "the sensible heat;" and gives as a reason that, in the actual experiments, the condensation must have taken place "under the same pressure, or nearly under the same pressure," as the evaporation. The question is not, *Did the condensation take place at a lower pressure than that of the entering steam?* but, *Did Regnault make the steam work an engine in passing through the calorimeter, or was there so much noise of steam rushing through it as to convert an appreciable portion of the total heat into external mechanical effect?* And a negative answer to this is a sufficient reason for adopting *with certainty* the opinion that the principle of his determination of the latent heat is correct.

of that paper, which shows the values of μ for the temperatures $\frac{1}{2}$, $1\frac{1}{2}$, $2\frac{1}{2}$...$230\frac{1}{2}$, or, the mean values of μ for each of the 230 successive Centigrade degrees of the air-thermometer above the freezing-point, as the basis of numerical applications of the theory. It may be added, that any experimental researches sufficiently trustworthy in point of accuracy, yet to be made, either on air or any other substance, which may lead to values of μ differing from those, must be admitted as proving a discrepancy between the true densities of saturated steam, and those which have been assumed *.

35.   Table II. of my former paper, which shows the values of $\int_0^t \mu dt$ for $t = 1$, $t = 2$, $t = 3$, ... $t = 231$, renders the calculation of the mechanical effect derivable from a given quantity of heat by means of a perfect engine, with any given range included between the limits 0 and 231, extremely easy; since the quantity to be divided by J† in the index of the exponential in the expression (8) will be found by subtracting the number in that table corresponding to the value of T, from that corresponding to the value of S.

36.   The following tables show some numerical results which have been obtained in this way, with a few (contained in the lower part of the second table) calculated from values of $\int_0^t \mu dt$

---

* I cannot see that any hypothesis, such as that adopted by Clausius fundamentally in his investigations on this subject, and leading, as he shows to determinations of the densities of saturated steam at different temperatures, which indicate enormous deviations from the gaseous laws of variation with temperature and pressure, is more probable, or is probably nearer the truth, than that the density of saturated steam does follow these laws as it is usually assumed to do.   In the present state of science it would perhaps be wrong to say that either hypothesis is more probable than the other [or that the rigorous truth of either hypothesis is probable at all].

† It ought to be remarked, that as the unit of force implied in the determinations of μ is the weight of a pound of matter at Paris, and the unit of force in terms of which J is expressed is the weight of a pound at Manchester, these numbers ought in strictness to be modified so as to express the values in terms of a common unit of force; but as the force of gravity at Paris differs by less than $\frac{1}{1000}$ of its own value from the force of gravity at Manchester, this correction will be much less than the probable errors from other sources, and may therefore be neglected.

estimated for temperatures above 230°, roughly, according to the rate of variation of that function within the experimental limits.

37. *Explanation of the Tables.*

Column I. in each table shows the assumed ranges.

Column II. shows ranges deduced by means of Table II. of the former paper, so that the value of $\int_{T}^{S} \mu dt$ for each may be the same as for the corresponding range shown in column I.

Column III. shows what would be the duty of a unit of heat if Carnot's theory required no modification (or the actual duty of a unit of heat with additions through the range, to compensate for the quantities converted into mechanical effect).

Column IV. shows the true duty of a unit of heat, and a comparison of the numbers in it with the corresponding numbers in column III. shows how much the true duty falls short of Carnot's theoretical duty in each case.

Column VI. is calculated by the formula

$$R = \epsilon^{-\frac{1}{1390} \int_{T}^{S} \mu dt},$$

where $\epsilon = 2{\cdot}71828$, and for $\int_{T}^{S} \mu dt$ the successive values shown in column III. are used.

Column IV. is calculated by the formula

$$W = 1390\,(1 - R)$$

from the values of $1 - R$ shown in column V.

38. Table of the Motive Power of Heat.

Range of temperatures.				III. Duty of a unit of heat through the whole range.	IV. Duty of a unit of heat supplied from the source.	V. Quantity of heat converted into mechanical effect.	VI. Quantity of heat wasted.
I.		II.					
S.	T.	S.	T.	$\int_T^S \mu dt.$ ft.-lbs	W. ft.-lbs.	$1-R.$	R.
1	0	31·08	30	4·960	4·948	·00356	·99644
10	0	40·86	30	48·987	48·1	·0346	·9654
20	0	51·7	30	96·656	93·4	·067	·933
30	0	62·6	30	143·06	136	·038	·902
40	0	73·6	30	188·22	176	·127	·873
50	0	84·5	30	232·18	214	·154	·846
60	0	95·4	30	274·97	249	·179	·821
70	0	106·3	30	316·64	283	·204	·796
80	0	117·2	30	357·27	315	·227	·773
90	0	128·0	30	396·93	345	·248	·752
100	0	138·8	30	435·69	374	·269	·731
110	0	149·1	30	473·62	401	·289	·711
120	0	160·3	30	510·77	427	·308	·692
130	0	171·0	30	547·21	452	·325	·675
140	0	181·7	30	582·98	476	·343	·657
150	0	192·3	30	618·14	499	·359	·641
160	0	203·0	30	652·74	521	·375	·625
170	0	213·6	30	686·80	542	·390	·610
180	0	224·2	30	720·39	562	·404	·596
190	0	190	0	753·50	582	·418	·582
200	0	200	0	786·17	600	·432	·568
210	0	210	0	818·45	619	·445	·555
220	0	220	0	850·34	636	·457	·542
230	0	230	0	881 87	653	·470	·530

39. Supplementary Table of the Motive Powers of Heat.

Range of temperatures.				III. Duty of a unit of heat through the whole range.	IV. Duty of a unit of heat supplied from the source.	V. Quantity of heat converted into mechanical effect.	VI. Quantity of heat wasted.
I.		II.					
S.	T.	S.	T.	$\int_T^S \mu dt.$ ft.-lbs.	W. ft.-lbs.	$1-R.$	R.
101·1	0	140	30	439·9	377	·271	·729
105·8	0	230	100	446·2	382	·275	·725
300	0	300	0	1099	757	·545	·455
400	0	400	0	1395	879	·632	·368
500	0	500	0	1690	979	·704	·296
600	0	600	0	1980	1059	·762	·238
∞	0	∞	0	∞	1390	1·000	·000

40. Taking the range 30° to 140° as an example suitable to the circumstances of some of the best steam-engines that have yet

been made (see Appendix to "Account of Carnot's Theory," Sec. v.),
we find in column III. of the supplementary table, 377 ft.-lbs. as
the corresponding duty of a unit of heat instead of 440, shown in
column III., which is Carnot's theoretical duty. We conclude
that the recorded performance of the Fowey-Consols engine in
1845, instead of being only 57½ per cent. amounted really to 67
per cent., or ⅔ of the duty of a perfect engine with the same range
of temperature; and this duty being ·271 (rather more than ¼)
of the whole equivalent of the heat used; we conclude further,
that $\frac{1}{5\cdot 49}$, or 18 per cent. of the whole heat supplied, was actually
converted into mechanical effect by that steam-engine.

41. The numbers in the lower part of the supplementary
table show the great advantage that may be anticipated from the
perfecting of the air-engine, or any other kind of thermo-dynamic
engine in which the range of the temperature can be increased
much beyond the limits actually attainable in steam-engines.
Thus an air-engine, with its hot part at 600°, and its cold part at
0° Cent., working with perfect economy, would convert 76 per
cent. of the whole heat used into mechanical effect; or working
with such economy as has been estimated for the Fowey-Consols
engine, that is, producing 67 per cent. of the theoretical duty
corresponding to its range of temperature, would convert 51 per
cent. of all the heat used into mechanical effect. [Note, of Dec. 30,
1881. A great advance towards realizing this principle is now
achieved in the gas-engine, of which the true dynamical economy
is believed to be already superior to that of the best modern compound steam-engine.]

42. It was suggested to me by Mr Joule, in a letter dated
December 9, 1848, that the true value of μ might be "inversely
as the temperatures from zero*;" and values for various temperatures calculated by means of the formula,

$$\mu = J\frac{E}{1 + Et}\dots\dots\dots\dots\dots\dots\dots(11),$$

* If we take $\mu = k\dfrac{E}{1+Et}$ where k may be any constant, we find

$$W = J\left(\frac{S-T}{\frac{1}{E}+S}\right)^{\frac{k}{J}};$$

which is the formula I gave when this paper was communicated. I have since
remarked, that Mr Joule's hypothesis implies essentially that the coefficient k must

were given for comparison with those which I had calculated from data regarding steam. This formula is also adopted by Clausius, who uses it fundamentally in his mathematical investigations. If μ were correctly expressed by it, we should have

$$\int_T^S \mu dt = J \log \frac{1 + ES}{1 + ET};$$

and therefore equations (1) and (2) would become

$$W = J \frac{S - T}{\frac{1}{E} + S} \quad \dots\dots\dots\dots\dots\dots\dots (12),$$

$$R = \frac{\frac{1}{E} + T}{\frac{1}{E} + S} \quad \dots\dots\dots\dots\dots\dots\dots\dots (13).$$

43. The reasons upon which Mr Joule's opinion is founded, that the preceding equation (11) may be the correct expression for Carnot's function, although the values calculated by means of it differ considerably from those shown in Table I. of my former paper, form the subject of a communication which I hope to have an opportunity of laying before the Royal Society previously to the close of the present session. [Part IV. §§ 61—80, below.]

PART III.

Applications of the Dynamical Theory to establish Relations between the Physical Properties of all Substances.

44. The two fundamental equations of the dynamical theory of heat, investigated above, express relations between quantities of heat required to produce changes of volume and temperature in any material medium whatever, subjected to a uniform pressure in all directions, which lead to various remarkable conclusions. Such

be as it is taken in the text, the mechanical equivalent of a thermal unit. Mr Rankine, in a letter dated March 27, 1851, informs me that he has deduced, from the principles laid down in his paper communicated last year to this Society, an approximate formula for the ratio of the maximum quantity of heat converted into mechanical effect to the whole quantity expended, in an expansive engine of any substance, which, on comparison, I find agrees exactly with the expression (12) given in the text as a consequence of the hypothesis suggested by Mr Joule regarding the value of μ at any temperature.—[April 4, 1851.]

of these as are independent of Joule's principle (expressed by
equation (2) of § 20), being also independent of the truth or
falseness of Carnot's contrary assumption regarding the permanence of heat, are common to his theory and to the dynamical
theory; and some of the most important of them * have been
given by Carnot himself, and other writers who adopted his
principles and mode of reasoning without modification. Other
remarkable conclusions on the same subject might have been
drawn from the equation $\dfrac{dM}{dt} - \dfrac{dN}{dv} = 0$, expressing Carnot's assumption (of the truth of which experimental tests might have
been thus suggested); but I am not aware that any conclusion
deducible from it, not included in Carnot's expression for the
motive power of heat through finite ranges of temperature, has yet
been actually obtained and published.

45. The recent writings of Rankine and Clausius contain
some of the consequences of the fundamental principle of the
dynamical theory (expressed in the first fundamental proposition
above) regarding physical properties of various substances; among
which may be mentioned especially a very remarkable discovery
regarding the specific heat of saturated steam (investigated also in
this paper in § 58 below), made independently by the two authors,
and a property of water at its freezing-point, deduced from the
corresponding investigation regarding ice and water under pressure
by Clausius; according to which he finds that, for each $\frac{1}{10}^{\circ}$ Cent.
that the solidifying point of water is lowered by pressure, its
latent heat, which under atmospheric pressure is 79, is diminished
by ·081. The investigations of both these writers involve fundamentally various hypotheses which may be or may not be found
by experiment to be approximately true; and which render it
difficult to gather from their writings what part of their conclusions, especially with reference to air and gases, depend merely on
the necessary principles of the dynamical theory.

46. In the remainder of this paper, the two fundamental
propositions, expressed by the equations

$$\frac{dM}{dt} - \frac{dN}{dv} = \frac{1}{J}\frac{dp}{dt} \dots\dots\dots(2) \text{ of § 20,}$$

---

* See above, § 22.

and
$$M = \frac{1}{\mu} \cdot \frac{dp}{dt} \quad \ldots\ldots\ldots\ldots\ldots(3) \text{ of } \S 21,$$

are applied to establish properties of the specific heats of any substance whatever; and then special conclusions are deduced for the case of a fluid following strictly the " gaseous laws " of density, and for the case of a medium consisting of parts in different states at the same temperature, as water and saturated steam, or ice and water.

47. In the first place it may be remarked, that by the definition of M and N in § 20, N must be what is commonly called the " specific heat at constant volume " of the substance, provided the quantity of the medium be the standard quantity adopted for specific heats, which, in all that follows, I shall take as the unit of weight.   Hence the fundamental equation of the dynamical theory, (2) of § 20, expresses a relation between this specific heat and the quantities for the particular substance denoted by M and p.   If we eliminate M from this equation, by means of equation (3) of § 21, derived from the expression of the second fundamental principle of the theory of the motive power of heat, we find

$$\frac{dN}{dv} = \frac{d\left(\frac{1}{\mu} \frac{dp}{dt}\right)}{dt} - \frac{1}{J} \frac{dp}{dt} \quad \ldots\ldots\ldots\ldots\ldots(14),$$

which expresses a relation between the variation in the specific heat at constant volume, of any substance, produced by an alteration of its volume at a constant temperature, and the variation of its pressure with its temperature when the volume is constant; involving a function, μ, of the temperature, which is the same for all substances.

48. Again, let K denote the specific heat of the substance under constant pressure.   Then, if dv and dt be so related that the pressure of the medium, when its volume and temperature are $v + dv$ and $t + dt$ respectively, is the same as when they are v and t, that is, if

$$0 = \frac{dp}{dv} dv + \frac{dp}{dt} dt;$$

we have
$$K dt = M dv + N dt.$$

Hence we find

$$M = \frac{-\dfrac{dp}{dv}}{\dfrac{dp}{dt}}(K - N)\ldots\ldots\ldots\ldots\ldots(15),$$

which merely shows the meaning in terms of the two specific heats, of what I have denoted by M. Using in this for M its value given by (3) of § 21, we find

$$K - N = \frac{\left(\dfrac{dp}{dt}\right)^2}{\mu \times -\dfrac{dp}{dv}}\ldots\ldots\ldots\ldots\ldots(16),$$

an expression for the difference between the two specific heats, derived without hypothesis from the second fundamental principle of the theory of the motive power of heat.

49. These results may be put into forms more convenient for use, in applications to liquid and solid media, by introducing the notation :—

$$\left.\begin{array}{l} \kappa = v \times -\dfrac{dp}{dv} \\[2mm] e = \dfrac{1}{\kappa}\dfrac{dp}{dt} \end{array}\right\}\ldots\ldots\ldots\ldots\ldots\ldots(17),$$

where κ will be the reciprocal of the compressibility, and e the coefficient of expansion with heat.

Equations (14), (16) and (3), thus become

$$\frac{dN}{dv} = \frac{d\left(\dfrac{\kappa e}{\mu}\right)}{dt} - \frac{\kappa e}{J}\ldots\ldots\ldots\ldots\ldots(18),$$

$$K - N = v\frac{\kappa e^2}{\mu}\ldots\ldots\ldots\ldots\ldots(19),$$

$$M = \frac{1}{\mu}.\kappa e\ldots\ldots\ldots\ldots\ldots(20);$$

the third of these equations being annexed to show explicitly the quantity of heat developed by the compression of the substance kept at a constant temperature. Lastly, if θ denote the rise in temperature produced by a compression from $v + dv$ to v before any heat is emitted, we have

$$\theta = \frac{1}{N} \cdot \frac{\kappa e}{\mu} \cdot dv = \frac{\kappa e}{\mu K - v\kappa e^2} \, dv \dots\dots\dots\dots(21).$$

50. The first of these expressions for θ shows that, when the substance contracts as its temperature rises (as is the case, for instance, with water between its freezing-point and its point of maximum density), its temperature would become lowered by a sudden compression. The second, which shows in terms of its compressibility and expansibility exactly how much the temperature of any substance is altered by an infinitely small alteration of its volume, leads to the approximate expression

$$\theta = \frac{\kappa e}{\mu K},$$

if, as is probably the case, for all known solids and liquids, e be so small that $e \cdot v\kappa e$ is very small compared with μK.

51. If, now, we suppose the substance to be a gas, and introduce the hypothesis that its density is strictly subject to the "gaseous laws," we should have, by Boyle and Mariotte's law of compression,

$$\frac{dp}{dv} = -\frac{p}{v} \dots\dots\dots\dots\dots\dots(22);$$

and by Dalton and Gay-Lussac's law of expansion,

$$\frac{dv}{dt} = \frac{Ev}{1 + Et} \dots\dots\dots\dots\dots\dots(23);$$

from which we deduce

$$\frac{dp}{dt} = \frac{Ep}{1 + Et}.$$

Equation (14) will consequently become

$$\frac{dN}{dv} = \frac{d \left\{ \dfrac{Ep}{\mu (1 + Et)} - \dfrac{p}{J} \right\}}{dt} \dots\dots\dots\dots(24),$$

a result peculiar to the dynamical theory and equation (16),

$$K - N = \frac{E^2 pv}{\mu (1 - Et)^2} \dots\dots\dots\dots\dots(25),$$

which agrees with the result of § 53 of my former paper.

If V be taken to denote the volume of the gas at the temperature $0°$ under unity of pressure, (25) becomes

$$K - N = \frac{E^2 V}{\mu (1 + Et)} \dots\dots\dots\dots\dots(26).$$

52. All the conclusions obtained by Clausius, with reference
to air or gases, are obtained immediately from these equations
by taking

$$\mu = J \frac{E}{1 + Et},$$

which will make $\frac{dN}{dv} = 0$, and by assuming, as he does, that N, thus
found to be independent of the density of the gas, is also independent of its temperature.

53. As a last application of the two fundamental equations of
the theory, let the medium with reference to which M and N are
defined consist of a weight $1 - x$ of a certain substance in one
state, and a weight x in another state at the same temperature,
containing more latent heat. To avoid circumlocution and to fix
the ideas, in what follows we may suppose the former state to be
liquid and the latter gaseous; but the investigation, as will be
seen, is equally applicable to the case of a solid in contact with the
same substance in the liquid or gaseous form.

54. The volume and temperature of the whole medium being,
as before, denoted respectively by v and t, we shall have

$$\lambda (1 - x) + \gamma x = v \dots\dots\dots\dots\dots(27),$$

if λ and γ be the volumes of unity of weight of the substance in
the liquid and the gaseous states respectively: and p, the pressure,
may be considered as a function of t, depending solely on the
nature of the substance. To express M and N for this mixed
medium, let L denote the latent heat of a unit of weight of the
vapour, c the specific heat of the liquid, and h the specific heat of
the vapour when kept in a state of saturation. We shall have

$$M dv = L \frac{dx}{dv} dv,$$

$$N dt = c (1 - x) dt + hx dt + L \frac{dx}{dt} dt.$$

Now, by (27), we have

$$(\gamma - \lambda) \frac{dx}{dv} = 1 \dots\dots\dots\dots\dots(28),$$

and

$$(\gamma - \lambda) \frac{dx}{dt} + (1 - x) \frac{d\lambda}{dt} + x \frac{d\gamma}{dt} = 0 \dots\dots\dots(29).$$

Hence

$$M = \frac{L}{\gamma - \lambda} \dots\dots\dots\dots\dots(30),$$

$$N = c(1 - x) + hx - L \frac{(1 - x) \frac{d\lambda}{dt} + x \frac{d\gamma}{dt}}{\gamma - \lambda} \dots\dots(31).$$

55.   The expression of the second fundamental proposition in this case becomes, consequently,

$$\mu = \frac{(\gamma - \lambda) \frac{dp}{dt}}{L} \dots\dots\dots\dots\dots(32),$$

which agrees with Carnot's original result, and is the formula that has been used (referred to above in § 31) for determining μ by means of Regnault's observations on steam.

56.   To express the conclusion derivable from the first fundamental proposition, we have, by differentiating the preceding expressions for M and N with reference to t and v respectively,

$$\frac{dM}{dv} = \frac{1}{\gamma - \lambda} \cdot \frac{dL}{dt} - \frac{L}{(\gamma - \lambda)^2} \cdot \frac{d(\gamma - \lambda)}{dt}$$

$$\frac{dN}{dt} = \left(h - c - L \frac{\frac{d\gamma}{dt} - \frac{d\lambda}{dt}}{\gamma - \lambda} \right) \frac{dx}{dv}$$

$$= \left\{ \frac{h - c}{\gamma - \lambda} - \frac{L}{(\gamma - \lambda)^2} \right\} \frac{d(\gamma - \lambda)}{dt}.$$

Hence equation (2) of § 20 becomes

$$\frac{\frac{dL}{dt} + c - h}{\gamma - \lambda} = \frac{1}{J} \frac{dp}{dt} \dots\dots\dots\dots(33).$$

Combining this with the conclusion (32) derived from the second fundamental proposition, we obtain

$$\frac{dL}{dt} + c - h = \frac{L\mu}{J} \quad\dotfill(34).$$

The former of these equations agrees precisely with one which was first given by Clausius, and the preceding investigation is substantially the same as the investigation by which he arrived at it. The second differs from another given by Clausius only in not implying any hypothesis as to the form of Carnot's function μ.

57. If we suppose μ and L to be known for any temperature, equation (32) enables us to determine the value of $\frac{dp}{dt}$ for that temperature; and thence deducing a value of dt, we have

$$dt = \frac{\gamma - \lambda}{\mu L} \, dp \quad\dotfill(35);$$

which shows the effect of pressure in altering the " boiling-point " if the mixed medium be a liquid and its vapour, or the melting-point if it be a solid in contact with the same substance in the liquid state. This agrees with the conclusion arrived at [see pp. 156—164 above] by my elder brother in his "Theoretical Investigation of the Effect of Pressure in Lowering the Freezing-Point of Water." His result, obtained by taking as the value for μ that derived from Table I. of my former paper for the temperature $0°$, is that the freezing-point is lowered by ·0075° Cent. by an additional atmosphere of pressure. Clausius, with the other data the same, obtains 00733° as the lowering of temperature by the same additional pressure, which differs from my brother's result only from having been calculated from a formula which implies the hypothetical expression $J\dfrac{E}{1 + Et}$ for μ. It was by applying equation (33) to determine $\dfrac{dL}{dt}$ for the same case that Clausius arrived at the curious result regarding the latent heat of water under pressure mentioned above (§ 45).

58. Lastly, it may be remarked that every quantity which appears in equation (33), except h, is known with tolerable ac-

curacy for saturated steam through a wide range of temperature; and we may therefore use this equation to find h, which has never yet been made an object of experimental research. Thus we have

$$- h = \frac{\gamma - \lambda}{J} \frac{dp}{dt} - \left(\frac{dL}{dt} + c \right).$$

For the value of γ the best data regarding the density of saturated steam that can be had must be taken. If for different temperatures we use the same values for the density of saturated steam (calculated according to the gaseous laws, and Regnault's observed pressure from $\frac{1}{1693\cdot5}$, taken as the density at 100°), the values obtained for the first term of the second member of the preceding equation are the same as if we take the form

$$- h = \frac{L\mu}{J} - \left(\frac{dL}{dt} + c \right)$$

derived from (34), and use the values of μ shown in Table I. of my former paper. The values of $-h$ in the second column in the following table have been so calculated, with, besides, the following data afforded by Regnault from his observations on the total heat of steam, and the specific heat of water

$$\frac{dL}{dt} + c = \cdot305.$$

$$L = 606\cdot5 + \cdot305t - (\cdot00002t^2 + \cdot000000t^3).$$

The values of $-h$ shown in the third column are those derived by Clausius from an equation which is the same as what (34) would become if $J \dfrac{E}{1 + Et}$ were substituted for μ.

t.	$-h$ according to Table I. of "Account of Carnot's Theory."	$-h$ according to Clausius.
0	1·863	1·916
50	1·479	1·465
100	1·174	1·133
150	0·951	0·879
200	0·780	0·676

59. From these results it appears, that through the whole range of temperatures at which observations have been made, the

value of h is negative; and, therefore, if a quantity of saturated vapour be compressed in a vessel containing no liquid water, heat must be continuously abstracted from it in order that it may remain saturated as its temperature rises; and conversely, if a quantity of saturated vapour be allowed to expand in a closed vessel, heat must be supplied to it to prevent any part of it from becoming condensed into the liquid form as the temperature of the whole sinks.   This very remarkable conclusion was first announced by Mr Rankine, in his paper communicated to this Society on the 4th of February last year.   It was discovered independently by Clausius, and published in his paper in Poggendorff's *Annalen* in the months of April and May of the same year.

60.   It might appear at first sight, that the well-known fact that steam rushing from a high-pressure boiler through a small orifice into the open air does not scald a hand exposed to it*, is inconsistent with the proposition, that steam expanding from a state of saturation must have heat given to it to prevent any part from becoming condensed; since the steam would scald the hand unless it were dry, and consequently above the boiling-point in temperature.   The explanation of this apparent difficulty, given in a letter which I wrote to Mr Joule last October [Art. XLVII. above], and which has since been published in the *Philosophical Magazine*, is, that the steam in rushing through the orifice pro-duces mechanical effect which is immediately wasted in fluid friction, and consequently reconverted into heat; so that the issuing steam at the atmospheric pressure would have to part with as much heat to convert it into water at the temperature $100°$ as it would have had to part with to have been condensed at the high pressure and then cooled down to $100°$, which for a

[* *Note added June 26, 1852, in Phil. Mag. reprint.*—At present I am inclined to believe that the rapidity of the current exercises a great influence on the sensation experienced in the circumstances, by causing the steam to mix with the surrounding air; for I have found that the hand suffers pain when exposed to the steam issuing from a common kettle, and dried by passing through a copper tube surrounded by red-hot coals or heated by lamps.   But although there may be uncertainty regarding the causes of the different sensations in the different circumstances, I believe there is no reason for doubting either the fact of the dryness of the steam issuing from a high-pressure boiler (except when there is "priming" to a considerable extent), or the correctness of the explanation of this fact which I have given in the letter referred to.]

T.   14

pound of steam initially saturated at the temperature t is, by Regnault's modification of Watt's law, ·305 $(t-100^\circ)$ more heat than a pound of saturated steam at 100° would have to part with to be reduced to the same state; and the issuing steam must therefore be above 100° in temperature, and dry.

PART IV.

[Note of Dec. 30, 1881. The experimental method suggested in this Article, was carried out practically by Mr Joule and the author in successive years from 1852 to 1856. Extracts from their joint papers describing their work are included below [Art. XLIX.] in the present reprint.]

On a Method of discovering experimentally the Relation between the Mechanical Work spent, and the Heat produced by the Compression of a Gaseous Fluid*.

61. The important researches of Joule on the thermal circumstances connected with the expansion and compression of air, and the admirable reasoning upon them expressed in his paper† "On the Changes of Temperature produced by the Rarefaction and Condensation of Air," especially the way in which he takes into account any mechanical effect that may be externally produced, or internally lost, in fluid friction, have introduced an entirely new method of treating questions regarding the physical properties of fluids. The object of the present paper is to show how, by the use of this new method, in connexion with the principles explained in my preceding paper, a complete theoretical view may be obtained of the phenomena experimented on by Joule; and to point out some of the objects to be attained by a continuation and extension of his experimental researches.

62. The Appendix to my "Account of Carnot's Theory"‡ contains a theoretical investigation of the heat developed by the

---

* From the *Transactions of the Royal Society of Edinburgh*, Vol. xx. part 2. April 17, 1851.

† *Philosophical Magazine*, May, 1845, Vol. xxvi. p. 369.

‡ *Transactions*, Vol. xvi. part 5.

compression of any fluid fulfilling the laws* of Boyle and Mariotte and of Dalton and Gay-Lussac. It has since been shown that that investigation requires no modification when the dynamical Theory is adopted, and therefore the formula obtained as the result may be regarded as being established for a fluid of the kind assumed, independently of any hypothesis whatever. We may obtain a corresponding formula applicable to a fluid not fulfilling the gaseous laws of density, or to a solid pressed uniformly on all sides, in the following manner.

63.   Let Mdv be the quantity of heat absorbed by a body kept at a constant temperature t, when its volume is increased from v to $v + dv$; let p be the uniform pressure which it experiences from without, when its volume is v and its temperature t; and let $p + \dfrac{dp}{dt} dt$ denote the value p would acquire if the temperature were raised to $t + dt$, the volume remaining unchanged. Then, by equation (3) of § 21 of my former paper, derived from Clausius's extension of Carnot's theory, we have

$$M = \frac{1}{\mu} \cdot \frac{dp}{dt} \quad \dots\dots\dots\dots\dots\dots\dots(a)\dagger,$$

where μ denotes Carnot's *function*, the same for all substances at the same temperature.

Now let the substance expand from any volume V to V', and, being kept constantly at the temperature t, let it absorb a quantity, H, of heat. Then

$$H = \int_V^{V'} Mdv = \frac{1}{\mu} \frac{d}{dt} \int_V^{V'} pdv \dots\dots\dots\dots(b).$$

But if W denote the mechanical work which the substance does in expanding, we have

$$W = \int_V^{V'} pdv \dots\dots\dots\dots\dots\dots(c),$$

and therefore

$$H = \frac{1}{\mu} \frac{dW}{dt} \dots\dots\dots\dots\dots\dots(d).$$

* To avoid circumlocution, these laws will, in what follows, be called simply the *gaseous laws*, or the *gaseous laws* of density.

† Throughout this paper, formulæ which involve no hypothesis whatever are marked with italic letters; formulæ which involve Boyle's and Dalton's laws are marked with Arabic numerals; and formulæ involving, besides, Mayer's hypothesis, are marked with Roman numerals.

This formula, established without any assumption admitting of doubt, expresses the relation between the heat developed by the compression of any substance whatever, and the mechanical work which is required to effect the compression, as far as it can be determined without hypothesis by purely theoretical considerations.

64.   The preceding formula leads to that which I formerly gave for the case of fluids subject to the gaseous laws; since for such we have

$$pv = p_0 v_0 (1 + Et)\dots\dots\dots\dots\dots\dots(1),$$

from which we deduce, by (c),

$$W = p_0 v_0 (1 + Et) \log \frac{V'}{V}\dots\dots\dots\dots\dots(2),$$

and $$\frac{dW}{dt} = Ep_0 v_0 . \log \frac{V'}{V} = \frac{E}{1 + Et} W\dots\dots\dots(3);$$

and therefore, by (d),

$$H = \frac{E}{\mu (1 + Et)} W\dots\dots\dots\dots\dots(4),$$

which agrees with equation (11) of § 49 of the former paper.

65.   Hence we conclude, that the heat evolved by any fluid fulfilling the gaseous laws is proportional to the work spent in compressing it at any given constant temperature; but that the quantity of work required to produce a unit of heat is not constant for all temperatures, unless Carnot's function for different temperatures vary inversely as $1 + Et$; and that it is not the simple mechanical equivalent of the heat, as it was unwarrantably* assumed by Mayer to be, unless this function have precisely the expression

$$\mu = J \cdot \frac{E}{1 + Et}\dots\dots\dots\dots\dots\dots(I.).$$

This formula was suggested to me by Mr Joule, in a letter dated December 9, 1848, as probably a true expression for μ, being required to reconcile the expression derived from Carnot's theory (which I had communicated to him) for the heat evolved in terms

---

* In violation of Carnot's important principle, that thermal agency and mechanical effect, or mechanical agency and thermal effect, cannot be regarded in the simple relation of cause and effect, when any other effect, such as the alteration of the density of a body, is finally concerned.

of the work spent in the compression of a gas, with the hypothesis that the latter of these is exactly the mechanical equivalent of the former, which he had adopted in consequence of its being, at least approximately, verified by his own experiments. This, which will be called Mayer's hypothesis, from its having been first assumed by Mayer, is also assumed by Clausius without any reason from experiment; and an expression for μ the same as the preceding, is consequently adopted by him as the foundation of his mathematical deductions from elementary reasoning regarding the motive power of heat. The preceding formulæ show, that if it be true at a particular temperature for any one fluid fulfilling the gaseous laws, it must be true for every such fluid at the same temperature.

66. Of the various experimental researches which might be suggested as suitable for testing Mayer's hypothesis, it appears from the preceding formula, that any which would give data for the determination of the values of μ through a wide range of temperatures would, with a single accurate determination of J, afford a complete test. Thus an experimental determination of the density of saturated steam for temperatures from $0°$ to $230°$ Cent. would complete the data, of which a part have been so accurately determined by Regnault, for the calculation of the values of μ between those wide limits, and would contribute more, perhaps, than any set of experimental researches that could at present be proposed, to advance the mechanical theory of heat.

67. The values of μ, given in Table I. of my "Account of Carnot's Theory," which were calculated from Regnault's observations on steam, with the assumption of $\frac{1}{1693\cdot5}$ (the maximum density of water being unity) for the density of saturated steam at $100°$ Cent., and of the gaseous laws for calculating it by means of Regnault's observed pressures, at other temperatures, are far from verifying equation (1), as appears from the Table of the values of $\frac{\mu(1+Et)}{E}$, given in the preceding paper, § 51; or as the following comparative Table shows :—

Col. 1. Temperature. t.	Col. 2. Values of μ according to assumed density of saturated steam. $[\mu]$.	Col. 3. Values of μ according to Joule's formula. $J\dfrac{E}{1+Et}$.	Col. 4. Values of μ according to modified assumption for density of saturated steam. $\dfrac{1717\cdot6}{1693\cdot5}\times[\mu]$.
0	4·967	5·087	5·038
10	4·832	4·908	4·901
20	4·703	4·740	4·769
30	4·578	4·584	4·643
40	4·456	4·438	4·519
50	4·337	4·300	4·399
60	4·221	4·171	4·281
70	4·114	4·050	4·172
80	4·013	3·935	4·070
90	3·921	3·827	3·977
100	3·833	3·724	3·887
110	3·753	3·627	3·806
120	3·679	3·535	3·731
130	3·611	3·447	3·662
140	3·546	3·364	3·596
150	3·487	3·284	3·536
160	3·432	3·209	3·481
170	3·382	3·136	3·430
180	3·335	3·067	3·382
190	3·289	3·001	3·336
200	3·247	2·937	3·293
210	3·208	2·876	3·254
220	3·171	2·818	3·216
230	3·135	2·762	3·179

where $[\mu]$ denotes the quantity tabulated for the temperatures 0°, 1°, 2°, ... 230° in Table I. of my "Account of Carnot's Theory;" and $[\sigma]$ denotes the density of saturated steam which was assumed in the calculation of that table, the values of $\dfrac{p}{\Pi}$ in the expression for it being obtained by dividing the numbers tabulated at the end of Regnault's eighth Mémoire by 760. The considerableness of the deviations from the gaseous laws which equation (II) indicates, is seen at once by comparing the numbers in column 2 with those in column 3 of the preceding table, and observing that the coefficient of $[\sigma]$ in (II) is, for each temperature shown in that table, obtained by dividing the corresponding number in column 2 by that in column 3. Column 4 shows what the values of μ would be if the density of saturated steam at 100° were $\frac{1}{1717\cdot6}$ instead of $\frac{1}{1693\cdot5}$, and, for other temperatures, varied according to the gaseous laws.

Mr Joule, when I pointed out these discrepancies to him in the year 1848, suggested that even between $0°$ and $100°$ the inaccuracy of the data regarding steam might be sufficient to account for them. I think it will be generally admitted that there can be no such inaccuracy in Regnault's part of the data, and there remains only the uncertainty regarding the density of saturated steam, to prevent the conclusion that μ cannot be expressed by $J\,\dfrac{E}{1+Et}$; so that Mayer's hypothesis would be confirmed if, and overturned unless, the density of saturated steam, instead of following the gaseous laws, were truly expressed by the equations

$$\left.\begin{aligned} \sigma &= \frac{\left(\dfrac{1}{E}+t\right)[\mu]}{J}\cdot[\sigma] \\[2ex] [\sigma] &= \frac{1}{1693\cdot5}\cdot\frac{1+E\times100}{1+Et}\cdot\frac{p}{\Pi} \end{aligned}\right\}\ \dots\dots\dots\text{(II.)}.$$

68. This subject has been very carefully examined by Clausius, who has indicated the great deviations from the gaseous laws of density that Mayer's hypothesis requires in saturated steam, and has given an empirical formula for the density of saturated steam founded on that hypothesis, and on Regnault's observations on the pressure and latent heat. In this direction theory can go no further, for want of experimental data; although, from what we know of gases and saturated vapours, it may be doubted whether such excessive deviations, in the case of steam, from the laws of a "perfect gas" are rendered probable by a hypothesis resting on no experimental evidence whatever [*].

69. To Joule we are indebted for a most important series of experimental researches on the relation between the thermal effects, the external mechanical effects, and the internal mechanical effects (*vis viva* destroyed by fluid friction) due to compressions

---

[*] Joule's experimental verification of Mayer's law for temperatures of from $50°$ to $60°$ Fahr. shows, if rigorously exact, that the density of saturated steam at about $10°$ Centigrade must be $\frac{1693\cdot5}{1717\cdot6}$ of what was assumed for it in the calculations of my former paper, but does not go towards indicating any deviation from the gaseous laws of variation in the density of saturated steam at different temperatures.

and expansions of air in various circumstances *. These researches afford actual tests, which, so far as they go, are verifications of the truth of Mayer's hypothesis for temperatures between 50° and 60° Fahr., founded on two distinct methods, either of which is perfect in principle, and might be made the foundation of experiments at any temperature whatever.

70.   The first of these methods consists simply in determining, by direct experiment, the heat evolved by the expenditure of a given amount of work in compressing air, and comparing it with the quantity of heat created by the same amount of work in Joule's original experiments on the heat developed by magneto-electricity, and by the friction of fluids in motion.

71.   The second method is especially remarkable, as affording in each experiment an independent test of the truth of Mayer's hypothesis for air at the temperature used, without requiring any knowledge of the absolute value of the mechanical equivalent of heat.   In Joule's actual experiments, the test is simply this:— the total external thermal effect is determined when air is allowed to expand, through a small orifice, from one vessel into another previously exhausted by an air-pump.   Here the first mechanical effect produced by the expanding gas is *vis viva* generated in the rushing of the air.   By the time equilibrium is established, all this mechanical effect has been lost in fluid friction (there being no appreciable mechanical effect produced externally in sound, which is the only external mechanical effect, other than heat, that can be produced by the motions of a fluid within a fixed rigid vessel); and no truth in physical science can be more certain, than that by the time thermal as well as mechanical equilibrium is established at the primitive temperature, the contents of the two vessels must have parted with just as much more heat than they would have parted with had the air in expanding pushed out a piston against an external resisting force, as is equivalent to the mechanical effect thus produced externally.   Hence if the two vessels and the tube connecting them be immersed (as they are in Joule's first set of experiments with this apparatus) in one vessel of water, and if, after time is allowed for the pressure and temperature of the air to become the same in the two vessels, the

* *Philosophical Magazine*, May, 1845.

water be found to have neither gained nor lost heat (it being understood, of course, that the air and all other matter external to the water are at an absolutely constant temperature during the experiment), then, for the temperature of the experiment, Mayer's hypothesis is perfectly confirmed; but any final elevation or depression of temperature in the water would show that the work due to the expansion is either greater than or less than the absolute equivalent of the heat absorbed.

72. Mr Joule's second experiment on the same apparatus, in which he examined separately the external thermal effects round each of the two vessels, and round a portion of the tube containing the small orifice (a stop-cock), has suggested to me a method which appears still simpler, and more suitable for obtaining an excessively delicate test of Mayer's hypothesis for any temperature. It consists merely in dispensing with the two vessels in Joule's apparatus, and substituting for them two long spirals of tube (instead of doing this for only one of the vessels, as Joule does in his third experiment with the same apparatus); and in forcing air continuously through the whole. The first spiral portion of the tube, up to a short distance from the orifice, ought to be kept as nearly as possible at the temperature of the atmosphere surrounding the portion containing the orifice, and serves merely to fix the temperature of the entering air. The following investigation shows what conclusions might be drawn by experimenting on the thermal phenomena of any fluid whatever treated in this manner.

73. Let p be the uniform pressure of the fluid in the first spiral, up to a short distance from the orifice, and let p' be the pressure a short distance from the orifice on the other side, which will be uniform through the second spiral. Let t be the constant external temperature, and let the air in both spirals be kept as closely as possible at the same temperature. If there be any elevation or depression of temperature of the fluid in passing through the orifice, it may only be after passing through a considerable length of the second spiral that it will again arrive sensibly at the temperature t; and the spiral must be made at least so long, that the fluid issuing from the open end of it, when accurately tested, may be found not to differ appreciably from the primitive temperature t.

74. Let H be the total quantity of heat emitted from the portion of the tube containing the orifice, and the second spiral, during the passage of a volume u through the first spiral, or of an equivalent volume u' through the parts of the second where the temperature is sensibly t. This will consist of two parts; one (positive) the heat produced by the fluid friction, and the other (negative) the heat emitted by that portion of the fluid which passes from one side to the other of the orifice, in virtue of its expansion. To find these two parts, let us first suppose the transference of the fluid to take place without loss of mechanical effect in fluid friction, as it would do if, instead of the partition with a small orifice, there were substituted a moveable piston, and if a volume u of fluid, on the side where the pressure is higher (p), were enclosed between that and another piston, and allowed to slide through the tube till the second piston should take the place of the first, and to expand till its volume should be u'. If we adopt the same notation with reference to the volume, v, of the substance between the pistons, kept at a constant temperature, t, as has been used uniformly in this and the preceding paper; we shall have, for the quantity of heat absorbed during the motion of the piston,

$$\int_u^u M dv;$$

or, by the second fundamental equation of the theory, (3) of § 21 of the preceding paper,

$$\frac{1}{\mu} \int_u^{u'} \frac{d\varpi}{dt}\, dv,$$

where ϖ denotes the actual pressure (intermediate between p and p') of the substance when its volume is v. Again, the work done by the pistons will be given by the equation

$$W = \int_u^u \varpi dv + pu - p'u' \dots\dots\dots\dots\dots\dots(e).$$

If now the transference of the substance from the one portion of the tube, where the pressure is p, to the other, where the pressure is p', take place through a small orifice, exactly that amount, W, of work will be lost as external mechanical effect, and will go to generate thermal *vis viva*. The quantity of heat thus produced will be

$$\frac{1}{J}\left\{\int_u^{u'}\varpi dv + pu - p'u'\right\}.$$

Hence the total quantity of heat emitted will be the excess of this above the amount previously found to be absorbed when the mechanical effect is all external; and therefore we have[*]

$$H = \frac{1}{J}\left\{\int_u^{u'}\varpi dv + pu' - p'u'\right\} - \frac{1}{\mu}\int_u^{u'}\frac{d\varpi}{dt}dv\ldots\ldots(f).$$

Whatever changes of temperature there may actually be of the air in or near the orifice, this expression will give rigorously the total quantity of heat emitted by that portion of tube which contains the orifice and the whole of the second spiral during the passage of a volume u through the first spiral, or u' through any portion of the second spiral where the temperature is sensibly t.

75. To apply this result to the case of a gas fulfilling the gaseous laws, we may put

$$pu = p'u'.$$

Hence (*e*) becomes

$$W = \int_u^{u'}\varpi dv = pu\log\frac{u'}{u} = p'u'\log\frac{p}{p'}\ldots\ldots\ldots(5),$$

and, by (3), we have

$$\frac{dW}{dt} = \frac{Epu}{1+Et}\log\frac{u'}{u} = \frac{EW}{1+Et}.$$

Hence the expression (*f*) for the heat emitted becomes

$$H = \left\{\frac{1}{J} - \frac{E}{\mu(1+Et)}\right\}W\ldots\ldots\ldots\ldots(6).$$

76. Lastly, if Mayer's hypothesis be fulfilled for the gas used in the experiment, the coefficient of W vanishes by (I), and therefore

$$H = 0\ldots\ldots\ldots\ldots\ldots\ldots(III).$$

---

[*] A more comprehensive investigation, [§§ 94—96 below,] including a proof of this result, is given in a subsequent communication (Royal Soc. Edinb. Dec. 15, 1851), constituting Part v. of the present series of articles, [§§ 81—96 below,] which will be republished in an early number of the *Philosophical Magazine*.

77. From equation (III) it follows, that if Mayer's hypothesis be true, there is neither emission nor absorption of heat, on the whole, required to reduce the temperature of the air after passing through the orifice to its primitive value, t. Hence, although no doubt those portions of the air in the intermediate neighbourhood of the orifice which are communicating, by their expansion, *vis viva* to those contiguous to them will be becoming colder, and those which are the means of occasioning the portions contiguous to them to lose *vis viva*, through fluid friction, will be becoming warmer at each instant; yet very near the orifice on each side, where the motion of the air is uniform, the temperature would be constantly equal to t. Hence the simplest conceivable test of the truth of Mayer's hypothesis would be, to try whether the temperature of the air is exactly the same on the two sides of the orifice. This might be done by very delicate thermometers adjusted in the tube at sufficient distances on each side of the orifice to be quite out of the *rush* which there is of air in the immediate neighbourhood of the orifice; but it might be done in a still more refined manner by means of a delicate galvanometer, and a small thermo-electric battery arranged so that one set of the solderings might be within the tube on the side of the entering current of air, and the other set within the tube on the side of the current from the orifice. The tube on each side of the orifice would need to be bent so as to bring two parts of it, at small distances from the orifice on each side, near enough one another to admit of the battery being so placed. The only difficulty I can perceive in the way of making the necessary arrangements is what might be experienced in fitting the two ends of the battery air-tight into the two parts of the tube. It first occurred to me that the little battery itself might be placed entirely within the tube, and the difference of pressure kept up in the two parts by the middle of the battery being fitted nearly air-tight in the tube by means of wax, or otherwise; but this arrangement would not be satisfactory, as portions of the bars of the battery, if not the ends themselves directly, would be altered in temperature, even if Mayer's hypothesis were rigorously true, on account of the rushing of the air among them. No part of the battery ought to be exposed to the rushing of the air in the neighbourhood of the orifice, and therefore the middle of the battery would have to be external to the tube, the ends being cemented into the tube by some indurating cement suffi-

ciently strong and compact to hold perfectly air-tight on the side
where the pressure is different from the atmospheric pressure.
By such means as these, I think a very satisfactory series of
experiments might easily be performed to test Mayer's hypothesis
for air through a very wide range of temperatures.

78. Should the differential method of experimenting just
described indicate any difference of temperature whatever on the
two sides of the orifice, Mayer's hypothesis would be shown to be
not exactly fulfilled, and, according as the air leaving the orifice is
found to be warmer or colder than the entering air, we should
infer that the heat absorbed, when air expands at a constant temperature, is less than or greater than the equivalent of the
mechanical effect produced by the expansion *.

79. Calorimetrical methods, like those used by Joule, might
then be followed for actually determining the heat emitted or
absorbed by the air in the neighbourhood of the orifice, or in the
second spiral, in acquiring the temperature of the air in the entering stream; and by careful experimenting, it is probable that
excessively accurate results might be thus obtained for a wide
range of temperature.

80. The result of each experiment would be a value of μ, in
terms of Joule's mechanical equivalent, to be calculated by the
following expression, derived from equations (5) and (6).

$$\mu = \frac{\dfrac{JE}{1+Et}}{1 - J \cdot \dfrac{H}{p'u' \log \dfrac{p}{p'}}} \quad \dots\dots\dots\dots\dots\dots\dots(7).$$

In the second member of this equation p' denotes the pressure
of the air through the second spiral, which would be the atmos-

---

* Note added in *Phil. Mag.* reprint. Experiments on the plan here suggested
have been recently made by Mr Joule and myself, and it has thus been ascertained
that the air leaves the *rapids* in the neighbourhood of the orifice at a lower temperature than it approached them, even if this temperature be as high as 170° F.;
and it follows that the heat absorbed is *greater* than the equivalent of the mechanical
effect of the expansion, even for so high a temperature, and probably for much
higher. See a paper published in the Supplement to this Volume of the *Philosophical Magazine*, in which these experiments are described.—Nov. 11, 1852.

pheric pressure, or excessively near it, if, as in Joule's third
experiment mentioned above (described in p. 378 of the volume *
containing his paper), the air leaving the second spiral be
measured by means of a pneumatic trough: p denotes the
pressure in the first spiral, which ought to be constant, and
must be carefully measured; u' denotes the volume of air which
leaves the apparatus in any time; and H denotes the quantity of
heat emitted in the same time.   The experiment might be continued for any length of time, and each one of these four quantities
might be determined with great accuracy, so that probably very
accurate direct results of observations might be obtained.   If so,
no way of experimenting could be better adapted than this to the
determination of Carnot's function, for different temperatures, in
terms of Joule's mechanical equivalent of heat.

PART V.

On the Quantities of Mechanical Energy contained in a Fluid in Different States, as to Temperature and Density †.

81.   A body which is either emitting heat, or altering its
dimensions against resisting forces, is doing work upon matter
external to it   The mechanical effect of this work in one case is
the excitation of thermal motions, and in the other the overcoming
of resistances.   The body must itself be altering in its circumstances, so as to contain a less store of work within it by an
amount precisely equal to the aggregate value of the mechanical
effects produced ; and conversely, the aggregate value of the
mechanical effects produced must depend solely on the initial
and final states of the body, and is therefore the same whatever
be the intermediate states through which the body passes, provided the *initial* and *final* states be the same.

82.   The total mechanical energy of a body might be defined
as the mechanical value of all the effect it would produce in heat
emitted and in resistances overcome, if it were cooled to the

* *Phil. Mag.*, Vol. xxvi.   May, 1845.
† From the *Transactions of the Royal Society of Edinburgh*, Vol. xx. Part 3;
read December 15, 1851.

utmost, and allowed to contract indefinitely or to expand indefinitely according as the forces between its particles are attractive
or repulsive, when the thermal motions within it are all stopped;
but in our present state of ignorance regarding perfect cold, and
the nature of molecular forces, we cannot determine this "total
mechanical energy" for any portion of matter, nor even can we
be sure that it is not infinitely great for a finite portion of
matter. Hence it is convenient to choose a certain state as
standard for the body under consideration, and to use the un-
qualified term, *mechanical energy*, with reference to this standard
state; so that the "mechanical energy of a body in a given
state" will denote the mechanical value of the effects the body
would produce in passing from the state in which it is given, to
the standard state, or the mechanical value of the whole agency
that would be required to bring the body from the standard state
to the state in which it is given.

83. In the present communication, a system of formulæ
founded on propositions established in the first part of my paper
on the Dynamical Theory of Heat, and expressing relations between the pressure of a fluid, and the thermal capacities and
mechanical energy of a given mass of it, all considered as functions
of the temperature and volume; and Carnot's function of the
temperature; are brought forward for the purpose of pointing
out the importance of making the *mechanical energy* of a fluid in
different states an object of research, along with the other elements
which have hitherto been considered, and partially investigated in
some cases.

84. If we consider the circumstances of a stated quantity (a
unit of matter, a pound, for instance) of a fluid, we find that its
condition, whether it be wholly in the liquid state or wholly
gaseous, or partly liquid and partly gaseous, is completely defined,
when its temperature, and the volume of the space within which
it is contained, are specified (§§ 20, 53,...56), it being understood, of course, that the dimensions of this space are so limited
that no sensible differences of density in different parts of the
fluid are produced by gravity. We shall therefore consider the
temperature, and the volume of unity of mass, of a fluid, as the
independent variables of which its pressure, thermal capacities,
and mechanical energy are functions. The volume and temperature

being denoted respectively by v and t, let e be the mechanical energy, p the pressure, K the thermal capacity under constant pressure, and N the thermal capacity in constant volume; and let M be such a function of these elements, that

$$K = N + \frac{\frac{dp}{dt}}{-\frac{dp}{dv}} M \dots\dots\dots\dots\dots\dots(1),$$

or (§§ 48, 20), such a quantity that

$$Mdv + Ndt \dots\dots\dots\dots\dots\dots\dots(2),$$

may express the quantity of heat that must be added to the fluid mass, to elevate its temperature by dt, when its volume is augmented by dv.

85. The mechanical value of the heat added to the fluid in any operation, or the quantity of heat added, multiplied by J (the mechanical equivalent of the thermal unit), must be diminished by the work done by the fluid in expanding against resistance, to find the actual increase of mechanical energy which the body acquires. Hence (de of course denoting the complete increment of e, when v and t are increased by dv and dt) we have

$$de = J(Mdv + Ndt) - pdv \dots\dots\dots\dots(3).$$

Hence, according to the usual notation for partial differential coefficients, we have

$$\frac{de}{dv} = JM - p \dots\dots\dots\dots\dots\dots(4),$$

$$\frac{de}{dt} = JN \dots\dots\dots\dots\dots\dots\dots(5).$$

Lastly, if we denote by μ, as formerly, Carnot's function of the temperature t, we have (§ 21)

$$\frac{dp}{dt} = \mu M \dots\dots\dots\dots\dots\dots\dots(6).$$

86. The use that may be made of these formulæ in investigations regarding the physical properties of any particular fluid must depend on the extent and accuracy of the general data belonging to the theory of the mechanical action of heat that are available. Thus, if nothing be known by experiment regarding the values of

J and μ, we may, in the first place, use equations (4) and (5), or the following deduced from them (§ 23), by eliminating e,

$$\frac{dp}{dt} = J\left(\frac{dM}{dt} - \frac{dN}{dv}\right) \dots\dots\dots\dots\dots\dots\dots(7),$$

and equation (6), as tests of the accuracy of experimental researches on the pressure and thermal capacities of a fluid, on account of the knowledge we have from theory that J is *certainly* an absolute constant, and that in all probability, if not with absolute certainty, we may regard μ as independent of v, and as the same for all fluids at the same temperature; and with experimental data of sufficient extent, we may use these equations as means of actually determining the values of J and μ. No other way than this has yet been attempted for determining μ; and if we except a conceivable, but certainly not at present practicable mode of determining this element by experiments on thermoelectric currents, no other way is yet known. Carnot's original determination of μ was effected by means of an expression equivalent to that of equation (6) applied to the case of a mass of air; and the determinations by Clapeyron, and those shown in Table I. of my "Account of Carnot's Theory," were calculated by the formula which is obtained when the same equation is applied to the case of a fluid mass, partly liquid and partly in the state of saturated vapour (§ 55).

87. As yet experiments have not been made on the pressure and thermal capacities of fluids to a sufficient extent to supply data for the evaluation, even in the roughest manner, of the expression given for J by equation (7); and it may be doubted whether such data can even be had with accuracy enough to give as exact a determination of this important element as may be effected by direct experiments on the generation of heat by means of friction. At present we may regard J as known, probably within $\frac{1}{300}$ of its own amount, by experiments of this kind.

88. The value of J being known, equations (4) and (5) may be used for determining the mechanical energy of a particular fluid mass in different states, from special experimental data regarding its pressure and thermal capacities, but not necessarily comprehending the values of each of these elements for all states of the fluid. The theory of the integration of functions of two

T.   15

independent variables will, when any set of data are proposed, make it manifest whether or not they are sufficient, and will point out the methods, whether of summation or of analytical integration, according to the forms in which the data are furnished, to be followed for determining the value of e for every value of v. Or the data may be such, that while the thermal capacities would be derived from them by differentiation, values of e may be obtained from them without integration. Thus, if the fluid mass consist of water and vapour of water at the temperature t, weighing in all one pound, and occupying the volume v^*, and if we regard the *zero* or "standard" state of the mass as being liquid water at the temperature $0°$, the mechanical energy of the mass in the given state will be the mechanical value of the heat required to raise the temperature of a pound of water from $0°$ to t, and to convert $\dfrac{v-\lambda}{\gamma-\lambda}$ of it into vapour, diminished by the work done in the expansion from the volume λ to the volume v; that is, we have

$$e = J\left(ct + L\frac{v-\lambda}{\gamma-\lambda}\right) - p(v-\lambda)\dots\dots\dots\dots(8).$$

The variables c, L, and p (which depend on t alone), in this expression have been experimentally determined by Regnault for all temperatures from $0°$ to $230°$; and when γ is also determined by experiments on the density of saturated steam, the elements for the determination of e in this case will be complete. The expressions investigated formerly for M and N in this case (§ 54) may be readily obtained by means of (4) and (5) of § 85, by the differentiation of (8).

89. If Carnot's function has once been determined by means of observations of any kind, whether on a single fluid or on different fluids, for a certain range of temperatures, then according

---

* The same notation is used here as formerly in § 54, viz. p is the pressure of saturated vapour at the temperature t, γ the volume, and L the latent heat of a pound of the vapour, λ the volume of a pound of liquid water, and c the mean thermal capacity of a pound of water between the temperatures 0 and t. A mass weighing a pound, and occupying the volume v, when at the temperature t, must consist of a weight $\dfrac{v-\lambda}{\gamma-\lambda}$ of vapour, and $\dfrac{\gamma-v}{\gamma-\lambda}$ of water.

to (6) of § 85, the value of $\dfrac{dp}{dt}/M$ for any substance whatever is known for all temperatures within that range. It follows that when the values of M for different states of a fluid have been determined experimentally, the law of pressures for all temperatures and volumes (with an arbitrary function of v to be determined by experiments on the pressure of the fluid at one particular temperature) may be deduced by means of equation (6); or conversely, which is more likely to be the case for any particular fluid, if the law of pressures is completely known, M may be deduced without further experimenting. Hence the second member of (4) becomes completely known, the equation assuming the following form, when, for M, its value according to (6) is substituted :—

$$\frac{de}{dv} = \frac{J}{\mu}\frac{dp}{dt} - p \dots\dots\dots\dots(4').$$

The integration of this equation with reference to v leads to an expression for e, involving an arbitrary function of t, for the determination of which more data from experiment are required. It would, for instance, be sufficient for this purpose to have the mechanical energy of the fluid for all temperatures when contained in a constant volume; or, what amounts to the same (it being now supposed that J is known), to have the thermal capacity of the fluid in constant volume for a particular volume and all temperatures. Hence we conclude, that when the elements J and μ belonging to the general theory of the mechanical action of heat are known, the mechanical energy of a particular fluid may be investigated without experiment, from determinations of its pressure for all temperatures and volumes, and its thermal capacity for any particular constant volume and all temperatures.

90.   For example, let the fluid be atmospheric air, or any other subject to the "gaseous" laws. Then if v_0 be the volume of a unit of weight of the fluid, and 0 the temperature, in the standard state from which the mechanical energy in any other state is reckoned, and if p_0 denote the corresponding pressure, we have

$$p = \frac{p_0 v_0}{v}(1 + Et), \quad \frac{dp}{dt} = \frac{p_0 v_0 E}{v},$$

and

$$\int_{v_0}^{v} \left(\frac{J}{\mu} \frac{dp}{dt} - p \right) dv = p_0 v_0 \left\{ \frac{JE}{\mu} - (1 + Et) \right\} \log \frac{v}{v_0}.$$

Hence if we denote by N_0 the value of N when $v = v_0$, whatever be the temperature, we have as the general expression for the mechanical energy of a unit weight of a fluid subject to the gaseous laws,

$$e = p_0 v_0 \left\{ \frac{JE}{\mu} - (1 + Et) \right\} \log \frac{v}{v_0} + J \int_0^t N_0 dt \ldots\ldots\ldots(9).$$

91.   Let us now suppose the mechanical energy of a particular fluid mass in various states to have been determined in any way, and let us find what results regarding its pressure and thermal capacities may be deduced.   In the first place, by integrating equation (4′), considered as a differential equation with reference to t for p, we find

$$p = \epsilon^{\frac{1}{J} \int_0^t \mu dt} \int_0^t \mu \frac{de}{dv} \epsilon^{-\frac{1}{J} \int_0^t \mu dt} dt + \psi(v) \epsilon^{\frac{1}{J} \int \mu dt} \ldots\ldots(10),$$

where $\psi(v)$ denotes a constant with reference to t, which may vary with v, and cannot be determined without experiment. Again, we have from (5), (4), and (1),

$$\left. \begin{array}{l} N = \dfrac{1}{J} \dfrac{de}{dt} \\[2em] K = \dfrac{1}{J} \dfrac{de}{dt} + \dfrac{1}{J} \left(\dfrac{de}{dv} + p \right) \dfrac{\frac{dp}{dt}}{-\dfrac{dp}{dv}} \end{array} \right\} \ldots\ldots\ldots\ldots\ldots(11).$$

From the first of these equations we infer, that, with a complete knowledge of the mechanical energy of a particular fluid, we have enough of data for determining for every state its thermal capacity in constant volume.   From equation (9) we infer, that with, besides, a knowledge of the pressure for all volumes and a particular temperature, or for all volumes and a particular series of temperatures, we have enough to determine completely the pressure, and consequently also, according to equation (11), to determine the two thermal capacities, for all states of the fluid.

92. For example, let these equations be applied to the case of a fluid subject to the gaseous laws. If we use for $\frac{de}{dv}$ its value derived from (9), in equation (10), we find

$$p = \frac{p_0 v_0}{v} (1 + Et) + \chi (v)\, \epsilon^{\frac{1}{J} \int \mu dt} \quad \ldots\ldots\ldots\ldots(12),$$

where $\chi (v)$, denoting an arbitrary function of v, is used instead of $\psi (v) - \frac{p_0 v_0}{v}$. We conclude that the same expression for the mechanical energy holds for any fluid whose pressure is expressed by this equation, as for one subject to the gaseous laws   Again, by using for $\frac{de}{dt}$ and $\frac{de}{dv}$, their values derived from (9), in equation (11), we have

$$N = N_0 + \frac{1}{J} p_0 v_0 \log \frac{v}{v_0} \frac{d \left\{ \frac{JE}{\mu} - (1 + Et) \right\}}{dt} \quad \ldots\ldots\ldots\ldots(13),$$

$$K = N_0 + \frac{1}{J} p_0 v_0 \log \frac{v}{v_0} \frac{d \left\{ \frac{JE}{\mu} - (1 + Et) \right\}}{dt} + \frac{E^2 p_0 v_0}{\mu (1 + Et)} \; ..(14).$$

The first of these equations shows, that unless Mayer's hypothesis be true, there is a difference in the thermal capacities in constant volume, of the same gas at the same temperatures for different densities, proportional in amount to the difference of the logarithms of the densities. The second, compared with the first, leads to an expression for the difference between the thermal capacities of a gas in constant volume, and under constant pressure, agreeing with results arrived at formerly. [Account of Carnot's Theory, Appendix III.; and Dynamical Theory of Heat, § 48.]

93. It may be that more or less information, regarding explicitly the pressure and thermal capacities of the fluid, may have been had as the data for determining the mechanical energy; but these converse deductions are still interesting, as showing how much information regarding its physical properties is comprehended in a knowledge of the mechanical energy of a fluid mass, and how useful a table of the values of this function for

different temperatures and volumes, or an empirical function of two variables expressing it, would be, whatever be the experimental data from which it is deduced. It is not improbable that such a table or empirical function, and a similar representation of the pressure, may be found to be the most convenient expression for results of complete observations on the compressibility, the law of expansion by heat, and the thermal capacities of a vapour or gas.

94. The principles brought forward in a former communication "On a Means of discovering experimentally, &c." (which is now referred to as Part IV. of a series of papers on the Dynamical Theory of Heat), may be expressed in a more convenient and in a somewhat more comprehensive manner than in the formulæ contained in that paper, by introducing the notations and principles which form the subject of the present communication. Thus, let t be the temperature and u the volume of a pound of air flowing gently in a pipe (under very high pressure it may be) towards a very narrow passage (a nearly closed stopcock, for instance), and let p be its pressure. Let t', u', and p' be the corresponding qualities of the air flowing gently through a continuation of the pipe, after having passed the "rapids" in and near the narrow passage. Let Q be the quantity of heat (which, according to circumstances, may be positive, zero, or negative) emitted by a pound of air during its whole passage from the former locality through the narrow passage to the latter; and let S denote the mechanical value of the sound emitted from the "rapids." The only other external mechanical effect besides these two produced by the air, is the excess (which, according to circumstances, may be negative, zero, or positive) of the work done by the air in pressing out through the second part of the pipe above that spent in pressing it in through the first; the amount of which, for each pound of air that passes, is of course $p'u' - pu$. Hence the whole mechanical value of the effects produced externally by each pound of the air from its own mechanical energy is

$$JQ + S + p'u' - pu \dots\dots\dots\dots\dots(15).$$

Hence if $\phi(v, t)$ denote the value of e expressed as a function of the independent variables v and t, so that $\phi(u, t)$ may express the mechanical energy of a pound of air before, and $\phi(u', t')$

the mechanical energy of a pound of air after passing the rapids, we have

$$\phi(u',\ t') = \phi(u,\ t) - \{JQ + S + p'u' - pu\}\ldots\ldots\ldots(16).$$

95. If the circumstances be arranged (as is always possible) so as to prevent the air from experiencing either gain or loss of heat by conduction through the pipe and stopcock, we shall have $Q = 0$; and if (as is perhaps also possible) only a mechanically inappreciable amount of sound be allowed to escape, we may take $S = 0$. Then the preceding equation becomes

$$\phi(u',\ t') = \phi(u,\ t) - (p'u' - pu)\ldots\ldots\ldots\ldots(17).$$

If by experimenting in such circumstances it be found that t' does not differ sensibly from t, Mayer's hypothesis is verified for air at the temperature t; and as $p'u'$ would then be equal to pu, according to Boyle and Mariotte's law, we should have

$$\phi(u',\ t) = \phi(u,\ t),$$

which is, in fact, the expression of Mayer's hypothesis, in terms of the notation for mechanical energy introduced in this paper. If, on the other hand, t' be found to differ from t*, let values of p, p', t, and t' be observed in various experiments of this kind, and, from the known laws of density of air, let u and u' be calculated. We then have, by an application of (13) to the results of each experiment, an equation showing the difference between the mechanical energies of a pound of air in two particular specified states as to temperature and density. All the particular equations thus obtained may be used towards forming, or for correcting, a table of the values of the mechanical energy of a mass of air at various temperatures and densities.

96. If, according to the plan proposed in my former communication (§ 72), the air on leaving the narrow passage be made to pass through a spiral pipe immersed in water in a calori-

---

* If the values of μ I have used formerly be correct, t' would be less than t for all cases in which t is lower than about 30° Cent.; but, on the contrary, if t be considerably above 30° Cent., t' would be found to exceed t. (See "Account of Carnot's Theory," Appendix II.) It may be shown, that if they are correct, air at the temperature 0° forced up with a pressure of ten atmospheres towards a small orifice, and expanding through it to the atmospheric pressure, would go down in temperature by about 4°·4; but that if it had the temperature of 100° in approaching the orifice, it would leave at a temperature about 5°·2 higher, provided that in each case there is no appreciable expenditure of mechanical energy on sound.

metrical apparatus, and be so brought back exactly to the primitive temperature t, we should have, according to Boyle and Mariotte's law, $p'u' - pu = 0$; and if H denote the value of Q in this particular case (or the quantity of heat measured by means of the calorimetric apparatus), the general equation (16) takes the form

$$\phi(u', t) = \phi(u, t) - (JH + S) \ldots\ldots\ldots\ldots(18).$$

If in this we neglect S, as probably insensible, and if we substitute for $\phi(u, t)$ and $\phi(u', t)$ expressions deduced from (9), we find

$$H = \left\{ \frac{1}{J} - \frac{E}{\mu(1 + Et)} \right\} pu \log \frac{u'}{u} \ldots\ldots\ldots\ldots(19),$$

which agrees exactly with the expression obtained by a synthetical process, founded on the same principles, in my former communication (§ 75).

PART VI.

Thermo-electric Currents.*

Preliminary §§ 97—101.   Fundamental Principles of General Thermo-dynamics recapitulated.

97. Mechanical action may be derived from heat, and heat may be generated by mechanical action, by means of forces either acting between contiguous parts of bodies, or due to electric excitation; but in no other way known, or even conceivable, in the present state of science. Hence thermo-dynamics falls naturally into two divisions, of which the subjects are respectively, *the relation of heat to the forces acting between contiguous parts of bodies,* and *the relation of heat to electrical agency.* The investigations of the conditions under which thermo-dynamic effects are produced, in operations on any fluid or fluids, whether gaseous or liquid, or passing from one state to the other, or to or from the solid state, and the establishment of universal relations between the physical properties of all substances in these different states, which have been given in Parts I.—V. of the present series of papers, belong to that first great division of thermo-dynamics—to be completed (as is intended for future communication to the Royal Society) by the extension of similar researches to the thermo-elastic properties of solids. [Note of Jan. 12, 1882. This proposed continuation of work on

---

* From the *Transactions* of the Royal Society of Edinburgh, vol. xxi. part I.; read 1st May, 1854.

thermo-dynamics was communicated not as first intended to the Royal Society of Edinburgh, but to the first Number of the *Quarterly Journal of Mathematics.* It is reprinted below as Part VII. of the present series of Articles on the Dynamical Theory of Heat.] The second division, or thermo-electricity, which may include many kinds of action as yet undiscovered, has hitherto been investigated only as far as regards the agency of heat in producing electrical effects in non-crystalline metals.   In a mechanical theory of electric currents, communicated to the Royal Society of Edinburgh, Dec. 15, 1851*, the application of the general laws of the dynamical theory of heat to this kind of agency was made, and certain universal relations precisely analogous to the thermo-elastic properties of fluids established in the previous treatment of the first division of the subject, were established between the thermo-electric properties of non-crystalline metals.   The object of the present communication is to extend the theory to the phænomena of thermo-electricity in crystalline metals; but as recent experimental researches on air have pointed out an absolute thermometric scale†,

---

* See *Proceedings* of that date, or *Philosophical Magazine*, first half-year, 1852 [reprinted as Note 1, appended to Art. xlviii. at the end of Part VII. below], where a sufficiently complete account of the investigations and principal results is given.

† That is a scale defined without reference to effects experienced by any particular kind of matter.   Such a scale, founded on general thermo-dynamic relations of heat and matter, and requiring reference to a particular thermometric substance only for defining the unit or degree, was, so far as I know, first proposed in a communication to the Cambridge Philosophical Society (*Proceedings*, May 1848, or *Philosophical Magazine*, October 1848, or Art. xxxix. above).   The particular thermometric assumption there suggested was, that a thermo-dynamic engine working to perfection, according to Carnot's criterion, would give the same work from the same quantity of heat, with its source and refrigerator differing by one degree of temperature in any part of the scale; the fixed points being taken the same as the 0^0 and 100^0 of the centigrade scale.   A comparison of temperature, according to this assumption, with temperature by the air thermometer, effected by the only data at that time afforded by experiment, namely Regnault's observations on the pressure and latent heat of saturated steam at temperatures of from 0^0 to 230^0 of the air thermometer, showed, as the nature of the assumption required, very wide discrepancies, even inconveniently wide between the fixed points of agreement.   A more convenient assumption has since been pointed to by Mr Joule's conjecture, that Carnot's function is equal to the mechanical equivalent of the thermal unit divided by the temperature by the air thermometer from its zero of expansion; an assumption which experiments on the thermal effects of air escaping through a porous plug, undertaken by him in conjunction with myself for the purpose of testing it (*Philosophical Magazine*, Oct. 1852), have shown to be not rigorously but very approximatively true.   More extensive and accurate experiments have given us data for a closer test (*Phil. Trans.*, June 1853), and in a joint communication by Mr Joule

the use of which in expressing the general laws of the dynamical theory of heat, both leads to a very concise mode of stating the principles, and shows the most convenient forms of the expressions brought forward in my former communication, the elementary theory of thermo-electricity in metals will be included in the investigations now communicated. I shall take the opportunity of introducing developments and illustrations, which, although communicated at the meeting of the Royal Society along with the original treatment of the subject, did not appear in the printed abstract; and I shall add some experimental conclusions which have since been arrived at, in answer to questions proposed in the former theoretical investigation.

98. Before entering on the treatment of the special subject, it is convenient to recall the fundamental laws of the dynamical theory of heat, and it is necessary to explain the thermometric assumption by which temperature is now to be measured.

The conditions under which heat and mechanical work are mutually convertible by means of any material system, subjected either to a continuous uniform action, or to a cycle of operations at the end of which the physical conditions of all its parts are the same as at the beginning, are subject to the following laws:—

Law I.—The material system must give out exactly as much energy as it takes in, either in heat or mechanical work.

Law II.—If every part of the action, and all its effects, be perfectly reversible, and if all the localities of the system by which

---

and myself to the Royal Society of London, to be made during the present session, we propose that the numerical measure of temperature shall be not founded on the expansion of air at a particular pressure, but shall be simply the mechanical equivalent of the thermal unit divided by Carnot's function. We deduce from our experimental results, a comparison between *differences on the new scale from the temperature of freezing water*, and *temperatures centigrade of* Regnault's *standard air thermometer*, which shows no greater discrepance than a few hundredths of a degree, at temperatures between the freezing- and boiling-points, and, through a range of 300^0 above the freezing-point, so close an agreement that it may be considered as perfect for most practical purposes. The form of assumption given below in the text as the foundation of the new thermometric system, without explicit reference to Carnot's function, is equivalent to that just stated, inasmuch as the formula for the action of a perfect thermo-dynamic engine, investigated in § 25, expresses (§ 42) that the heat used is to the heat rejected in the proportion of the temperature of the source to the temperature of the refrigerator, if Carnot's function have the form there given as a conjecture, and now adopted as the definition of temperature.

heat is either emitted or taken in, be at one or other of two temperatures, the aggregate amount of heat taken in or emitted at the higher temperature, must exceed the amount emitted or taken in at the lower temperature, always in the same ratio when these temperatures are the same, whatever be the particular substance or arrangement of the material system, and whatever be the particular nature of the operations to which it is subject.

99. *Definition of temperature* and *general thermometric assumption*.—If two bodies be put in contact, and neither gives heat to the other, their temperatures are said to be the same; but if one gives heat to the other, its temperature is said to be higher.

The temperatures of two bodies are proportional to the quantities of heat respectively taken in and given out in localities at one temperature and at the other, respectively, by a material system subjected to a complete cycle of perfectly reversible thermodynamic operations, and not allowed to part with or take in heat at any other temperature: or, the absolute values of two temperatures are to one another in the proportion of the heat taken in to the heat rejected in a perfect thermo-dynamic engine working with a source and refrigerator at the higher and lower of the temperatures respectively.

100. *Convention for thermometric unit, and determination of absolute temperatures of fixed points in terms of it.*

Two fixed points of temperature being chosen according to Sir Isaac Newton's suggestion, by particular effects on a particular substance or substances, the difference of these temperatures is to be called unity, or any number of units or degrees as may be found convenient. The particular convention is, that the difference of temperatures between the freezing- and boiling-points of water under standard atmospheric pressure shall be called 100 degrees. The determination of the absolute temperatures of the fixed points is then to be effected by means of observations indicating the economy of a perfect thermo-dynamic engine, with the higher and the lower respectively as the temperatures of its source and refrigerator. The kind of observation best adapted for this object was originated by Mr Joule, whose work in 1844* laid the founda-

* "On the Changes of Temperature occasioned by the Rarefaction and Condensation of Air," see *Proceedings* of the Royal Society, June 1844; or, for the paper in full, *Phil. Mag.*, May 1845.

tion of the theory, and opened the experimental investigation; and it has been carried out by him, in conjunction with myself, within the last two years, in accordance with the plan proposed in Part IV.* of the present series. The best result, as regards this determination, which we have yet been able to obtain is, that the temperature of freezing water is 273·7 on the absolute scale; that of the boiling-point being consequently 373·7†. Further details regarding the new thermometric system will be found in a joint communication to be made by Mr Joule and myself to the Royal Society of London before the close of the present session.

101. A corollary from the second general law of the dynamical theory stated above in § 98, equivalent to the law itself in generality, is, that if a material system experience a continuous action, or a complete cycle of operations, of a perfectly reversible kind, the quantities of heat which it takes in at different temperatures are subject to a linear equation, of which the coefficients are the corresponding values of an absolute function of the temperature. The thermometric assumption which has been adopted is equivalent to assuming that this absolute function is the reciprocal of the temperature; and the equation consequently takes the form

$$\frac{H_t}{t} + \frac{H_{t'}}{t'} + \frac{H_{t''}}{t''} + \&c. = 0,$$

if t, t', &c. denote the temperatures of the different localities where there is either emission or absorption of heat, and $\pm H_t$, $\pm H_{t'}$, $\pm H_{t''}$, &c. the quantities of heat taken in or given out in those localities respectively. To prove this, conceive an engine emitting a quantity H_t of heat at the temperature t, and taking in the corresponding quantity $\frac{t'}{t} H_t$ at the temperature t'; then an engine emitting the quantity $\frac{t'}{t} H_t + H_{t'}$ at t', and taking in the corresponding quantity $t''\left(\frac{H_t}{t} + \frac{H_{t'}}{t'}\right)$ at the temperature t''; another

* "On a Method of discovering experimentally the Relation between the Heat Produced and the Work Spent in the Compression of a Gas." *Trans. R.S.E.*, April 1851; or *Phil. Mag.* 1852, second half-year.

† [Note of Dec. 1881. Later results show that these numbers are more accurately 273·1 and 373·1. Article on Heat by the author, *Encyc. Brit.*; also published separately under the title "Heat," Edinburgh, Black, 1880.]

emitting $t'' \left(\dfrac{H_t}{t} + \dfrac{H_{t'}}{t'} \right) + H_t$ at t'', and taking in the corresponding

quantity $t''' \left(\dfrac{H_t}{t} + \dfrac{H_{t'}}{t'} + \dfrac{H_{t''}}{t''} \right)$ at t'''; and so on. Considering $n-2$
such engines as forming one system, we have a material system
causing, by reversible operations, an emission of heat amounting
to H_t at the temperature t, $H_{t'}$ at the temperature t',... and $H_{t^{(n-2)}}$
at $t^{(n-2)}$; and taking in $t^{(n-1)} \left(\dfrac{H_t}{t} + \dfrac{H_{t'}}{t'} + ... + \dfrac{H_{t^{(n-2)}}}{t^{(n-2)}} \right)$ at the temperature $t^{(n-1)}$. Now this system, along with the given one, constitutes
a complex system, which causes on the whole neither absorption
nor emission of heat at the temperatures t, t', &c., or at any other
temperatures than $t^{(n-1)}$, $t^{(n)}$; but gives rise to an absorption or
emission equal to

$$\pm \left[t^{(n-1)} \left(\frac{H_t}{t} + \frac{H_{t'}}{t'} + ... + \frac{H_{t^{(n-2)}}}{t^{(n-2)}} \right) + H_{t^{(n-1)}} \right]$$

at $t^{(n-1)}$, and an emission or absorption equal to $\pm H_{t^{(n)}}$ at $t^{(n)}$. This
complete system fulfils the criterion of reversibility, and, having
only two temperatures at localities where heat is taken in or given
out, is therefore subject to Law II.; that is, we must have

$$H_{t^{(n)}} = - \frac{t^{(n)}}{t^{(n-1)}} \left[t^{(n-1)} \left(\frac{H_t}{t} + \frac{H_{t'}}{t'} + ... + \frac{H_{t^{(n-2)}}}{t^{(n-2)}} \right) + H_{t^{(n-1)}} \right],$$

which is the same as

$$\frac{H_t}{t} + \frac{H_{t'}}{t'} + ... + \frac{H_{t^{(n-1)}}}{t^{(n-1)}} + \frac{H_{t^{(n)}}}{t^{(n)}} = 0 ....(1).$$

This equation may be considered as the mathematical expression
of the second fundamental law of the dynamical theory of heat.
The corresponding expression of the first law is

$$W + J(H_t + H_{t'} + ... + H_{t^{(n-1)}} + H_{t^{(n)}}) = 0 ....(2),$$

where W denotes the aggregate amount of work spent in producing the operations, and J the mechanical equivalent of the
thermal unit.

§§ 102—106. *Initial examination of Thermo-dynamic circum-stances regarding Electric Currents in Linear Conductors.*

102. Peltier's admirable discovery, that an electric current in
a metallic circuit of antimony and bismuth produces cold where it

passes from bismuth to antimony, and heat where it passes from
antimony to bismuth, shows how an evolution of mechanical effect,
by means of thermo-electric currents, involves transference of heat
from a body at a higher temperature to a body at a lower temperature, and how a reverse thermal effect may be produced, by
thermo-electric means, from the expenditure of work.   For if a
galvanic engine be kept in motion doing work, by a thermo-electric
battery of bismuth and antimony, the current by means of which
this is effected passing, as it does, from bismuth to antimony
through the hot junctions, and from antimony to bismuth through
the cold junctions, must cause absorption of heat in each of the
former, and evolution of heat in each of the latter; and to sustain
the difference of temperature required for the excitation of the
electromotive force, even were there no propagation of heat by
conduction through the battery, it would be necessary continually,
during the existence of the current, to supply heat from a source
to the hot junctions, and to draw off heat from the cold junctions
by a refrigerator:—Or, if work be spent to turn the engine faster
than the rate at which its inductive reaction balances the electro-
motive force of the battery, there will be a reverse current sent
through the circuit, producing absorption of heat at the cold junctions, and evolution of heat at the hot junctions, and consequently
effecting the transference of some heat from the refrigerator to the
source.

103.   We see, then, that in Peltier's phænomenon we have a
reversible thermal agency of exactly the kind supposed in the
second law of the dynamical theory of heat.   Before, however, we
can apply either this or the first law, we must consider other
thermal actions which are involved in the circumstances of a
thermo-electric current; and with reference to the second law, we
shall have to examine whether there are any such of an essentially
irreversible kind.

104.   It is to be remarked, in the first place, that a current
cannot pass through a homogeneous conductor without generating
heat in overcoming resistance.   This effect, which we shall call the
frictional generation of heat, has been discovered by Joule to be
produced at a rate proportional to the square of the strength of
the current; and, taking place equally with the current in one
direction, or in the contrary, is obviously of an irreversible kind.

Any other thermal action that can take place must depend on the heterogeneity of the circuit, and must be of a kind reversible with the current.

105.   Now if in an unbroken circuit with an engine driven by a thermo-electric current, the strength of the current be infinitely small, compared with what it would be were the engine held at rest, or, which is the same, if the engine be kept at some such speed that its inductive electromotive force may fall short of, or may exceed, by only an infinitely small fraction of itself, the amount required to balance the thermal electromotive force of the battery, there will be only an infinitely small fraction of the work done by the current in the former case, or of the work done in turning the engine in the latter, wasted on the frictional generation of heat through the electric circuit.   In these circumstances, it is clear, that whatever mechanical effect would be produced in any time by the engine from a direct current of a certain strength, an equal amount of work would have to be spent in forcing it to move faster and keeping up an equal reverse current for the same length of time; and as the direct and reverse currents would certainly produce equal and opposite thermal effects at the junctions, and elsewhere in all actions depending on heterogeneousness of the circuit, it appears that, were there no propagation of heat through the battery by ordinary conduction, Carnot's criterion of a perfect thermo-dynamic engine would be completely fulfilled, and a definite relation, the same as that which has been already investigated (§ 25) by considering expansive engines fulfilling the same criterion, would hold between the operative thermal agency and the mechanical effect produced.   It appears extremely probable that this relation does actually subsist between the *part of the thermal agency which is reversed with the current* and the mechanical effect produced by the engine, and that the ordinary conduction of heat through the battery takes place independently of the electrical circumstances.   The following proposition is therefore assumed as a fundamental hypothesis in the theory at present laid before the Royal Society.

106.   *The electromotive forces produced by inequalities of temperature in a circuit of different metals, and the thermal effects of electric currents circulating in it, are subject to the laws which*

*would follow from the general principles of the dynamical theory of
heat if there were no conduction of heat from one part of the circuit
to another.*

In adopting this hypothesis, it must be distinctly understood
that it is only a hypothesis, and that, however probable it may
appear, experimental evidence in the special phænomena of thermo-
electricity is quite necessary to prove it.   Not only are the conditions prescribed in the second law of the dynamical theory not
completely fulfilled, but the part of the agency which does fulfil
them, is in all known circumstances of thermo-electric currents excessively small in proportion to the agency inseparably accompanying it and essentially violating those conditions.   Thus, if the
current be of the full strength which the thermal electromotor
alone can sustain against the resistance in its circuit, the whole
mechanical energy of the thermo-electric action is at once spent
in generating heat in the conductor,—an essentially irreversible
process.   The whole thermal agency immediately concerned in the
current, even in this case when the current is at the strongest, is
(from all we know of the magnitude of the thermo-electric force
and absorptions and evolutions of heat) probably very small in comparison with the transference of heat from hot to cold by ordinary
conduction through the metal of the circuit.   It might be imagined,
that by choosing, for the circuit, materials which are good conductors of electricity and bad conductors of heat, we might diminish
indefinitely the effect of conduction in comparison with the
thermal effects of the current; but unfortunately we have no such
substance as a *non-conductor* of heat.   The metals which are the
worst conductors of heat are, nearly in the same proportion, the
worst conductors of electricity; and all other substances appear to
be comparatively very much worse conductors of electricity than
of heat; stones, glass, dry wood, and so on, being, as compared
with metals, nearly perfect non-conductors of electricity, and yet
possessing very considerable conducting powers for heat.   It is
true we may, as has been shown above, diminish without limit the
waste of energy by frictional generation of heat in the circuit, by
using an engine to do work and react against the thermal electro-
motive force; but, as we have also seen, this can only be done by
keeping the strength of the current very small compared with
what it would be if allowed to waste all the energy of the electro-

motive force on the frictional generation of heat, and it therefore
requires a very slow use of the thermo-electric action. At the
same time it does not in any degree restrain the dissipation of
energy by conduction, which is always going on, and which will
therefore bear even a much greater proportion to the thermal
agency electrically spent than in the case in which the latter was
supposed to be unrestrained by the operation of the engine. By
far the greater part of the heat taken in at all, then, in any
thermo-electric arrangement is essentially dissipated, and there
would be no violation of the great natural law expressed in Carnot's principle if the small part of the whole action, which is reversible, gave a different, even an enormously different, and either
a greater or a less, proportion of heat converted into work to heat
taken in than that law requires in all completely reversible processes. Still the reversible part of the agency, in the thermo-
electric circumstances we have supposed, is in itself so *perfect*,
that it appears in the highest degree probable it may be found to
fulfil independently the same conditions as the general law would
impose on it if it took place unaccompanied by any other thermal
or thermo-dynamic process.

§§ 107—111. *Mathematical expression of the Thermo-dynamic
circumstances of Currents in Linear Conductors.*

107. In a heterogeneous metallic conductor, the whole heat
developed in a given time will consist of a quantity generated
frictionally, increased or diminished by the quantities produced or
absorbed in the different parts by action depending on heterogeneity of the circuit. The former, according to the law discovered by Joule, may be represented by a term $B\gamma^2$, in which B
denotes a constant depending only on the resistance of the circuit.
The latter, being reversible with the current, may be assumed, at
least for infinitely feeble currents, to be, in a given conductor, proportional simply to the strength of the current; and hence the
whole quantity of heat evolved in a given time must be expressible
by a term of the form $-A\gamma$; where A, whether it varies with γ
or not, has a finite positive or negative value when γ is infinitely
small. Hence the whole heat developed in any portion of a
heterogeneous metallic conductor in a unit of time must be expressible by the formula

$$-A\gamma + B\gamma^2,$$

T.   16

where B is essentially positive; but A may be positive, negative, or zero, according to the nature of the different parts of the conducting arc. It may be assumed with great probability, that the quantities A and B are absolutely constant for a given conductor with its different parts at given constant temperatures; and that when the temperatures of the different parts of a conductor are kept as nearly constant as possible with currents of different strengths passing through it, the quantities A and B can only depend on γ, inasmuch as it may be impossible to prevent the interior parts of the conductor from varying in temperature, and so changing in their resistance to the conduction of electricity, or in their thermo-electric properties. In the present paper, accordingly, A and B are assumed to depend solely on the nature and thermal circumstances of the conductor, and to be independent of γ; but the investigations and conclusions would be applicable to cases of action with sufficiently feeble currents, probably to all currents due solely to the thermal electromotive force, even if A and B were in reality variable, provided the limiting values of these quantities for infinitely small values of γ be used.

108. Let us consider a conductor of any length and form, but of comparatively small transverse dimensions, composed of various metals at different temperatures, but having portions at its two extremities homogeneous and at the same temperature. These terminal portions will be denoted by E and E', and will be called the *principal electrodes*, or *the electrodes of the principal conductor*; the conductor itself being called the *principal conductor* to distinguish it from others, either joining its extremities or otherwise circumstanced, which we may have to consider again.

Let an electromotive force be made to act continuously and uniformly between these electrodes; as may be done, for instance, by means of a metallic disc included in the circuit touched by electrodes at its centre and a point of its circumference, and made to rotate between the poles of a powerful magnet, an arrangement equivalent to the "engine" spoken of above. Let the amount of this electromotive force be denoted by P, to be regarded as positive, when it tends to produce a current from E through the principal conductor, to E'. Let the absolute strength of the current, which in these circumstances passes through the principal conductor, be denoted by γ, to be considered as positive if in the direction of P when positive.

109.   Then $P\gamma$ will be the amount of work done by the electro-motive force in the unit of time.   As this work is spent wholly in keeping up a uniform electric current in the principal conductor, it must be equal to the mechanical equivalent of the heat gene-rated, since no other effect is produced by the current.   Hence if $-A\gamma + B\gamma^2$ be, in accordance with the preceding explanations, the expression for the heat developed in the conductor in the unit of time by the current γ, and if J, as formerly, denote the mechanical equivalent of the thermal unit, we have

$$P\gamma = J(-A\gamma + B\gamma^2)\dots\dots\dots\dots\dots\dots(3),$$

which is the expression for the particular circumstances of the first fundamental law of the dynamical theory of heat.

Hence, by dividing by γ, we have

$$P = J(-A + B\gamma)\dots\dots\dots\dots\dots\dots\dots(4),$$

from which we deduce

$$\gamma = \frac{P + JA}{JB}\dots\dots\dots\dots\dots\dots\dots\dots(5).$$

These equations show that, according as P is greater than, equal to, or less than $-JA$, the value of γ is positive, zero, or negative; and that, in any of the circumstances, the strength of the actual current is just the same as that of the current which an electromotive force equal to $P + JA$ would excite in a homo-geneous metallic conductor having JB for the absolute numerical measure of its galvanic resistance.   Hence we conclude:—

(1)   That in all cases in which the value of A is finite, there must be an intrinsic electromotive force in the principal conductor, which would itself produce a current if the electrodes E, E' were put in contact with one another, and which must be balanced by an equal and opposite force, JA, applied either by means of a perfect non-conductor, or some electromotor, placed between E and E', in order that there may be electrical equilibrium in the principal conductor.

And (2), That JB, which cannot vanish in any case, is the

absolute numerical measure of the galvanic resistance of the principal conductor itself*.

It appears, therefore, that the whole theory of thermo-electric force in linear conductors is reduced to a knowledge of the circumstances on which the value of the coefficient A, in the expression $-A\gamma + B\gamma^2$ for the heat developed throughout any given conductor, depends.

110.   To express the second general law, we must take into account the temperatures of the different localities of the circuit in which heat is evolved or absorbed, when the current is kept so feeble (by the action of the electromotive force P against the thermo-electric force of the system) as to render the frictional generation of heat insensible.   Denoting then by $\alpha_t\gamma$ the heat absorbed in all parts of the circuit which are at the temperature t, by the action of a current of infinitely small strength γ: so that the term $-A\gamma$, expressing the whole heat generated not frictionally throughout the principal conductor in any case, will be the sum of all such terms with their signs changed, or

$$A\gamma = \Sigma\alpha_t\gamma,$$

which gives

$$\Sigma\alpha_t = A \dots\dots\dots\dots\dots\dots\dots(6);$$

and denoting by F the value of the electromotive force required to balance the thermo-electric tendency; we have

$$F = J\Sigma\alpha_t \dots\dots\dots\dots\dots\dots\dots(7).$$

---

* This conclusion was first given by Joule in his first paper, which was communicated to the Royal Society, December 17, 1840, " On the Production of Heat by Voltaic Electricity " (see *Proceedings* of that date).   The paper was published in the *Philosophical Magazine*, vol. XIX. p. 260.   See also " On the Calorific Effects of Magneto-electricity, and the Mechanical Value of Heat," by the same author (*Phil. Mag.* vol. XXIII. 1843), where the principles of mechanical action in the electric generation of heat are more fully developed.

The conclusion stated in the text was also given by Helmholtz in his *Erhaltung der Kraft*, Berlin, 1847 (translated in Taylor's New Scientific Memoirs).   It was given by the author of the present paper with various numerical applications regarding the electromotive forces of electro-chemical arrangements and the resistances of metallic conductors in absolute units, in two papers in the *Philosophical Magazine*, December 1851, " On the Mechanical Theory of Electrolysis " [Art. LIII. below], and " On Applications of the Principle of Mechanical Effect to the Measurement of Electro-motive Forces, and of Galvanic Resistances, in Absolute Units." [Art. LIV. below.]

The second general law, as expressed above in equation (1), ap-plied to the present circumstances, gives immediadely

$$\Sigma \frac{\alpha t \gamma'}{t} = 0 \dots\dots\dots\dots\dots\dots(8);$$

or, since γ is the same for all terms of the sum,

$$\Sigma \frac{\alpha_t}{t} = 0 \dots\dots\dots\dots\dots\dots\dots(9).$$

111. Of these equations, (7) and (3), from which (7) is derived, involve no hypothesis whatever, but merely express the application of a great natural law—discovered by Joule for every case of thermal action, whether chemical, electrical, or mechanical—to the electrical circumstances of a solid linear conductor having in any way the property of experiencing reverse thermal effects from infinitely feeble currents in the two directions through it.   Equa-tion (9) expresses the hypothetical application of the second general law discussed above in § 106.   The two equations, (7) and (9), express all the information that can be derived from the general dynamical theory of heat, regarding the special thermal and elec-trical energies brought into action by inequalities of temperature, or by the independent excitation of a current in a solid linear conductor, whether crystalline or not.   The condition that the cir-cuit is to be linear, being merely one of convenience in the initial treatment of the subject, may of course be removed by supposing linear conductors to be put together so as to represent the circum-stances of a solid conductor of electricity, with any distribution of electric currents whatever through it; and we may therefore regard these two equations as the fundamental equations of the mechani-cal theory of thermo-electric currents.   To work out the theory for crystalline or non-crystalline conductors, it is necessary to consider all the conditions which determine the generation or absorption of heat in different parts of the circuit, whatever be the properties of the metals of which it is formed.   This we may now proceed to do; first for non-crystalline, and after that for crystalline metals.

§§ 112—124. *General Equations of Thermo-electric Currents in non-crystalline Linear Conductors.*

112. The only reversible thermal effect of electric currents which experiment has yet demonstrated, is that which Peltier has

discovered in the passage of electricity from one metal to another. Besides this, we may conceive that in one homogeneous metal formed into a conductor of varying section, different thermal effects may be produced by a current in any part, according as it passes in the direction in which the section increases, or in the contrary direction; and with greater probability we may suppose, that a current in a conductor of one metal unequally heated may produce different thermal effects, according as it passes from hot to cold, or from cold to hot. But Magnus has shown by careful experiments, that no application of heat can sustain a current in a circuit of one homogeneous non-crystalline metal, however varying in section; and from this it is easy to conclude, by equations (7) and (9), that there can be no reversible thermal effect due to the passage of a current between parts of a homogeneous metallic conductor having different sections. Now it is clear that no circumstances, except those which have just been mentioned, can possibly give rise to different thermal effects in any part of a linear conductor of the same or of different metals, uniformly or non-uniformly heated, provided none of them be crystalline; and we have therefore at present nothing in the sum Σa_t, besides the terms depending on the passage of electricity from one metal to another, which certainly exist, and terms which may possibly be discovered (§ 121 below), depending on its passage from hot to cold, or from cold to hot, in the same metal.

113.   Let the principal conductor consist of n different metals, in all $n+1$ parts, of which the first and last are of the same metal, and have their terminal portions (which we have called the electrodes E and E') at the same temperature T_0. Let T_1, T_2, T_3, &c. denote the temperatures of the different junctions in order, and let Π_1, Π_2, Π_3, &c. denote the amounts (positive or negative) of heat absorbed at them respectively by a positive current of unit strength during the unit of time. Let $\gamma\sigma_1 dt, \gamma\sigma_2 dt, \gamma\sigma_3 dt$, &c. denote the quantities of heat evolved in each of the different metals in the unit of time by a current of infinitely small strength, γ, passing from a locality at temperature $t + dt$ to a locality at temperature t. Without hypothesis, but by an obvious analogy, we may call the elements σ_1, σ_2, &c. the *specific heats of electricity in the different metals*, since they express the quantities of heat absorbed or evolved by the unit of current electricity in passing from cold to hot, or

from hot to cold, between localities differing by a degree of temperature in each metal respectively. It is easily shown (as will be
seen by the treatment of the subject to follow immediately) that
if the values of σ_1, σ_2, &c. depend either on the *section* of the
conductor, or on the rate of variation of temperature along it, or
on any other variable differing in different parts of the conductor,
except the temperature, a current might be maintained by the
application of heat to a homogeneous metallic conductor. We
may therefore at once assume them to be, if not invariable, absolute functions of the temperature. From this it follows, that if ϕt
denote any function of t, the value of the sum $\int \phi t \sigma dt$ for any
conducting arc of homogeneous metal depends only on the temperatures of its extremities, and therefore the parts of the sums $\Sigma \alpha_t$
and $\Sigma \alpha_t/t$ corresponding to the successive metals in the principal
conductor, are respectively

$$-\int_{T_1}^{T_0} \sigma_1 dt, \quad -\int_{T_2}^{T_1} \sigma_2 dt, \ldots \quad -\int_{T_n}^{T_{n-1}} \sigma_n dt, \quad -\int_{T_0}^{T_n} \sigma_1 dt,$$

and

$$-\int_{T_1}^{T_0} \frac{\sigma_1}{t} dt, \quad -\int_{T_2}^{T_1} \frac{\sigma_2}{t} dt \ldots \quad -\int_{T_n}^{T_{n-1}} \frac{\sigma_n}{t} dt, \quad -\int_{T_0}^{T_n} \frac{\sigma_1}{t} dt.$$

Hence the general equations (7) and (9) become

$$F = J \left\{ \Pi_1 + \Pi_2 + \ldots + \Pi_n - \int_{T_1}^{T_0} \sigma_1 dt - \int_{T_2}^{T_1} \sigma_2 dt - \ldots \right.$$
$$\left. - \int_{T_n}^{T_{n-1}} \sigma_n dt - \int_{T_n}^{T_n} \sigma_1 dt \right\} \ldots \ldots (10),$$

$$\frac{\Pi_1}{T_1} + \frac{\Pi_2}{T_2} + \ldots + \frac{\Pi_n}{T_n} - \int_{T_1}^{T_0} \frac{\sigma_1}{t} dt - \int_{T_2}^{T_1} \frac{\sigma_2}{t} dt - \ldots - \int_{T_n}^{T_{n-1}} \frac{\sigma_n}{t} dt$$
$$- \int_{T_0}^{T_n} \frac{\sigma_1}{t} dt = 0 \ldots \ldots \ldots (11),$$

which are the fundamental equations of thermo-electricity in non-
crystalline conductors. In these, along with the equation

$$\gamma = \frac{P+F}{JB} \ldots \ldots \ldots \ldots (12),$$

which shows the strength of the current actually sustained in the
conductor when an independent electromotive force, P, is applied
between the principal electrodes E, E', we have a full expression

of the most general circumstances of thermo-electric currents in linear conductors of non-crystalline metals.

114. The special qualities of the metals of a thermo-electric circuit must be investigated experimentally before we can fix the values of Π_1, Π_2, &c., and σ_1, σ_2, &c. for any particular case. The relation between these quantities expressed in the general equation (11) having, as we have seen, a very high degree of probability, not merely as an approximate law, but as an essential truth, may be used as a guide, but must be held provisionally until we have sufficient experimental evidence in its favour. The first fundamental equation (10) admits of no doubt whatever as to its universal application, and we shall see (§ 123*) that it leads to most remarkable conclusions from known experimental facts.

The general principles are most conveniently applied by restricting the number of metals referred to in the general equations to two; a case which we accordingly proceed to consider.

115. Let the principal conductor consist of two metals, one constituting the middle, and the other the two terminal portions. Let the junctions of these portions next the terminals E, E' be denoted by A, A' respectively, and let their temperatures be T, T'. Let also $\Pi(T)$, $-\Pi(T')$ be the quantities of heat absorbed at them per second by a current of unit strength. We should have

$$\Pi(T) = \Pi(T'),$$

if the temperatures were equal, since the Peltier phænomenon consists, as we have seen, of equal quantities of heat evolved or absorbed, according to the direction of a current crossing the junction of two different metals; and if these quantities be not actually equal, we may consider them as particular values of a function Π of the temperature, which depends on the particular relative thermo-electric quality of the two metals. Accordingly, the preceding notation is reduced to $n = 2$, $T_1 = T$, $T_2 = T'$, $\Pi_1 = \Pi(T)$, $\Pi_2 = -\Pi(T')$; and we have

$$\int_{T_1}^{T_0} \sigma_1 dt + \int_{T_2}^{T_1} \sigma_2 dt + \int_{T_0}^{T_2} \sigma_1 dt = \int_T^{T'} (\sigma_1 - \sigma_2) dt,$$

* See *Proceedings of the Royal Society*, May 4, 1854; *Phil. Mag.* July 1854, p. 63 [Art. LI. below].

and similarly for the integral involving $1/t$.   Hence the general equations become

$$F = J\left\{ \Pi(T) - \Pi(T') + \int_{T'}^{T} (\sigma_1 - \sigma_2)\, dt \right\} \dots\dots\dots(13),$$

$$\frac{\Pi(T)}{T} - \frac{\Pi(T')}{T'} + \int_{T'}^{T} \frac{\sigma_1 - \sigma_2}{t}\, dt = 0 \dots\dots\dots\dots(14).$$

If in the latter equation we substitute t for T, and differentiate with reference to this variable, we have as an equivalent equation,

$$\frac{d(\Pi/t)}{dt} + \frac{\sigma_1 - \sigma_2}{t} = 0 \dots\dots\dots\dots\dots(15),$$

or

$$\sigma_1 - \sigma_2 = \frac{\Pi}{t} - \frac{d\Pi}{dt} \dots\dots\dots\dots\dots(16).$$

This last equation leads to a remarkably simple expression for the electromotive force of a thermo-electric pair, solely in the terms of the Peltier evolution of heat at any temperature intermediate between the temperatures of its junctions; for we have only to eliminate by means of it $(\sigma_1 - \sigma_2)$ from (13), to find

$$F = J \int_{T'}^{T} \frac{\Pi}{t}\, dt \dots\dots\dots\dots\dots(17).$$

116.   Let us first apply these equations to the case of a thermo-electric pair, with the two junctions kept at temperatures differing by an infinitely small amount τ.   In this case we have

$$\Pi(T) - \Pi(T') = \frac{d\Pi}{dt}\, \tau,$$

$$\int_{T'}^{T} (\sigma_1 - \sigma_2)\, dt = (\sigma_1 - \sigma_2)\tau;$$

and equation (13) becomes

$$F = J\left\{ \frac{d\Pi}{dt} + \sigma_1 - \sigma_2 \right\} \tau \dots\dots\dots\dots(18).$$

If we make use of (16) in this, we have

$$F = J \frac{\Pi}{t}\, \tau \dots\dots\dots\dots\dots(19).$$

The first of these expressions for the electromotive force involves no hypothesis, but only the general principle of equivalence of heat and work.   Its agreement with any experimental results is only to be looked on as a verification of the accuracy of the experiments, and can add nothing to the certainty of the part of the theory from which it is deduced. On the other hand, it would be extremely important to test the second expression (19) by direct experiment, and so confirm or correct the only doubtful part of the theory. The way to do so would be to determine in absolute measure the electromotive force, F, due to a small difference of temperature, τ, in any thermo-electric pair, and to determine, in known thermal units, the amount of the Peltier effect at a junction of the two metals with a current of strength measured in electrodynamic units, as we should then, by these determinations, be able from direct experiments to find the values of the two members separately which appear equated in (19). As yet no observations have been made which lead directly or indirectly to the evaluation of the second member of (19) in any case, but I hope before long to succeed in carrying out a plan I have formed for this object. Neither have any observations been made yet which give in any case a determination of the first member; but they may easily be accomplished by any person who possesses a conductor of which the resistance has been determined in absolute measure. Mr Joule having kindly put me in possession of the silver wire on which his observations of the electrical generation of heat, in 1845, were made with currents measured by a tangent galvanometer used by him about the same time in experimenting on the electrolysis of sulphate of copper and sulphate of zinc, I hope to be able to complete the test of the theoretical result without difficulty, in any case in which I may succeed in determining the amount of the Peltier thermal effect.

117. In the mean time it is interesting to form an estimate, however rough, of the absolute values of the thermo-electric elements in any case in which observations that have been made afford, directly or indirectly, the requisite data. This I have done for copper and bismuth, and copper and iron, in the manner shown in the following explanation, which was communicated in full to the Royal Society of Edinburgh when the theory was first brought forward in 1851, although only the part enclosed in double quotation marks was printed in the *Proceedings*.

118. Example 1. *Copper and Bismuth.*—"'Failing direct data, the absolute value of the electromotive force in an element

of copper and bismuth, with its two junctions kept at the temperatures 0° and 100° Cent., may be estimated indirectly from Pouillet's
comparison of the strength of the current it sends through a copper
wire 20ᵐ long and 1 millim. in diameter, with the strength of a
current decomposing water at an observed rate, by means of the
determinations by Weber and others, of the specific resistance of
copper and the electro-chemical equivalent of water, in absolute
units. The specific resistances of different specimens of copper
having been found to differ considerably from one another, it is
impossible, without experiments on the individual wire used by
M. Pouillet, to determine with much accuracy the absolute resistance of his circuit; but the author has estimated it on the hypothesis that the specific resistance of its substance is $2\frac{1}{4}$ British
units. Taking ·02 as the electro-chemical equivalent of water in
British absolute units, the author has thus found 16,300 as the
electromotive force of an element of copper and bismuth, with the
two junctions at 0° and 100° respectively. About 154 of such elements would be required to produce the same electromotive force
as a single cell of Daniell's—if, in Daniell's battery, the whole
chemical action were electrically efficient[*]. A battery of 1000
copper and bismuth elements, with the two sets of junctions at 0°
and 100° C., employed to work a galvanic engine, if the resistance
in the whole circuit be equivalent to that of a copper wire of about
100 feet long and about one-eighth of an inch in diameter, and if
the engine be allowed to move at such a rate as by inductive
reaction to diminish the strength of the current to the half of
what it is when the engine is at rest, would produce mechanical
effect at the rate of about one-fifth of a horse-power. The electro-
motive force of a copper and bismuth element, with its two
junctions at 0° and 1°, being found by Pouillet to be about $\frac{1}{100}$ of

---

[*] M. Jules Regnauld has since found experimentally, that 165 copper-bismuth
elements balance the electromotive force of a single cell of Daniell's (see *Comptes
Rendus*, Jan. 9, 1854, or *Bibliothèque Universelle de Genève*, March 1854), a result
agreeing with the estimate quoted in the text more closely than the uncertainty and
indirectness of the data on which that estimate was founded would have justified us
in expecting. The comparison of course affords no test of the thermo-electric
theory; and only shows that, as far as the observations of Weber and others
alluded to render Pouillet's available for determining the absolute electromotive
force of a copper-bismuth element, the absolute electromotive force of a single cell
of Daniell's, obtained by multiplying it by the number found by M. Regnauld,
agrees with that which I first gave on the hypothesis of all the chemical action
being electrically efficient (*Phil. Mag.* Dec. 1851), and so confirms this hypothesis.

the electromotive force when the junctions are at 0° and 100, must be about 163. The value of Θ_0''' [*i.e.* in terms of the notation now used, $\Pi(273\cdot7)$, or the value of $\Pi(t)$, for the freezing-point] "'for copper and bismuth, or the quantity of heat absorbed in a second of time by a current of unit strength in passing from bismuth to copper, when the temperature is kept at 0°C., must therefore be 163/160·16, or very nearly equal to the quantity required to raise the temperature of a grain of water from 0° to 1°C.'"

119.   Example 2.   *Copper and Iron.*—"By directing the electromotive force of one copper and bismuth element against that of a thermo-electric battery of a variable number of copper and iron wire elements in one circuit, I have found, by a galvanometer included in the same circuit, that when the range of temperature in all the thermo-electric elements is the same, and not very far at either limit from the freezing-point of water, the current passes in the direction of the copper-bismuth agency when only three, and in the contrary direction when four or more of the copper-iron elements are opposed to it. Hence the electromotive force of a copper-bismuth element is between three and four times that of a copper-iron element with the same range of temperature, a little above the freezing-point of water. The electromotive force of a copper-iron element, with its two junctions at 0° and 1°C. respectively, must therefore be something greater than one-fourth of the number found above for copper-bismuth with the same range of temperature, that is, something more than forty British absolute units, and we may consequently represent it by $m \times 40$, where $m > 1$. We have then by the equation expressing the application of Carnot's principle [equation (19) of § 116],

$$\Theta_0\mu = \Theta_0\frac{J}{273\cdot7} = m \times 40,$$

whence*

$$\Theta_0 = \tfrac{1}{4}m \text{ nearly} \dots\dots\dots\dots\dots\dots(a).$$

---

* The value of J now used being $32\cdot2 \times 1390 = 44{,}758$, which is the equivalent of the unit of heat in "absolute units" of work. The "absolute unit of force" on which this unit of work is founded, and which is generally used in magnetic and electro-magnetic expressions, is the force which acting on the unit of matter (one grain) during the unit of time (one second), generates a unit of velocity (one foot per second). The "absolute unit of work" is the work done by the absolute unit of force in acting through the unit of space (one foot).

"Now, by the principle of mechanical effect, we have

$$F_0^{280} = J\left(\int_0^{280} \vartheta dt - \Theta_0\right);$$

if F_0^{280} denote the electromotive force of a copper-iron element of which the two junctions are respectively $0°$ and $280°$ C., and ϑdt the quantity of heat absorbed per second by a current of unit strength, in passing in copper from a locality at temperature t to a locality at $t + dt$, and in iron from a locality at $t + dt$ to a locality at t [*]; since the Peltier generation of heat between copper and iron at their neutral point, $280°$, vanishes[†], and therefore the only absorption of heat is that due to the electric convection expressed by $\int \vartheta dt$; while there is evolution of heat amounting to Θ_0 at the cold junction, and of mechanical effect by the current amounting to F units of work. If we estimate the value of F_0^{280} as half what it would be were the electromotive force the same for all equal differences of temperature as for small differences near the freezing-point[‡], that is, if we take $F_0^{280} = \frac{1}{2} \times 40m \times 280$, the preceding equation becomes

$$140 \times m \times 40 = J\left(\int_0^{280} \vartheta dt - \Theta_0\right).$$

But we found

$$m \times 40 = \mu\Theta_0.$$

Hence

$$\int_0^{280} \vartheta dt = \Theta_0\left(1 + 140\frac{\mu}{J}\right) = \Theta_0(1 + \tfrac{140}{2727}) = \tfrac{3}{2}\Theta_0 \text{ nearly...}(b);$$

or, according to (a),

$$\int_0^{280} \vartheta dt = \tfrac{3}{8}m \dots\dots\dots\dots(c);$$

results, of which (b) shows how the difference of the aggregate amount of the theoretically indicated convective effect in the two metals is related to the Peltier effect at the cold junction; and (c) shows that its absolute value is rather more than one-third of a thermal unit per second per unit strength of current.

* That is, if ϑ denote the algebraic excess of the specific heat of electricity in copper, above the specific heat of electricity in iron, according to the terms more recently introduced.
† See *Proceedings of the Royal Society*, May, 1854, *Phil. Mag.*, July, 1854. Instead of $240°$ conjectured from Regnauld's observation when these details were first published, $280°$ is now taken as a closer approximation to the neutral point of copper and iron.
‡ See *Phil. Mag.* for July 1854, p. 63.

120   If the specific heats of current electricity either vanished or were equal in the different metals, we should have by (15) and (16),

$$\frac{\Pi}{t} = \text{constant} \dots\dots\dots\dots\dots\dots\dots(20),$$

and
$$F = J\frac{\Pi}{t}(T - T') \dots\dots\dots\dots\dots\dots(21),$$

or the Peltier thermal effect at a junction of two metals would be proportional to the absolute temperature at which it takes place, and the electromotive force in a circuit of any two metals would vary in the simple ratio of the difference of temperature on the new absolute scale between their junctions[*]. Whatever thermometric system be followed, the second of these conclusions would require the same law of variation of electromotive force with the temperatures of the junctions in every pair of metals used as a thermo-electric element.

121.   Before the existence of a convective effect of electricity in an unequally heated metal had even been conjectured, I arrived at the preceding conclusions by a theory in which the Peltier effect was taken as the only thermal effect reversible with the current in a thermo-electric circuit, and found them at variance with known facts which show remarkably different laws of electromotive force in thermo-electric pairs of different metals. I therefore inferred, that, besides the Peltier effect, there must be other reversible thermal effects; and I showed that these can be due to no other cause than the inequalities of temperature in single metals in the circuit. A convective effect of electricity in an unequally heated conductor of one metal was thus first demonstrated by theoretical reasoning; but only the difference of the amount of this effect produced by currents of equal strength in different metals, not its quality or its absolute value in any one metal, could be inferred from the data of thermo-electric force alone. The case of a thermo-electric circuit of copper and iron, being that

---

[*] When the theory was first communicated to the Royal Society of Edinburgh, I stated these conclusions with reference to temperature by the air thermometer, and therefore in terms of Carnot's absolute function of the temperature, not simply, as now, in terms of absolute temperature. At the same time I gave, as consequences of Mayer's hypothesis, the same statement in terms of air-thermometer temperatures, as is now made absolutely. See *Proceedings*, Dec. 15, 1851; or *Philosophical Magazine*, 1852, first half-year [Note 1 appended to Art. xlviii., Part VII., below].

which first forced on me the conclusion that an electric current must produce different effects according as it passes from hot to cold, or from cold to hot, in an unequally heated metal, was taken as an example in my first communication of the theory to this Society*; and the two metals, copper and iron, were made the subjects of a consequent experimental investigation, to ascertain the quality of the anticipated property in each of them separately. The application of the general reasoning to this particular case, and the answers which I have derived by experiment to the question which it raises, are described in the following extract from a Report communicated to the Royal Society of London, March 31, and published in the *Proceedings* for May 4, 1854, and in the *Philosophical Magazine* for July, 1854." [This extract, which constituted §§ 122—133 of "Dynamical Theory of Heat," is not reprinted here because the paper from which it was taken is reprinted in full, as the first Part of Art. LI. below.]

§§ 134, 135. *Inserted September* 15, 1854.

134. A continuation of the experiments has shown many remarkable variations of order in the thermo-electric series. The following Table exhibits the results of observations to determine neutral points for different pairs of metals; the number at the head of each column being the temperature Centigrade at which the two metals written below it are thermo-electrically neutral to one another; and the lower metal in each column being that which *passes the other from bismuth towards antimony* as the temperature rises.

14^0 C.	$-12^0{\cdot}2.$	$-1^0{\cdot}5.$	$8^0{\cdot}2.$	$36^0.$	$38^0.$	$44^0.$	$44^0.$	$64^0.$	$99^0.$	$121^0.$	$130^0.$	$162^0{\cdot}5.$	$237^0.$	$280^0.$
	P_1	P_1	P_1	P_2	P_2	P_2	Lead	P_1	P_1	P_1	P_1	Iron	Iron	Iron
Brass	Cadmium	Silver	Zinc	Lead	Brass	Tin	Brass	Copper	Brass	Lead	Tin	Cadmium	Silver	Copper

I also found that brass becomes neutral to copper, and copper becomes neutral to silver, at some high temperatures, estimated at from 800^0 to 1400^0 Cent. in the former case, and from 700^0 to 1000^0 in the latter, being a little below the melting-point of silver. The following diagram exhibits the results graphically, constructed on the principle of drawing a line through the letters corresponding

---

* See *Proceedings of the Royal Society of Edinburgh*, Dec. 15, 1851; or *Phil. Mag.* 1852, first half-year [reprinted as Note 1 appended to Art. XLVIII., Part VII., below].

to any one of the metallic specimens in a table such as that of § 130*, and arranging the spaces so that each line shall be as nearly straight as possible, if not exactly so.

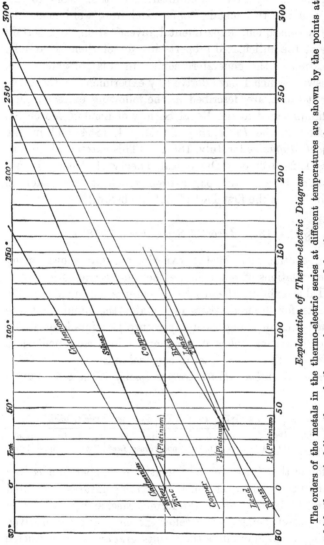

Explanation of Thermo-electric Diagram.

The orders of the metals in the thermo-electric series at different temperatures are shown by the points at which the vertical lines through the numbers expressed by the temperatures Centigrade are cut by the horizontal and oblique lines named for the different metallic specimens.

P_1, P_2, P_3 denote three platinum wires found to differ thermo-electrically, and used as standards.

The object to be aimed at in perfecting a thermo-electric diagram, and perhaps approximately attained to (conjecturally) in the

* *Phil. Mag.* for July 1854, p. 67 [Art. LI. below].

preceding, is to make the ordinates of the lines (which will in general be curves) corresponding to the different metallic specimens, be exactly proportional to their *thermo-electric powers**, with reference to a standard metal (P_3 in the actual diagram).

135. Judging by the eye from the diagram, as regards the convective agency of electricity in unequally heated conductors, I infer that the different metals are probably to be ranked as follows, in order of the values of the specific heat of electricity in them :—

Specific Heat of Vitreous Electricity.

In   Cadmium .... Positive.

...   Brass ....

...   Copper ....

... $\left\{\begin{array}{l}\text{Lead, Tin,}\\ \text{Silver,}\end{array}\right\}$ .... Positive, zero, or negative.

...   Platinum .... Probably negative.

...   Iron .... Negative.

Zinc probably stands high, certainly above platinum.

136. A very close analogy subsists between the thermo-dynamical circumstances of an electrical current in a circuit of two metals, and those of a fluid circulating in a closed rectangular tube, consisting of two vertical branches connected by two horizontal branches. Thus if, by the application of electromotive force in one case, or by the action of pistons in the other, a current be instituted, and if at the same time the temperature be kept uniform throughout the circuit, heat will be evolved and absorbed at the two junctions respectively in the former case, and heat will be evolved in one and absorbed in the other of the vertical branches of the tube in the latter case, in consequence of the variations of pressure experienced by the fluid in moving through those parts of the circuit. If the temperature of one junction of the electrical circuit be raised above that of the other, and if the temperature of one vertical branch of the tube containing fluid be raised above that of the other, a current will in each case be occasioned without any other motive appliance. If the current be directed to do work with all its energy, by means of an engine in each case, there will be a conversion of heat into mechanical effect, with perfectly ana-

---

* See § 140, below.

logous relations as to absorption and evolution of heat in different
parts of the circuit, provided the engine worked by the fluid current
be arranged to pass the fluid through it, from or to either of the
vertical branches of the tube, without variation of temperature. If
σ_1 and σ_2 denote the specific heat of unity of mass of the fluid
under the constant pressures at which it exists in the lower and
upper horizontal branches of the tube in the second case; $\Pi(T)$,
$\Pi(T')$ the quantities of heat evolved and absorbed respectively by
the passage of a unit mass of fluid through the two vertical
branches kept at the respective temperatures T, T'; and if F
denote the work done by a unit mass of the fluid in passing
through the engine; the fundamental equations obtained above
with reference to the thermo-electric circumstances, may be at once
written down for the case of the ordinary fluid as the expression
of the two fundamental laws of the dynamical theory of heat, both
of which are applicable to this case, without any uncertainty such
as that shown to be conceivable as regards the application of the
second law to the case of a thermo-electric current.   The two
equations thus obtained are equivalent to the two general equations
given in §§ 20 and 21 of the first part of this series of papers, as
the expressions of the fundamental laws of the dynamical theory
of heat applied to the elasticity and expansive properties of fluids.
In fact, when we suppose the ranges of both temperature and
pressure in the circulating fluid to be infinitely small, the equation
$F = J \int_{T'}^{T} \frac{\Pi}{t}\, dt$, reduced to the notation formerly used, and modified
by changing the independent variables from (t, p) to (t, v), becomes

$$M = \frac{t}{J} \frac{dp}{dt},$$

which is the same as (3) of § 21; and a combination of this with

$$\frac{d}{dt}\left(\frac{\Pi}{t}\right) = \frac{\sigma_2 - \sigma_1}{t}$$

gives

$$\frac{dM}{dt} - \frac{dN}{dv} = \frac{1}{J}\frac{dp}{dt},$$

which is identical with (2) of § 20. It appears, then, that the
consideration of the case of fluid motion here brought forward as
analogous to thermo-electric currents in non-crystalline linear con-

ductors, is sufficient for establishing the general thermo-dynamical equations of fluids; and consequently the universal relations [XLVIII. Part VII. below] among specific heats, elasticities, and thermal effects of condensation or rarefaction, derived from them in Part III., are all included in the investigation at present indicated. Not going into the details of this investigation, because the former investigation, which is on the whole more convenient, is fully given in Parts I. and III., I shall merely point out a special application of it to the case of a liquid which has a temperature of maximum density, as for instance water.

137.   In the first place, it is to be remarked that if the two vertical branches be kept at temperatures a little above and below the point of maximum density, no current will be produced; and therefore if T_0 denote this temperature, the equation $F = \int \dfrac{\Pi}{t}\, dt$ gives $\Pi(T_0) = 0$. Again, if one of the vertical branches be kept at T_0, and the other be kept at a temperature either higher or lower, a current will set, and always in the same direction. Hence $\int_{T_0}^{T} \dfrac{\Pi}{t}\, dt$ has the same sign, whether T be greater or less than T_0 and consequently $\Pi(t)$ must have contrary signs for values of above and below T_0; which, by attending to the signs in the general formulæ, we see must be such as to express evolution of heat by the actual current in the second vertical branch when its temperature is below, and absorption when above T_0. As the current in each case ascends in this vertical branch, we conclude that a slight diminution of pressure causes evolution or absorption of heat in water, according as its temperature is below or above that of maximum density; or conversely, that when water is suddenly compressed, it becomes colder if initially below, or warmer if initially above, its temperature of maximum density. This conclusion from general thermo-dynamic principles was first, so far as I know, mentioned along with the description of an experiment to prove the lowering of the freezing-point of water by pressure communicated to the Royal Society of Edinburgh in January 1850 * [Art. XLV. above]. The quantitative expression for the effect, which was given in § 50 of Part III., may be derived with ease from the

---

* See *Proceedings* of that date, or *Philosophical Magazine*, Aug. 1850.

considerations now brought forward. The other thermo-dynamic equation

$$\frac{\sigma_2 - \sigma_1}{t} = \frac{d}{dt}\left(\frac{\Pi}{t}\right)$$

shows that the specific heat of the water must be greater in the upper horizontal branch than in the lower, or that the specific heat of water under constant pressure is increased by a diminution of the pressure. The same conclusion, and the amount of the effect, are also implied in equations (18) and (19) of Part III. We may arrive at it without referring to any of the mathematical formulæ, merely by an application of the general principle of mechanical effect, when once the conclusion regarding the thermal effects of condensation or rarefaction is established; exactly as the conclusion regarding the specific heats of electricity in copper and in iron was first arrived at*. For if we suppose one vertical branch to be kept at the temperature of maximum density (corresponding to the neutral point of the metals in the corresponding thermo-electric case), and the other at some lower temperature, a current will set downwards through the former branch, and upwards through the latter. This current will cause evolution of heat, in consequence of the expansion of the fluid, in the branch through which it rises, but will cause neither absorption nor evolution in the other vertical branch, since in it the temperature is that of the maximum density. There will also be heat generated in various parts by fluid friction. There must then be, on the whole, absorption of heat in the horizontal branches, because otherwise there would be no source of energy for the heat constantly evolved to be drawn from. But heat will be evolved by the fluid in passing in the lower horizontal branch from hot to cold; and therefore, exactly to the extent of the heat otherwise evolved, this must be over-compensated by the heat absorbed in the upper horizontal branch by the fluid passing from cold to hot. On the other hand, if one of the vertical branches be kept above the temperature of maximum density and the other at this point, the fluid will sink in the latter, causing neither absorption nor evolution of heat, and rise in the former, causing absorption; and therefore more heat must be evolved by the fluid passing from hot to cold in the upper horizontal branch than is absorbed by it in

* *Proceedings of the Royal Society of Edinburgh*, Dec. 15, 1851; or extract of *Proceedings of the Royal Society*, May 1854.

passing from cold to hot in the lower. From either case we infer that the specific heat of the water is greater in the upper than in the lower branch. The analogy with the thermo-electric circumstances of two metals which have a neutral point is perfect algebraically in all particulars. The proposition just enunciated corresponds exactly to the conclusion arrived at formerly, that if one metal passes another in the direction from bismuth towards antimony in the thermo-electric scale, the specific heat of electricity is greater in the former metal than in the latter; this statement holding algebraically, even in such a case as that of copper and iron, where the specific heats are of contrary sign in the two metals, although the existence of such contrary effects is enough to show how difficult it is to conceive the physical circumstances of an electric current as physically analogous to those of a current of fluid in one direction.

§§ 138—140. *General Lemma, regarding relative thermo-electric properties of Metals, and multiple combinations in a Linear Circuit.*

138. The general equation (11), investigated above, shows that *the aggregate amount of all the thermal effects produced by a current, or by any system of currents, in any solid conductor or combination of solid conductors must be zero, if all the localities in which they are produced are kept at the same temperature.*

COR. 1. If in any circuit of solid conductors the temperature be uniform from a point P through all the conducting matter to a point Q, both the aggregate thermal actions and the electro-motive force are totally independent of this intermediate matter, whether it be homogeneous or heterogeneous, crystalline or non-crystalline, linear or solid, and is the same as if P and Q were put in contact. The importance of this simple and elementary truth in thermo-electric experiments of various kinds is very obvious. It appears to have been overlooked by many experimenters, who have scrupulously avoided introducing extraneous matter (as solder) in making thermo-electric junctions, and who have attempted to explain away Cumming's and Becquerel's remarkable discovery of thermo-electric inversions, by referring the phænomena observed to coatings of oxide formed on the metals at their surfaces of contact.

COR. 2. If $\Pi(A, B)$, $\Pi(B, C)$, $\Pi(C, D)$,... $\Pi(Z, A)$ denote the amounts of the Peltier absorption of heat per unit of strength of current per unit of time, at the successive junctions of a circuit of metals, A, B, C,...Z, A, we must have

$$\Pi(A, B) + \Pi(B, C) + ... + \Pi(Z, A) = 0.$$

Thus if the circuit consist of three metals,

$$\Pi(A, B) + \Pi(B, C) + \Pi(C, A) = 0;$$

from which, since $\Pi(C, A) = -\Pi(A, C)$, we derive

$$\Pi(B, C) = \Pi(A, C) - \Pi(A, B).$$

139. Now, by (19) above, the electromotive force in an element of the two metals (A, B), tending from B to A through the hot junction, for an infinitely small difference of temperature τ, and a mean absolute temperature t, is $\dfrac{J\Pi(A, B)}{t}\,\tau$, and so for every other pair of metals. Hence, if $\phi(A, B)$, $\phi(B, C)$, &c. denote the quantities by which the infinitely small range τ must be multiplied to get the electromotive forces of elements composed of successive pairs of the metals in the same thermal circumstances, we have

$$\phi(A, B) + \phi(B, C) + ... + \phi(Z, A) = 0;$$

and for the case of three metals,

$$\phi(B, C) = \phi(A, C) - \phi(A, B).$$

Since the thermo-electric force for any range of temperature is the sum of the thermo-electric forces for all the infinitely small ranges into which we may divide the whole range (being, as proved above, equal to $\int_{T}^{T'} \phi\,dt$), in the case of each element, the theorem expressed by these equations is true of the thermo-electric forces in the single elements for *all ranges* of temperature, provided the absolute temperatures of the hot and cold junctions be the same in the different elements. The second equation, by successive applications of which the first may be derived, is the simplest expression of a theorem which was, I believe, first pointed out and experimentally verified by Becquerel in researches described in the second volume of his *Traité d'Electricité*.

140. For brevity, we shall call what has been denoted by $\phi(B, C)$ *the thermo-electric relation* of the metal B to the metal C;

we shall call a certain metal (perhaps copper or silver) the standard metal; and if A be the standard metal, we shall call $\phi(A, B)$ the *thermo-electric power of the metal B.* The theorem expressed by the last equation may now be stated thus: *The thermo-electric relation between two metals is equal to the difference of their thermo-electric powers,* which is nearly identical with Becquerel's own statement of his theorem.

§§ 141—146.　*Elementary Explanations in Electro-kinematics and Electro-mechanics.*

141.　When we confined our attention to electric currents flowing along linear conductors, it was only necessary to consider in each case the *whole strength of the current,* and the longitudinal electromotive force in any part of the circuit, without taking into account any of the transverse dimensions of the conducting channel. In what follows, it will be frequently necessary to consider distributions of currents in various directions through solid conductors, and it is therefore convenient at present to notice some elementary properties, and to define various terms, adapted for specifications of systems of electric currents and electromotive forces distributed in any manner whatever throughout a solid.

142.　It is to be remarked, in the first place, that any portion of a solid traversed by current electricity may be divided, by tubular surfaces coinciding with lines of electric motion, into an infinite number of channels or conducting arcs, each containing an independent linear current. The *strength* of a linear current being, as before, defined to denote the quantity of electricity flowing across any section in the unit of time, we may now define the *intensity of the current* at any point of a conductor as the strength of a linear current of infinitely small transverse dimensions through this point, divided by the area of a normal section of its channel. The elementary proposition of the composition of motions, common to the kinematics of ordinary fluids and of electricity, shows that the superposition of two systems of currents in a body gives a resultant system, of which the intensity and direction at any point are represented by the diagonal of a parallelogram described upon lines representing the intensity and direction of the component systems respectively. Hence we may define the components, along

three lines at right angles to one another, of the intensity of
electric current through any point of a body, as the products of
the intensity of the current at that point into the cosines of the
inclination of its direction to those three lines respectively; and
we may regard the specification of a distribution of currents
through a body as complete, when the components parallel to
three fixed rectangular axes of reference of the intensity of the
current at every point are given.

143.   The term electromotive force has been applied in what
precedes, consistently with the ordinary usage, to the whole force
urging electricity through a linear conducting arc.   When a current
is sustained through a conducting arc by energy proceeding from
sources belonging entirely to the remainder of the circuit, the
electromotive force may be considered as applied from without to
its extremities; and in all such cases it may be measured—electro-
statically, by determining in any way the difference of potential
between two conducting bodies insulated from one another and
put in metallic communication with the extremities of the conducting arc;—or electro-dynamically, by applying to these points
the extremities of another linear conductor of infinitely greater
resistance (practically, for instance, a long fine wire used as a
galvanometer coil), and determining the strength of the current
which branches into it when it is so applied.   These tests may of
course be regarded as giving either the amount of the electromotive
force with which the remainder of the circuit acts on, or the
whole of the electromotive force efficient in, the passive conducting
arc first considered.   On the other hand, the electromotive force
acting in the portion from which the energy proceeds is not itself
determined by such tests, but is equal to the whole electromotive
force of the sources contained in it, diminished by the reaction of
the force which is measured in the manner just explained.   The
same tests applied to any two points whatever of a complete conducting circuit, however the sources of energy are distributed
through it, show simply the electromotive force acting and reacting
between the two parts into which the circuit might be separated
by breaking it at these points.   In some cases, for instance some
cases of thermo-electric action which we shall have to consider*,
these tests would give a zero indication to whatever two points of

---

* For one of these see § 167 below.

a circuit through which a current is actually passing they are applied, and would therefore show that there is no electric action and reaction between different parts of the circuit, but that each part contains intrinsically the electromotive force required to sustain the current through it at the existing rate. An actual test of the electromotive force of sources contained in any part of a linear conductor is defined, with especial reference to the circumstances of thermo-electricity, in the following statement :—

144. DEF. The actual intrinsic electromotive force in any part of a linear conducting circuit is the difference of potential which it produces in two insulated conductors of a standard metal at one temperature, when its extremities are connected with them by conducting arcs of the same metal, and are insulated from the remainder of the circuit.

The electromotive force so defined may be determined, either by determining by some electrostatical method the difference of potentials in the two conductors of standard metal mentioned in the definition, or by measuring the strength of the current produced in a conducting arc of the standard metal of infinitely greater resistance than the given conducting arc, applied to connect its extremities when insulated from the remainder of its own circuit.

145. With reference to the distribution of electromotive force through a solid, the following definitions are laid down :—

DEF. 1. The intrinsic electromotive force of a linear conductor at any point is the actual intrinsic electromotive force in an infinitely small arc through this point divided by its length.

DEF. 2. The efficient electromotive force at any point of a linear conducting circuit is the sum of the actual intrinsic electromotive force in an infinitely small arc, and the electromotive force produced by the remainder of the circuit on its extremities, divided by its length.

DEF. 3. The intrinsic electromotive force in any direction, at any point in a solid, is the electromotive force that would be experienced by an infinitely thin conducting arc of standard metal, applied with its extremities to two points in a line with this direction, in an infinitely small portion insulated all round from the rest of the solid, divided by the distance between these points.

DEF. 4. The electromotive force efficient at any point of a solid, in any direction, is the difference of the electromotive forces that would be experienced by an infinitely thin conducting arc of standard metal, with its extremities applied to two points infinitely near one another in this direction, divided by the distance between the points, in the two cases separately of the solid being left unchanged, and of an infinitely small portion of it containing these points being insulated from the remainder.

146.   Principle of the superposition of thermo-electric action. It may be assumed as an axiom, that each of any number of coexisting systems of electric currents produces the same reversible thermal effect in any locality as if it existed alone.

§§ 147—155.   *On Thermo-electric Currents in Linear Conductors of Crystalline Substance.*

147.   The general characteristic of crystalline matter is, that physical agencies, having particular directions in the space through which they act, and depending on particular qualities of the substance occupying that space, take place with different intensities in different directions if the substance be crystalline.   Substances not naturally crystalline may have the crystalline characteristic induced in them by the action of some directional agency, such as mechanical strain or magnetization, and may be said to be inductively crystalline.   Or again, minute fragments of non-crystalline substances may be put together so as to constitute solids, which on a large scale possess the general characteristic of homogeneous crystalline substances; and such bodies may be said to possess the crystalline characteristic by structure, or to be structurally crystalline.

148.   As regards thermo-electric currents, the characteristic of crystalline substance must be, that bars cut from it in different directions would, when treated thermo-electrically as linear conductors, be found in different positions in the thermo-electric series; or that two bars cut from different directions in the substance would be thermo-electrically related to one another like different metals.   This property has been experimentally demonstrated by Svanberg for crystals of bismuth and antimony; and there can be no doubt but that other natural metallic crystals will be found to possess it.   I have myself observed, that the thermo-electric properties of copper and iron wires are affected by alternate

tension and relaxation in such a manner as to leave no doubt but that a mass of either metal, when compressed or extended in one direction, possesses different thermo-electric relations in different directions. Fragments of different metals may be put together so as to form solids, possessing by structure the thermo-electric characteristic of a crystal, in an infinite variety of ways. Thus, a structure consisting of thin layers alternately of two different metals, possesses obviously the thermo-electric qualities of a crystal with an axis of symmetry. I have investigated the thermo-electric properties in all directions of such a structure in terms of the conducting powers for heat and electricity, and the thermo-electric powers, of the two metals of which it is composed; and bars made up of alternate layers of copper and iron, one with the layers perpendicular, another with the layers oblique, and a third with the layers parallel to the length, illustrating the theoretical results which were communicated along with this paper, were exhibited to the Royal Society. The principal advantage of considering metallic structures with reference to the theory of thermo-electricity is, as will be seen below, that we are so enabled to demonstrate the possibility of crystalline thermo-electric qualities of the most general conceivable type, and are shown how to construct solids (whether or not natural crystals may be ever found) actually possessing them.

149. The following two propositions with reference to thermo-electric effects in a particular case of crystalline matter are premised to the unrestricted treatment of the subject, because they will serve to guide us as to the nature of the agencies for which the general mathematical expressions are to be investigated.

PROP. I. If a bar of crystalline substance, possessing an axis of thermo-electric symmetry, has its length oblique to this axis, a current of electricity sustained in it longitudinally will cause evolution of heat at one side and absorption of heat at the opposite side, all along the bar, when the whole substance is kept at one temperature.

PROP. II. If the two sides of such a bar be kept at different temperatures, and a homogeneous conducting arc be applied to points of the ends which are at the same temperature, a current will be produced along the bar, and through the arc completing the circuit.

150. For proving these propositions, it will be convenient to investigate fully the thermo-electric agency experienced by a bar cut obliquely from a crystalline substance possessing an axis of symmetry, when placed longitudinally in a circuit of which the remainder is composed of the standard metal, and kept with either its sides or its ends unequally heated. Let θ and ϕ denote the thermo-electric powers of two bars cut from the given substance in directions parallel and perpendicular to its axis of symmetry respectively. Let us suppose the actual bar to be of rectangular section, with two of its opposite sides perpendicular to the plane of its length and the axis of symmetry of its substance. Let a longitudinal section in this plane be represented by the accompanying diagram; let OA or any line parallel to it be the direction of the axis of symmetry through any point; and let ω denote the inclination of this line to the length of the bar. Let the breadth of the two opposite sides of the bar perpendicular to the plane of the diagram be denoted by a, and in the plane of the diagram by b. The area of the transverse section of the bar will be ab; and therefore if γ denote the strength, and i the intensity, of the current in it, we have (§ 142)

$$i = \frac{\gamma}{ab}.$$

151. We may suppose the current, itself parallel to the length of the bar, and in the direction from left to right of the diagram, to be resolved (§ 142) at any point P at the side of the bar into two components in directions parallel and perpendicular to OA, of

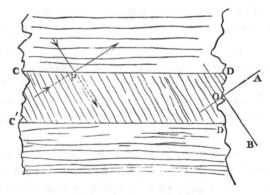

which the intensities will be $i \cos \omega$ and $i \sin \omega$ respectively. The former of these components may be supposed to belong to a system of currents crossing the bar in lines parallel to OA, and passing out of it across the side CD into a conductor of the standard metal; and the latter, to a system of currents entering the bar across CD from the same conductor of standard metal, and crossing it in lines perpendicular to OA. The resultant current in the supposed standard metal beside the bar will clearly be parallel to the length, and can therefore (this metal being non-crystalline) produce no effect influencing the thermal agency at the side of the bar or within it. The inclinations of the currents to a perpendicular to the separating plane of the two metals being respectively $90^\circ - \omega$ and ω, their strengths per unit of area of this plane, obtained by multiplying their intensities by the cosine of those angles respectively, will be each equal to

$$i \cos \omega \sin \omega.$$

Hence the absorptions of heat which they will produce at the surface of separation of the metals per unit of area per second will be

$$-\frac{1}{J} i \cos \omega \sin \omega \, t\theta, \text{ and } \frac{1}{J} i \cos \omega \sin \omega \, t\phi,$$

respectively. According to the general principle of the superposition of thermo-electric actions stated above, the sum of these is the rate of absorption of heat per unit of surface when the two systems of currents coexist. But the resultant of these systems is simply the given longitudinal current in the bar, with no flow either out of it or into it across any of its sides. Hence a simple current of intensity i, parallel to the sides of the bar, causes absorption of heat at the side CD amounting to

$$\frac{1}{J} i \cos \omega \sin \omega \, t \, (\phi - \theta),$$

per unit of area per second; and the same demonstration shows that an equal amount of evolution must be produced at the opposite side $C'D'$ These effects take place quite independently of the matter round the bar, since the metal carrying electric currents which we supposed to exist at the sides of the bar in the course of the demonstration, can exercise no influence on the phænomena.

152. If l denotes the length of the bar, the area of each of the sides perpendicular to the plane of the diagram will be la; and therefore the absorption over the whole of the side CD, and

the evolution over the whole of the other side $C'D'$, per second, will be

$$\frac{1}{J}ila \cos \omega \sin \omega t\,(\phi - \theta),$$

or
$$\frac{1}{J}\gamma\,\frac{l}{b}\cos \omega \sin \omega t\,(\phi - \theta).$$

It is obvious that there can be neither evolution nor absorption of heat at the two other sides.

153.   An investigation similar to that which has just been completed, shows that if the actual current enter from a conductor of the standard metal at one end of the bar, and leave it by a conductor of the same metal at its other end, the absorption and evolution of heat at these ends respectively will amount to

$$\frac{1}{J}\gamma\,(t\theta \cos^2\omega + t\phi \sin^2\omega)$$

per second.

154.   Let us now suppose the two sides CD, $C'D'$ to be kept at uniform temperatures, T, T', and the two ends to be kept with equal and similar distributions of temperatures, whether a current is crossing them or not.   Then if a current of strength γ be sent through the bar from left to right of the diagram, in a circuit of which the remainder is the standard metal, there will be reversible thermal action, consisting of the following parts, each stated per unit of time.

(1)   Absorption amounting to $\Omega(T)\dfrac{l}{b}\gamma$, in a locality at the temperature T.

(2)   Evolution amounting to $\Omega(T')\dfrac{l}{b}\gamma$, in a locality at the temperature T'.

(3)   Absorption amounting to $\Pi\gamma$ at one end (that beyond CC').

and   (4)   Evolution amounting to $\Pi\gamma$ at the other end;

where, for brevity, $\Omega(T)$ and $\Omega(T')$ are assumed to denote the values of $\dfrac{t}{J}(\phi - \theta)\sin \omega \cos \omega$ at the temperatures T and T'; and Π the mean value of $\dfrac{t}{J}(\theta \cos^2\omega + \phi \sin^2\omega)$ for either end of the bar.

The contributions towards the sums appearing in the general thermo-dynamic equations which are due to these items of thermal agency are as follows:

$$[\Omega(T) - \Omega(T')]\frac{l}{b}\gamma \quad \text{towards } \Sigma H_t,$$

and

$$\left[\frac{\Omega(T)}{T} - \frac{\Omega(T')}{T'}\right]\frac{l}{b}\gamma \quad \text{towards } \Sigma\frac{H_t}{t};$$

the thermal agencies at the ends disappearing from each sum in consequence of their being mutually equal and opposite, and similarly distributed through localities equally heated. Now when every reversible thermal effect is included, the value of $\Sigma\dfrac{H_t}{t}$ must be zero, according to the second general law. Hence either

$$\frac{\Omega(T)}{T} - \frac{\Omega(T')}{T'}$$

must vanish, or there must be a reversible thermal agency not yet taken into account. But probably $\dfrac{\Omega(T)}{T} - \dfrac{\Omega(T')}{T'}$ may not vanish, that is $\dfrac{\Omega}{t}$ may vary with the temperature, for natural crystals; and it certainly does vary with the temperature for metallic combinations structurally crystalline (for instance, for a bar cut obliquely from a solid consisting of alternate layers of copper and iron, the value of Ω decreases to zero as the temperature is raised from an ordinary atmospheric temperature up to about 280⁰, and has a contrary sign for higher temperatures). Hence in general there must be another reversible thermal agency, besides the agencies at the ends and at the sides of the bar which we have investigated. This agency must be in the interior; and since the substance is homogeneous, and uniformly affected by the current, the new agency must be uniformly distributed through the length, as different points of the same cross section can only differ in virtue of their different circumstances as to temperature. If there were no variation of temperature, there could be no such effect anywhere in the interior of the bar; and therefore if dt denote the variation of temperature in an infinitely small space dx across the bar in the plane of the diagram, and χ an unknown element, constant or a

function of the temperature, depending on the nature of the substance, we may assume

$$i\chi \frac{dt}{dx}$$

as the amount of absorption, per unit of the volume of the bar
due to a current of intensity i, by means of the new agency. The
whole amount in a lamina of thickness dx, length l, and breadth a
perpendicular to the plane of the diagram, is therefore

$$i\chi \frac{dt}{dx} \, aldx,$$

or

$$\gamma \frac{l}{b} \chi dt.$$

As there cannot possibly be any other reversible thermal agency
to be taken into account, we may now assume

$$\Sigma H_t = \gamma \frac{l}{b} \left\{ [\Omega(T) - \Omega(T')] + \int_{T'}^{T} \chi dt \right\} \quad \ldots\ldots\ldots(22),$$

$$\Sigma \frac{H_t}{t} = \gamma \frac{l}{b} \left\{ \frac{\Omega(T)}{T} - \frac{\Omega(T')}{T'} + \int_{T'}^{T} \frac{\chi}{t} \, dt \right\} \quad \ldots\ldots\ldots(23).$$

The second general law, showing that $\Sigma \dfrac{H_t}{t}$ must vanish, gives by
the second of these equations,

$$\frac{\Omega(T)}{T} - \frac{\Omega(T')}{T'} + \int_{T'}^{T} \frac{\chi}{t} \, dt = 0 \ldots\ldots\ldots\ldots(24).$$

Substituting in place of T, t, and differentiating with reference to
this variable, we have as an equivalent equation,

$$\frac{\chi}{t} = - \frac{d}{dt}\left(\frac{\Omega}{t} \right) \ldots\ldots\ldots\ldots\ldots(25),$$

and using this in (22), we have

$$\Sigma H_t = \gamma \frac{l}{b} \int_{T'}^{T} \frac{\Omega}{t} \, dt \ldots\ldots\ldots\ldots(26).$$

This expresses the full amount of heat taken in through the agency
of the current γ, of which the mechanical equivalent is therefore
the work done by the current. Hence (according to principles
fully explained above, §§ 109, 110) the thermal circumstances must

actually cause an electromotive force F, of which the amount is given by the equation

$$F = J\frac{l}{b}\int_{T'}^{T}\frac{\Omega}{t}\,dt\ldots\ldots\ldots\ldots\ldots\ldots\ldots(27),$$

to act along the bar from left to right of the diagram; which will produce a current unless balanced by an equal and contrary re-action. This result both establishes Proposition II., enunciated above in § 149, and shows the amount of the electromotive force producing the stated effect in terms of T and T', the temperatures of the two sides of the bar, the obliquity of the bar to the crystal-line axis of symmetry, and the thermo-electric properties of the substance; since if θ and ϕ denote its thermo-electric powers along the axis of symmetry, and along lines perpendicular to this axis, at the temperature t, and ω the inclination of this axis to the length of the bar when the substance is at the temperature t, we have

$$\Omega = \frac{t}{J}(\phi - \theta)\sin\omega\cos\omega\ldots\ldots\ldots\ldots(28).$$

155.   By an investigation exactly similar to that of § 115, which had reference to non-crystalline linear conductors, we deduce the following expression for the electromotive force, when the ends of the bar are kept at temperatures T, T' from the terminal thermal agency Π, of a current investigated in § 153:

$$F = J\int_{T'}^{T}\frac{\Pi}{t}\,dt\ldots\ldots\ldots\ldots\ldots\ldots\ldots(29),$$

where

$$\Pi = \frac{t}{J}(\theta\cos^2\omega + \phi\sin^2\omega)\ldots\ldots\ldots\ldots(30).$$

§§ 156—170.   *On the Thermal Effects and the Thermo-electric Excitation of Electrical Currents in Homogeneous Crystalline Solids.*

156.   The Propositions I. and II., investigated above, suggest the kind of assumptions to be made regarding the reversible thermal effects of currents in uniformly heated crystalline solids, and the electromotive forces induced by any thermal circumstances which cause inequalities of temperature in different parts. The formulæ expressing these agencies in the particular case which we have now investigated, guide us to the precise forms required to express those assumptions in the most general possible manner.

T.   18

157.   Let us first suppose a rectangular parallelepiped (a, b, c) of homogeneous crystalline conducting matter, completely surrounded by continuous metal of the standard thermo-electric quality touching it on all sides, to be traversed in any direction by a uniform electric current, of which the intensity components parallel to the three edges of the parallelepiped are h, i, j, and to be kept in all points at a uniform temperature t.   Then taking ϕ, θ, ψ to denote the thermo-electric powers of bars of the substance cut from directions parallel to the edges of the parallelepiped, quantities which would be equal to one another in whatever directions those edges are if the substance were non-crystalline; and θ', θ'', ϕ', ϕ'', ψ', ψ'' other elements depending on the nature of the substance with reference to the directions of the sides of the parallelepiped, to which the name of thermo-electric obliquities may be given, and which must vanish for every system of rectangular planes through the substance if it be non-crystalline, we may assume the following expression for the reversible thermal effects of the current:

$$\left. \begin{aligned} Q_{(b,\,c)} &= bc\,\frac{t}{J}\,(h\theta + i\phi'' + j\psi') \\[2mm] Q_{(c,\,a)} &= ca\,\frac{t}{J}\,(h\theta' + i\phi + j\psi'') \\[2mm] Q_{(a,\,b)} &= ab\,\frac{t}{J}\,(h\theta'' + i\phi' + j\psi) \end{aligned} \right\} \quad \ldots\ldots\ldots\ldots\ldots(31),$$

where $Q_{(b,\,c)}$, $Q_{(c,\,a)}$, $Q_{(a,\,b)}$ denote quantities of heat absorbed per second at the sides by which positive current components enter, or quantities evolved in the same time at the opposite sides.   Hence if the opposite sides be kept at different temperatures, currents will pass, unless prevented by the resistance of surrounding matter; and the electromotive forces by which these currents are urged in directions parallel to the three edges of the parallelepiped have the following expressions, in which ua, vb, and wc denote the difference of temperature between corresponding points in the pairs of sides bc, ca, and ab respectively; reckoned positive, when the temperature increases in the direction of positive components of current:

$$\left. \begin{aligned} E &= -a(u\theta + v\theta' + w\theta'') \\ F &= -b(u\phi'' + v\phi + w\phi') \\ G &= -c(u\psi' + v\psi'' + w\psi) \end{aligned} \right\} \quad \ldots\ldots\ldots\ldots\ldots(32).$$

The negative signs are prefixed, in order that positive values of the electromotive components may correspond to forces in the direction assumed for positive components of current.

158.   The most general conceivable elementary type of crystalline thermo-electric properties is expressed in the last equations, along with the equations (31) by which we arrived at them; and we shall see that every possible case of thermo-electric action in solids of whatever kind may be investigated by using them with values, and variations it may be, of the coefficients ϕ, θ, &c., suitable to the circumstances.   It might be doubted, indeed, whether these nine coefficients can be perfectly independent of one another; and indeed it might appear very probable that they are essentially reducible to six independent coefficients, from the extraordinary nature of certain conclusions which we shall show can only be obviated by supposing

$$\theta' = \phi'', \quad \theta'' = \psi', \text{ and } \phi' = \psi''.$$

Before going on to investigate any consequences from the unrestricted fundamental equations, I shall prove that it is worth while to do so, by demonstrating that a metallic structure may be actually made, which, when treated on a large scale as a continuous solid, according to the electric and thermal conditions specified for the substance with reference to which the equations (31) and (32) have been applied, shall exhibit the precise electric and thermal properties respectively expressed by those sets of equations with nine arbitrarily prescribed values for the coefficients θ, ϕ, &c.

159.   Let two zigzag linear conductors of equal dimensions, each consisting of infinitely short equal lengths of infinitely fine straight wire alternately of two different metals, forming right angles at the successive junctions, be placed in perpendicular planes, and not touching one another at any point, but with a common straight line joining the points of bisection of the small

Cold.

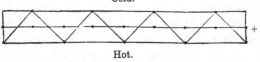

Hot.

straight parts of each conductor.   Let an insulating substance be moulded round them so as to form a solid bar of square section,

just containing the two zigzags imbedded in it in planes parallel
to its sides. Although this substance is a non-conductor of electricity, we may suppose it to have enough of conducting power for
heat, or the wires of the electric conductors to be fine enough,
that the conduction of heat through the bar when it is unequally
heated may be sensibly the same as if its substance were homogeneous throughout, and consequently that the electric conductors
take at every point the temperatures which the bar would have
at the same point if they were removed. Let an infinite number
of such bars, equal and similar, and of the same substance, be
constructed; and let a second system of equal and similar bars be
constructed with zigzag conductors of different metals from the
former; and a third with other different metals; the sole condition
imposed on the different zigzag conductors being that the two in
each bar, and those in the bars of different systems, exercise the
same resistance against electric conduction. Let an infinite number
of bars of the first set be laid on
a plane parallel to one another,
with intervals between every two
in order, equal to the breadth of
each. Lay perpendicularly across
them an infinite number of bars
of the second system similarly disposed relatively to one another;
place on these again bars of the
first system, constituting another
layer similar and parallel to the
first; on this, again, a layer similar
and parallel to the second; and

so on till the thickness of the superimposed layers is equal to the
length of each bar. Then let an infinite number of the bars of
the third system be taken and pushed into the square prismatic
apertures perpendicular to the plane of the layers; the cubical
hollows which are left (not visible in the diagram) being previously
filled up with insulating matter, such as that used in the composition of the bars. Let the complex solid cube thus formed be
coated round its sides with infinitely thin connected sheets of the
standard metal, so thin that the resistance to the conduction of
electricity along them is infinitely great, compared to the resistance
to conduction experienced by a current traversing the interior of

the cube by the zigzag linear conductors imbedded in it. (For instance, we may suppose the resistance of four parallel sides of the cube to be as great as, or greater than, the resistance of each one of the zigzag linear conductors.) Let an infinite number of such cubes be built together, with their structural directions preserved parallel, so as to form a solid, which, taken on a large scale, shall be homogeneous. A rectangular parallelepiped, *abc*, of such a solid, with its sides parallel to the sides of the elementary cubes, will present exactly the thermo-electric phenomena expressed above by the equations (31) and (32), provided the thermo-electric powers ϖ_1, ϖ_1', ϖ_1'', ϖ_1''', ϖ_2, ϖ_2', ϖ_2'', ϖ_2''', and ϖ_3, ϖ_3', ϖ_3'', ϖ_3''' of the metals used in the three systems fulfil the following conditions:

$$\left.\begin{aligned}
&\tfrac{1}{4}(\varpi_1 + \varpi_1' + \varpi_1'' + \varpi_1''') = \theta, \\
&\tfrac{1}{4}(\varpi_1 - \varpi_1') = \theta', \quad \tfrac{1}{4}(\varpi_1'' - \varpi_1''') = \theta'', \\
&\tfrac{1}{4}(\varpi_2 + \varpi_2' + \varpi_2'' + \varpi_2''') = \phi, \\
&\tfrac{1}{4}(\varpi_2 - \varpi_2') = \phi', \quad \tfrac{1}{4}(\varpi_2'' - \varpi_2''') = \phi'', \\
&\tfrac{1}{4}(\varpi_3 + \varpi_3' + \varpi_3'' + \varpi_3''') = \psi, \\
&\tfrac{1}{4}(\varpi_3 - \varpi_3') = \psi', \quad \tfrac{1}{4}(\varpi_3'' - \varpi_3''') = \psi''
\end{aligned}\right\} \quad \ldots\ldots\ldots\ldots(33).$$

160. To prove this, let us first consider the condition of a bar of any of the three systems, taken alone, and put in the same thermal circumstances ás those in which each bar of the same system exists in the compound mass. If, for instance, we take a bar of the first system, we must suppose the temperature to vary at the rate u per unit of space along its length; at the rate v across it, perpendicularly to two of its sides; and at the rate w across it, perpendicularly to its other two sides. If l be its length, and e the breadth of each side, its ends will differ in temperature by ul; corresponding points in one pair of its sides by ve, and corresponding points in the other pair of sides by we. Now it is easily proved that the longitudinal electromotive force (that is, according to the definition, the electromotive force between conductors of the standard metal connected with its ends) would, with no difference of temperatures between its sides, and the actual difference ul between its ends, be equal to $\tfrac{1}{2}(\varpi_1 + \varpi_1')ul$, if only the first of the zigzag conductors existed imbedded in the bar, or equal to $\tfrac{1}{2}(\varpi_1'' + \varpi_1''')ul$, if only the second; and since the two have equal resistances to conduction, and are connected by a little square disc of the standard metal, it follows that the longi-

tudinal electromotive force of the actual bar, with only the longitudinal variation of temperature, is

$$\tfrac{1}{4}(\varpi_1 + \varpi_1{}' + \varpi_1{}'' + \varpi_1{}''')ul.$$

Again, with only the lateral variation ve, we have in one of the zigzags a little thermo-electric battery, of a number of elements amounting to the greatest integer in $l/2e$, which is sensibly equal to $l/2e$, since the value of this is infinitely great; the electromotive force of each element is $(\varpi_1 - \varpi_1{}')ve$; and therefore the whole electromotive force of the zigzag is

$$\frac{l}{2e} \times (\varpi - \varpi_1{}')ve, \text{ or } \tfrac{1}{2}l \times (\varpi_1 - \varpi_1{}')v.$$

This battery is part of a complete circuit with the little terminal squares and the other zigzag, and therefore its electromotive force will sustain a current in one direction through itself, and in the contrary through the second zigzag; but since the resistances are equal in the two zigzags, and those of the terminal connexions may be neglected, just half the electromotive force of the first zigzag, being equal to the action and reaction between the two parts of the circuit, must remain ready to act between conductors applied to the terminal discs of the standard metal. In the circumstances now supposed, the second zigzag is throughout at one temperature, and therefore has no intrinsic electromotive force; and the resultant intrinsic electromotive force of the bar is therefore

$$\tfrac{1}{4}l\,(\varpi_1 - \varpi_1{}')\,v.$$

Similarly, if there were only the lateral variation we of temperature in the bar, we should find a resultant longitudinal electromotive force equal to

$$\tfrac{1}{4}l\,(\varpi_1{}'' - \varpi_1{}''')w.$$

If all the three variations of temperature are maintained simultaneously, each will produce its own electromotive force as if the others did not exist, and the resultant electromotive force due to them all will therefore be

$$\tfrac{1}{4}l\,\{(\varpi_1 + \varpi_1{}' + \varpi_1{}'' + \varpi''')u + (\varpi_1 - \varpi_1{}')v + (\varpi_1{}'' - \varpi_1{}''')\,w\}.$$

This being the electromotive force of each bar of the first system in any of the cubes composing the actual solid, must be

the component electromotive force of each cube in the direction
to which they are parallel, and therefore

$$\tfrac{1}{4}a\left\{(\varpi_1 + \varpi_1{}' + \varpi_1{}'' + \varpi_1{}''')u + (\varpi_1 - \varpi_1{}')v + (\varpi_1{}'' - \varpi_1{}''')w\right\}$$

must be the component electromotive force of the entire parallelepiped in the same direction. Similar expressions give the component electromotive forces parallel to the edges b and c of the
solid, which are similarly produced by the bars of the second and
third systems, and we infer the proposition which was to be
proved.

161. COR. By choosing metals of which the thermo-electric
relations, both to the standard metal and to one another, vary,
we may not only make the nine coefficients have any arbitrarily
given values for a particular temperature, but we may make them
each vary to any extent with a given change of temperature.

162. For the sake of convenience in comparing the actual
phenomena of thermo-electric force in different directions presented by an unequally heated crystalline solid, let us now, instead
of a parallelepiped imbedded in the standard metal, consider an
insulated sphere of the crystalline substance, with sources of heat
and cold applied at its surface, so as to maintain a uniform variation of temperature in all lines perpendicular to the parallel
isothermal planes. Let the rate of variation of temperature per
unit of length, perpendicular to the isothermal surfaces, be q, and
let the cosines of the inclinations of this direction to the three
rectangular directions in the substance to which the edges of the
parallelepiped first considered were parallel, and which we shall
now call the lines of reference, be l, m, n respectively. Then if we
take

$$ql = u, \quad qm = v, \quad qn = w,$$

the substance of the sphere will be in exactly the same thermal
condition as an equal spherical portion of the parallelepiped; and
it is clear that the preceding expressions for the component electro-
motive forces of the parallelepiped will give the electromotive forces
of the sphere between the pairs of points at the extremities of
diameters coinciding with the rectangular lines of reference, if we
take each of the three quantities, a, b, c, equal to the diameter of
the sphere. Calling this unity, then, we have

$$
\left.
\begin{aligned}
-E &= u\theta + v\theta' + w\theta'' \\
-F &= u\phi'' + v\phi + w\phi' \\
-G &= u\psi' + v\psi'' + w\psi
\end{aligned}
\right\} \quad \ldots\ldots\ldots\ldots\ldots(34).
$$

According to the definition given above (§ 144, Def. 3), it appears that these quantities, E, F, G, are the three components of *the intrinsic electromotive force at any point in the substance*, whether the portion of it we are considering be limited and spherical, or rectangular, or of any other shape, or be continued to any indefinite extent by homogeneous or heterogeneous solid conducting matter with any distribution of temperature through it. The component electromotive force P, along a diameter of the sphere inclined to the rectangular lines of reference, at angles whose cosines are l, m, n is of course given by the equation

$$
P = El + Fm + Gn \ldots\ldots\ldots\ldots\ldots\ldots(35),
$$

which may also be employed to transform the general expressions for the components of the electromotive force to any other lines of reference.

163.   A question now naturally presents itself: Are there three principal axes at right angles to one another in the substance, possessing properties of symmetry with reference to the thermo-electric qualities, analogous to those which have been established for the dynamical phenomena of a solid rotating about a fixed point, and for electrostatical and for magnetic forces, in natural crystals or in substances structurally crystalline as regards electric or magnetic induction? The following transformation, suggested by Mr Stokes's paper on the *Conduction of Heat in Crystals**, in which a perfectly analogous transformation is applied to the most general conceivable equations expressing flux of heat in terms of variations of temperature along rectangular lines of reference in a solid, will show the nature of the answer†.

---

* *Cambridge and Dublin Mathematical Journal*, Nov. 1851; also *Mathematical and Physical Papers*. Stokes. Vol. II.

† [Note of March 3, 1882. In this paper Stokes explained the rotatory quality expressed by the general equations of thermal conduction, with nine distinct arbitrary constants, as follows:—

"Conceive an elastic solid to be fixed at the origin, and to expand alike in all directions and at all points with a velocity of expansion unity, so that a particle which at the end of time t is situated at a distance r from the origin, at the end of the time $t + dt$ is situated at a distance $r(1 + dt)$. Conceive this solid to turn with

164. The direction cosines of the line of greatest thermal variation, or the perpendicular to the isothermal planes, are

$$\frac{u}{q}, \ \frac{v}{q}, \ \frac{w}{q},$$

where q, denoting the rate of variation of temperature in the direction of that line, is given by the equation

$$q = (u^2 + v^2 + w^2) \quad \text{.....................(36)}.$$

Taking these values for l, m, n, in the preceding general expression for the electromotive force in any direction, we find

$$P = \frac{1}{q} \{ \theta u^2 + \phi v^2 + \psi w^2 + (\phi' + \psi'')vw + (\psi' + \theta'')wu + (\theta' + \phi'')uv \},$$

the negative sign being omitted on the understanding that P shall be considered positive when the electromotive force is from hot to cold in the substance. This formula suggests the following changes in the notation expressing the general thermo-electric coefficients:—

$$\left. \begin{array}{llll} \phi' + \psi'' = 2\theta_1, & \psi' + \theta'' = 2\phi_1, & \theta' + \phi'' = 2\psi_1 \\ -\phi' + \psi'' = 2\zeta, & -\psi' + \theta'' = 2\eta, & -\theta' + \phi'' = 2\vartheta \end{array} \right\} \quad \text{........(37)},$$

an angular velocity ω equal to $\sqrt{(\omega'^2 + \omega''^2 + \omega'''^2)}$, about an axis whose direction cosines are $\omega'\omega^{-1}$, $\omega''\omega^{-1}$, $\omega'''\omega^{-1}$. The direction of motion of any particle will represent the direction of the flow of heat in what we may still call the *auxiliary solid*, from whence the direction of the flow of heat in the given solid will be obtained by merely conceiving the whole figure differently magnified or diminished in three rectangular directions.

"This *rotatory* sort of motion of heat, produced by the mere diffusion from the source outwards, certainly seems very strange, and leads us to think, independently of the theory of molecular radiation, that the expressions for the flux with six arbitrary constants only, namely the expressions (8), or the equivalent expressions (7), are the most general possible."

It appeared to me that the assumption made in this last clause was conceivably invalidated by the existence of crystals (such as those that have the "Pyroelectric" quality) with difference of crystalline form between two ends, the "Dipolar" difference as I called it (§ 168 below) : and that it certainly was inapplicable to matter in a magnetic field (compare §§ 168—171 below), whether the matter be solid or fluid, or ferro-magnetic or diamagnetic ; that on the contrary we might expect to actually find the rotatory quality in thermal conduction under magnetic influence. The same considerations are of course applicable to electrical conduction, and accordingly (§ 176, below) I would not admit less than nine coefficients for electrical conduction. Hall's recent great discovery shows that the hypothesis which 28 years ago I refused to admit, was incorrect, and proves the rotatory quality to exist for electrical conduction through metals in the magnetic field.]

which reduce the general equations, and the formula itself which suggests them, to

$$\left.\begin{array}{l} -E = \theta u + \psi_1 v + \phi_1 w + (\eta w - \Im v) \\ -F = \psi_1 u + \phi v + \theta_1 w + (\Im u - \zeta w) \\ -G = \phi_1 u + \theta_1 v + \psi w + (\zeta v - \eta u) \end{array}\right\} \dots\dots\dots\dots(38),$$

$$P = \frac{1}{q} \left(\theta u^2 + \phi v^2 + \psi w^2 + 2\theta_1 vw + 2\phi_1 wu + 2\psi_1 uv \right) \dots\dots(39).$$

165. The well-known process of the reduction of the general equation of the second degree shows that three rectangular axes may be determined for which the coefficients θ_1, ϕ_1, ψ_1 in these expressions vanish, and for which, consequently, the equations become

$$\left.\begin{array}{l} -E = \theta u + (\eta w - \Im v) \\ -F = \phi v + (\Im u - \zeta w) \\ -G = \psi w + (\zeta v - \eta u) \end{array}\right\} \dots\dots\dots\dots(40),$$

$$P = \frac{1}{q} \left(\theta u^2 + \phi v^2 + \psi w^2 \right) \dots\dots\dots\dots:(41).$$

166. The law of transformation of the binomial terms $(\eta w - \Im v)$ &c. in these expressions is clearly, that if ρ denote a quantity independent of the lines of reference, and expressing a specific thermoelectric quality of the substance, which I shall call its thermoelectric rotatory power, and if λ, μ, ν denote the inclinations of a certain axis fixed in the substance, which I shall call its axis of thermo-electric rotation, to any three rectangular lines of reference, then the values of $\zeta, \eta, \Im$ for these lines of reference are as follows:

$$\zeta = \rho \cos \lambda, \quad \eta = \rho \cos \mu, \quad \Im = \rho \cos \nu.$$

If i denote the inclination of the direction $\left(\dfrac{u}{q}, \dfrac{v}{q}, \dfrac{w}{q} \right)$, in which the temperature varies most rapidly, to the axis of thermo-electric rotation, and if α, β, γ denote the angles at which a line perpendicular to the plane of this angle i is inclined to the axes of reference, we have

$$\left.\begin{array}{l} \eta w - \Im v = \rho q \sin i \cos \alpha \\ \Im u - \zeta w = \rho q \sin i \cos \beta \\ \zeta v - \eta u = \rho q \sin i \cos \gamma \end{array}\right\} \dots\dots\dots\dots(42).$$

Hence we see that the last terms of the general formula for the component electromotive forces along the lines of reference express the components of an electromotive force acting along a line perpendicular both to the axis of thermo-electric rotation, and to the direct line from hot to cold in the substance, and equal in magnitude to the greatest rate of variation of temperature perpendicular to that axis, multiplied by the coefficient ρ.

167.   Or again, if we consider a uniform circular ring of rectangular section, cut from any plane of the substance inclined at an angle λ to a plane perpendicular to the axis of thermo-electric rotation, and if the temperature of the outer and inner cylindrical surfaces of this ring be kept each uniform, but different from one another, so that there may be a constant rate of variation, q, of temperature in the radial direction, but no variation either tangentially or in the transverse direction perpendicular to the plane of the ring, we find immediately, from (42), that the last terms of the general expressions indicate a tangential electromotive force, equal in value to $\rho q \cos \lambda$, acting uniformly all round the ring. This tangential force vanishes if the plane of the ring contains the axis of thermo-electric rotation, and is greatest when the ring is in a plane perpendicular to the same axis.

168.   The peculiar quality of a solid expressed by these terms would be destroyed by cutting it into an infinite number of plates of equal infinitely small thickness, inverting every second plate, and putting them all together again into a continuous solid, in planes perpendicular to the axis of thermo-electric rotation; a process which would clearly not in any way affect the thermo-electric relations expressed by the first term of the general expressions for the components of electromotive force ; and it is therefore of a type, to which also belongs the rotatory property with reference to light discovered by Faraday as induced by magnetization in transparent solids, which I shall call dipolar, to distinguish it from such a rotatory property with reference to light as that which is naturally possessed by many transparent liquids and solids, and which may be called an isotropic rotatory property.   The axis of thermo-electric rotation, since the agency distinguishing it as a line also distinguishes between the two directions in it, may be called a dipolar axis; so may the axis of rotation of a rotating rigid body*,

* [*Added, Liverpool, Sept.* 27, 1854.]—As is perfectly illustrated by M. Foucault's beautiful experiment of a rotating solid, placing its axis parallel to that of the

or the direction of magnetization of a magnetized element of matter; and its general type is obviously different from that of a principal axis of inertia of a rigid body, or a principal axis of magnetic inductive capacity in a crystal, or a line of mechanical tension in a solid; any of which may be called an isotropic axis.

169.   The general directional properties expressed by the first terms of the second members of (40) are perfectly symmetrical regarding the three rectangular lines of reference, and are of a type so familiar that they require no explanation here. We conclude that every substance has three principal isotropic axes of maximum and minimum properties regarding thermo-electric power, which are at right angles to one another; but that it is only for a particular class of conceivable substances that the thermo-electric properties are entirely symmetrical with reference to these axes; all substances for which the rotatory power, ρ, does not vanish, having besides a dipolar axis of thermo-electric rotation which may be inclined in any way to them.

170.   These principal isotropic axes lose distinction from all other directions in the solid when the thermo-electric powers along them (the values of the coefficients θ, ϕ, ψ) are equal; but a rotatory property, distinguishing a certain line as a dipolar axis, may still exist. By § 159, we see how metallic structures possessing any of these properties (for instance, having equal thermo-electric power in all directions, and possessing a given rotatory power, ρ, in a given direction about a given system of parallel lines) may be actually made. [*Added, March* 3, 1882. To adapt the general statement and notation of § 159 and the notation of § 166 to this case, suppose for example OX to be the resultant axis of the thermo-electric rotation. We must have

$$\left. \begin{aligned} \varpi_1 = \varpi_1{}' = \varpi_1{}'' = \varpi_1{}''' = \varpi_2{}'' = \varpi_2{}''' = \varpi_3 = \varpi_3{}' = \theta \\ -\varpi_2 = -\varpi_3{}''' = \omega_2{}' = \omega_3{}'' = 2\rho \end{aligned} \right\} \dots (42').$$

To simplify the subject however and to aid a thorough understanding of the anticipated thermo-electric rotatory quality, instead of following the directions of § 159 begin afresh thus. Prepare a little thermo-electric battery, such as that used in the

earth's, and so turned that it may itself be rotating in the same direction as the earth; which the meeting of the British Association just concluded has given me an opportunity of witnessing.

old thermo-multiplier of Melloni, with bismuth and antimony, or with the more familiar metals copper and iron as illustrated in the annexed diagram, in which the shaded areas represent sections of square iron plates, and the plain black lines sections of square copper plates, except two half-squares on the left-hand side below and right-hand side above. These iron and copper squares or rectangles are soldered together at their upper and lower edges,

Fig. 1.

Cold.

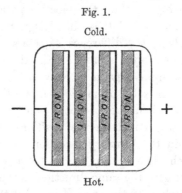

Hot.

as shown in the diagram (Fig. 1). The bounding line of the diagram represents in section a sheet copper cube with rounded edges, to the middles of two opposite sides of which are soldered the copper half-squares on the left and right of the diagram (Fig. 1). All interstices are filled up with plaster of Paris so as to form a solid cube coated with thin sheet copper; and having the thermo-electric arrangement firmly embedded in its interior. The thickness of the sheet copper is to be made different in the several sides of the cube so as to fulfil the condition that for the three directions perpendicular to the three pairs of sides, it is

Fig. 2.

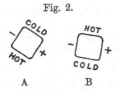

A   B

isotropic in respect to both electrical and thermal conduction of the mass as a whole, consisting as it does of copper, iron, and plaster of Paris. Imagine now millions of millions of such cubes, tossed as dice out of a dice-box, and all consolidated together by plaster of Paris. The resulting solid taken as a whole on a large scale, will be perfectly isotropic in respect to all physical qualities, and there-fore of necessity destitute of the thermo-electric rotatory quality. Now looking at the designation " − hot + cold" marked round the cube of Fig. 1, understand that it means that if the side marked hot is at any time at a higher temperature than the side marked cold, when the cube is insulated and undisturbed by any external electric influence, the electric potential on the side marked + will be higher than the electric potential on the side marked −; and understand that the designation does not mean that there actually is the difference of temperature or potential indicated, but it is merely a symbol for the thermo-electric quality of the cube. Now from a point of view at a very great distance look all through our solid of millions of millions of cubes, and referring to Fig. 2 when the "− hot + cold" are seen, as in the cube A, situated successively in the positions reached by a hand travelling round in the contrary direction to the motion of the hands of a watch, leave it as it is. When it is seen with "− hot + cold" lying in the same direction as that of the motion of the hands of a watch, as in the cube B, invert it through 180° round any axis taken at random through its centre perpendicular to your line of vision. Do this for every cube, whether as in the diagram the four thermo-electrically efficient sides are parallel to the line of vision or whether they are oblique to it: invert it in the manner described if the symbols − hot + cold lie in the order which would be indicated by the hands of a watch with its face seen directly or obliquely; leave it as it is if the symbols are seen to lie in the opposite direction. The result will be a solid fulfilling precisely the description of § 170. It will be in fact isotropic in respect to every physical quality except that it has the thermo-electric rotatory power, with the supposed line of vision for axis.]

171. [*Added, July* 1854.] It is far from improbable that a piece of iron in a state of magnetization, which I have, since § 147 was written, ascertained to possess different thermo-electric proper-ties in different directions, may also possess rotatory thermo-electric power[*], distinguishing its axis of magnetization, which is essen-

---

* [*Added, Sept.* 13, 1854.]—By an experiment made to test its existence, which has given only negative results, I have ascertained that this "rotatory power," if it exists in inductively magnetized iron at all, must be very small in comparison with the amount by which the thermo-electric power in the direction of magnetization differs from the thermo-electric power of the same metal not magnetized.

tially, in its magnetic character, dipolar, as thermo-electrically dipolar also.

§§ 172—181. *On the general equations of Thermo-electric Action in any homogeneous or heterogeneous crystallized or non-crystallized solid.*

172. Let t denote the absolute temperature at any point, x, y, z of a solid. Let θ, ϕ, ψ, θ', ϕ', ψ', θ'', ϕ'', ψ'' be the values of the nine thermo-electric coefficients for the substance at this point, quantities which may vary from point to point, either by heterogeneousness of the solid, or in virtue of non-uniformity of its temperature. Let h, i, j be the components of the intensity of electric current through the same point (x, y, z).

173. Then, applying equations (31) of § 157 to infinitely small, contiguous, rectangular parallelepipeds in the neighbourhood of the point (x, y, z), and denoting by $H dx dy dz$ the resultant reversible absorption of heat occasioned by the electric current across the infinitely small element $dx dy dz$, we find

$$H = \frac{t}{J} \left\{ \frac{d}{dx} (h\theta + i\phi'' + j\psi') + \frac{d}{dy} (h\theta' + i\phi + j\psi'') + \frac{d}{dz} (h\theta'' + i\phi' + j\psi) \right\} \quad (43$$

174. By the analysis of discontinuous functions, this expression may be applied not only to homogeneous or to continuously varying heterogeneous substances, but to abrupt transitions from one kind of substance to another. Still it may be convenient to have formulæ immediately applicable to such cases, and therefore I add the following expression for the reversible thermal effect in any part of the bounding surfaces separating the given solid from a solid of the standard metal in contact with it:

$$Q = \frac{t}{J} \left\{ p(h\theta + i\phi'' + j\psi') + q(h\theta' + i\phi + j\psi'') + r(h\theta'' + i\phi' + j\psi) \right\} .. (44),$$

where Q denotes the quantity of heat absorbed per second per unit of surface at a point of the bounding surface, and (p, q, r) the direction cosines of a normal to the surface at the same point.

175. Equations (34) give explicitly the intrinsic electromotive force at any point of the solid when the distribution of temperature is given; but we must take into account also the reaction proceeding from the surrounding matter, to get the efficient electromotive

force determining the current through any part of the body. This reaction will be the electrostatical resultant force due to accumulations of electricity at the bounding surface and in the interior of the conducting mass throughout which the electrical circuits are completed. Hence if V denote the electrical potential at (x, y, z) due to these accumulations, the components of the reactional electromotive force are

$$-\frac{dV}{dx}, \quad -\frac{dV}{dy}, \quad -\frac{dV}{dz};$$

and the components of the efficient electromotive force in the solid are therefore

$$E - \frac{dV}{dx}, \quad F - \frac{dV}{dy}, \quad G - \frac{dV}{dz},$$

where E, F, G are given by the following equations, derived from (34) by substituting for u, v, w their values dt/dx, dt/dy, dt/dz, in terms of the notation now introduced:—

$$\left. \begin{aligned} -E &= \frac{dt}{dx}\,\theta \ + \frac{dt}{dy}\,\theta' \ + \frac{dt}{dz}\,\theta'' \\ -F &= \frac{dt}{dx}\,\phi'' + \frac{dt}{dy}\,\phi \ + \frac{dt}{dz}\,\phi' \\ -G &= \frac{dt}{dx}\,\psi' + \frac{dt}{dy}\,\psi'' + \frac{dt}{dz}\,\psi \end{aligned} \right\} \ \ldots\ldots\ldots\ldots\ldots(45).$$

176.   The body, being crystalline, probably possesses different electrical conductivities in different directions, and the relation between current and electromotive force cannot, without hypothesis*, be expressed with less than nine coefficients. These, which we shall call the coefficients of electric conductivity, we shall denote by κ, λ, &c.; and we have the following equations, expressing by means of them the components of the intensity of electric current in terms of the efficient electromotive force at any point of the solid:—

---

* [Note of March 3, 1882. See footnote on § 163 above. Hall's discovery in terms of the notation of formula (46), is simply that $\lambda' - \mu''$, $\mu' - \kappa''$, $\kappa' - \lambda''$, are not each of them equal to zero for a metal in a magnetic field.]

$$
\left.
\begin{aligned}
h &= \kappa \left(E - \frac{dV}{dx}\right) + \kappa' \left(F - \frac{dV}{dy}\right) + \kappa'' \left(G - \frac{dV}{dz}\right) \\
i &= \lambda'' \left(E - \frac{dV}{dx}\right) + \lambda \left(F - \frac{dV}{dy}\right) + \lambda' \left(G - \frac{dV}{dz}\right) \\
j &= \mu' \left(E - \frac{dV}{dx}\right) + \mu'' \left(F - \frac{dV}{dy}\right) + \mu \left(G - \frac{dV}{dz}\right)
\end{aligned}
\right\} \ldots\ldots(46).
$$

These equations (45) and (46), with

$$
\frac{dh}{dx} + \frac{di}{dy} + \frac{dj}{dz} = 0 \ldots\ldots\ldots\ldots\ldots\ldots(47),
$$

which expresses that as much electricity flows out of any portion of the solid as into it, in any time (in all seven equations,) are sufficient to determine the seven functions E, F, G, V, h, i, j, for every point of the solid, subject to whatever conditions may be prescribed for the bounding surface, and so to complete the problem of finding the motion of electricity across the body in its actual circumstances; provided the values of dt/dx, dt/dy, dt/dz are known, as they will be when the distribution of temperature is given. We may certainly, in an electrical problem such as this, suppose the temperature actually given at every point of the solid considered, since we may conceive thermal sources distributed through its interior to make the temperature have an arbitrary value at every point.

177. Yet practically the temperature will, in all ordinary cases, follow by conduction from given thermal circumstances at the surface. The equations of motion of heat, by which, along with those of thermo-electric force, such problems may be solved, are as follows:—(1), three equations,

$$
\left.
\begin{aligned}
\zeta &= - \left(k \frac{dt}{dx} + k' \frac{dt}{dy} + k'' \frac{dt}{dz}\right) \\
\eta &= - \left(l'' \frac{dt}{dx} + l \frac{dt}{dy} + l' \frac{dt}{dz}\right) \\
\vartheta &= - \left(m' \frac{dt}{dx} + m'' \frac{dt}{dy} + m \frac{dt}{dz}\right)
\end{aligned}
\right\} \ldots\ldots\ldots\ldots(48),
$$

to express the components ζ, η, ϑ of the "flux of heat" at any point of the solid, in terms of the variations of temperature dt/dx, dt/dy, dt/dz multiplied by coefficients k, l, m, k', &c., which may be called the nine coefficients of thermal conductivity of the substance; and (2) the single equation

$$\frac{d\zeta}{dx} + \frac{d\eta}{dy} + \frac{d\vartheta}{dz}$$

$$= -\frac{t}{J}\left\{\frac{d}{dx}(h\theta + i\phi'' + j\psi') + \frac{d}{dy}(h\theta' + i\phi + j\psi'') + \frac{d}{dz}(h\theta'' + i\phi' + j\psi)\right\}$$

$$+ \frac{1}{J}\left\{h\left(E - \frac{dV}{dx}\right) + i\left(F - \frac{dV}{dy}\right) + j\left(G - \frac{dV}{dz}\right)\right\} \dots\dots\dots(49),$$

of which the first member expresses the rate at which heat flows out of any part of the solid per unit of volume; and the second member, to which it is equated, the resultant thermal agency (positive when there is, on the whole, evolution at xyz) produced by the electric currents.

178. The general treatment of these eleven equations, (45), (46), (47), (48), (49), leads to two non-linear partial differential equations of the second order and degree for the determination of the functions t and V.

179. It may be remarked, however, that the second term of the second member of (49), when the prefixed negative sign is removed, expresses the frictional generation of heat by currents through the solid, and will therefore, when the electromotive forces in action are solely thermo-electric, be very small, even in comparison with the reversible generation and absorption of heat in various parts of the body, provided the differences of temperature between these different localities are small fractions of the temperature, on the absolute scale from its zero. Excepting, then, cases in which there are wide ranges (for instance, of 53°C. or more) of temperature, the second principal term of the second member of (49) may be neglected, and the partial differential equations to which t and V are subject will become linear; so that one of the unknown functions may be readily eliminated, and a linear equation of the fourth order obtained for the determination of the other.

180. Further, it may be remarked that probably in most, if not in all known cases, the reversible as well as the frictional thermal action of the currents, when excited by thermo-electric force alone, is very small in comparison with that of conduction, perhaps quite

insensible. (See above, § 106.) Hence, except when more powerful electromotive forces than the thermo-electric forces of the solid
itself, and of its relation to the matter touching it round its surface, act to drive currents through it, we may possibly in all,
certainly in many cases, neglect the entire second member of (49)
without sensible loss of accuracy; and we then have a differential
equation of the second order for the determination of the temperature in the interior of the body, simply from ordinary conduction,
according to the conditions imposed on its surface. To express
these last conditions generally, a superficial application of the
three equations (48) with their nine independent coefficients is
required.

181. When t is either given or determined in any way, the
solution of the purely electrical problem is, as was remarked above,
to be had from the seven equations (45), (46), and (47). These
lead to a single partial differential equation of the second order
for the determination of V through the interior, subject to conditions as to electromotive force and electrical currents across the
surface, for the expression of which superficial applications of (45)
and (46) will be required. When V is determined, the solution
of the problem is given by (45) and (46), expressing respectively
the electromotive force and the motion of electricity through the
solid.

PART VII.*

§§ 182—210.   *On the Thermo-elastic, Thermomagnetic, and Pyro-
electric properties of Matter.*

[*Phil. Mag.* Jan. 1878.]

182. A body which is either emitting heat, or altering its
dimensions against resisting forces, is doing work upon matter
external to it. The mechanical effect of this work in one case

* [(Note, written in Sept. 1877 for the Reprint in *Phil. Mag.*) This paper
is in the main a reprint from an article which appeared under the title "On
the Thermo-elastic and Thermomagnetic Properties of Matter, Part I.," in April
1855, in the first number of the *Quarterly Journal of Mathematics*, but which was
confined to the thermo-elastic part of the subject. The continuation, in which
it was intended to make a similar application of thermodynamic principles to mag-

is the excitation of thermal motions, and in the other the over-coming of resistances. The body must itself be altering in its circumstances, so as to contain a less store of energy within it, by an amount precisely equal to the aggregate value of the mechanical effects produced; and conversely, the aggregate value of the mechanical effects produced must depend solely on the initial and final states of the body, and is therefore the same, whatever be the intermediate states through which the body passes, provided the *initial* and *final* states be the same.

183.   The *total intrinsic energy* of a body might be defined as the mechanical value of all the effect it would produce, in heat emitted and in resistances overcome, if it were cooled to the utmost, and allowed to contract indefinitely or to expand indefinitely according as the forces between its particles are attractive or repulsive, when the thermal motions within it are all stopped; but in our present state of ignorance regarding perfect cold, and the nature of molecular forces, we cannot determine this "total intrinsic energy" for any portion of matter; nor even can we be sure that it is not infinitely great for a finite portion of matter. Hence it is convenient to choose a certain state as standard for the body under consideration, and to use the un-qualified term *intrinsic energy* with reference to this standard state; so that the "intrinsic energy of a body in a given state" will denote the mechanical value of the effects the body would produce in passing from the state in which it is given, to the standard state—or, which is the same, the mechanical value of the whole agency that would be required to bring the body from the standard state to the state in which it is given.

netic induction, was never published or written; but the results which it should have contained were sufficiently indicated in a short article on "Thermomagnetism," which I wrote at the request of my friend and colleague the late Professor J. P. Nichol for the second edition of his *Cyclopædia*, published in 1860, and which I include in the present reprint. The addition of "Pyro-Electric," which I now make to the title of the former article, is justified by another short quotation from the second edition of Nichol's *Cyclopædia* (article "Thermo-Electric, Division I.—Pyro-Elec-tricity, or Thermo-Electricity of Nonconducting Crystals"), and a short addition, now written and published for the first time, in which the same thermodynamic principles are applied to this form of thermo-electric action.

Several additions both in the shape of text and footnote are appended in the course of the reprint, These are all distinguished by being enclosed in brackets, [ ].]

184. In Part V.* (§§ 81—96, above) of a series of papers on
the Dynamical Theory of Heat, communicated to the Royal
Society of Edinburgh, a system of formulæ founded on propositions established in Part I.† (§§ 7—23 above), of the same series of
papers, and expressing, for a given fluid mass, relations between its
pressure, its thermal capacities, its intrinsic energy (all considered
as functions of its temperature and volume), and Carnot's function
of the temperature, were brought forward for the purpose of pointing out the importance of making the *intrinsic energy* of a fluid in
different states an object of research along with the other elements
which have hitherto been considered, and partially investigated
in some cases.   In the present communication a similar mode
of treatment, extended to include solid bodies, unmagnetic [and
unelectrified], or magnetized [or electrified] in any way, is shown
to lead to the most general possible theory of elasticity, whether
of solids or fluids, and to point out various thermodynamic properties of solids and various thermal effects of magnetism [and of
electricity] not hitherto discovered.

SECTION I.—*Elasticity of Solids or Fluids not subjected
to Magnetic Force.*

185.   Let $x, y, z, \xi, \eta, \zeta$ be six independent variables expressing
the mechanical condition of a homogeneous solid mass, homogeneously strained in any way‡, and let t be its temperature; and (in
accordance with the preceding explanations) let e denote its
intrinsic energy, reckoned from a certain "standard state" defined
by particular values, $x_0, y_0, z_0, \xi_0, \eta_0, \zeta_0, t_0$, on which its physical
condition depends.   Thus, if ϕ denotes a certain function depending on the nature of the substance, and vanishing for the values
$x_0, y_0, \dots t_0$ of the independent variables, we have

$$e = \phi (x, y, z, \xi, \eta, \zeta, t) \dots\dots\dots\dots(1);$$

and a knowledge of the function ϕ [with besides a knowledge of

---

* *Trans. Roy. Soc. Edinb.*, December 15, 1851.

† *Ibid.*, March 17, 1851.

‡ The terms *a strain*, or *to strain*, are used simply with reference to alterations
of dimensions or form in a solid—the forces by which "a strain" is produced being
called the *straining tensions or pressures*, or sometimes merely *the tensions or pressures*, to which the solid is subjected.   This distinction of terms is adopted in
accordance with the expressions used by Mr Rankine in his paper on the Elasticity
of Solids (*Cambridge and Dublin Mathematical Journal*, February 1851).

w for one particular temperature*] comprehends all the thermo-
elastic qualities of the solid.

186.   Now let us suppose the body to be strained so as to pass
from the mechanical state $(x_0, y_0, z_0, \xi_0, \eta_0, \zeta_0)$ to $(x, y, z, \xi, \eta, \zeta)$
while it is constantly kept at the temperature t; and let H denote
the quantity of heat that must be supplied to it during this
process to prevent its temperature from being lowered (a quantity
which of course is zero, or negative, for such strains as cause no
thermal effects, or which cause positive evolutions of heat).   Let
the body be brought back to its mechanical condition $(x_0, y_0, z_0,$
$\xi_0, \eta_0, \zeta_0)$ through the same or any other of all the infinitely
varied successions of states by which it may be made to pass from
one to the other of the two which have been named, its temperature being kept always at t.   Then, by the second Fundamental
Law of the Dynamical Theory of Heat (§ 98, above) (see *Trans.
Roy. Soc. Edinb.*, May 1, 1854, p. 126), we must have (§ 101,
above)

$$\frac{H}{t} + \frac{H'}{t'} = 0,$$

and therefore $H' = -H$.

187.   We conclude that the quantity of heat absorbed by the
body in being strained from one state to another at the same
temperature is quite independent of the particular succession of
states through which it is made to pass, provided it has throughout
the same temperature.   Hence we must have

$$H = \psi (x, y, z, \xi, \eta, \zeta, t) - \psi (x_0, y_0, z_0, \xi_0, \eta_0, \zeta_0, t)...(2),$$

where ψ denotes a function of the variables.   Now the mechanical
value of the heat taken in by the body while it passes from one
condition to the other, together with the work spent in compelling
it to do so, constitutes the whole augmentation of mechanical
energy which it experiences; so that if ϵ denote this augmentation—that is, if

$$\epsilon = \phi (x, y, z, \xi, \eta, \zeta, t) - \phi (x_0, y_0, z_0, \xi_0, \eta_0, \zeta_0, t) ...(3),$$

and if w denote the work done by the applied forces and J the
mechanical equivalent of the thermal unit, we have

$$\epsilon = w + JH ....(4).$$

[See equations (10), (11) of § 188 below.]

From this we conclude that the work required to strain the body from one to another of two given mechanical states, keeping it always at the same temperature, is independent of the particular succession of mechanical states through which it is made to pass, and is always the same when the initial and final states are the same. This theorem was, I believe, first given by Green (as a consequence of the most general conceivable hypothesis that could be framed to explain the mutual actions of the different parts of a body on which its elasticity depends), who inferred from it that there cannot be 36, but only 21, independent coefficients [or "moduluses"] of elasticity, with reference to axes chosen arbitrarily in any solid whatever. It is now demonstrated as a particular consequence of the Second General Thermodynamic Law. It might at first sight be regarded as simply a consequence of the general principle of mechanical effect; but this would be a mistake, fallen into from forgetting that heat is in general evolved or absorbed when a solid is strained in any way; and the only absurdity to which a denial of the proposition could lead would be the possibility of a self-acting machine going on continually drawing heat from a body surrounded by others at a higher temperature, without the assistance of any at a lower temperature, and performing an equivalent of mechanical work.

188.   The full expression of the Second Thermodynamic Law for the circumstances of elastic force is, as is shown in the passage referred to above (§ 101, above) (*Trans. Roy. Soc. Edinb.*, May 1, 1854, p. 126), that if H_t, H'_t, &c., denote the quantities of heat emitted from a body when at temperatures* t, t' respectively, during operations changing its physical state in any way, the sum $\Sigma \dfrac{H_t}{t}$ must vanish for any cycle of changes, if each is of a perfectly

---

* Reckoned on the absolute thermodynamic scale, according to which "temperature" is defined § 99 above as the mechanical equivalent of the thermal unit divided by "Carnot's function." In a paper "On the Thermal Effects of Fluids in Motion" by Mr Joule and myself [Art. XLIX. below], communicated to the Royal Society last June [1854], and since published in the *Philosophical Transactions*, it is shown that temperature on the absolute thermodynamic scale does not differ sensibly from temperature on the ordinary scale of the air-thermometer, except by the addition of a constant number, which we find to be about 273·7 for the Centigrade scale. Thus, on the system now adopted, the temperature of melting ice is 273·7, that of boiling water is 373·7, and differences of temperature are sensibly the same as on an ordinary standard Centigrade thermometer.

reversible character, and if at the end of all the body is brought back to its primitive state in every respect. Let us consider, for instance, the following cycle, which obviously fulfils both conditions.

(I.) Let the body, initially in the state $(x_0, y_0, z_0, \xi_0, \eta_0, \zeta_0, t)$, be raised in temperature from t to t', its form and dimensions being maintained constant.

(II.) Let it be strained from the state $(x_0, y_0, z_0, \xi_0, \eta_0, \zeta_0)$ to the state $(x, y, z, \xi, \eta, \zeta)$, while its temperature is kept always at t'.

(III.) Let it be lowered in temperature from t' to t, its form and dimensions being retained.

(IV.) Let it be brought back to the mechanical state $(x_0, y_0, z_0, \xi_0, \eta_0, \zeta_0)$, while its temperature is kept constantly at t.

The quantities of heat taken in by the body in these successive operations are respectively :—

(I.) $\dfrac{1}{J} \{\phi\, (x_0, y_0, z_0, \xi_0, \eta_0, \zeta_0, t') - \phi\, (x_0, y_0, z_0, \xi_0, \eta_0, \zeta_0, t)\}$,

because the difference of the whole mechanical energies is simply the mechanical value of the heat taken in or emitted in all cases in which no work is either done on the body or received by it in virtue of the action of applied forces ;

(II.) $\psi\, (x, y, z, \xi, \eta, \zeta, t') - \psi\, (x_0, y_0, z_0, \xi_0, \eta_0, \zeta_0, t')$,

according to the notation expressed by equation (2) above ;

(III.) $-\dfrac{1}{J} \{\phi\, (x, y, z, \xi, \eta, \zeta, t') - \phi\, (x, y, z, \xi, \eta, \zeta, t)\}$,

and

(IV.) $-\{\psi\, (x, y, z, \xi, \eta, \zeta, t) - \psi\, (x_0, y_0, z_0, \xi_0, \eta_0, \zeta_0, t)\}$.

If we suppose $t' - t$ to be infinitely small, these expressions become respectively, in accordance with the previous notation :—

(I.) $\dfrac{1}{J} \dfrac{de_0}{dt} (t' - t)$,

where e_0 denotes the value of e for $(x_0, y_0, z_0, \xi_0, \eta_0, \zeta_0, t)$;

(II.)
$$H + \frac{dH}{dt}(t' - t);$$

(III.)
$$-\frac{1}{J}\frac{de}{dt}(t' - t);$$

(IV.)
$$-H.$$

Hence we have

$$\Sigma \frac{Ht}{t} = \frac{\frac{1}{J}\frac{de_0}{dt}(t' - t)}{\frac{1}{2}(t + t')} + \frac{H + \frac{dH}{dt}(t' - t)}{t'}$$

$$+ \frac{-\frac{1}{J}\frac{de}{dt}(t' - t)}{\frac{1}{2}(t + t')} + \frac{-H}{t};$$

or, since $e - e_0$ is what we have denoted by ϵ,

$$\Sigma \frac{Ht}{t} = (t' - t)\left\{\frac{d}{dt}\left(\frac{H}{t}\right) - \frac{1}{J}\frac{d\epsilon}{tdt}\right\};$$

and the expression of the Second Thermodynamic Law becomes

$$\frac{d}{dt}\left(\frac{H}{t}\right) - \frac{1}{J}\frac{d\epsilon}{dt} = 0 \ \dots\dots\dots\dots\dots(5).$$

Eliminating ϵ from this by (4), we have

$$H = -\frac{t}{J}\frac{dw}{dt} \ \dots\dots\dots\dots\dots\dots(6);$$

and, eliminating H,
$$\epsilon = w - t\frac{dw}{dt} \ \dots\dots\dots\dots\dots\dots(7).$$

This is equivalent to
$$e = e_0 + w - t\frac{dw}{dt} \ \dots\dots\dots\dots\dots(8);$$

or, if N_0 denote the specific heat of the mass at any temperature t, when kept constantly in the mechanical state $(x_0, y_0, z_0, \xi_0, \eta_0, \zeta_0)$,

$$e = J\int_{t_0}^{t} N_0 dt + w - t\frac{dw}{dt} \ \dots\dots\dots\dots(9),$$

an expression which shows how the "intrinsic energy" of the body may be determined from observations giving w as a function of the seven independent variables, and N_0 as a function of the temperature, for a particular set of values of the geometrical elements. Conversely, by (5) we have

$$H = \frac{t}{J}\int \frac{d\epsilon}{t\,dt}.\,dt \quad\ldots\ldots\ldots\ldots(10);$$

and by (6) and (7), or simply by (4),

$$w = \epsilon - JH \quad\ldots\ldots\ldots\ldots(11);$$

which show how H and w may be determined for all temperatures from a knowledge of the intrinsic energy of the body, and of [one of] those functions themselves for a particular temperature.

189.   Let K denote the specific heat of the body at any temperature t, when it is allowed or compelled to vary in form and dimensions with the temperature, according to any fixed law—that is, when each of the variables x, y, z, ξ, η, ζ is a given function of t; and let N denote what this becomes in the particular case of each of these elements being maintained at a constant value; or, which is the same, let N be the specific heat of the body at any temperature when maintained at constant dimensions $(x, y, z, \xi, \eta, \zeta)$.   We have

$$JN = \frac{de}{dt} \quad\ldots\ldots\ldots\ldots(12),$$

$$JK = \frac{de}{dt} + \frac{d}{dx}(JH)\frac{dx}{dt} + \frac{d}{dy}(JH)\frac{dy}{dt} + \frac{d}{dz}(JH)\frac{dz}{dt}$$
$$+ \frac{d}{d\xi}(JH)\frac{d\xi}{dt} + \frac{d}{d\eta}(JH)\frac{d\eta}{dt} + \frac{d}{d\zeta}(JH)\frac{d\zeta}{dt} \quad\ldots\ldots(13).$$

Since JH is equal to $e - w$, this expression may be modified as follows:—If D denote the total differential of e,

$$JK = \frac{De}{dt} - \left(\frac{dw}{dx}\frac{dx}{dt} + \frac{dw}{dy}\frac{dy}{dt} + \frac{dw}{dz}\frac{dz}{dt}\right.$$
$$\left. + \frac{dw}{d\xi}\frac{d\xi}{dt} + \frac{dw}{d\eta}\frac{d\eta}{dt} + \frac{dw}{d\zeta}\frac{d\zeta}{dt}\right) \quad\ldots\ldots\ldots(13\,bis).$$

190.   These equations may be applied to any kind of matter; and they express all the information that can be derived, from the general thermodynamic principles, regarding the relations between thermal and mechanical effects produced by condensations,

rarefactions, or distortions of any kind. For the case of a fluid they become reduced at once to the forms investigated specially for fluids in my previous communications. Thus, if the mass considered be one pound of any kind of fluid, we may take one of the six variables x, y, z, ξ, η, ζ as the volume v, which it is made to occupy in any particular condition, and the remaining five will not affect its physical properties, and will therefore disappear from all the preceding equations; and the state of the fluid will be completely defined by the values of the two independent variables v, t. Then, if p denote the pressure, we must have

$$\frac{dw}{dv} = -p;$$

since $-pdv$ is the work done upon a fluid in compressing it under pressure p from a volume v to a volume $v + dv$. But from (9) we have

$$\frac{de}{dv} = \frac{dw}{dv} - t \frac{d}{dt} \frac{dw}{dv},$$

and, therefore, $$\frac{de}{dv} = t \frac{dp}{dt} - p \dots\dots\dots\dots\dots\dots(14),$$

and $$\frac{d(JH)}{dv} = t \frac{dp}{dt} \dots\dots\dots\dots\dots(14 \, bis).$$

Hence (13) becomes

$$JK = \frac{de}{dt} + t \frac{dp}{dt}\left(\frac{dv}{dt}\right),$$

where $\left(\dfrac{dv}{dt}\right)$ expresses the assumed relation between the naturally independent variables v, t. If this be such that the pressure is constant, we have

$$\frac{dv}{dt} = \frac{-dp/dt}{dp/dv};$$

and, K being now the specific heat under constant pressure, we have finally

$$JK = \frac{de}{dt} + \frac{t\,(dp/dt)^2}{-(dp/dv)} \dots\dots\dots\dots\dots(15).$$

191. These equations (14) and (15), together with the unmodified equation (12), which retains the same form in all cases, express the general thermodynamic relations between the intrinsic

energy, the pressure, and the specific heats of a fluid. If we eliminate e, we have

$$JK - JN = \frac{t(dp/dt)^2}{-dp/dv} \quad \dots\dots\dots\dots\dots\dots\dots(16),$$

and

$$\frac{d}{dv}(JN) = t\frac{d^2p}{dt^2} \quad \dots\dots\dots\dots\dots\dots\dots\dots(17),$$

which are the equations used to express those relations in a recent paper by Mr Joule and myself, "On the Thermal Effects of Fluids in Motion*" [Art. XLIX., below].

If, instead of Jdt/t, we substitute μdt, considering μ as a function of Carnot's function of the temperature, they become identical with the two fundamental equations (14) and (16) given in Part III. of my first communication "On the Dynamical Theory of Heat†" [§§ 47, 48, above].

192. To apply the preceding equations to a body possessing rigidity, it is necessary to take the form as well as the bulk into account, and therefore to retain, besides the temperature, six independent variables to express those elements. There is, of course, an infinite variety of ways in which the form and bulk of a homogeneously strained body may be expressed by means of six independent variables. Thus the lengths (three variables) and the mutual inclinations (three variables) of the edges of a parallelepiped enclosing always same portion of the solid in all states of strain (which of course always remains a parallelepiped, provided the strain is homogeneous throughout the solid), may be chosen for the independent variables; or we may choose the six elements of an ellipsoid enclosing always the same portion of the solid (which will always remain an ellipsoid however the solid be strained, provided it is strained homogeneously). Thus, let us actually take for x, y, z the lengths of three conterminous edges OX, OY, OZ of a certain parallelepiped of the solid, and for ξ, η, ζ the angles between the planes meeting in these edges respectively, the parallelepiped being so chosen that it becomes strained into a cube of unit dimensions, when the solid is in the particular state at which we wish to investigate its thermo-elastic properties.

---

* *Transactions of the Royal Society*, June 15, 1854.

† *Transactions of the Royal Society of Edinburgh*, March 17, 1851.

193.   If then we take

$$x_0 = 1, \qquad y_0 = 1, \qquad z_0 = 1,$$

$$\xi_0 = \tfrac{1}{2}\pi, \quad \eta_0 = \tfrac{1}{2}\pi, \quad \zeta_0 = \tfrac{1}{2}\pi,$$

and if we suppose x, y, z, ξ, η, ζ to differ infinitely little from $x_0, y_0, z_0, \xi_0, \eta_0, \zeta_0$ respectively, the actual state $(x, y, z, \xi, \eta, \zeta)$ will be one in which the body is strained from the state $(x_0, y_0, z_0, \xi_0, \eta_0, \zeta_0)$ by the edges of the cube being elongated by $x - x_0$, $y - y_0$, $z - z_0$, and the angles meeting in three conterminous edges receiving augmentations of $\xi - \xi_0$, $\eta - \eta_0$, $\zeta - \zeta_0$. It is clear, since the altered angles differ each infinitely little from a right angle, that the strains represented by $\xi - \xi_0$, $\eta - \eta_0$, $\zeta - \zeta_0$ involve no change of volume, and are simple deformations, each of a perfectly definite kind, in the planes YOZ, ZOX, XOY respectively, and that the change of volume due to the six coexistent strains is actually an infinitely small augmentation amounting to

$$x - x_0 + y - y_0 + z - z_0.$$

194.   Considering still $x - x_0$ &c. as each very small, we have the following development by Maclaurin's theorem, the zero suffixes to the differential coefficients being used for brevity to denote the values of the different coefficients at $(x_0, y_0, z_0, \xi_0, \eta_0, \zeta_0)$.

$$w = \left(\frac{dw}{dx}\right)_0 (x - x_0) + \left(\frac{dw}{dy}\right)_0 (y - y_0) + \left(\frac{dw}{dz}\right)_0 (z - z_0)$$

$$+ \left(\frac{dw}{d\xi}\right)_0 (\xi - \xi_0) + \left(\frac{dw}{d\eta}\right)_0 (\eta - \eta_0) + \left(\frac{dw}{d\zeta}\right)_0 (\zeta - \zeta_0)$$

$$+ \tfrac{1}{2}\left\{ \left(\frac{d^2w}{dx^2}\right)_0 (x - x_0)^2 + \left(\frac{d^2w}{dy^2}\right)_0 (y - y_0)^2 + \left(\frac{d^2w}{dz^2}\right)_0 (z - z_0)^2 \right.$$

$$+ \left(\frac{d^2w}{d\xi^2}\right)_0 (\xi - \xi_0)^2 + \left(\frac{d^2w}{d\eta^2}\right)_0 (\eta - \eta_0)^2 + \left(\frac{d^2w}{d\zeta^2}\right)_0 (\zeta - \zeta_0)^2$$

$$+ 2\left(\frac{d^2w}{dy\,dz}\right)_0 (y - y_0)(z - z_0) + 2\left(\frac{d^2w}{dz\,dx}\right)_0 (z - z_0)(x - x_0)$$

$$\left. + 2\left(\frac{d^2w}{dx\,dy}\right)_0 (x - x_0)(y - y_0)\right\}$$

$$+ 2 \left(\frac{d^2w}{d\eta\, d\zeta}\right)_0 (\eta - \eta_0)(\zeta - \zeta_0) + 2 \left(\frac{d^2w}{d\zeta d\xi}\right)_0 (\zeta - \zeta_0)(\xi - \xi_0)$$

$$+ 2 \left(\frac{d^2w}{d\xi d\eta}\right)_0 (\xi - \xi_0)(\eta - \eta_0)$$

$$+ 2 \left(\frac{d^2w}{dx d\xi}\right)_0 (x - x_0)(\xi - \xi_0) + 2 \left(\frac{d^2w}{dy d\eta}\right)_0 (y - y_0)(\eta - \eta_0)$$

$$+ 2 \left(\frac{d^2w}{dz d\zeta}\right)_0 (z - z_0)(\zeta - \zeta_0)$$

$$+ 2 \left(\frac{d^2w}{dx d\eta}\right)_0 (x - x_0)(\eta - \eta_0) + 2 \left(\frac{d^2w}{dx d\zeta}\right)_0 (x - x_0)(\zeta - \zeta_0)$$

$$+ 2 \left(\frac{d^2w}{d\zeta dy}\right)_0 (y - y_0)(\zeta - \zeta_0) + 2 \left(\frac{d^2w}{dy d\xi}\right)_0 (y - y_0)(\xi - \xi_0)$$

$$+ 2 \left(\frac{d^2w}{dz d\xi}\right)_0 (z - z_0)(\xi - \xi_0) + 2 \left(\frac{d^2w}{dz d\eta}\right)_0 (z - z_0)(\eta - \eta_0) + \&\text{c. (18)}.$$

195. According to the system of variables[*] which we have adopted, as set forth in § 193, when $x - x_0$ &c. are each infinitely small, x increasing corresponds to a motion of all the particles in a plane at a distance unity from YOZ, in directions perpendicular to this plane, through a space numerically equal to the increment of x; ξ increasing corresponds to a motion of all the particles at a distance unity from XOY, in directions parallel to YO, through a space equal to the increment of ξ, or to a motion of all the particles at a distance unity from XOZ, in directions parallel to ZO, through a space equal to the increment of ξ, or to two such motions superimposed, through any spaces respectively, amounting together to a quantity equal to the increment of ξ. Similar statements apply to the effects of variations of the other four variables. Hence, if P, Q, R denote the normal components of the superficial tensions experienced respectively by the three pairs of opposite faces of the unit cube of the solid in the state of strain in which we are considering it, and if S, T, U be the components, along the planes of the faces, of the actual tensions, taken in order of symmetry, so that S denotes the component, perpendicular to the edge opposite to OX, of the superficial tension in either of the

---

[*] [A method of generalized stress and strain components is fully developed in "Elements of a Mathematical Theory of Elasticity," first published in the *Transactions of the Royal Society* for April, 1856, and embodied in an article on "Elasticity," about to be published in the *Encyclopædia Britannica*.]

faces meeting in that edge (which are equal for these two faces, or else the cube would not be in equilibrium, but would experience the effect of a couple in a plane perpendicular to OX), and T and U denote components, perpendicular respectively to OY and OZ, of the superficial tensions of the pairs of faces meeting in those edges, the work done on the parallelepiped during an infinitely small strain in which the variables become augmented by dx, dy, &c. respectively will be

$$Pdx + Qdy + Rdz + Sd\xi + Td\eta + Ud\zeta.$$

Hence, if the portion of matter of which the intrinsic energy is denoted by e, and to which the notation ϵ, w, &c. applies, be the matter within the parallelepiped referred to, we have

$$\left. \begin{array}{l} \dfrac{dw}{dx} = P, \quad \dfrac{dw}{dy} = Q, \quad \dfrac{dw}{dz} = R \\[2mm] \dfrac{dw}{d\xi} = S, \quad \dfrac{dw}{d\eta} = T, \quad \dfrac{dw}{d\zeta} = U \end{array} \right\} \quad \ldots\ldots\ldots\ldots(19).$$

196.   Using the development of w expressed by (18), we derive from these equations the following expressions for the six component tensions:—

$$\left. \begin{aligned} P &= \left(\frac{dw}{dx}\right)_0 \\ &+ \left(\frac{d^2w}{dx^2}\right)_0 (x - x_0) + \left(\frac{d^2w}{dx\,dy}\right)_0 (y - y_0) + \left(\frac{d^2w}{dx\,dz}\right)_0 (z - z_0) \\ &+ \left(\frac{d^2w}{dx\,d\xi}\right)_0 (\xi - \xi_0) + \left(\frac{d^2w}{dx\,d\eta}\right)_0 (\eta - \eta_0) + \left(\frac{d^2w}{dx\,d\zeta}\right)_0 (\zeta - \zeta_0); \\[2mm] Q &= \left(\frac{dw}{dy}\right)_0 \\ &+ \left(\frac{d^2w}{dx\,dy}\right)_0 (x - x_0) + \left(\frac{d^2w}{dy^2}\right)_0 (y - y_0) + \left(\frac{d^2w}{dy\,dz}\right)_0 (z - z_0) \\ &+ \left(\frac{d^2w}{dy\,d\xi}\right)_0 (\xi - \xi_0) + \left(\frac{d^2w}{dy\,d\eta}\right)_0 (\eta - \eta_0) + \left(\frac{d^2w}{dy\,d\zeta}\right)_0 (\zeta - \zeta_0); \\[2mm] R &= \left(\frac{dw}{dz}\right)_0 \\ &+ \left(\frac{d^2w}{dz\,dx}\right)_0 (x - x_0) + \left(\frac{d^2w}{dz\,dy}\right)_0 (y - y_0) + \left(\frac{d^2w}{dz^2}\right)_0 (z - z_0) \\ &+ \left(\frac{d^2w}{dz\,d\xi}\right)_0 (\xi - \xi_0) + \left(\frac{d^2w}{dz\,d\eta}\right)_0 (\eta - \eta_0) + \left(\frac{d^2w}{dz\,d\zeta}\right)_0 (\zeta - \zeta_0); \end{aligned} \right\} \quad (20).$$

$$
\begin{aligned}
S = {} & \left(\frac{dw}{d\xi}\right)_0 \\
& + \left(\frac{d^2w}{d\xi dx}\right)_0 (x - x_0) + \left(\frac{d^2w}{d\xi dy}\right)_0 (y - y_0) + \left(\frac{d^2w}{d\xi dz}\right)_0 (z - z_0) \\
& + \left(\frac{d^2w}{d\xi^2}\right)_0 (\xi - \xi_0) + \left(\frac{d^2w}{d\xi d\eta}\right)_0 (\eta - \eta_0) + \left(\frac{d^2w}{d\xi d\zeta}\right)_0 (\zeta - \zeta_0); \\[6pt]
T = {} & \left(\frac{dw}{d\eta}\right)_0 \\
& + \left(\frac{d^2w}{d\eta dx}\right)_0 (x - x_0) + \left(\frac{d^2w}{d\eta dy}\right)_0 (y - y_0) + \left(\frac{d^2w}{d\eta dz}\right)_0 (z - z_0) \\
& + \left(\frac{d^2w}{d\xi d\eta}\right)_0 (\xi - \xi_0) + \left(\frac{d^2w}{d\eta^2}\right)_0 (\eta - \eta_0) + \left(\frac{d^2w}{d\zeta d\eta}\right)_0 (\zeta - \zeta_0); \\[6pt]
U = {} & \left(\frac{dw}{d\zeta}\right)_0 \\
& + \left(\frac{d^2w}{d\zeta dx}\right)_0 (x - x_0) + \left(\frac{d^2w}{d\zeta dy}\right)_0 (y - y_0) + \left(\frac{d^2w}{d\zeta dz}\right)_0 (z - z_0) \\
& + \left(\frac{d^2w}{d\zeta d\xi}\right)_0 (\xi - \xi_0) + \left(\frac{d^2w}{d\zeta d\eta}\right)_0 (\eta - \eta_0) + \left(\frac{d^2w}{d\zeta^2}\right)_0 (\zeta - \zeta_0)
\end{aligned} \right\} \quad (21).
$$

197. These equations express in the most general possible manner the conditions of equilibrium of a solid in any state of strain whatever at a constant temperature. They show how the straining forces are altered with any infinitely small alteration of the strain. If we denote by P_0 &c. the values of P &c. for the state $(x_0, y_0, z_0, \xi_0, \eta_0, \zeta_0, t)$, the values of $P - P_0$, $Q - Q_0$, $R - R_0$, $S - S_0$, $T - T_0$, $U - U_0$ given by these equations as linear functions of the strains $(x - x_0)$, $(y - y_0)$, $(z - z_0)$, $(\xi - \xi_0)$, $(\eta - \eta_0)$, $(\zeta - \zeta_0)$, with twenty-one coefficients, express the whole tensions required to apply these strains to the cube, if the condition of the solid when the parallelepiped is exactly cubical is a condition of no strain, and in this case become (if single letters are substituted for the co-efficients $(d^2w/dx^2)_0$ &c.) identical with the equations of equilibrium of an elastic solid subjected to infinitely small strains, which have been given by Green, Cauchy, Haughton, and other writers. Many mathematicians and experimenters have endeavoured to show that in actual solids there are certain essential relations between these twenty-one coefficients [or moduluses] of elasticity. Whether or not it may be true that such relations do hold for natural crystals, it is quite certain that an arrangement of actual pieces of matter may be made, constituting a homogeneous whole when considered

on a large scale (being, in fact, as homogeneous as writers adopting the atomic theory in any form consider a natural crystal to be),
which shall have an arbitrarily prescribed value for each one of
these twenty-one coefficients.   No one can legitimately deny for all
natural crystals, known and unknown, any property of elasticity, or
any other mechanical or physical property, which a solid composed
of natural bodies artificially put together may have in reality.   To
do so is to assume that the infinitely inconceivable structure of the
particles of a crystal is essentially restricted by arbitrary conditions
imposed by mathematicians for the sake of shortening the equations
by which their properties are expressed.   It is true experiment
might, and does, show particular values for the coefficients for particular bodies; but I believe even the collation of recorded experimental investigations is enough to show bodies violating every
relation that has been imposed ; and I have not a doubt that an
experiment on a natural crystal, magnetized if necessary, might be
made to show each supposed relation violated.   Thus it has been
shown, first I believe by Mr Stokes, that the relation which the
earlier writers supposed to exist between rigidity and resistance to
compression is not verified, because experiments on the torsion of
wires of various metals, rods of india-rubber, &c. indicate, on the
whole, less rigidity than would follow, according to that relation, from their resistance to compression, and less in different proportions for different metals.   It is quite certain that india-rubber,
jelly of any kind (ever so stiff), and gutta-percha, are all of them
enormously less rigid in proportion to their resistance to compression than glass or the metals ; and they are all certainly substances
which may be prepared so as to be at least as homogeneous as rods,
wires, bars, or tubes of metals.   From some experiments communicated to me by Mr Clerk Maxwell, which he has made on iron
wire by flexure and torsion, it appears highly probable that iron is
more rigid in proportion to its resistance to compression than M.
Wertheim's experiments on brass and glass show these bodies to be.

[198.   Since the publication of this paper, the same conclusion
as to the relative qualities of iron and brass has been arrived at by
Everett (*Transactions of the Royal Society*, 1865 and 1866) as a
result of fresh experiments made by himself on these substances
in the Physical Laboratory of the University of Glasgow :—but
an opposite conclusion with reference to two specimens of flint
glass upon which he experimented, and which both showed greater

rigidity in proportion to compressibility than either his own experiments or those of others had shown for iron or any other substance accurately experimented on. Far beyond these specimens of glass, with respect to greatness of rigidity in proportion to compressibility, is cork; which though not hitherto accurately experimented on, and though no doubt very variable in its elastic quality, shows obviously a very remarkable property, on which its use for corking bottles depends, viz. that a column of it compressed endwise does not swell out sidewise to any sensible degree, if at all. It is easy to construct a model elastic solid, on the plan suggested above, which shall actually show lateral shrinking when compressed longitudinally, and lateral swelling when pulled out longitudinally. The false theory, referred to above as having been first proved to be at fault by Stokes, gives for every kind of solid $\frac{1}{4}$ as the ratio of the lateral shrinking to the longitudinal elongation when a rod is pulled out lengthwise. (Compare Thomson and Tait's *Natural Philosophy*, § 685, or *Encyclopædia Britannica*, Art. Elasticity, § 48.) The following Table shows how different are the values of that ratio determined by experiment on several real solids.]

Substances.	Authority.	Ratio of lateral shrinking or swelling to longitudinal extension or shortening under the influence of push or pull on both ends of a column of the substance.
Cork.............................	General experience and some accurate measurements of diameter of a cork under various degrees of end-pressure, producing shortenings from small amounts up to as much as $\frac{1}{5}$ of the original length............	0
Specimens of "crystal" glass	Wertheim..................	·33
A specimen of flint glass.	Everett (1865)	·26
Another specimen of flint glass	Everett (1866)	·23
A specimen of brass.......	Wertheim	·34
Drawn brass rod............	Everett (1866)	·47
Copper	W. Thomson...............	from ·40 to ·23
Iron	Clerk Maxwell	·27
Steel	Kirchhoff	·29
Cast steel...................	Everett (1866)	·31
Vulcanized india-rubber..	Joule	Less than ·5 by an exceedingly small amount.

199.   The known fact that [many] gelatinous bodies, and the nearly certain fact that most bodies of all kinds, when their temperatures are raised, become less rigid to a much more marked extent than that of any effect on their compressibilities, are enough to show that neither the relation first supposed to exist, nor any other constant relation between compressibility and rigidity, can hold even for one body at different temperatures.

200.   Again, some of the relations which have been supposed to exist lead to three principal axes of elasticity.   Many natural crystals do certainly exhibit perfect symmetry of form with reference to three rectangular axes, and therefore probably possess all their physical properties symmetrically with reference to those axes; but as certainly many, and among them some of the best known, of natural crystals do not exhibit symmetry of form with reference to rectangular axes, and possess the mechanical property of resisting fracture differently in different directions, without symmetry about any three rectangular axes—for instance, Iceland spar, which has three planes of greatest brittleness ("cleavage-planes"), inclined at equal angles to one another and to a common axis (the "optic-axis" of the crystal).   If, as probably must be the case, the elastic properties within the limits of elasticity have correspondence with the mechanical properties on which the brittleness in different directions depends, the last-mentioned class of crystals cannot have three principal axes of elasticity at right angles to one another.   It will be an interesting inquiry (*Encyclopædia Britannica*, Art. Elasticity, § 41) to examine thoroughly the various directional properties of an elastic solid represented by the different coefficients (of which the entire number may of course be reduced from twenty-one to eighteen by a choice of axes), or by various combinations of them.

201.   The general thermodynamic principles expressed above in the equations (6), (8), (12), and (13) enable us to determine the relations between the evolution of heat or cold by strains of any kind effected on an elastic solid, the variation of its elastic forces with temperature, and the differences and variations of its specific heats.   Thus (6) gives at once, when the development of w expressed by (18) is used, and for $(dw/dx)_0$, &c. are substituted P, &c., which are infinitely nearly equal to them, the following expression for the heat absorbed by an infinitely small straining, namely from $(x_0, y_0, z_0, \xi_0, \eta_0, \zeta_0)$ to $(x, y, z, \xi, \eta, \zeta)$:—

$$H = \frac{t}{J}\left\{ -\frac{dP}{dt}(x-x_0) - \frac{dQ}{dt}(y-y_0) - \frac{dR}{dt}(z-z_0) \right.$$

$$\left. -\frac{dS}{dt}(\xi-\xi_0) - \frac{dT}{dt}(\eta-\eta_0) - \frac{dU}{dt}(\zeta-\zeta_0) \right\}\ldots(22).$$

202. We conclude that cold is produced whenever a solid is strained by opposing, and heat when it is strained by yielding to, any elastic force of its own, the strength of which would diminish if the temperature were raised—but that, on the contrary, heat is produced when a solid is strained against, and cold when it is strained by yielding to, any elastic force of its own, the strength of which would increase if the temperature were raised. When the stress is a pressure, uniform in all directions, fluids may be included in the statement. Thus we may conclude as certain:—

(1) That a cubical compression of any elastic fluid or solid in an ordinary condition would cause an evolution of heat; but that, on the contrary, a cubical compression would produce cold in any substance, solid or fluid, in such an abnormal state that it would contract if heated, while kept under constant pressure.

(2) That if a wire already twisted be suddenly twisted further, always, however, within its limits of elasticity, cold will be produced; and that if it be allowed suddenly to untwist, heat will be evolved from itself (besides heat generated externally by any work allowed to be wasted, which it does in untwisting). For I suppose it is certain that the torsive rigidity of every wire is diminished by an elevation of temperature.

(3) That a spiral spring suddenly drawn out will become lower in temperature, and will rise in temperature when suddenly allowed to draw in. [This result has since been experimentally verified by Joule (" Thermodynamic Properties of Solids," *Trans. Roy. Soc.*, June, 1858), and the amount of the effect found to agree with that calculated, according to the preceding thermodynamic theory, from the amount of the weakening of the spring which he found by experiment.]

(4) That a bar or rod or wire of any substance with or without a weight hung on it, or experiencing any degree of end thrust, to begin with, becomes cooled if suddenly elongated by end pull or by diminution of end thrust, and warmed if suddenly shortened by end thrust or by diminution of end pull: except abnormal cases, in which, with constant end pull or end thrust, elevation of tempera-

ture produces shortening; in every such case pull or diminished thrust produces elevation of temperature, thrust or diminished pull lowering of temperature.

(4') That an india-rubber band suddenly drawn out (within its limits of elasticity) produces cold, and that, on the contrary, when allowed to contract, heat will be evolved from it. For it is certain that an india-rubber band with a weight suspended by it will expand in length if the temperature be raised. [Alas for over-confident assertion! This is not true—at all events not true *in general* for either natural or vulcanized india-rubber, but only true for india-rubber in somewhat exceptional circumstances. It was founded on the supposition that india-rubber becomes less rigid when raised in temperature, which, besides seeming to be expectable for solids generally, seemed to be experimentally proved for india-rubber by the familiar stiffness of common india-rubber in very cold weather. My original supposition is in fact correct for india-rubber which has become rigid by being kept at rest at a low temperature for some time. In this condition india-rubber was found by Joule to be cooled when suddenly stretched, and heated when the stretching weight was removed; and therefore, when in this condition, it is certain, from the thermodynamic principle, that a band of the substance bearing a weight will expand in length if the temperature is raised, and shrink when the temperature is lowered. But the very piece of india-rubber in which Joule found a cooling effect by pull when its temperature was 5° C., gave him a heating effect by pull, and a cooling effect on withdrawal of pull, when the temperature was 15° C. Joule experimented also on vulcanized india-rubber, and with it always found a heating effect when the substance was pulled out, and a cooling effect when it was allowed to shrink back. I pointed out to him that therefore, by thermodynamic theory, a vulcanized india-rubber band, when stretched by a constant weight of sufficient amount hung on it, must, when heated, pull up the weight, and when cooled, allow the weight to descend. This is an experiment which any one can make with the greatest ease by hanging a few pounds weight on a common india-rubber band, and taking a red-hot coal, in a pair of tongs, or a red-hot poker, and moving it up and down close to the band. The way in which the weight rises when the red-hot body is near, and falls when it is removed, is quite startling. Joule experimented on the amount of shrinking per degree of elevation of temperature, with different weights hung on a band of vulcanized india-rubber, and found that they closely agreed with the amounts calculated by my theory from the heating effects of pull, and cooling effects of ceasing to pull, which he had observed in the same piece of india-rubber. Joule's experiments leave the statements of the following paragraph (5) true for common india-rubber at 5° C., but reverse it for common india-rubber at higher temperatures and for vulcanized india-rubber at all ordinary temperatures—that is to say, leave it applicable to these substances with "pull" substituted for "push" throughout.]

(5) We may conclude as highly probable, that pushing a column of india-rubber together longitudinally while leaving it free at its sides will cause the evolution of heat, when the force by which its ends are pushed together falls short of a certain limit; but that, on the contrary, if this force exceeds a certain limit, cold will be produced by suddenly increasing the force a very little, so as to contract the column further. For I suppose it is certain that a column of india-rubber with no weight, or only a small weight on its top, will expand longitudinally when its temperature is raised; but it appears to me highly probable that if the weight on the top of the column exceed a certain limit, the diminished rigidity of the column will allow it to descend when the temperature is raised. [This second change we now know to be contrary to the true state of the case; for we have seen that the rigidity of vulcanized india-rubber is augmented by elevation of temperature.]

203. The specific heat of an elastic solid homogeneously strained under given pressures or tensions will be obtained by finding the differential coefficients of $x, y, z, \xi, \eta, \zeta$ with reference to t, so as to make P, Q, R, S, T, U each remain constant or vary in a given manner—that is to say, by finding the coefficients of expansion in various dimensions for the body with an infinitely small change of its temperature, and using these in (3) above.

204. The elastic properties of such a crystal as is frequently found in natural specimens of garnet—a regular rhombic dodecahedron—must, if they correspond to the crystalline form, be symmetrical with reference to six axes in the substance perpendicular to the six pairs of opposite faces of the dodecahedron, or to the six edges of a regular tetrahedron related to the dodecahedron in a determinate manner (having for its corners four of the eight

trihedral corners of the dodecahedron); and yet they may differ, and in all probability they do differ, in different directions through the crystal. The relations among the coefficients of elasticity, according to the system of independent variables used in the preceding paper, which are required to express such circumstances, may be investigated by choosing for the normal cube a cube with faces perpendicular to the lines joining the three pairs of opposite tetrahedral corners of the dodecahedron. The choice of the normal cube makes all the coefficients vanish except nine, and makes these nine related one to another as follows:

$$\left(\frac{d^2w}{dx^2}\right)_0 = \left(\frac{d^2w}{dy^2}\right)_0 = \left(\frac{d^2w}{dz^2}\right)_0 = \lambda + 2\mu,$$

$$\left(\frac{d^2w}{d\xi^2}\right)_0 = \left(\frac{d^2w}{d\eta^2}\right)_0 = \left(\frac{d^2w}{d\zeta^2}\right)_0 = \mu + \kappa,$$

and $$\left(\frac{d^2w}{dy\,dz}\right)_0 = \left(\frac{d^2w}{dz\,dx}\right)_0 = \left(\frac{d^2w}{dx\,dy}\right)_0 = \lambda;$$

where λ, μ, κ are three independent coefficients, introduced merely for the sake of comparison with M. Lamé's notation. In different natural crystals of the cubical system, such as fluor-spar, garnet, &c., it is probable that the three coefficients here left have different relations with one another. The body would, as is known, be, in its elastic qualities, perfectly isotropic if, and not so unless, the further relation

$$\left(\frac{d^2w}{dx^2}\right)_0 = \left(\frac{d^2w}{dy\,dz}\right)_0 + 2\left(\frac{d^2w}{d\xi^2}\right)_0$$

were fulfilled. Hence the quantity κ in the preceding formulæ expresses the crystalline quality which I suppose to exist in the elasticity of a crystal of the cubic class.

205. The fact of there being a system of six axes of symmetry in cubic crystals (the diagonals of sides of the cube), has suggested to me a system of independent variables, symmetrical with respect to those six axes, which I believe may be found extremely convenient in the treatment of a mechanical theory of crystallography, and which, so far as I know, has not hitherto been introduced for the expression of a state of strain in an elastic solid. It is simply the *six edges of a tetrahedron enclosing always the same part of the solid*, a system of variables which might be used in all expressions connected with the theory of the elasticity of solids. To apply it to express the elastic properties of a crystal of the cubical class, let

the tetrahedron be chosen with its edges parallel to the six lines which are lines of symmetry when the solid is unstrained. In any state of strain let x, y, z be the lengths of three edges lying in one plane, and ξ, η, ζ those of the three others (which meet in a point). Let x_0, y_0, z_0, ξ_0, η_0, ζ_0 denote the values (equal among themselves) of these variables for the unstrained state, and let w be the work required to bring any portion of the solid (whether the tetrahedron itself or not is of no consequence) from the unstrained state to the state (x, y, z, ξ, η, ζ) while kept at a constant temperature. The relations among the coefficients of elasticity according to this system of variables, to express perfect symmetry with reference to the six axes, will clearly be :—

$$\left(\frac{d^2w}{dx^2}\right)_0 = \left(\frac{d^2w}{dy^2}\right)_0 = \left(\frac{d^2w}{dz^2}\right)_0 = \left(\frac{d^2w}{d\xi^2}\right)_0 = \left(\frac{d^2w}{d\eta^2}\right)_0 = \left(\frac{d^2w}{d\zeta^2}\right)_0 = \varpi ;$$

$$\left(\frac{d^2w}{dy\,dz}\right)_0 = \left(\frac{d^2w}{dz\,dx}\right)_0 = \left(\frac{d^2w}{dx\,dy}\right)_0 = \left(\frac{d^2w}{d\eta\,d\zeta}\right)_0 = \left(\frac{d^2w}{d\zeta\,d\xi}\right)_0$$

$$= \left(\frac{d^2w}{d\xi\,d\eta}\right)_0 = \left(\frac{d^2w}{dx\,d\eta}\right)_0 = \left(\frac{d^2w}{dx\,d\zeta}\right)_0 = \left(\frac{d^2w}{dy\,d\zeta}\right)_0$$

$$= \left(\frac{d^2w}{dy\,d\xi}\right)_0 = \left(\frac{d^2w}{dz\,d\xi}\right)_0 = \left(\frac{d^2w}{d\zeta\,d\eta}\right)_0 = \sigma ;$$

$$\left(\frac{d^2w}{dx\,d\xi}\right)_0 = \left(\frac{d^2w}{dy\,d\eta}\right)_0 = \left(\frac{d^2w}{dz\,d\zeta}\right)_0 = \omega ,—$$

where ϖ, σ, ω denote three independent coefficients of elasticity for the substance. The definition of the new system of variables may be given as simply, and in some respects more conveniently, by referring to the dodecahedron, whose faces are perpendicular to the edges of the tetrahedron. Thus the six variables x, y, z, ξ, η, ζ may be taken to denote respectively the mutual distances of the six pairs of parallel faces of the rhombohedron into which the regular dodecahedron is altered when the solid is strained in any manner. Thus, if the portion of the solid considered be the dodecahedron itself, and of such dimensions that when it is in its normal state the area of each face is unity, the values of $d\omega/dx$ &c., denoted, as in the preceding paper, by P, Q, R, S, T, U, are normal tensions (reckoned, as usual, per unit of area) on surfaces in the solid parallel to the faces of the dodecahedron, which compounded give the actual straining force to which the solid is subjected. The coefficients denoted above by ϖ, σ, ω are such as to give the following expressions for the component straining tensions in terms of the strains:—

$$P = \varpi\,(x - x_0) + \omega\,(\xi - \xi_0) + \sigma\,(y - y_0 + z - z_0 + \eta - \eta_0 + \zeta - \zeta_0)\,;$$
$$Q = \varpi\,(y - y_0) + \omega\,(\eta - \eta_0) + \sigma\,(z - z_0 + x - x_0 + \zeta - \zeta_0 + \xi - \xi_0)\,;$$
$$R = \varpi\,(z - z_0) + \omega\,(\zeta - \zeta_0) + \sigma\,(x - x_0 + y - y_0 + \xi - \xi_0 + \eta - \eta_0)\,;$$
$$S = \varpi\,(\xi - \xi_0) + \omega\,(x - x_0) + \sigma\,(y - y_0 + z - z_0 + \eta - \eta_0 + \zeta - \zeta_0)\,;$$
$$T = \varpi\,(\eta - \eta_0) + \omega\,(y - y_0) + \sigma\,(z - z_0 + x - x_0 + \zeta - \zeta_0 + \xi - \xi_0)\,;$$
$$U = \varpi\,(\zeta - \zeta_0) + \omega\,(z - z_0) + \sigma\,(x - x_0 + y - y_0 + \xi - \xi_0 + \eta - \eta_0).$$

206. The three quantities, ϖ, ω, σ, or the three coefficients of elasticity according to the new system of independent variables, will express, by their different relative values, the elastic properties of all crystals of the cubical class. For a perfectly isotropic body, a particular numerical relation, which I have not yet determined, must hold between ϖ, ρ, and σ; and two independent coefficients of elasticity will remain. To determine this relation, and to find the formulæ of transformation from one set of variables to another on the new system, or from the new system to the ordinary system (that which was used in the preceding portion of this paper), or *vice versâ*, may be interesting objects of enquiry.

GLASGOW COLLEGE, *March* 10, 1855.

207. *Extracted from Nichol's "Cyclopædia of the Physical Sciences,"* second edition, 1860. *Thermomagnetism.* (1) *Experimental Facts.*—Gilbert found that if a piece of soft iron between the poles of a magnet be raised to a bright red heat it loses all its ordinary indications of magnetism, and it only retains (Faraday, *Exp. Res.* 2344—2347) slight traces of the paramagnetic character. Nickel loses its magnetic inductive capacity very rapidly as its temperature rises about 635° Fahr., and has very little left at the temperature of boiling oil. Cobalt loses its inductive capacity at a far higher temperature than that of either, near the melting-point of copper. Of the three metals, iron remains nearly constant, nickel falls gradually, and *cobalt actually rises* in inductive capacity as the temperature is raised from 0° to 300° Fahr. (Faraday, *Exp. Res.* 3428; *Phil. Trans.* Nov. 1855). Cobalt, of course, must have a maximum inductive capacity at some temperature intermediate between 300° Fahr. and the temperature of melting copper. Crystals, when their

temperatures are raised, have their magnetic inductive capacities in different directions of the crystalline substance rendered less unequal, and in general to a very marked degree. Thus Faraday found the difference of inductive capacities in different directions in a crystal of bismuth (a diamagnetic crystal) reduced to less than half when the temperature was raised from 100° to 280°. In carbonate of iron (a paramagnetic crystal) the difference of inductive capacities in different directions was reduced to one third when the temperature was raised from 70° to 289° Fahr., and was tripled when the temperature was again brought down to 70° (*Exp. Res.* 3400 and 3411).

(2) *Thermodynamic Relations.*—The theory of the mutual convertibility of heat and mechanical work in reversible operations when applied to these phenomena proves:—1. That a piece of soft iron at a moderate or low red heat, when drawn gently away from a magnet experiences a cooling effect, and when allowed to approach a magnet experiences a heating effect; that nickel at ordinary temperatures, and cobalt at high temperatures, within some definite range below that of melting copper, experience the same kind of effects when subjected to similar magnetic operations. 2. That cobalt at ordinary atmospheric temperatures, and at all temperatures upwards to its temperature of maximum inductive capacity, experiences a cooling effect when allowed to approach a magnet slowly, and a heating effect when drawn away. 3. That a crystal in a magnetic field experiences a cooling effect when its axis of greatest paramagnetic or of least diamagnetic inductive capacity is turned round from a position along to a position across the lines of force, and a heating effect when such a motion is reversed.

208. [Let there be three rectangular axes fixed relatively to the moveable body, whether soft iron, or copper, or a crystal in a magnetic field, and, considering the whole magnetic motive* on the body, reduce it, after the manner of Poinsot, to three component forces along the magnetic axes and three couples round these axes. Let P, Q, R be the force-components, and S, T, U the couple-components thus obtained, which we must

---

* [In dynamics the want is keenly felt of an expression for a system of forces acting on a body: adopting a suggestion of my brother, Professor James Thomson, the word "motive" is used in the text to supply this want.]

suppose to be known functions of t, the temperature. Equation (22) of § 201 above gives H, the quantity of heat which must be supplied to prevent the body from becoming cooler when it is moved through infinitesimal spaces $x - x_0$, $y - y_0$, $z - z_0$ in the directions of the three axes, and turned through infinitesimal angles $\xi - \xi_0$, $\eta - \eta_0$, $\zeta - \zeta_0$ round the same axes. The lowering of temperature which it experiences if heat is neither given to it nor taken from it is equal to $\dfrac{H}{C}$, where C denotes the whole capacity for heat of the body, or the product of its mass or bulk by its specific heat per unit of mass or per unit of bulk. If the directions of x, y, z and ξ, η, ζ are such that P, Q, R, S, T, U are positive, then for iron and nickel, and for cobalt at temperatures above that of its maximum inductive capacity,

$$ -\frac{dP}{dt}, \quad -\frac{dQ}{dt}, \quad -\frac{dR}{dt}, \quad -\frac{dS}{dt}, \quad -\frac{dT}{dt}, \quad -\frac{dU}{dt} $$

are positive, and therefore the substance experiences a cooling effect when it is moved in such a manner as to require work to be done against magnetic force ; and the reverse is the case for cobalt at ordinary temperatures.]

209. *Extracted from Nichol's " Cyclopædia of the Physical Sciences,"* second edition, 1860.—The most probable account that can be given of the pyro-electric quality of dipolar crystals is, that these bodies intrinsically possess the same kind of *bodily electro-polarization* which Faraday, in his *Experimental Researches*, has clearly proved to be temporarily produced in solid and liquid non-conductors, and that they possess this property to different degrees at different temperatures.

The inductive action exercised by this electro-polar state of the substance, on the matter touching the body all round, in-duces a superficial electrification which perfectly balances its electric force on all points in the external matter ; but when the crystal is broken in two across its electric axis, the two parts exhibit as wholes contrary electrifications, not only by the free electro-polarities on the fractured surfaces discovered by Canton, but by the induced electrifications on the whole surface, belonging to the old state of electric equilibrium, and gradually lost by slow conduction, while a new superficial distribution of electricity on each fragment is acquired which ultimately masks all external symptoms of electric excitement. When the temperature of the substance is changed, its electro-polarization changes simultaneously, while the masking superficial electrification follows the change only by slow degrees—more or less slow according to the greater or less resistance offered to electric conduction in the substance or along its surface.

[210. If the preceding explanation of pyro-electricity be true, it must follow that a pyro-electric crystal moved about in an electric field will experience cooling effects or heating effects calculable by formula (22) of § 201, with the same notation for the electric subject as that of § 208 for magnetism. Thus the effects will be the same for a crystal at the same temperature whatever be the electrification of its surface. Thus it is remarkable that, in virtue of the wholly latent electric polarity of a seemingly neutral pyro-electric crystal (that is to say, a crystal at the surface of which there is an electrification neutralizing for external space the force due to its internal electric polarity), the same cooling and heating effects will be produced by moving it in an electric field, as similar motions would produce in a similar crystal which, by having been heated in hot water, dried at the high temperature, and cooled, is in a state of pyro-electric excitement.

YACHT 'LALLA ROOKH,'
    LARGS, *Sept.* 13, 1877.]

APPENDIX. (NOTES I., II., III.)

NOTE I.

[From *Proceedings of the Royal Society of Edinburgh*, Dec. 1851.]

On a Mechanical Theory of Thermo-Electric Currents.

It was discovered by Peltier that heat is absorbed at a surface of contact of bismuth and antimony in a compound metallic conductor, when electricity traverses it from the bismuth to the antimony, and that heat is generated when electricity traverses it in the contrary direction. This fact, taken in connection with Joule's law of the electrical generation of heat in a homogeneous metallic conductor, suggests the following assumption, which is the foundation of the theory at present laid before the Royal Society.

When electricity passes in a current of uniform strength γ through a heterogeneous linear conductor, no part of which is permitted to vary in temperature, the heat generated in a given time is expressible by the formula

$$A\gamma + B\gamma^2,$$

where A, which may be either positive or negative, and B, which is essentially positive, denote qualities independent of γ.

The fundamental equations of the theory are the following :—

$$F\gamma = J\left(\gamma \Sigma a_t + B\gamma^2\right) \dots\dots\dots\dots\dots\dots\dots\dots\dots(a),$$

$$\Sigma a_t = \Sigma a_t \left(1 - \epsilon^{-\frac{1}{J}\int_T^t \mu dt}\right) \dots\dots\dots\dots\dots\dots\dots(b),$$

where F denotes the electromotive force (considered as of the same sign with γ, when it acts in the direction of the current) which must act to produce or to permit the current γ to circulate uniformly through the conductor; J the mechanical equivalent of the thermal unit; $a_t\gamma$ the quantity of heat evolved in the unit of time in all parts of the conductor which are at the temperature t when γ is infinitely small; μ " Carnot's function*" of the temperature t; T the temperature of the coldest part of the circuit; and Σ a summation including all parts of the circuit.

The first of these equations is a mere expression of the equivalence, according to the principles established by Joule, of the work, $F\gamma$†, done in a unit of time by the electromotive force, to the heat developed, which, in the circumstances, is the sole effect produced. The second is a consequence of the first and of the following equation :—

$$\phi \cdot \gamma = \mu \Sigma a_t \gamma \cdot (t - T) \dots\dots\dots\dots\dots\dots\dots\dots\dots(c),$$

where ϕ denotes the electromotive force when γ is infinitely small, and when the temperatures in all parts of the circuit are infinitely

---

* The values of this function, calculated from Regnault's observations, and the hypothesis that the density of saturated steam follows the "gaseous laws," for every degree of temperature from 0^0 to 230^0 cent., are shown in Table I. of the author's "Account of Carnot's Theory" [Art. XLI., above], *Transactions*, vol. XVI., p. 541.

† See *Philosophical Magazine*, Dec. 1851, "On Applications of the Principle of Mechanical Effect," &c. [Art. LIV., below].

nearly equal. This latter equation is an expression, for the present circumstances, of the proposition* (first enunciated by Carnot and first established in the dynamical theory by Clausius) that the obtaining of mechanical effect from heat, by means of a perfectly reversible arrangement, depends in a definite manner on the transmission of a certain quantity of heat from one body to another at a lower temperature. There is a degree of uncertainty in the present application of this principle, on account of the conduction of heat that must necessarily go on from the hotter to the colder parts of the circuit; an agency which is not reversed when the direction of the current is changed. As it cannot be shown that the thermal effect of this agency is infinitely small, compared with that of the electric current, unless γ be so large that the term $B\gamma^2$, expressing the thermal effect of another irreversible agency, cannot be neglected, the conditions required for the application of Carnot and Clausius's principle, according to the demonstrations of it which have been already given, are not completely fulfilled: the author therefore considers that at present this part of the theory requires experimental verification.

1.   A first application of the theory is to the case of antimony and bismuth; and it is shown that the fact discovered by Seebeck is, according to equation (c), a consequence of the more recent discovery of Peltier referred to above,—a partial verification of the only doubtful part of the theory being thus afforded.

2.   If $\Theta\gamma$ denote the quantity of heat evolved, [or $-\Theta\gamma$ the quantity absorbed,] at the surface of separation of two metals in a compound circuit, by the passage of a current of electricity of strength γ across it, when the temperature t is kept constant; and if ϕ denote the electromotive force produced in the same circuit by keeping the two junctions at temperatures t and t', which differ from one another by an infinitely small amount, the magnitude of this force is given by the equation

$$\phi = \Theta\mu\,(t' - t) \dots\dots\dots\dots\dots\dots (d),$$

and its direction is such, that a current produced by it would cause the absorption of heat at the hotter junction, and the evolution of heat at the colder. A complete experimental verification of this conclusion would fully establish the theory.

* "Dynamical Theory of Heat" [§ 9, above] (*Transactions*, vol. xx., part ii.), Prop. II., &c.

3. If a current of electricity, passing from hot to cold, or from cold to hot, in the same metal produced the same thermal effects; that is, if no term of Σa_t depended upon variation of temperature from point to point of the same metal; we should have, by equation (a),

$$\phi = J \frac{d\Theta}{dt} (t' - t); \text{ and therefore, by } (d), \frac{d\Theta}{dt} = \frac{1}{J}\Theta\mu.$$

From this we deduce

$$\Theta = \Theta_0 \, \epsilon^{\frac{1}{J}\int_0^t \mu dt}; \text{ and } \phi = (t' - t)\, \mu\Theta_0 \, \epsilon^{\frac{1}{J}\int_0^t \mu dt}$$

A table of the values of $\dfrac{\phi}{\Theta_0 (t' - t)}$ for every tenth degree from 0 to 230 is given, according to the values of μ*, used in the author's previous papers; showing, that if the hypothesis just mentioned were true, the thermal electromotive force corresponding to a given very small difference of temperatures would, for the same two metals, increase very slowly, as the mean absolute temperature is raised.   Or, if Mayer's hypothesis, which leads to the expression $\dfrac{JE}{1 + Et}$ for μ, were true, the electromotive force of the same pair of metals would be the same, for the same difference of temperatures, whatever be the absolute temperatures.   Whether the values of μ previously found were correct or not, it would follow, from the preceding expression for ϕ, that the electromotive force of a thermo-electric pair is subject to the same law of variation, with the temperatures of the two junctions, whatever be the metals of which it is composed.   This result being at variance with known facts, the hypothesis on which it is founded must be false; and the author arrives at the remarkable conclusion, that *an electric current produces different thermal effects, according as it passes from hot to cold, or from cold to hot, in the same metal.*

4. If $\ni (t' - t)$ be taken to denote the value of the part of Σa_t which depends on this circumstance, and which corresponds to all parts of the circuit of which the temperatures lie within an

---

* The unit of force adopted in magnetic and electro-magnetic researches, being that force which, acting on a unit of matter, generates a unit of velocity in the unit of time, the values of μ and J used in this paper are obtained by multiplying the values used in the author's former papers, by 32·2.

infinitely small range t to t'; the equations to be substituted for the preceding are

$$\phi = J \frac{d\Theta}{dt} (t' - t) + J\vartheta (t' - t) \quad\dots\dots\dots\dots\dots\dots\dots(e),$$

and therefore, by (d),

$$\frac{d\Theta}{dt} + \vartheta = \frac{1}{J} \Theta\mu \quad\dots\dots\dots\dots\dots\dots\dots\dots\dots\dots(f).$$

5.   The following expressions for F, the electromotive force in a thermo-electric pair, with the two junctions at temperatures S and T differing by any finite amount, are then established in terms of the preceding notations, with the addition of suffixes to denote the particular values of Θ for the temperatures of the junctions.

$$\left. \begin{aligned} F &= \int_T^S \mu\Theta dt = J \left\{ \Theta_s - \Theta_T + \int_T^S \vartheta dt \right\} \\ &= J \left\{ \Theta_s (1 - \epsilon^{-\frac{1}{J}\int_T^S \mu dt}) + \int_T^S \vartheta \, (1 - \epsilon^{-\frac{1}{J}\int_T^t \mu dt}) \, dt \right\} \end{aligned} \right\} \dots\dots\dots(g).$$

6.   It has been shown by Magnus, that no sensible electromotive force is produced by keeping the different parts of a circuit of one homogeneous metal at different temperatures, however different their sections may be.   It is concluded that for this case $\vartheta = 0$; and therefore that, for a thermo-electric element of two metals, we must have

$$\vartheta = \Psi_1 (t) - \Psi_2 (t),$$

where Ψ_1 and Ψ_2 denote functions depending solely on the qualities of the two metals, and expressing the thermal effects of a current passing through a conductor of either metal, kept at different uniform temperatures in different parts.   Thus, with reference to the metal to which Ψ_1 corresponds, if a current of strength γ pass through a conductor consisting of it, the quantity of heat *absorbed* in any infinitely small part PP' is $\Psi_1 (t) (t' - t) \gamma$, if t and t' be the temperatures at P and P' respectively, and if the current be in the direction from P to P'.   An application to the case of copper and iron is made, in which it is shown that, if Ψ_1, and Ψ_2 refer to these metals respectively, if S be a certain temperature defined below (which, according to Regnault's observations, cannot differ much from 240 cent.), and if T be any lower temperature;

we have

$$\int_T^S \{\Psi_1(t) - \Psi_2(t)\} dt = \Theta_T + \frac{1}{J} F,$$

since the experiments made by Becquerel lead to the conclusion, that at a certain high temperature iron and copper change their places in the thermo-electric series (a conclusion which the author has experimentally verified), and if this temperature be denoted by S, we must consequently have $\Theta^S = 0$.

The quantities denoted by Θ_T and F in the preceding equation being both positive, it is concluded that, *when a thermo-electric current passes through a piece of iron from one end kept at about 240° cent., to the other end kept cold, in a circuit of which the remainder is copper, including a long resistance wire of uniform temperature throughout or an electro-magnetic engine raising weights, there is heat evolved at the cold junction of the copper and iron, and (no heat being either absorbed or evolved at the hot junction) there must be a quantity of heat absorbed on the whole in the rest of the circuit. When there is no engine raising weights, in the circuit, the sum of the quantities evolved, at the cold junction, and generated in the " resistance wire," is equal to the quantity absorbed on the whole in the other parts of the circuit. When there is an engine in the circuit, the sum of the heat evolved at the cold junction and the thermal equivalent of the weights raised, is equal to the quantity of heat absorbed on the whole in all the circuit except the cold junction.*

7.   An application of the theory to the case of a circuit consisting of several different metals, shows that if

$$\phi(A, B), \ \phi(B, C), \ \phi(C, D), \ldots\ldots\ldots\ldots\ldots\ldots\phi(Z, A)$$

denote the electromotive forces in single elements, consisting respectively of different metals taken in order, with the same absolute temperatures of the junctions in each element, we have

$$\phi(A, B) + \phi(B, C) + \phi(C, D) \ldots\ldots\ldots\ldots\ldots\ldots+ \phi(Z, A) = 0,$$

which expresses a proposition, the truth of which was first pointed out and experimentally verified by Becquerel. A curious experimental verification of this proposition (so far as regards the signs of the terms of the preceding equation) was made by the author, with reference to certain specimens of platinum wire, and iron and copper wires. He had observed that the platinum wire, with iron wires bent round its ends, constituted a less powerful thermo-

electric element than an iron wire with copper wires bent round
its ends, for temperatures within atmospheric limits. He tried, in
consequence, the platinum wire with copper wires bent round its
ends, and connected with the ends of a galvanometer coil; and he
found that, with temperatures within atmospheric limits, a current
passed from the copper to the platinum through the hot junction,
and concluded that, in the thermo-electric series

$$+$$
$$\text{Antimony, Iron,} \left\{ \begin{array}{c} \text{Copper,} \\ \text{Platinum,} \end{array} \right\} \text{Bismuth,}$$

this platinum wire must, at ordinary temperatures, be between
iron and copper. He found that the platinum wire retained the
same properties after having been heated to redness in a spirit-
lamp and cooled again; but with temperatures above some limit
itself considerably below that of boiling water, he found that the
iron and platinum constituted a more powerful thermo-electric
element than the iron and copper; and he verified that for such
temperatures, in the platinum and copper element the current
was from the platinum to the copper through the hot junction,
and therefore that the copper now lay between the iron and the
platinum of the series, or in the position in which other observers
have generally found copper to lie with reference to platinum. A
second somewhat thinner platinum wire was found to lie invariably
on the negative side of copper, for all temperatures above the
freezing point; but a third, still thinner, possessed the same
property as the first, although in a less marked degree, as the
superior limit of the range of temperatures for which it was positive towards copper was lower than in the case of the first wire.
By making an element of the first and third platinum wire, it was
found that the former was positive towards the latter, as was to be
expected.

In conclusion, various objects of experimental research regarding thermo-electric forces and currents are pointed out, and methods
of experimenting are suggested. It is pointed out that, failing
direct data, the absolute value of the electromotive force in an
element of copper and bismuth, with its two junctions kept at the
temperatures 0° and 100° cent., may be estimated indirectly from
Pouillet's comparison of the strength of the current it sends
through a copper wire 20 metres long and 1 millimetre in diameter,

XLVIII.]   ON THE DYNAMICAL THEORY OF HEAT.   **323**

with the strength of a current decomposing water at an observed rate ; by means of determinations by Weber, and of others, of the specific resistance of copper and the electro-chemical equivalent of water, in absolute units. The specific resistances of different specimens of copper having been found to differ considerably from one another, it is impossible, without experiments on the individual wire used by M. Pouillet, to determine with much accuracy the absolute resistance of his circuit, but the author has estimated it on the hypothesis that the specific resistance of its substance is $2\frac{1}{4}$ British units. Taking ·02 as the electro-chemical equivalent of water in British absolute units, the author has thus found 16300 as the electromotive force of an element of copper and bismuth, with the two junctions at $0°$ and $100°$ respectively. About 154 of such elements would be required to produce the same electromotive force as a single cell of Daniell's ; if, in Daniell's battery, the whole chemical action were electrically efficient. A battery of 1000 copper and bismuth elements, with the two sets of junctions at $0°$ and $100°$ cent., employed to work a galvanic engine, if the resistance in the whole circuit be equivalent to that of a copper wire of about 100 feet long and about one-eighth of an inch in diameter, and if the engine be allowed to move at such a rate as by inductive reaction to diminish the strength of the current to the half of what it is when the engine is at rest, would produce mechanical effect at the rate of about one fifth of a horse-power. The electromotive force of a copper and bismuth element, with its two junctions at $0°$ and $1°$, being found by Pouillet to be about $\frac{1}{100}$ of the electromotive force when the junctions are at $0°$ and $100°$, must be about 163. The value of Θ_0 for copper and bismuth, according to these results (and to the value $160\cdot16$ of μ at $0°$), or the quantity of heat absorbed in a second of time by a current of unit strength in passing from bismuth to copper, when the temperature is kept at $0°$, is $\frac{163}{160\cdot16}$, or very nearly equal to the quantity required to raise the temperature of a grain of water from $0°$ to $1°$ cent.

NOTE II.

[From *Proceedings of the Royal Society of Edinburgh*, May, 1854.]

A MECHANICAL THEORY OF THERMO-ELECTRIC CURRENTS IN CRYSTALLINE SOLIDS.

[Abstract of Part VI. (§§ 97—181) above.]

In this paper the Mechanical Theory of Thermo-electric Currents in linear conductors of non-crystalline substance, first communicated to the Royal Society of Edinburgh December 15, 1851 [Note I. of this appendix], is extended to solids of any form and of crystalline substance.

It is proved, that if a solid be such that bars cut from it in different directions have different thermo-electric powers relatively to one another, or to other linear conductors, forming part of a circuit, there must, for every bar cut from it, except in certain particular directions (principal thermo-electric axes), be a new thermo-electric quality, of a kind quite distinct from any hitherto known ; giving rise to a reciprocal thermo-dynamic action, which consists of *a difference in temperature at the sides of the bar causing a current to flow longitudinally, when the two ends, being at the same temperature, are connected by a uniformly heated conductor; and a current through the bar causing an absorption and evolution of heat at its two sides, when these are kept at the same temperature.*

The most general conceivable thermo-electric relations of a crystalline solid, or body possessing, inductively or structurally, different physical properties in different directions, are next examined. It is shown how a metallic structure may be actually made up of pieces of different non-crystalline metals, which, taken on a large scale compared with the dimensions of the heterogeneous elements of which it is composed, will be found to exhibit the most general type of thermo-electric directional relations indicated by the abstract investigation; and it is inferred that it would be wrong to limit the general expressions by any particular assumption, even if we only discover simpler types of thermo-electric relations in natural crystals.

The general equations determining the thermo-electric currents in any naturally, inductively, or structurally crystalline solid; resulting either from a completely specified distribution of temperature through it; or from given external appliances of heat, on which, and on the thermo-electric currents themselves, the distribution of heat through the interior will depend; are investigated.

Certain particular applications of the general equations are also made; and the thermo-electric properties of metallic structures (laid before the Society as solids actually possessing the properties referred to), are investigated.

The paper in which this extension of the theory is described, includes a more developed account of the theory of thermo-electric currents in non-crystalline conductors, formerly communicated, than has been hitherto printed; with a simplification in the fundamental equations introduced without hypothesis, by the adoption of a thermometric assumption proposed as the foundation of an absolute scale of temperature, in consequence of thermo-dynamic experiments on air recently made by Mr Joule and the author. It also includes a brief outline of some experimental investigations undertaken to answer questions proposed in the former theoretical communication, and suggested by various considerations which occurred in the course of the research, and by the new part of the theory now communicated to the Royal Society.

NOTE III.

[From the *Cambridge and Dublin Mathematical Journal*, Nov. 1853.]

NOTE (A) ON THE MECHANICAL ACTION OF HEAT, AND (B) THE SPECIFIC HEAT OF AIR *.

(A) *Synthetical Investigation of the Duty of a Perfect Thermo-Dynamic Engine founded on the Expansions and Condensations of a Fluid, for which the gaseous laws hold and the ratio (k) of the specific heat under constant pressure to the specific heat in constant volume is constant; and modification of the result by the assumption of* MAYER'S *hypothesis* †.

Let the source from which the heat is supplied be at the temperature S, and let T denote the temperature of the coldest body that can be obtained as a refrigerator. A cycle of the following four operations, *being reversible in every respect*, gives, according to Carnot's principle, first demonstrated for the Dynamical Theory by Clausius, the greatest possible statical mechanical effect that can be obtained in these circumstances from a quantity of heat supplied from the source.

(1) Let a quantity of air contained in a cylinder and piston, at the temperature S, be allowed to expand to any extent, and let heat be supplied to it to keep its temperature constantly S.

(2) Let the air expand farther, without being allowed to take heat from or to part with heat to surrounding matter, until its temperature sinks to T.

(3) Let the air be allowed to part with heat so as to keep its temperature constantly T, while it is compressed to such an extent that at the end of the fourth operation the temperature may be S.

---

* Extracted from the *Philosophical Transactions* (Part II. 1852), being a Note added to a paper by Mr Joule on the Air Engine.

† That is, that the heat evolved when air is compressed and kept at constant temperature, is the thermal equivalent of the work spent in the compression —(Addition, April, 1853). Experiments recently made by Mr Joule and myself have shown that this hypothesis is so nearly true for atmospheric air, that it may be used in all calculations in which the deviations from "the gaseous laws" of compression and expansion are not taken into account. See *Philosophical Magazine*, Oct. 1852.

(4) Let the air be farther compressed, and prevented from either gaining or parting with heat, till the piston reaches its primitive position.

The amount of mechanical effect gained on the whole of this cycle of operations will be the excess of the mechanical effect obtained by the first and second above the work spent in the third and fourth. Now if P and V denote the primitive pressure and volume of the air, and if P_1 and V_1, P_2 and V_2, P_3 and V_3, P_4 and V_4, denote the pressure and volume respectively, at the ends of the four successive operations, we have by the gaseous laws, and by Poisson's formula and a conclusion from it quoted above[*], the following expressions :—

Mechanical effect obtained by the first operation $= PV \log \dfrac{V_1}{V}$.

Mechanical effect obtained by the second operation

$$= P_2 V_2 \cdot \frac{1}{k-1} \cdot \left\{ \left(\frac{V_2}{V_1}\right)^{k-1} - 1 \right\}.$$

Work spent in the third operation

$$= P_3 V_3 \log \frac{V_2}{V_3}.$$

Work spent in the fourth operation

$$= P_3 V_3 \cdot \frac{1}{k-1} \left\{ \left(\frac{V_3}{V_4}\right)^{k-1} - 1 \right\}.$$

"To find the work necessary to compress a given mass of air to a given fraction of its volume, when no heat is permitted to leave the air; let P, V, T be the primitive pressure, volume, and temperature, respectively; let p, v, and t be the pressure, volume, and temperature at any instant during the compression; and let P', V', and T' be what they become when the compression is concluded. Then if k denote the ratio of the specific heat of air at constant pressure to the specific heat of air kept in a space of constant volume, and if, as appears to be nearly, if not rigorously true, k be constant for varying temperatures and pressures, we shall have by the investigation in Miller's 'Hydrostatics' (Edit. 1835, p. 22)—

$$\frac{1+Et}{1+ET} = \left(\frac{V}{v}\right)^{k-1}$$

But $$\frac{pv}{PV} = \frac{1+Et}{1+ET},$$

therefore $$pv = PV \left(\frac{V}{v}\right)^{k-1}$$

Now, according to the gaseous laws, we have

$$P_1V_1 = PV, \quad P_2V_2 = P_1V_1\frac{1+ET}{1+ES},$$

and $\quad P_3V_3 = P_2V_2;$ and (since $V_4 = V$) $P_4 = P.$

Also, by Poisson's formula,

$$\left(\frac{V_2}{V_1}\right)^{k-1} = \left(\frac{V_3}{V}\right)^{k-1} = \frac{1+ES}{1+ET}.$$

By means of these we perceive that the work spent in the fourth operation is equal to the mechanical effect gained in the second; and we find, for the whole gain of mechanical effect (denoted by M), the expressions

$$M = (PV - P_3V_3)\log\frac{V_1}{V} = PV\log\frac{V_1}{V}\cdot\frac{E(S-T)}{1+ES}.$$

All the preceding formulæ are founded on the assumption of the gaseous laws and the constancy of the ratio (k) of the specific heat under constant pressure to the specific heat in constant volume, for the air contained in the cylinder and piston, and involve no other hypothesis*. If now we add the assumption of Mayer's hypothesis, which for the actual circumstances is $PV\log\frac{V_1}{V} = JH$, H denoting the heat abstracted by the air from the surrounding matter in the first operation, and J the mechani-

Now the work done in compressing the mass from volume v to volume $v - dv$ will be pdv, or by what precedes,

$$PV \cdot V^{k-1}\frac{dv}{v^k}.$$

Hence by the integral calculus we readily find, for the work, W, necessary to compress from V to V',

$$W = PV \cdot \frac{1}{k-1}\left\{\left(\frac{V}{V'}\right)^{k-1} - 1\right\}."$$

* From the sole hypothesis that k is constant for one fluid fulfilling the gaseous laws and having E for its coefficient of expansion, I find it follows, as a necessary consequence, that Carnot's function would have the form $\frac{JE}{1+Et+C}$; where C denotes an unknown absolute constant, and t the temperature measured by a thermometer founded on the equable expansions of that gas. From this it follows, that for such a gas subjected to the four operations described in the text, we must have

$$PV\log\frac{V_1}{V} = JH\frac{1+ES}{1+ES+C}, \text{ and consequently, } \mu = JH\frac{E(S-T)}{1+ES+C},$$

which is Mr Rankine's general formula.

cal equivalent of a thermal unit, we have

$$M = JH \cdot \frac{E(S-T)}{1+ES} \, .$$

The investigation of this formula given in my paper on the Dynamical Theory of Heat, shows that it would be true for every perfect thermo-dynamic engine, if Mayer's hypothesis were true for a fluid subject to the gaseous laws of pressure and density, whether, for such a fluid (did it exist), k were constant or not.

It was first obtained by using, in the formula

$$M = JH\epsilon^{-\frac{1}{J}\int_{T}^{S}\mu dt} \, ,$$

which involves no hypothesis, the expression

$$\mu = \frac{J}{1/E + t}$$

for Carnot's function, which Mr Joule had suggested to me, in a letter dated December 9, 1848, as the expression of Mayer's hypothesis, in terms of the notation of my "Account of Carnot's Theory*" [Art. XLI. above]. Mr Rankine† has arrived at a formula agreeing with it (with the exception of a constant term in the denominator, which, as its value is unknown, but probably small, he neglects in the actual use of the formula), as a consequence of the fundamental principles assumed in his Theory of Molecular Vortices (*Phil. Mag.*, Dec. 1851), when applied to any fluid whatever, subjected to a cycle of four operations satisfying Carnot's criterion of reversibility (being in fact precisely analogous to those described above, and originally invented by Carnot); and he thus establishes Carnot's law as a consequence of the equations of the mutual conversion of heat and expansive power, which had been given in the first section of his paper on the Mechanical Action of Heat‡.

---

* Royal Society of Edinburgh, Jan. 2, 1849, *Transactions*, Vol. XVI. pt. 5.
† "On the Economy of Heat in Expansive Engines." Royal Society of Edinburgh, April 21, 1851, *Transactions*, Vol. XX. part 2.
‡ Royal Society of Edinburgh, Feb. 4, 1850, *Transactions*, Vol. XX. pt. 1.

(B) *Note on the Specific Heats of Air.*

Let N be the specific heat of unity of weight of any fluid at the temperature t, kept within constant volume, v; and let kN be the specific heat of the same fluid mass, under constant pressure, p. Without any other assumption than that of Carnot's principle, the following equation is demonstrated in my paper* on the "Dynamical Theory of Heat" [§ 48 above],

$$kN - N = \frac{\left(\frac{dp}{dt}\right)^2}{\mu \times -\frac{dp}{dv}},$$

where μ denotes the value of Carnot's function, for the temperature t, and the differentiations indicated are with reference to v and t considered as independent variables, of which p is a function. If the fluid be subject to Boyle's and Mariotte's law of compression, we have

$$\frac{dp}{dv} = -\frac{p}{v};$$

and if it be subject also to Gay-Lussac's law of expansion,

$$\frac{dp}{dt} = \frac{p}{1 + Et}.$$

Hence, for such a fluid,

$$kN - N = \frac{E^2 pv}{\mu (1 + Et)^2} \dagger.$$

In the case of dry air these laws are fulfilled to a very high degree of approximation, and, for it, according to Regnault's observations,

$$\frac{pv}{1 + Et} = 26215, \quad E = \cdot 00366$$

(a British foot being the unit of length, and the weight of a British pound at Paris, the unit of force).

* Royal Society of Edinburgh, Mar. 17, 1851, *Transactions*, Vol. xx. pt. 2.

† This equation expresses a proposition first demonstrated by Carnot. See "Account of Carnot's Theory" [Art. xli. above], Appendix III. Page 148 (*Transactions*, Royal Society of Edinburgh, Vol. xvi. part 5).

We have consequently, for dry air,

$$kN - N = \frac{26215 E^2}{\mu\,(1 + Et)}\quad\dots\dots\dots\dots\dots(1).$$

Now it is demonstrated, without any other assumption than that of Carnot's principle, in my "Account of Carnot's Theory" (Appendix III.) [Art. XLI. above], that

$$\frac{E}{\mu\,(1 + Et)} = \frac{H}{W},$$

if W denote the quantity of work that must be spent in compressing a fluid subject to the gaseous laws, to produce H units of heat when its temperature is kept at t.   Hence

$$kN - N = 26215 E \times \frac{H}{W} = 95\cdot947 \times \frac{H}{W}\dots\dots(2).$$

If we adopt the values of μ shown in Table I. of the "Account of Carnot's Theory" [Art. XLI. above], depending on no uncertain data except the densities of saturated steam at different temperatures, which, for want of accurate experimental data, were derived from the value 1693·5 for the density of saturated vapour at 100°, by the assumption of the "gaseous laws" of variation with temperature and pressure; we find 1357 and 1369 for the values of $\frac{E}{\mu\,(1 + Et)}$ at the temperature 0 and 10° respectively; and hence, for these temperatures,

$$\left.\begin{array}{l} (t = 0)\quad kN - N = \dfrac{95\cdot947}{1357} = \cdot07071 \\[2mm] (t = 10°)\ kN - N = \dfrac{95\cdot947}{1369} = \cdot07008 \end{array}\right\}\dots\dots\dots\dots(a).$$

Or, if we adopt Mayer's hypothesis, according to which $\frac{W}{H}$ is equal to the mechanical equivalent of the thermal unit*, we have $\frac{W}{H} = 1390$; and hence, for all temperatures,

$$kN - N = \frac{95\cdot947}{1390} = \cdot06903\dots\dots\dots\dots\dots(a').$$

---

* The number 1390, derived from Mr Joule's experiments on the friction of fluids, cannot differ by $\frac{1}{100}$, and probably does not differ by $\frac{1}{300}$, of its own value, from the true value of the mechanical equivalent of the thermal unit.

The very accurate observations which have been made on the velocity of sound in air, taken in connection with the results of Regnault's observations on its density, &c., lead to the value 1·410 for k, which is probably true in three if not in four of its figures. Now, k being known, the preceding equations enable us to determine the absolute values of the two specific heats (kN, and N) according to the hypotheses used in (a) and (a') respectively; and we thus find,

	Specific heat of air under constant pressure (kN).	Specific heat of air in constant volume (N).
for $t=0$,	·2431	·1724,
for $t=10$,	·2410	·1709,

according to the tabulated values of Carnot's function.

Or, for all temperatures, ·2374 1684,
according to Mayer's hypothesis.

By the adoption of hypotheses involving that of Mayer, and taking 1389·6 and 1·4 as the values of J and k, respectively, Mr Rankine finds ·2404 and ·1717 as the values of the two specific heats.

Hence it is probable that the values of the specific heat of air under constant pressure, found by Suermann (·3046), and by De la Roche and Berard (·2669), are both considerably too great; and the true value, to two significant figures, is probably ·24.

GLASGOW COLLEGE, *Feb.* 19, 1852.

POSTSCRIPT.

In a paper communicated to the Royal Society, along with the above (March, 1852), Mr Joule described a new experimental determination of the specific heat of air under constant atmospheric pressure, which gave ·23 as a mean result, but he used ·2389 as probably nearer the truth, correcting certain tables, calculated from De la Roche and Berard's result, which he had given in his paper on the Air Engine. M. Regnault has just published (*Comptes Rendus*, April 18, 1853) the results of experimental researches on the specific heat of air, by which he finds that for all temperatures from −30° to +225° centigrade, and for all pressures from one up to ten atmospheres, the specific heat of air is from ·237 to ·2379, and thus both pushes to a minuter degree of accuracy the direct confirmation which the theoretical results published by Mr Rankine and myself first obtained from Mr Joule's experiments, and justifies Mr Joule in the number he actually used in his calculations.

GLASGOW COLLEGE, *April*, 1853.

ART. XLIX. ON THE THERMAL EFFECTS OF FLUIDS IN MOTION.
By J. P. JOULE and W. THOMSON.

PRELIMINARY.

ON THE THERMAL EFFECTS EXPERIENCED BY AIR IN RUSHING
THROUGH SMALL APERTURES*.

[*Phil. Mag.* 2nd half year, 1852.]

THE hypothesis that the heat evolved from air compressed and
kept at a constant temperature is mechanically equivalent to the
work spent in effecting the compression, assumed by Mayer as the
foundation for an estimate of the numerical relation between
quantities of heat and mechanical work, and adopted by
Holtzmann, Clausius, and other writers, was made the subject of
an experimental research by Mr Joule†, and verified as at least
approximately true for air at ordinary atmospheric temperatures.
A theoretical investigation, founded on a conclusion of Carnot's‡,
which requires no modification§ in the dynamical theory of heat,
also leads to a verification of Mayer's hypothesis within limits of
accuracy as close as those which can be attributed to Mr Joule's
experimental tests.   But the same investigation establishes the
conclusion, that that hypothesis cannot be rigorously true except
for one definite temperature within the range of Regnault's experiments on the pressure and latent heat of saturated aqueous
vapour, unless the density of the vapour both differs considerably
at the temperature 100° Cent. from what it is usually supposed to
be, and for other temperatures and pressures presents great discrepancies from the gaseous laws.   No experiments, however,

* Communicated by the Authors; having been read to the British Association
at Belfast, Sept. 3, 1852.
† *Phil. Mag.* May 1845, p. 375, " On the Changes of Temperature produced by
the Rarefaction and Condensation of Air."
‡ *Transactions of the Royal Society of Edinburgh* (April, 1849), Vol. XVI. part 5,
"Appendix to Account of Carnot's Theory," §§ 46—51.
§ *Trans. Royal Soc. Edinb.* (March, 1851), Vol. XX. part 2, or *Phil. Mag.* Aug.
1852, "On the Dynamical Theory of Heat," § 30.

which have yet been published on the density of saturated aqueous vapour are of sufficient accuracy to admit of an unconditional statement of the indications of theory regarding the truth of Mayer's hypothesis, which cannot therefore be considered to have been hitherto sufficiently tested either experimentally or theoretically. The experiments described in the present communication were commenced by the authors jointly in Manchester last May. The results which have been already obtained, although they appear to establish beyond doubt a very considerable discrepancy from Mayer's hypothesis for temperatures from 40° to 170° Fahr., are far from satisfactory; but as the authors are convinced that, without apparatus on a much larger scale, and a much more ample source of mechanical work than has hitherto been available to them, they could not get as complete and accurate results as are to be desired, they think it right at present to publish an account of the progress they have made in the inquiry.

The following brief statement of the proposed method, and the principles on which it is founded, is drawn from §§ 77, 78 of Part IV. of the series of articles on the Dynamical Theory of Heat [Art. XLVIII. above] republished in this Magazine from the *Transactions of the Royal Society of Edinburgh* in 1851* (Vol. XX. part 2, pp. 296, 297).

Let air be forced continuously and as uniformly as possible, by means of a forcing-pump, through a long tube, open to the atmosphere at the far end, and nearly stopped in one place so as to leave, for a short space, only an extremely narrow passage, on each side of which, and in every other part of the tube, the passage is comparatively very wide; and let us suppose, first, that the air in rushing through the narrow passage is not allowed to gain any heat from, nor (if it had any tendency to do so) to part with any to, the surrounding matter. Then, if Mayer's hypothesis were true, the air after leaving the narrow passage would have exactly the same temperature as it had before reaching it. If, on the contrary, the air experiences either a cooling or a heating effect in the circumstances, we may infer that the heat produced by the fluid friction in the rapids, or, which is the same, the thermal equivalent of the work done by the air in expanding from its state of high pressure on one side of the narrow passage to the state of atmospheric pressure which it has after passing the rapids,

---

* See also "Dynamical Theory of Heat," part 5, *Trans. Roy. Soc. Edinb.* 1852.

is in one case less, and in the other more, than sufficient to
compensate the cold due to the expansion; 'and the hypothesis in
question would be disproved.

The apparatus consisted principally of a forcing-pump of
$10\frac{1}{2}$ inches stroke and $1\frac{3}{8}$ internal diameter, worked by a hand-
lever, and adapted to pump air, through a strong copper vessel*
of 136 cubic inches capacity (used for the purpose of equalizing
the pressure of the air), into one end of a spiral leaden pipe
24 feet long and $\frac{5}{16}$ths of an inch in diameter, provided with a
stop-cock at its other end. The spiral was in all the experiments
kept immersed in a large water-bath.

In the first series of experiments, the temperature of the bath
was kept as nearly as possible the same as that of the surrounding
atmosphere; and the stop-cock, which was kept just above the
surface of the water, had a vulcanized india-rubber tube tied to its
mouth. The forcing-pump was worked uniformly, and the stop-
cock was kept so nearly closed as to sustain a pressure of from two
to five atmospheres within the spiral. A thermometer placed in
the vulcanized india-rubber tube, with its bulb near the stop-cock,
always showed a somewhat lower temperature than another placed
in the water-bath†; and it was concluded that the air had
experienced a cooling effect in passing through the stop-cock.

To diminish the effects which might be anticipated from the
conduction of heat through the solid matter round the narrow
passage, a strong vulcanized india-rubber tube, a few inches long,
and of considerably less diameter than the former, was tied on
the mouth of the stop-cock in place of that one which was
removed, and tied over the mouth of the narrower. The stop-cock
was now kept wide open, and the narrow passage was obtained by

* This and the forcing-pump are parts of the apparatus used by Mr Joule in his
original experiments on air. See *Phil. Mag.* May, 1845.

† When the forcing-pump is worked so as to keep up a uniform pressure in the
spiral, and the water of the bath is stirred so as to be at a uniform temperature
throughout, this temperature will be, with almost perfect accuracy, the temperature
of the air as it approaches the stop-cock. It is to be remarked, however, that when,
by altering the aperture of the stop-cock, or the rate of working the pump, the
pressure within the spiral is altered, even although not very suddenly, the air
throughout the spiral, up to the narrow passage, alters in temperature on account
of the expansion or condensation which it is experiencing, and there is an immediate
corresponding alteration in the temperature of the stream of air flowing from *the
rapids*, which produces often a most sensible effect on the thermometer in the
issuing stream.

squeezing the double india-rubber tube by means of a pair of wooden pincers applied to compress the inner tube very near its end, through the other surrounding it. The two thermometers were placed, one, as before, in the bath, and the other in the wide india-rubber tube, with its bulb let down so as to be close to the end of the narrower one within. It was still found that, the forcing-pump being worked as before, when the pincers were applied so as to keep up a steady pressure of two atmospheres or more in the spiral, the thermometer placed in the current of air flowing from the narrow passage showed a lower temperature than that of the air in the spiral, as shown by the other. Sometimes the whole of the narrow india-rubber tube, the wooden pincers, and several inches of the wider tube containing the thermometer, were kept below the surface of the bath, and still the cooling effect was observed; and this even when hot water, at a temperature of about 150° F., was used, although in this case the observed cooling effect was less than when the temperature of the bath was lower.

As it was considered possible that the cooling effects observed in these experiments might be due wholly or partly to the air reaching the thermometer-bulb before it had lost all the *vis viva* produced by the expansion in the narrow passage, and consequently before the full equivalent of heat had been produced by the friction, and as some influence (although this might be expected to diminish the cooling effect) must have been produced by the conduction of heat through the solid matter round the air, especially about the narrow passage, an attempt was made to determine the whole thermal effect by means of a calorimetrical apparatus applied externally. For this purpose the india-rubber tubes were removed, and the stop-cock was again had recourse to for producing the narrow passage. A piece of small block-tin tube, about 10 inches long, was attached to the mouth of the stop-cock, and was bent into a spiral, as close round the stop-cock as it could be conveniently arranged. A portion of the block-tin pipe was unbent from the principal spiral, and was bent down so as to allow the stop-cock to be removed from the water-bath, and to be immersed with the exit spiral in a small glass jar filled with water. The forcing-pump was now worked at a uniform rate, with the stop-cock nearly closed, for a quarter of an hour, and then nearly open for a quarter of an hour, and so on for several

alternations. The temperatures of the water in the large bath
and in the glass jar were observed at frequent stated intervals
during these experiments ; but, instead of there being any cooling
effect discovered when the stop-cock was nearly closed, there was
found to be a slight elevation of temperature during every period
of the experiments, averaging nominally ·06525° F. for four
periods of a quarter of an hour when the stop-cock was nearly
closed, and 06533° when it was wide open, or, within the limits of
the accuracy of the observations, ·065° in each case ; a rise due, no
doubt, to the rising temperature of the surrounding atmosphere
during the series of experiments. Hence the results appear at
first sight only negative ; but it is to be remarked that, the
temperature of the bath having been on an average $3\frac{1}{2}$° F.
lower than that of the water in the glass jar, the natural rise of
temperature in the glass jar must have been somewhat checked
by the air coming from the principal spiral ; and had there been
no cooling effect due to rushing through the stop-cock when it
was nearly closed, would have been more checked when the stop-
cock was wide open than when it was nearly closed, as the same
number of strokes of the pump must have sent considerably more
air through the apparatus in one case than in the other. A
cooling effect on the whole, due to the rushing through the nearly
closed stop-cock, is thus indicated, if not satisfactorily proved.

Other calorimetric experiments were made with the stop-cock
immersed in water in one glass jar, and the air from it, conducted
by a vulcanized india-rubber tube, to flow through a small spiral
of block-tin pipe immersed in a second glass jar of equal capacity ;
and it was found that the water in the jar round the stop-cock
was cooled, while that in the other, containing the exit spiral, was
heated, during the working of the pump, with the stop-cock
nearly closed, and a pressure of about three atmospheres in the
principal spiral. The explanation of this curious result is clearly,
that the water round the stop-cock supplied a little heat to the
air in the first part of the rapids, where it has been cooled by
expansion and has not yet received all the heat of the friction,
and that the heat so obtained, along with the heat produced by
friction throughout the rapids, raises the temperature of the air a
little above what it would have had if no heat had been gained
from without ; so that about the end of the rapids the air has a
temperature a little above that of the surrounding water, and is

led, under the protection of the india-rubber tube, to the exit
spiral with a slightly elevated temperature. This is what would
necessarily happen in any case of an arrangement such as that
described, if Mayer's hypothesis were strictly true; but then the
quantity of heat emitted to the water in the second glass jar, from
the air in passing through the exit spiral, would be exactly equal
to that taken by conduction through the stop-cock from the water
in the first. In reality, according to the discrepancy from Mayer's
hypothesis, which the other experiments described in this communication appear to establish, there must have been somewhat
more heat taken in by conduction through the stop-cock than
was emitted by it in flowing through the exit spiral; but the
experiments were not of sufficient accuracy, and were affected by
too many disturbing circumstances, to allow this difference to be
tested.

To obtain a decisive test of the discrepancy from Mayer's
hypothesis, indicated by the experiments which have been described, and to obtain either comparative or absolute determinations of its amount for different temperatures, some alterations in
the apparatus, especially with regard to the narrow passage and
the thermometer for the temperature of the air flowing from it,
were found to be necessary by Mr Joule, who continued the
research alone, and made the experiments described in what
follows.

A piece of brass piping, *a* (see the accompanying sketch drawn
half the actual size), was soldered to the termination of the leaden
spiral, and a bit of calf-skin leather, *b*, having been tightly bound
over its end, it was found that the natural pores of the leather
were sufficient to allow of a uniform and conveniently rapid flow
of air from the receiver. By protecting the end over which the
leather diaphragm was bound with a piece of vulcanized india-
rubber tube, *c*, the former could be immersed to the depth of
about two inches in the bath of water. A small thermometer*,
having a spherical bulb ⅛th of an inch in diameter, was placed
within the india-rubber tube, the bulb being allowed to rest on the
central part of the leather diaphragm†.

---

* We had two of these thermometers, one of which had Fahrenheit's, the other
an arbitrary scale.

† The bulb was kept in this position for convenience sake, but it was ascertained
that the effects were not perceptibly diminished when it was raised ¼ of an inch
above the diaphragm.

In making the experiments, the pump was worked at a uniform rate until the pressure of the air in the spiral and the

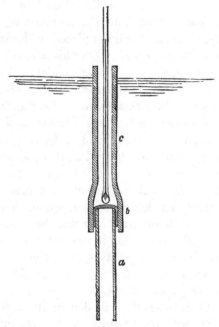

temperature of the thermometer had become sensibly constant. The water of the bath was at the same time constantly stirred, and by various devices kept as uniform as possible during each series of experiments. The temperature of the stream of air having been observed, the same thermometer was immediately plunged into the bath to ascertain its temperature, the difference between the two readings giving of course the cooling effect of the rushing air.

According to theory*, the cooling effect for a given tempera-

* See "Account of Carnot's Theory," Appendix II. [Art. XLI. above], *Trans. Roy. Soc. Edinb.* Jan. 1849, Vol. XVI. p. 566; and "Dynamical Theory," § 75 [Art. XLVIII. above], *Trans. Roy. Soc. Edinb.* April, 1851, Vol. XX. p. 296; or *Phil. Mag.* Dec. 1852. The numbers shown in the table of § 51 of the former paper being used in the formula of § 75 of the latter, and 1390 being used for *J*, we find (according to the numerical data used formerly for deriving numerical results from the theory) how much heat would have to be added to each pound of the issuing stream of air to bring it back to the temperature it had when approaching the narrow passage; and this number, divided by ·24, the specific heat of air under constant pressure, would be the depression of temperature (in Centigrade degrees) actually experienced by the air when no heat is communicated to it in or after the rapids.

ture would be independent of the kind of aperture and of the copiousness of the stream, and would be simply proportional to the logarithm of the pressure, if the insulation of the current against gain or loss of heat from the surrounding matter were perfect, and if the thermometer be so placed in the issuing stream as to be quite out of the *rapids*. On this account the values of the cooling effect divided by the logarithm of the pressure were calculated, and are shown in the last columns of the tables of results given below. When this was done for the first two series of experiments, the discrepancies (see columns 5 of the first two of the tables given below) were found to be so great, and especially among the results of the different experiments for the higher temperature of 160° F., all made with the pressure and other circumstances as nearly as possible the same, so irregular, that great uncertainty was felt as to the numerical results, which must obviously have been much affected by purely accidental circumstances. At the same time it was noticed, that in the case of Series 1, in which the temperature of the bath was always as nearly as possible that of the atmosphere, and different pressures were used, the discrepancies showed a somewhat regular tendency of the value of the cooling effect divided by the logarithm of the pressure to increase with the pressure; which was probably owing to the circumstance that the stream was more copious, and that less of the cooling effect was lost (as some probably was in every case) by the conduction of heat from without, the higher the pressure under which the air approached the narrow passage. Hence in all the subsequent experiments the quantity of air pumped through per second was noted.

The following Tables show the results obtained from ten series of experiments conducted in the manner described:

Series 1.

Col. 1.	Col. 2.	Col. 3.	Col. 4.	Col. 5*.
Quantity of air pumped in cubic inches per second.	Temperature of bath.	Pressure of air in atmospheres.	Cooling effect.	Cooling effect divided by logarithm of pressure.
A.	T.	P.	D.	$\dfrac{D}{\log P}$.
Not noted.	61°	1·79	0·5°	1·98°
Not noted.	61	2·64	0·9	2·13
Not noted.	61	2·9	0·7	1·51
Not noted.	61	3·22	1·5	2·95
Not noted.	61	3·4	1·4	2·64
Not noted.	61	3·61	1·4	2·51
Not noted.	61	3·61	1·3	2·33
Not noted.	61	3·61	1·4	2·51
Not noted.	61	3·84	1·5	2·57
Not noted.	61	4·11	1·7	2·77
Mean......	61	......	......	2·39

Series 2.

Not noted.	160	2·64	0·264	0·62
Not noted.	160	2·64	0·396	0·94
Not noted.	160	2·64	0·66	1·56
Not noted.	160	2·64	0·528	1·25
Not noted.	160	2·64	0·66	1·56
Mean	160	2·64	0·502	1·18

Series 3.

5·6	170·8	3·61	0·396	0·71
5·6	170·8	4·11	0·528	0·86
5·6	170·8	4·11	0·66	1·08
5·6	170·8	4·11	0·726	1·18
5·6	170·8	4·26	0·66	1·05
8·4	170·8	4·78	0·858	1·26
8·4	170·8	4·98	0·858	1·23
Mean 6·4	170·8	4·28	0·67	1·05

Series 4.

5·6	37·8	3·4	0·8	1·51
5·6	38·8	3·4	1·1	2·07
5·6	37·9	3·61	0·6	1·08
5·6	44·4	3·04	1·1	2·28
5·6	45·3	3·04	0·9	1·86
5·6	46·3	3·04	1·0	2·07
Mean 5·6	41·75	......	......	1·81

* The true value of $\dfrac{D}{\log P}$ for any particular temperature would be the depression of temperature that would be experienced by air approaching the narrow passage at that temperature and under ten atmospheres of pressure, since P is measured in atmospheres, and the *common* logarithm is taken.

Table (*continued*).

Series 5.

Col. 1. Quantity of air pumped in cubic inches per second. A.	Col. 2. Temperature of bath. T.	Col. 3. Pressure of air in atmospheres. P.	Col. 4. Cooling effect. D.	Col. 5. Cooling effect divided by logarithm of pressure. $\frac{D}{\log P}$.
8·4	46·8	3·84	1·2	2·06
8·4	38·7	4·11	1·8	2·93
8·4	39·3	4·11	1·8	2·93
Mean 8·4	41·6	......	......	2·64

Series 6.

11·2	39·7	4·4	1·7	2·64
11·2	40·9	4·4	1·9	2·95
11·2	41·9	4·4	1·5	2·33
11·2	43	4·4	1·5	2·33
Mean 11·2	41·38	4·4	1·65	2·56

Series 7.

1·4	64·1	1·9	0·3	1·08
1·4	64·2	1·87	0·45	1·65
1·4	64·0	1·9	0·4	1·43
1·4	64·2	1·9	0·5	1·79
1·4	64·3	1·9	0·45	1·61
Mean 1·4	64·16	1·894	0·42	1·51

Series 8.

2·8	64·2	2·41	0·5	1·31
2·8	64·3	2·41	0·5	1·31
2·8	64·5	2·41	0·5	1·31
2·8	64·7	2·41	0·7	1·83
2·8	64·7	2·41	0·6	1·57
Mean 2·8	64·48	2·41	0·56	1·46

Series 9.

5·6	64·6	2·9	0·8	1·73
5·6	64·7	2·9	0·8	1·73
5·6	64·8	3·04	0·8	1·66
5·6	65·0	2·97	0·7	1·48
Mean 5·6	64·775	......	......	1·65

Series 10.

11·2	65·	4·11	1·2	1·95
11·2	65·1	4·11	1·3	2·12
11·2	65·1	4·11	1·4	2·28
Mean 11·2	65·06	4·11	1·3	2·12

The numbers in the last column of any one of these tables show, by their discrepancies, how much uncertainty there must be in the results on account of purely accidental circumstances.

The following table is arranged, with double argument of temperature and of quantity of air passing per second, to show a comparison of the means of the different series (Series 3 being divided into two, one consisting of the first five experiments, and the other of the remaining two).

Table of Mean Values of $\dfrac{D}{\log P}$ in different Series of Experiments.

		Quantity of air passing per second.				
		1·4	2·8	5·6	8·4	11·2
Temperature of bath.	$41\frac{1}{2}°$	......	......	1·81	2·64	2·56
	$64\frac{1}{2}$	1·51	1·46	1·65	......	2·12
	171	......	......	·98	1·25	......

The general increase of the numbers from left to right in this table shows that very much of the cooling effect must be lost on account of the insufficiency of the current of air. This loss might possibly be diminished by improving the thermal insulation of the current in and after the rapids; but it appears probable that it could be reduced sufficiently to admit of satisfactory observations being made, only by using a much more copious current of air than could be obtained with the apparatus hitherto employed.

The decrease of the numbers from the upper to the lower spaces, especially in the one complete vertical column (that under the argument 5·6), shows that the cooling effect is less to a remarkable degree for the higher than for the lower temperatures. Even from 41° to 65° F. the diminution is most sensible; and at 171° the cooling effect appears to be only about half as much as at 41°.

The best results for the different temperatures are probably those shown under the arguments 8·4 and 11·2, being those obtained from the most copious currents; but it is probable that they all fall considerably short of the true values of $\dfrac{D}{\log P}$ for the actual temperatures; and we may consider it as perfectly established by the experiments described above, that *there is a final cooling effect produced by air rushing through a small aperture at any temperature up to* 170° F., and that *the amount of this cooling effect decreases as the temperature is augmented.* Now according to the theoretical views on this subject brought forward in the papers on "Carnot's Theory" [Art. XLI. above], and "On the Dynamical Theory of Heat" [Art. XLVIII. above], already referred to, a cooling effect was expected for low temperatures; and the amount of this effect was expected to be the *less* the *higher* the temperature; expectations which have therefore been perfectly confirmed by experiment. But since the excess of the heat of compression above the thermal equivalent of the work was, in the theoretical investigation, found to diminish to zero* as the temperature is raised to about 33° Cent., or 92° Fahr., and to be negative for all higher temperatures, a *heating* instead of a *cooling* effect would be found for such a temperature as 171° F., if the data regarding saturated steam used in obtaining numerical results from the theory were correct. All of these data except the *density* had been obtained from Regnault's very exact experimental determinations; and we may consequently consider it as nearly certain, that the true values of the density of saturated aqueous vapour differ considerably from those which were assumed. Thus, if the error is to be accounted for by the *density* alone, the fact of there being any cooling effect in the air experiments at 171° Fahr. (77° Cent.) shows that the density of saturated aqueous vapour at that temperature must be greater than it was assumed to be in the ratio of something more than 1416 to 1390, or must be more than 1·019 of what it was assumed to be : and, since the experiments render it almost if not absolutely certain, that even at 100° Cent. air rushing through a small aperture would produce a final cooling effect, it is probable that the density of steam at the ordinary boiling-point, instead of being about $\frac{1}{1693\cdot5}$, as it is generally supposed to be, must be something more than $\frac{1430\cdot6}{1390}$ of this; that is, must exceed $\frac{1}{1645}$.

With a view to ascertain what effect would be produced in the case of the air rushing violently against the thermometer-bulb, the leather diaphragm was now perforated with a fine needle, and

---

* See the table in § 51 of the "Account of Carnot's Theory" [Art. XLI. above], from which it appears that the element tabulated would have the value 1390, or that of the mechanical equivalent of the thermal unit, at about 33° Cent.

the bulb placed on the orifice so as to cause the air to rush between the leather and the sides of the bulb. With this arrangement the following results were obtained :—

Series 11.

A.	T.	P.	D.	$\frac{\log P}{D}$.
11·2	64°	3·22	3·5°	6·90°
11·2	64	3·31	3·5	6·73
11·2	64	3·61	3·8	6·82
11·2	64	2·30	4·0	11·05
11·2	64	3·31	6·1	11·73
11·2	64	2·58	4·7	11·41
11·2	64	4·78	5·3	7·80
11·2	64	1·9	4·0	14·34
Mean 11·2	64	......	......	9·60

The great irregularities in the last column of the above table are owing to the difficulty of keeping the bulb of the thermometer in exactly the same place over the orifice. The least variation would occasion an immediate and considerable change of temperature; and when the bulb was removed to only $\frac{1}{4}$ of an inch above the orifice, the cooling effects were reduced to the amount observed when the natural pores alone of the leather were employed. There can be no doubt but that the reason why the cooling effects experienced by the thermometer-bulb were greater in these experiments than in the former is, that in these it was exposed to the current of air in localities in which a sensible portion of the mechanical effect of the work done by the expansion had not been converted into heat by friction, but still existed in the form of *vis viva* of fluid motion. Hence this series of experiments confirms the theoretical anticipations formerly published* regarding the condition of the air in *the rapids* caused by flowing through a small aperture.

* See "Dynamical Theory," § 77 [Art. xlviii. above], *Trans. Royal Soc. Edinb.* April, 1851; or *Phil. Mag.* Dec. 1852.

[From *Transactions of the Royal Society*, June, 1853.]

PART I.

IN a paper communicated to the Royal Society, June 20, 1844, "On the Changes of Temperature produced by the Rarefaction and Condensation of Air*," Mr Joule pointed out the dynamical cause of the principal phenomena, and described the experiments upon which his conclusions were founded. Subsequently Professor Thomson pointed out that the accordance discovered in that investigation between the work spent and the mechanical equivalent of the heat evolved in the compression of air may be only approximate, and in a paper communicated to the Royal Society of Edinburgh in April, 1851, "On a Method of discovering experimentally the relation between the Mechanical Work spent, and the Heat produced by the compression of a Gaseous Fluid†" [Art. XLVIII. above, §§ 61—80], proposed the method of experimenting adopted in the present investigation, by means of which we have already arrived at partial results‡. This method consists in forcing the compressed elastic fluid through a mass of porous non-conducting material, and observing the consequent change of temperature in the elastic fluid. The porous plug was adopted instead of a single orifice, in order that the work done by the expanding fluid may be immediately spent in friction, without any appreciable portion of it being even temporarily employed to generate ordinary *vis viva*, or being devoted to produce sound. The non-conducting material was chosen to diminish as much as possible all loss of thermal effect by conduction, either from the air on one side to the air on the other side of the plug, or between the plug and the surrounding matter.

A principal object of the researches is to determine the value of μ, Carnot's function. If the gas fulfilled perfectly the laws of compression and expansion ordinarily assumed, we should have§

$$\frac{1}{\mu} = \frac{1/E + t}{J} + \frac{K\delta}{Ep_0u_0 \log P},$$

---

* *Philosophical Magazine*, May, 1845, p. 369.

† *Transactions of the Royal Society, Edinburgh*, April 21, 1851.

‡ *Philosophical Magazine*, Dec. 1852, p. 481.

§ Dynamical Theory of Heat, (equation 7), § 80 [Art. XLVIII. above], *Transactions of the Royal Society of Edinburgh*, April 21, 1851.

where J is the mechanical equivalent of the thermal unit; $p_0 u_0$ the product of the pressure in pounds on the square foot into the volume in cubic feet of a pound of the gas at $0°$ Cent.; P is the ratio of the pressure on the high pressure side to that on the other side of the plug; δ is the observed cooling effect; t the temperature Cent. of the bath, and K the thermal capacity of a pound of the gas under constant pressure equal to that on the low pressure side of the gas.   To establish this equation it is only necessary to remark that $K\delta$ is the heat that would have to be added to each pound of the exit stream of air, to bring it to the temperature of the bath, and is the same (according to the general principle of mechanical energy) as would have to be added to it in passing through the plug, to make it leave the plug with its temperature unaltered.   We have therefore $K\delta = -H$, in terms of the notation used in the passage referred to.

On the above hypothesis (that the gas fulfils the laws of compression and expansion ordinarily assumed) $\delta/\log P$ would be the same for all values of P; but Regnault has shown that the hypothesis is not rigorously true for atmospheric air, and our experiments show that $\delta/\log P$ increases with P.   Hence, in reducing the experiments, a correction must be first applied to take into account the deviations, as far as they are known, of the fluid used, from the gaseous laws, and then the value of μ may be determined.   The formula by which this is to be done is the following (Dynamical Theory of Heat [Art. XLVIII. above], equation (f), § 74, or equation (17), § 95, and (8), § 88)—

$$\frac{1}{\mu} = \frac{\frac{1}{J}\{w - (p'u' - pu)\} + K\delta}{\frac{dw}{dt}},$$

where $$w = \int_u^{u'} p\, dv,$$

u and u' denoting the volumes of a pound of the gas at the high pressure and low pressure respectively, and at the same temperature (that of the bath), and v the volume of a pound of it at that temperature, when at any intermediate pressure p.   An expression for w for any temperature may be derived from an empirical formula for the compressibility of air at that temperature, and between the limits of pressure in the experiment.

The apparatus, which we have been enabled to provide by the assistance of a grant from the Royal Society, consists mainly of a pump, by which air may be forced into a series of tubes acting at at once as a receiver of the elastic fluid, and as a means of communicating to it any required temperature; nozzles, and plugs of porous material being employed to discharge the air against the bulb of a thermometer.

The pump a, fig. 1, consists of a cast-iron cylinder of 6 inches internal diameter, in which a piston, fig. 2, fitted with spiral

Fig. 1.

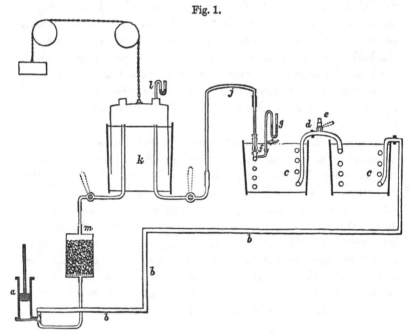

metallic packing (of antifriction metal), works by the direct action of the beam of a steam-engine through a stroke of 22 inches. The pump is single-acting, the air entering at the base of the cylinder during the up-stroke, and being expelled thence into the receiving tubes by the down-stroke. The governor of the steam-engine limits the number of complete strokes of the pump to 27 per minute. The valves, fig. 3, consist of loose spheres of brass 0·6 of an inch in diameter, which fall by their own gravity over orifices 0·45 of an inch diameter. The cylinder and valves in connection with it are immersed in water to prevent the wear and tear which might arise from a variable or too elevated temperature.

Wrought-iron tubing, bb, fig. 1, of 2 inches internal diameter, conducts the compressed air horizontally a distance of 6 feet,

thence vertically to an elevation of 18 feet, where another length
of 23 feet conveys it to the copper tubing, *cc*; the junction being

Fig. 3.

Fig. 2.

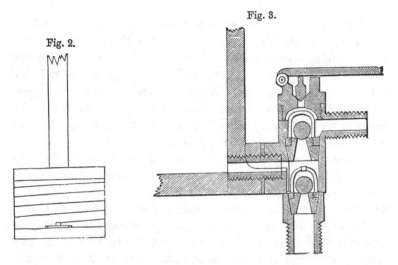

effected by means of a coupling-joint.   The copper tubing, which
is of 2 inches internal diameter and 74 feet in length, is arranged
in two coils, each being immersed in a wooden vessel of 4 feet

diameter, from the bottom and sides
of which it is kept at a distance of
6 inches.   The coils are connected
by means of a coupling-joint *d*, near
which a stop-cock, *e*, is placed, in
order to let a portion of air escape
when it is wanted to reduce the pressure. The terminal coil has a flange,
f, to which any required nozzle
may be attached by means of screw-
bolts. Near the flange, a small pipe,
g, is screwed, at the termination of
which a calibrated glass tube bent
(as shown in fig. 4), and partly filled
with mercury, is tightly secured. A

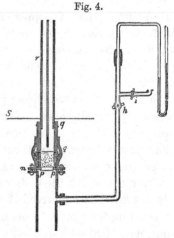

Fig. 4.

stop-cock at *h*, and another in a small branch pipe at *i*, permit the
air at any time to be let off, so as to examine the state of the gauge
when uninfluenced by any except atmospheric pressure. The branch
pipe is also employed in collecting a small portion of air for chemical

analysis during each experiment. A pipe, j, is so suspended, that by means of india-rubber junctions, a communication can readily be made to convey the air issuing from the nozzle into the gas-meter, k, which has a capacity of 40 cubic feet, and is carefully graduated by calibration. A bent glass tube, l, inserted in the top of the meter, and containing a little water, indicates the slight difference which sometimes exists between the pressure of air in the meter and that of the external atmosphere. When required, a wrought-iron pipe, m, 1 inch in diameter, is used to convey the elastic fluid from the meter to the desiccating apparatus, and thence to the pump so as to circulate through the entire apparatus.

We have already pointed out the different thermal effects to be anticipated from the rushing of air from a single narrow orifice. They are *cold*, on the one hand, from the expenditure of heat in labouring force to communicate rapid motion to the air by means of expansion; and *heat*, on the other, in consequence of the *vis viva* of the rushing air being reconverted into heat. The two opposite effects nearly neutralize each other at 2 or 3 inches distance from the orifice, leaving however a slight preponderance of cooling effect; but close to the orifice the variations of temperature are excessive, as will be made manifest by the following experiments.

A thin plate of copper, having a hole of $\frac{1}{20}$th of an inch diameter, drilled in the centre, was bolted to the flange, an india-rubber washer making the joint air-tight. At the ordinary velocity of the pump the orifice was sufficient to discharge the whole quantity of air when its pressure arrived at 124 lbs. on the square inch. When however lower pressures were tried, the stop-cock e was kept partially open. The thermometer used was one with a spherical bulb 0·15 of an inch in diameter. Holding it as close to the orifice as possible without touching the metal, the following observations were made at various pressures, the temperature of the water in which the coils were immersed being 22° Cent. The air was dried and deprived of carbonic acid by passing it, previous to entering the pump, through a vessel 4½ feet long and 20 inches diameter, filled with quick-lime.

Total pressure of the air in lbs. on the square inch.	Temperature Centigrade.	Depression below temperature of bath.
124	8·58	13·42
72	11·65	10·35
31	16·25	5·75

The heating effect was exhibited as follows :—The bulb of the thermometer was
inserted into a piece of conical gutta percha
pipe in such a manner that an extremely
narrow passage was allowed between the
interior surface of the pipe and the bulb.
Thus armed, the thermometer was held, as
represented by fig. 5, at half an inch distance
from the orifice, when the following results were obtained :—

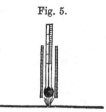

Fig. 5.

Total pressure of the air in lbs. on the square inch.	Temperature Centigrade.	Elevation above temperature of bath.
124	45·75	23·75
71	39·23	17·23
31	26·2	4·20

It must be remarked, that the above recorded thermal effects
are not to be taken as representing the maximum results to be
derived from the rushing air at the pressures named. The determination of these, in the form of experiment above given, is
prevented by several circumstances. In particular it must be
observed, that the cooling effects must have been reduced in consequence of the heat evolved by the friction of the rushing air
against the bulb of the thermometer. The heating effects, resulting as they do from the absorption and conversion into heat
of the *vis viva* of the rushing air, depend very much upon the
narrowness of the space between the thermometer and gutta
percha pipe. We intend further on to return to this subject,
but in the mean time will mention three forms of experiment
whereby the heating effect is very strikingly and instructively
exhibited.

Experiment 1.—The finger and thumb are brought over the
orifice, as represented in fig. 6, so that by gradually closing them the stream of air is pinched.
It is found that the effort to close the finger and
thumb is opposed by considerable force, which
increases with the pressure applied. At the
same time a strong tremulous motion is felt and
a shrill noise is heard, whilst the heat produced in five or six
seconds necessitates the termination of the experiment.

Fig. 6.

Experiment 2.—Fig. 7. The finger is placed over the orifice
Fig. 7.   and pressed until a thin stratum of air escapes
between the copper-plate and the finger. In
this case the burning heat of the rushing air is
equally remarkable in spite of the proximity
of the finger to the cold metal.

Experiment 3.—Fig. 8. A piece of thick india-rubber is
Fig. 8.   pressed by the finger over the narrow orifice so
as to allow a thin stream of air to rush between
the india-rubber and the plate of copper. In
this case the india-rubber is speedily raised to a
temperature which prevents its being handled comfortably.

We have now adduced enough to illustrate the immense and
sudden changes of temperature which exist in the "rapids" of
a current of air, changes which point out the necessity of employing a porous plug, in order that when the air arrives at the
thermometer its state may be reduced to a uniform condition.
Fig. 9.   Figs. 4 and 9 represent our first arrangement for
the porous plug, where n is a brass casting with
flange to bolt to the copper tube. It has eight
studs, o, and eight holes, pp, drilled into the
inner part of the flange. The studs and holes
furnish the means of securing the porous material
(in the present instance of cotton wool) in its place, by binding
it down tightly with twine. Immediate contact between the
cotton and metal is prevented by the insertion of a piece of
india-rubber tubing; qqq are three pieces of india-rubber tube
inserted within each other, the inner one communicating with
a glass tube r, through which the divisions of the thermometer
may be seen, and which serves to convey the air to the meter.
In the experiments about to be given, the thermometer was in
immediate contact with the cotton plug as represented in the
figure, and the nozzle was immersed in the bath up to the line s.
The weight of the cotton wool in the dry state was 251 grs., its
specific gravity 1·404, and being compressed into a space 1½ inch
in diameter and 1·9 inch long, the opening left for the passage
of air must have been equal in volume to a pipe of 1·33 of an
inch diameter.

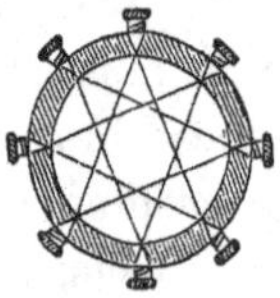

First series of experiments. Atmospheric air dried and deprived of carbonic acid by quick-lime. Gauge 73·6; barom. 30·04 = 14·695 lbs. pressure per square inch.

Gauge.	Total pressure in lbs. per square inch.	Cubic inches of air passed per minute reduced to atmospheric pressure.	Temperature of bath * ascertained by Thermometer No. 1, in Centigrade degrees.	Temperature of the issuing air, ascertained by Thermometer No. 2.	Cooling effect.
37·5 37·5 38 }37·7 37·8 38	35·854	12703	445 445·5 445·9 }445·6 = 18°·2676 446	414 414 414·6 }414·35 = 17°·8298 414·8	0°·4378
38 38 38 }37·9 37·8 38	35·647	12703	446·1 446·6 446·8 }446·65 = 18·3128 447·1	415·4 416 416·8 }416·45 = 17·9295 417·6	0·3833
38 37·75 37·5 }37·69 37·5	35·866	12703	447·2 447·5 447·8 }447·62 = 18·3545 448	418 418·2 418·4 }418·15 = 18·0110 418	0·3435

A Liebig tube containing sulphuric acid, specific gravity 1·8, gained 0·03 of a grain by passing through it, during the experiment, 100 cubic inches of air.

The observations above tabulated were made at intervals of two or three minutes. It will be observed that the cooling effect appeared to be greater at the commencement than at the termination of the series. This may be attributed in a great measure to the drying of the cotton, which was found to contain at least 5 per cent. of moisture after exposure to the atmosphere. There was also another source of interference with the accuracy of the results owing to a considerable oscillation of pressure arising from the action of the pump. We had remarked that when the number of strokes of the engine was suddenly reduced from twenty-seven to twenty-five per minute, a depression of the thermometer equal to some hundredths of a degree Cent. took place, a circumstance evidently owing to the entire mass of air in the coils and cotton plugs suffering dilatation without allowing time for the escape of the consequent thermal effect. Hence it was found absolutely essential

* By varying the temperature of the water in which the coils were immersed, it was found that the temperature of the water surrounding the first coil exercised no perceptible influence, the temperature of the rushing air being entirely regulated by that of the terminal coil. However, the precaution was taken of keeping both coils at nearly the same temperature.

T.   23

to keep the pump working at a perfectly uniform rate. For a similar reason it was also most important to prevent the oscillations of pressure due to the action of the pump, particularly as it appeared obvious that the heat evolved by the sudden increase of pressure, on the admission of a fresh supply of air from the pump, would arrive at the thermometer in a larger proportion than the cold produced by the subsequent gradual dilatation. In fact, on making an experiment in which the air was kept at a low pressure, by opening a stop-cock provided for the purpose, the oscillations of pressure amounting to $\frac{1}{20}$th of the whole, it was found that an apparent heating effect, equal to $0°\cdot2$ Cent., was produced instead of a small cooling effect.

It became therefore necessary to obviate the above source of error, and the method first employed with that view, was to place a diaphragm of copper with a hole in its centre $\frac{1}{4}$th of an inch in diameter at the junction between the iron and copper pipes. The oscillation being thus reduced, so as to be hardly perceptible, we made the following observations.

Second series of experiments. Atmospheric air dried and deprived of carbonic acid by quick-lime. Gauge $73\cdot75$; barometer $30\cdot162 = 14\cdot755$ lbs. pressure per square inch; thermometer $19°\cdot3$ Cent.

Gauge.	Total pressure in lbs. per square inch.	Cubic inches of air issuing per minute at atmospheric pressure.	Temperature of bath by Thermometer No. 1, degrees Centigrade.	Temperature of issuing air by Thermometer No. 2, degrees Centigrade.	Cooling effect.
39 38·6 38·5 38·5 } 38·65	36·069	11796	467 467 467 467·1 } 467·02 = 19·186	434·6 435 435 435 } 434·9 = 18·810	0·377
38·5 38·8 38·8 38·75 } 38·79	35·912	11796	467·1 467·2 467·2 467·3 } 467·2 = 19·194	435·1 435·4 435·6 435·4 } 435·37 = 18·832	0·362
38·8 38·8 38·8 38·8 } 38·8	35·900	11796	467·3 467·4 467·4 467·4 } 467·37 = 19·202	435·6 435·8 435·9 436 } 435·82 = 18·854	0·348

Suspecting that particles of the sperm oil employed for lubricating the pump were carried mechanically to the cotton plug and interfered with the results, we now substituted a box with perforated caps, filled with cotton wool, for the diaphragm used in the last series. With this arrangement the pressure was kept as uniform as with the other, and all solid and liquid particles were kept back by filtration.

Third series of experiments. Atmospheric air dried and deprived of carbonic acid by quick-lime*, and filtered through cotton. Gauge 73·7; thermometer 21°·7 Cent.; barometer 30·10 = 14·71 lbs. on the square inch.

Time of observation.	Gauge.	Total pressure in lbs. per square inch.	Cubic inches of air issuing per minute at atmospheric pressure.	Temperature of bath by Thermometer No. 1, in degrees Centigrade.	Temperature of the issuing air by Thermometer No. 2, in degrees Centigrade.	Cooling effect.
m 3 6 9 12	39 39·1 39·5 39·2 }39·2	34·410	11784	357·7 357·8 358 358·2 }357·92 = 14°·506	337·35 337·8 338 338·4 }337·89 = 14°·183	0°·323
15 16 18 21	39·1 39·35 39·1 39·2 }39·19	34·418	11784	358·7 358·9 359·1 359·2 }358·97 = 14·552	338·8 338·7 339 338·9 }338·85 = 14·230	0·322
23 25 28 30	39·2 39·1 39·2 39·2 }39·18	34·426	11784	359·4 359·7 359·8 360 }359·72 = 14·584	339·25 339·8 339·7 340 }339·69 = 14·270	0·314
32 34 36 38	39·5 39·3 39·25 39·3 }39·34	34·279	11784	360·1 360·2 360·4 360·4 }360·27 = 14·607	340 340·2 340·4 340·4 }340·25 = 14·296	0·311

The stop-cock for reducing pressure being now partially opened, the observations were continued as follows :—

---

* The use of quick-lime as a desiccating agent was suggested to us by Mr Thomas Ransome. It answered its purpose admirably after it had fallen a little by use, so as to be finely subdivided. The perfection of its action was shown by the desiccating cylinder remaining, after having been used two hours, cold at the lower part, while the upper part for about 9 inches was made very hot. The analysis of the air passed during the third series of experiments showed that one of the Liebig tubes had gained no weight whatever ; and in one instance we have observed that the sulphuric acid of 1·8 specific gravity, actually lost weight, apparently indicating that the air dried by quick-lime was able to remove water from acid of that density.

Time of observation.	Gauge.	Total pressure in lbs. per square inch.	Temperature of bath by Thermometer No. 1, in degrees Centigrade.	Temperature of the issuing air by Thermometer No. 2, in degrees Centigrade.	Cooling effect.
h m					
50	55·1		361·7	344·	
52	55·1		361·9	344·8	
54	55·1		361·9	345·3	
55	55·1		361·9	345·8	
57	55·1		362·1	346·0	
59	55·1	} 55·12 22·876	362·3 } 362·26 = $14\overset{o}{\cdot}693$	346·4 } 346·19 = $14\overset{o}{\cdot}579$	$0\overset{o}{\cdot}114$
1 1	55·1		362·4	346·9	
3	55·1		362·7	347·2	
5	55·1		362·7	347·6	
7	55·3		363	347·9	
11	54·3		363·3	348·9	
13	54·4		363·3	348·9	
15	54·4		363·5	349·2	
17	54·7		363·7	349·4	
19	54·5	} 54·51 23·217	363·9 } 363·82 = 14·760	350· } 349·74 = 14·749	0·011
20	54·5		364·1	350·	
22	54·6		364·2	350·3	
24	54·6		364·2	350·4	
26	54·6		364·2	350·6	
30	54·6		375·	356·4	
32	54·6		375·4	358·2	
33	54·2		375·4	359·4	
35	54·3		375·5	359·8	
37	54·4	} 54·38 23·277	375·8 } 375·7 = 15·270	360· } 360 = 15·238	0·032
39	54·6		375·7	360·1	
40	54·3		375·8	360·3	
42	54·5		376·	360·4	

During the above experiment 100 cubic inches of the air was slowly passed through two Liebig tubes containing sulphuric acid, specific gravity 1·8. The first tube gained 0·006 of a grain, the second remained at exactly the same weight.

P.S.   Oct. 14, 1853.—The apparently anomalous results contained in the last Table have been fully explained, and shown to depend on the alteration of pressure which took place towards the beginning of the interval of time from 42^m to 50^m, by subsequent researches which we hope soon to lay before the Royal Society.

[From the *Transactions of the Royal Society*, June, 1854.]

PART II.

IN the last experiment related in our former paper*, in which
a low pressure of air was employed, a considerable variation of the
cooling effect was observed, which it was necessary to account for
in order to ascertain its influence on the results. We therefore
continued the experiments at low pressures, trying the various
arrangements which might be supposed to exercise influence over
the phenomena. We had already interposed a plug of cotton wool
between the iron and copper pipes, which was found to have the
very important effect of equalizing the pressure, besides stopping
any solid or liquid particles driven from the pump, and which
has therefore been retained in all the subsequent experiments.
Another improvement was now effected by introducing a nozzle
constructed of boxwood, instead of the brass one previously used.
This nozzle is represented by fig. 1, Plate IV., in which *aa* is a
brass casting which bolts upon the terminal flange of the copper
piping, *bb* is a turned piece of boxwood screwing into the above,
having two ledges for the reception of perforated brass plates, the
upper plate being secured in its place by the turned boxwood *cc*,
which is screwed into the top of the first piece. The space enclosed by the perforated plates is 2·72 inches long and an inch and
a half in diameter, and being filled with cotton, silk, or other
material more or less compressed, presents as much resistance to
the passage of the air as may be desired. A tin can, *d*, filled with
cotton wool, and screwing to the brass casting, serves to keep the
water of the bath from coming in contact with the boxwood
nozzle.

In the following experiments, made in order to ascertain the
variations in the cooling effect above referred to, the nozzle was
filled with 382 grs. of cotton wool, which was sufficient to keep up
a pressure of about 34 lbs. on the inch in the tubes, when the
pump was working at the ordinary rate. By opening the stop-cock
in the main pipe this pressure could be further reduced to about
22 lbs. by diminishing the quantity of air arriving at the nozzle.

---

* *Transactions of the Royal Society*, June, 1853.

By shutting and opening the stop-cock we had therefore the means of producing a temporary variation of pressure, and of investigating its effect on the temperature of the air issuing from the nozzle. In the first experiments the stop-cock was kept open for a length of time, until the temperature of the rushing air became pretty constant; it was then shut for a period of $3\frac{3}{4}$, $7\frac{1}{2}$, 15, 30 or 60 seconds, then reopened. The oscillations of temperature thus produced are laid down upon the Chart No. 1, in which the ordinates of the curves represent the temperatures according to the scale of thermometer C, each division corresponding to 0·0477 of a degree Centigrade. The divisions of the horizontal lines represent intervals of time equal to a quarter of a minute. The horizontal black lines show the temperature of the bath in each experiment.

The effect upon the pressure of the air produced by shutting the stop-cock during various intervals of time, is given in the following Table :—

Stop-cock shut for	5ˢ.	15ˢ.	30ˢ.	1ᵐ.	2ᵐ.
m s					
Initial pressure..............	22·35	22·35	22·35	22·35	22·35
Pressure after 0 5	24·92	24·92	24·92	24·92	24·92
Pressure after 0 15	23·07	28·46	28·46	28·46	28·46
Pressure after 0 30	22·43	23·38	30·84	30·84	30·84
Pressure after 0 45	22·35	22·5	24·27	32·03	32·03
Pressure after 1 0	22·35	22·43	22·83	32·79	32·79
Pressure after 1 15	......	22·35	22·45	24·54	33·08
Pressure after 1 30	......	22·35	22·35	22·83	33·25
Pressure after 1 45	......	......	22·35	22·43	33·33
Pressure after 2 0	......	......	......	22·35	33·41
Pressure after 2 15	......	......	......	22·35	24·54
Pressure after 2 30	......	......	......	......	22·54
Pressure after 2 45	......	......	......	......	22·40
Pressure after 3 0	......	......	......	......	22·35

The last column gives also the effect occasioned by the permanent shutting or opening of the stop-cock, 33·41 lbs. being nearly equal to the pressure when the stop-cock has been closed for a long time.

In the next experiments, the opposite effect of opening the stop-cock was tried, the results of which are laid down on Chart No. 2.

The effect upon the pressure of the air produced by opening the stop-cock during the various intervals of time employed in the experiments, is exhibited in the next Table :—

Stop-cock opened for	$3\frac{3}{4}$ˢ.	$7\frac{1}{2}$ˢ.	15ˢ.	30ˢ.	1ᵐ.
m s					
Initial pressure.............	34·37	34·37	34·37	34·37	34·37
Pressure after .... 0 $3\frac{3}{4}$	29·57	29·57	29·57	29·57	29·57
Pressure after .... 0 $7\frac{1}{2}$	......	27·43	27·43	27·43	27·43
Pressure after 0 15	22·47	30·41	25·15	25·15	25·15
Pressure after 0 30	33·5	32·47	30·41	23·23	23·23
Pressure after 0 45	33·94	33·5	32·4	29·4	22·9
Pressure after 1 0	34·1	34·1	33·5	32·13	22·76
Pressure after 1 15	34·2	34·3	33·94	33·24	28·82
Pressure after 1 30	34·33	34·37	34·14	33·90	31·44
Pressure after 1 45	34·37	34·37	34·30	34·14	32·9
Pressure after 2 0	34·37	......	34·37	34·33	33·66
Pressure after 2 15	......	......	......	34·37	34·06
Pressure after 2 30	......	......	......	......	34·20
Pressure after 2 45	......	......	......	......	34·37

The remarkable fluctuations of temperature in the issuing stream accompanying such changes of pressure, and continuing to be very perceptible in the different cases for periods of from 3 or 4 minutes up to nearly half an hour after the pressure had become sensibly uniform, depend on a complication of circumstances, which appear to consist of (1) the change of cooling effect due to the instantaneous change of pressure; (2) a heating or cooling effect produced instantaneously by compression or expansion in all the air flowing towards and entering the plug, and conveyed through the plug to the issuing stream; and (3) heat or cold communicated by contact from the air on the high-pressure side, to the metals and boxwood, and conducted through them to the issuing stream.

The first of these causes may be expected to influence the issuing stream instantaneously on any change in the stop-cock; and after fluctuations from other sources have ceased, it must leave a permanent effect in those cases in which the stop-cock is permanently changed. But after a certain interval the reverse agency of the second cause, much more considerable in amount, will begin to affect the issuing stream, will soon preponderate over the first, and (always on the supposition that this convection is uninfluenced by conduction of any of the materials) will affect it with all the variations, undiminished in amount, which the air entering the plug experiences, but behind time by a constant interval equal to the time occupied by as much air as is equal in thermal capacity to the cotton of the plug, in passing through the apparatus*; this, in the experiments with the stop-cock shut,

* To prove this, we have only to investigate the convection of heat through a

would be very exactly a quarter of a minute; but it appears to prismatic solid of porous material, when a fluid entering it with a varying tempera-

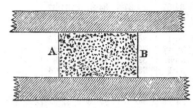

ture is forced through it in a continuous and uniform stream. Let AB be the porous body, of length a and transverse section S; and let a fluid be pressed continuously through it in the direction from A to B, the temperature of this fluid as it enters at A being an arbitrary function $F(t)$ of the time. Then if v be the common temperature of the porous body and fluid passing through it, at a distance x from the end A, we have

$$\sigma \frac{dv}{dt} = k \frac{d^2v}{dx^2} - \frac{\theta}{S} \frac{dv}{dx} \quad \dots \dots \dots (1);$$

if k be the conducting power of the porous solid for heat (the solid surrounding it being supposed to be an infinitely bad conductor, or 'the circumstances to be other-wise arranged, as is practicable in a variety of ways, so that there may be no lateral conduction of heat), σ the thermal capacity of unity of its bulk, and θ the thermal capacity of as much of the fluid as passes in the unit time. Now if, as is probably the case in the actual circumstances, conduction through the porous solid itself is insensible in its influence as compared with the convection of the fluid, this equation will become approximately

$$\sigma \frac{dv}{dt} = - \frac{\theta}{S} \frac{dv}{dx} \quad \dots \dots \dots (2),$$

which, in fact, expresses rigorously the effect of the second cause mentioned in the text if alone operative.

If F denote any arbitrary function, and if θ be supposed to be constant, the general integral of this equation is—

$$v = F \left(t - \frac{\sigma S}{\theta} x \right) \quad \dots \dots \dots (3);$$

and if the arbitrary function be chosen to express by $F(t)$ the given variation of temperature where the fluid enters the porous body, we have the particular solution of the proposed problem. We infer from it that, at any distance x in the porous body from the entrance, the temperature will follow the same law and extent of variation as at the entrance, only later in time by an interval equal to $\sigma Sx/\theta$. We conclude that the variations of temperature in the issuing stream due to the second cause alone, in the actual circumstances, are equal and similar to those of the air entering the plug, but later in time by $\sigma Sa/\theta$. In this expression, the numerator, σSa, denotes simply the thermal capacity of the whole plug. The plug, in the actual experiments, having consisted of 382 grains of cotton, of which the thermal capacity is about 191 times that of a grain of water, and (when the stop-cock was closed) the air having been pumped through at the rate, per second, of 50 grains, of which the capacity is twelve times that of a grain of water, the value of $\sigma Sa/\theta$ must have been $\frac{191}{12}$ seconds, or about a quarter of a minute. When the stop-cock was open, an unknown quantity of air escaped through it, and therefore the value of $\sigma Sa/\theta$ must have been somewhat greater. The variation which the value of θ must have experienced when the stop-cock was opened or closed in the course of an experi-ment, or even merely in consequence of the change of pressure following the initial opening or closing of the stop-cock, makes the circumstances not such as in any of

have averaged more nearly one-third of a minute in the varying
circumstances of the actual experiments, since our observations (as
may be partially judged from the preceding charts) showed us
with very remarkable sharpness, in each case about twenty
seconds after the shutting or opening of the stop-cock, the commencement of the heating or cooling effect on the issuing stream,
due to the sudden compression or rarefaction instantaneously produced in the air on the other side of the plug.

The entering air will, very soon after its pressure ceases to
vary, be reduced to the temperature of the bath by the excellent
conducting action of the spiral copper pipe through which it
passes; and, consequently, twenty seconds or so later, the issuing
stream can experience no further fluctuations in temperature except by the agency depending on the third cause.

That the third cause may produce very considerable effects
is obvious, when we think how great the variations of temperature
must be to which the surfaces of the solid materials in the neighbourhood of the plug on the high-pressure side are subjected
during the sudden changes of pressure: and that the heat consequently taken in or emitted by these bodies may influence the
issuing stream perceptibly for a quarter or a half hour after the
changes of pressure from which it originated have ceased, is quite
intelligible on account of the slowness of conduction of heat
through the wood and metals, when we take into account the
actual dimensions of the parts of the apparatus round the plug.
It is not easy, however, to explain all the fluctuations of temperature which have been observed after the pressure had become
constant in the different cases.   Those shown in the first set of
diagrams are just such as might be expected from the alternate
heating and cooling which the solids must have experienced at
their surfaces on the high-pressure side, and which must be conducted through so as to affect the issuing stream after a considerable time; but the great elevations of temperature shown in the
second set of diagrams, which correspond to cases when the
pressure was temporarily or permanently *diminished,* are not, so
far as we see, explained by the causes we have mentioned, and the
circumstances of these cases require further examination.

the cases to correspond rigorously to the preceding solution; which, notwithstanding, represents the general nature of the convective effect nearly enough for the explanation in the text.

When we had thus examined the causes of the fluctuations of temperature in the issuing air, the precautions to prevent their injurious effect upon the accuracy of the determinations of the cooling effect in the passage of air through the porous plug became evident. These were simply to render the action of the pump as uniform as possible, and to commence the record of observations only after one hour and a half or two hours had elapsed from the starting of the pump. The system then adopted was to observe the thermometers in the bath and stream of air, and the pressure-gauge every two minutes or minute and a half; the means of which observations are recorded in the columns of the Tables. In some instances the air previous to passing into the pump was transmitted through a cylinder which had been filled with quick-lime. But since by previous use its power of absorbing water had been considerably deteriorated, a portion of the air was always transmitted through a Liebig tube containing asbestos moistened with sulphuric acid or chloride of zinc. The influence of a small quantity of moisture in the air is trifling, but will hereafter be examined. That of the carbonic acid contained by the atmosphere was, as will appear in the sequel, quite inappreciable. It will be proper to observe that the thermometers by which the temperature of the bath and issuing air was ascertained, were repeatedly compared together to avoid any error which might arise from the alteration of their fixed points from time to time.

TABLE I.—Experiments with a plug consisting of 191 grains of cotton wool.

1.	2.	3.	4.	5.	6.	7.	8.
Number of observations from which the results in Columns 4, 6, and 7, are obtained.	Cubic inches passed through the nozzle per minute.	Water in 100 grains of air, in grains.	Pressure in lbs. on the square inch.	Atmospheric pressure.	Temperature of the bath.	Temperature of the issuing air.	Cooling effect in Cent. degrees.
20	10822	0·51	21·326	14·400	20·295°	20·201°	0·094°
20	10998	0·30	21·239	14·252	16·740	16·615	0·125
10	Not observed.	0·56	20·446	14·609	17·738	17·622	0·116
10	10769	0·66	20·910	14·772	16·039	15·924	0·115
10	10769	0·66	20·934	14·775	16·065	15·967	0·098
10	10769	0·66	20·995	14·779	16·084	15·984	0·100
10	10769	0·66	20·933	14·782	16·081	15·974	0·107
Mean	......	0·57	20·969	14·624	17·006	16·898	0·108

In each, excepting the first of the seven experiments above recorded, the air was passed through the quick-lime cylinder.

In the next experiments the nozzle was filled with 382 grains of cotton wool. The intermediate stop-cock was however partly opened, in order that by discharging a portion of the air before its arrival at the nozzle, the pressure might not be widely different from that employed in the last series. In all excepting the last experiment recorded in the following Table, the cylinder of lime was dispensed with.

TABLE II.—Experiments with a smaller quantity of air passed through a plug consisting of 382 grs. of cotton wool.

1.	2.	3.	4.	5.	6.	7.	8.
Number of observations from which the results in Columns 4, 6, and 7, are obtained.	Cubic inches passed through the nozzle per minute.	Water in 100 grains of air, in grains.	Pressure in lbs. on the square inch.	Atmospheric pressure.	Temperature of the bath.	Temperature of the issuing air.	Cooling effect in Cent. degrees.
20	3865	0·59	22·614	14·513	20·363	20·224	0·139
30	3960	0·73	22·818	14·514	19·853	19·769	0·084
20	Not observed.	0·56	22·818	14·604	20·481	20·407	0·074
45	3125	0·65	22·296	14·590	20·584	20·313	0·271
20	Not observed.	1·23	22·000	14·518	18·636	18·476	0·160
36	Not observed.	1·20	22·616	14·520	20·474	20·336	0·138
50	Not observed.	1·36	22·582	14·518	20·485	20·325	0·160
Mean	......	0·90	22·678	14·540	20·125	19·979	0·146

TABLE III.—Experiments in which the entire quantity of air propelled by the pump was passed through a plug consisting of 382 grains of cotton wool. The cylinder of lime was not employed.

1.	2.	3.	4.	5.	6.	7.	8.
Number of observations from which the results in Columns 4, 6, and 7, are obtained.	Cubic inches passed through the nozzle per minute.	Water in 100 grains of air, in grains.	Pressure in lbs. on the square inch.	Atmospheric pressure.	Temperature of the bath.	Temperature of the issuing air.	Cooling effect in Cent. degrees.
7	11766	0·56	36·625	14·583	19·869	19·535	0·334
10	Not observed.	0·56	35·671	14·790	20·419	20·098	0·321
10	Not observed.	0·36	35·772	14·504	16·096	15·730	0·366
10	Not observed.	0·36	35·872	14·504	16·104	15·721	0·383
10	Not observed.	0·36	36·026	14·504	16·232	15·869	0·363
Mean	......	0·44	35·993	14·577	17·744	17·390	0·354

In the next series of experiments the air was passed through a plug of silk, formed by rolling a silk handkerchief into a cylindrical shape, and then screwing it into the nozzle. The silk weighed 580 grains, and the small quantity of cotton wool placed on the side next the thermometer in order to equalize the stream of air more completely, weighed 15 grains. The stop-cock was partly opened as in the experiments of Table II., in order to reduce the pressure to that obtained by passing the full quantity of air propelled by the pump through a more porous plug. The cylinder of lime was employed.

TABLE IV.—Experiments in which a smaller quantity of air was passed through a plug consisting of 580 grains of silk.

1.	2.	3.	4.	5.	6.	7.	8.
Number of observations from which the results in Columns 4, 6, and 7, are obtained.	Cubic inches passed through the nozzle per minute.	Water in 100 grains of air, in grains.	Pressure in lbs. on the square inch.	Atmospheric pressure.	Temperature of the bath.	Temperature of the issuing air.	Cooling effect in Cent. degrees.
10	3071	0·18	33·168	14·727	18·882	18·524	0·358
10	Not observed.	0·18	33·024	14·732	18·884	18·536	0·348
10	Not observed.	0·14	33·820	14·660	19·066	18·686	0·380
10	Not observed.	0·14	33·226	14·650	19·068	18·695	0·373
Mean	......	0·16	33·309	14·692	18·975	18·610	0·365

TABLE V.—Experiments in which the entire quantity of air propelled by the pump was passed through the silk plug. The cylinder of lime was employed in all excepting the first two experiments.

1.	2.	3.	4.	5.	6.	7.	8.
Number of observations from which the result in Columns 4, 6, and 7, are obtained.	Cubic inches passed through the nozzle per minute.	Water in 100 grains of air, in grains.	Pressure in lbs. on the square inch.	Atmospheric pressure.	Temperature of the bath.	Temperature of the issuing air.	Cooling effect in Cent. degrees.
10	7594	0·40	53·722	14·580	17·585	16·903	0·682
10	Not observed.	0·40	53·530	14·580	17·628	16·954	0·674
10	Not observed.	0·32	53·317	14·563	17·993	17·318	0·675
10	Not observed.	0·32	53·317	14·568	18·027	17·357	0·670
10	7742	0·11	55·797	14·615	17·822	17·063	0·759
10	Not observed.	0·11	54·074	14·611	17·813	17·079	0·734
10	Not observed.	0·11	55·720	14·608	17·808	17·082	0·726
10	Not observed.	0·11	56·174	14·605	17·796	17·058	0·738
Mean	......	0·23	54·456	14·591	17·809	17·102	0·707

In order to obtain a greater pressure, a plug was formed of silk "waste" compressed very tightly into the nozzle.

TABLE VI.—Experiments in which the air, after passing through the cylinder of lime, was forced through a plug consisting of 740 grains of silk.

1.	2.	3.	4.	5.	6.	7.	8.
Number of observations from which the results in Columns 4, 6, and 7, are obtained.	Cubic inches passed through the nozzle per minute.	Water in 100 grains of air, in grains.	Pressure in lbs. on the square inch.	Atmospheric pressure.	Temperature of the bath.	Temperature of the issuing air.	Cooling effect in Cent. degrees.
10	Not observed.	0·19	79·852	14·777	17·050	15·884	1·166
10	Not observed.	0·19	80·133	14·782	17·066	15·913	1·153
10	Not observed.	0·19	79·870	14·787	17·079	15·945	1·134
10	5650	0·19	80·013	14·793	17·083	15·967	1·116
10	Not observed.	0·15	79·814	14·960	16·481	15·338	1·143
10	Not observed.	0·15	80·274	14·957	16·489	15·374	1·115
10	Not observed.	0·15	79·903	14·953	16·505	15·392	1·113
10	5378	0·15	77·867	14·950	16·521	15·428	1·093
10	Not observed.	0·14	78·214	14·638	12·851	11·770	1·081
10	Not observed.	0·14	78·245	14·638	12·877	11·800	1·077
10	Not observed.	0·14	78·180	14·638	12·885	11·824	1·061
10	Not observed.	0·14	78·633	14·638	12·905	11·839	1·066
Mean	......	0·16	79·250	14·793	15·483	14·373	1·110

In the foregoing experiments the pressure of the air on its exit from the plug was always exactly equal to the atmospheric pressure. To ascertain the effect of an alteration in the pressure of the exit air, we now enclosed a long siphon barometer within the glass tube (fig. 10). The upper part of this tube was surmounted with a cap, furnished with a stop-cock, by partially closing which the air at its exit could be brought to the required pressure. The influence of pressure in raising the mercury in the thermometer by compressing its bulb, was ascertained by plunging the instrument into a bottle of water within the glass tube, and noting the amount of the sudden rise or fall of the quicksilver on a sudden augmentation or reduction of pressure. It was found that the pressure equal to that of 17 inches of mercury, raised the indication by 0°·09 ; which quantity was therefore subtracted after the usual reduction of the thermometric scale.

TABLE VII.—Experiments with the plug consisting of 740 grains of silk. Pressure of the exit air increased. Cylinder of lime used.

1.	2.	3.	4.	5.	6.	7.	8.
Number of observations from which the results in Columns 4, 6, and 7, are obtained.	Cubic inches passed through the nozzle per minute.	Water in 100 grains of air, in grains.	Pressure in lbs. on the square inch.	Pressure of the exit air.	Temperature of the bath.	Temperature of the issuing air.	Cooling effect in degrees Cent. degrees.
10	Not observed.	0·14	82·982	23·093	12·673	11·612	1·061
10	Not observed.	0·14	82·510	22·878	12·713	11·676	1·037
10	Not observed.	0·14	81·895	22·798	12·755	11·725	1·030
10	Not observed.	0·14	80·630	22·488	12·795	11·792	1·003
Mean	Estimated at 5400	0·14	82·004	22·814	12·734	11·701	1·033

With reference to the experiments in Table VII. it may be remarked, that the cooling effect must be the excess of that which would have been obtained had the air been only resisted by the atmospheric pressure in escaping from the plug, above the cooling effect that would be found in an experiment with the temperature of the bath and the pressure of the entering air the same as the temperature and pressure of the exit air in the actual experiment, and the air issuing at atmospheric pressure. Hence, since two or three degrees of difference of temperature in the bath would not sensibly alter the cooling effect in any of the experiments on air, the cooling effect in an experiment in which the pressure of the exit air is increased, must be sensibly equal to the difference of the cooling effects in two of the ordinary experiments, with the high pressures the same as those used for the entering and issuing air respectively, and the low pressure that of the atmosphere in each case; a conclusion which is verified by the actual results, as the comparison given below shows.

The results recorded in the foregoing Tables are laid down on Chart No. 3, in which the horizontal lines represent the excess of the pressure of the air in the receiver over that of the exit air as found by subtracting the fifth from the fourth columns of the Tables, and the vertical lines represent the cooling effect in tenths of a degree Centigrade. It will be remarked that the

line drawn through the points of observation is nearly straight,
indicating that the cooling effect is, approximately at least, proportional to the excess of pressure, being about ·018° per pound
on the square inch of difference of pressure. Or we may arrive
at the same conclusion by dividing the cooling effect (δ) by the
difference of pressures $(P - P')$ in the different experiments.
We thus find, from the means shown in the different tables,—

$$\text{Table (I.)} \quad \frac{\delta}{P - P'} = ·0170$$

(II.)	·0179
(III.)	·0165
(IV.)	·0196
(V.)	·0177
(VI.)	·0172
(VII.)	·0174
Mean	·0176

*On the Cooling Effects experienced by Carbonic Acid in passing
through a porous Plug.*

The position of the apparatus gave us considerable practical
facilities in experimenting with carbonic acid. A fermenting
tun 10 feet deep and 8 feet square was filled with wort to a depth
of 6 feet. After the fermentation had been carried on for about
forty hours, the gas was found to be produced in sufficient quantity
to supply the pump for the requisite time. The carbonic acid
was conveyed by a gutta-percha pipe, and passed through two
glass vessels surrounded by ice in order to condense the greater
portion of vapours. In the succeeding experiment the total
quantity of liquid so condensed was 300 grains, which having a
specific gravity of ·9965, was composed of 10 grains of alcohol
and 290 grains of water. On analysing a portion of the gas
during the experiment by passing it through a tube containing
chloride of zinc, it was found to contain 0·733 gr. of water to
100 grs. of carbonic acid.

TABLE VIII.—Carbonic acid forced through a plug of 382 grs.
of cotton wool. Mean barometric pressure 29·45 inches,
equivalent to 14·399 lbs. Gauge under atmospheric pressure
151. The pump was placed in connexion with the pipe immersed in carbonic acid at $10^h 55^m$.

1.	2.	3.	4.	5.	6.
Time of observation.	Volume percentage of carbonic acid.	Pressure-gauge; mean pressure in lbs. on the square inch.	Temperature of the bath, by indications of thermometer.	Temperature of the issuing gas, by indications of thermometer.	Cooling effect in Cent. degrees
h m					
10 47	0	79·0	486·0	198·5	
49	0	79·0	486·0	198·5	
53	0	79·6	486·0	198·2	
57		85·2	486·0	195·0	
58		86·0	486·0	186·0	
59		85·0	486·0	188·6	
11 0	95·51	85·0	486·0	188·5	
2		86·4	486·0	187·6	
4		86·7	486·0	187·8	
6		86·6	486·0	188·9	
9	95·51	86·6	486·0	188·9	
13		84·0	486·0	188·65	
14		84·2	486·0	188·1	
15	95·51 ⟩94·89	84·4 ⟩84·906 = 32·989 lbs.	486·0 ⟩486·00 = 20·001	188·0 ⟩188·36 = 18·611	1·390
19		84·5	486·0	188·0	
22		84·1	486·0	188·1	
24		84·6	486·0	188·3	
25	93·03	84·2	486·0	188·5	
28		84·1	486·0	188·6	
32		83·2	486·0	188·9	
33		83·8	486·0	188·9	
35	86·92	84·0	486·0	189·0	
40		83·8	486·0	189·6	
41		83·9	486·0	189·7	
43		85·0	485·9	189·9	
45	79·37 ⟩80·61	86·0 ⟩84·245 = 33·286	485·9 ⟩485·94 = 19·998	190·4 ⟩190·1 = 18·787	1·211
49		84·6	485·9	190·8	
51		84·5	485·9	190·8	
53		83·9	485·9	190·6	
55	75·65	83·6	485·9	190·6	
12 0		83·6	485·9	190·8	
2		83·0	485·7	190·8	
5	70·68	82·7	485·7	190·9	
9	⟩68·82	82·7 ⟩82·783 = 33·960	485·4 ⟩485·52 = 19·980	190·8 ⟩191·07 = 18·884	1·096
13		82·9	485·4	191·1	
15	66·96	82·7	485·5	191·3	
21		82·7	485·4	191·5	
23		82·8	485·4	191·55	
25	65·72	82·9	485·4	191·6	
28		82·9	485·4	191·7	
33		82·2	485·4	191·8	
35	63·23	82·3	485·4	191·7	
40		81·9	485·3	191·65	
44		81·9	485·2	191·6	
45	63·23 ⟩63·85	82·1 ⟩82·986 = 33·864	485·2 ⟩485·18 = 19·966	191·6 ⟩191·82 = 18·959	1·007
52		82·4	485·0	191·65	
55	62·0	83·9	485·0	192·0	
1 2		84·1	485·0	192·1	
5	63·23	84·9	485·0	192·1	
11		85·4	485·0	192·3	
15	65·72	82·1	484·9	192·1	

TABLE IX.—Carbonic acid forced through a plug consisting of 191 grs. of cotton wool. Mean barometric pressure 29·6 inches, equivalent to 14·472 lbs. Gauge under atmospheric pressure 150·6. Pump placed in connexion with the pipe immersed in carbonic acid at $10^h\ 38^m$.

1. Time of observation.	2. Volume percentage of carbonic acid.	3. Pressure-gauge, and pressure in lbs. on the square inch equivalent thereto.	4. Indication of thermometer. Temperature of the bath.	5. Indication of thermometer. Temperature of the issuing gas.	6. Cooling effect in Cent. degrees.
h m					
10 40		123·0	461·5	189·5	
42		123·1	461·6	187·6	
44		123·1⌉	461·6⌉	187·25⌉	
50	90·51				
53		123·0	461·75	187·5	
55		123·2	461·75	187·45	
57	94·58	123·0 ⌡122·91 = 20·43	461·75 ⌡461·78 = 18°·962	187·55 ⌡187·49 = 18°·522	0·44
59		122·9	461·8	187·55	
11 0	93·65				
1		122·6	461·9	187·55	
3		122·6⌋	461·95⌋	187·6⌋	
5		122·6⌉	462·0⌉	187·55⌉	
7		122·5	462·0	188·1	
9		122·8	462·0	188·1	
10	81·86				
11		122·1	462·0	188·4	
15	76·27	121·6 ⌡121·91 = 20·682	462·2 ⌡462·11 = 18·976	188·4 ⌡188·35 = 18·609	0·367
17		121·7	462·15	188·4	
19		121·6	462·2	188·55	
20	70·68	121·7	462·2	188·7	
21		121·3	462·2	188·65	
25		121·2⌋	462·2⌋	188·7⌋	

In the above, as well as in the next series, the carbonic acid contained 0·35 per cent. of water.

TABLE X.—Experiment in which carbonic acid was forced through a plug consisting of 580 grs. of silk. Mean barometric pressure 29·56, equivalent to 14·452 lbs. Gauge under atmospheric pressure 150·8. Pump placed in connexion with the pipe immersed in carbonic acid at 12^h 53^m. Quantity of gas forced through the plug about 7170 cubic inches per minute.

1.	2.	3.	4.	5.	6.
Time of observation.	Volume percentage of carbonic acid.	Pressure-gauge, and pressure in lbs. on the square inch equivalent thereto.	Indication of thermometer. Temperature of the bath.	Indication of thermometer. Temperature of the issuing gas.	Cooling effect in Cent. degrees
h s					
12 42	0	52·2	464·2	185·6	
44	0	52·2	464·35	185·5	
46	0	52·2 lbs.	464·4	185·5	
49	0	52·2 }52·2 = 55·454	464·35 }464·34=19·072	185·55 }185·53=18·323	0·749
50	0	52·2	464·35	185·55	
52	0	52·2	464·4	185·5	
54		56·0	464·55	179·0	
57		55·7	464·65	166·3	
1 0	95·51	56·0	464·3	165·0	
5		56·0	464·55	165·0	
7		56·0	464·5	165·0	
9		56·0	464·4	164·9	
10	96·0 }94·85	}55·92 = 51·7	}464·47=19·077	165·0 =16·256	2·821
11		55·8	464·6	164·9	
13		55·6	464·55	164·8	
17		56·0	464·5	165·0	
20	93·03	56·0	464·4	165·4	
24		55·5	464·6	166·0	
25		55·7	464·6	166·3	
27		56·1	464·6	166·8	
30	85·92	56·0 }55·94=51·68	464·7 }464·71=19·088	167·9 }167·8 =16·538	2·550
35		56·1	464·8	168·9	
36		56·1	464·8	169·1	
38		56·1	464·9	169·6	

In the above experiment, as well as in those of the adjoining Tables, the sudden diminution of pressure on connecting the pump with the receiver containing carbonic acid, is in perfect accordance with the discovery by Professor GRAHAM of the superior facility with which that gas may be transmitted through a porous body compared with an equal volume of atmospheric air.

TABLE XI.—Experiment in which carbonic acid was forced through a plug consisting of 740 grs. of silk. Mean barometric pressure 30·065, equivalent to 14·723 lbs. on the inch. Gauge under atmospheric pressure 145·65. Pump placed in connexion with the pipe immersed in carbonic acid at $11^h 37^m$. Per-centage of moisture in the carbonic acid 0·15.

1.	2.	3.	4.	5.	6.
Time of observation.	Volume percentage of carbonic acid.	Pressure-gauge, and pressure in lbs. on the square inch equivalent thereto.	Indication of thermometer. Temperature of the bath.	Indication of thermometer. Temperature of the issuing gas.	Cooling effect in Cent. degrees.
h m					
11 28		35·5	318·9	117·9	
30		35·1	318·95	118·0	
32		35·6	318·95	118·0	
34		35·2	318·95	117·9	
36		35·2	318·95	117·73	
37		36·0	318·95	117·5	
38		36·2	318·95	112·0	
39	95·51	36·6		94·0	
43		36·9	319·03	83·95	
45	95·51	37·0		83·6	
47		37·1		83·0	
50	95·51 ⎤	37·0 ⎤	319·05 ⎤	82·6 ⎤	
53		37·0		82·4	
55	95·51	37·0	319·15	82·35	
57	⎬ 95·51	37·0 ⎬ 37·0 = 75·324 lbs.	⎬ 319·17 = 12°·844	82·3 ⎬ 82·62 = 7°·974	4°·87
12 0	95·51	37·0		82·7	
2		37·0	319·3	83·0	
5	95·51 ⎦	37·0 ⎦	⎦	83·0 ⎦	

In order to ascertain the cooling effect due to pure carbonic acid, we may at present neglect the effect due to the small quantity of watery vapour contained by the gas; and as the cooling effects observed in the various mixtures of atmospheric air and carbonic acid appear nearly consistent with the hypothesis that the specific heats of the two elastic fluids are for equal volumes equal to one another, and that each fluid experiences in the mixture the same absolute thermo-dynamic effect as if the other were removed, we may for the present take the following estimate of the cooling effects due to pure carbonic acid, at the various temperatures and pressures employed, calculated by means of this hypothesis from the observations in which the per-centage of carbonic acid was the greatest, and in fact so great, that a considerable error in the correction for the common air would scarcely affect the result to any sensible extent.

	Temperature of the bath.	Excess of pressure, $P - P'$	Cooling effect, δ.	Cooling effect divided by excess of pressure.
From Table IX....	18·962	5·958	0·459	·0770
From Table VIII....	20·001	18·590	1·446	·0778
From Table X....	19·077	37·248	2·938	·0789
From Table XI....	12·844	60·601	5·049	·0833
	Mean 17·721			Mean of first three ·0779
				Mean of all......... ·0793

We shall see immediately that the temperature of the bath makes a very considerable alteration in the cooling effect, and we therefore select the first three results, obtained at nearly the same temperature, in order to indicate the effect of pressure. On referring to Chart No. 3, it will be remarked that these three results range themselves almost accurately in a straight line. Or, by looking to the numbers in the last column, we arrive at the same conclusion.

Cooling Effect* experienced by Hydrogen in passing· through a porous Plug.

Not having been able as yet to arrange the large apparatus so as to avoid danger in using this gas in it, we have contented ourselves for the present with obtaining a determination by the help of the smaller force-pump employed in our preliminary experiments. The hydrogen, after passing through a tube filled with fragments of caustic potash, was forced, at a pressure of 68·4 lbs. on the inch, through a piece of leather in contact with the bulb of a small thermometer, the latter being protected from the water of the bath by a piece of india-rubber tube. At a temperature of about 10° Cent., a slight cooling effect was observed, which was found by repeated trials to be 0°·076. The pressure of the atmosphere being 14·7 lbs., it would appear that the cooling effect experienced by this gas is only one-thirteenth of

---

* [Note of March 29, 1882. We afterwards found that it was not a cooling effect, but a heating effect, that was experienced by Hydrogen: very small at low temperatures; and somewhat larger but still very small at temperatures of about 90° Cent. See Part IV. and Abstracts below.]

that observed with atmospheric air. We state this result with
some reserve, on account of the imperfection of such experiments
on a small scale, but there can be no doubt that the effect of
hydrogen is vastly inferior to that of atmospheric air.

Influence of Temperature on the Cooling Effect.

By passing steam through pipes plunged into the water of
the bath, we were able to maintain it at a high temperature
without any considerable variation. The passage of hot air
speedily raised the temperature of the stem of the thermometer,
as well as of the glass tube in which it was enclosed ; but never-
theless the precaution was taken of enclosing the whole in a
tin vessel, by means of which water in constant circulation with
the water of the bath was kept within one or two inches of the
level of the mercury in the thermometer. The bath was completely covered with a wooden lid, and the water kept in constant
and vigorous agitation by a proper stirrer.

TABLE XII.– Experiment in which—1st, air; 2nd, carbonic acid ;
3rd, air dried by quicklime was forced through a plug consisting of 740 grs. of silk. Mean barometric pressure 30·015,
equivalent to 14·68 lbs. on the inch. Gauge under the atmospheric pressure 150. Per-centage of moisture in the carbonic acid 0·31. Pump placed in connexion with the pipe
immersed in carbonic acid at 11^h 24^m. Disconnected and
attached to the quicklime cylinder at 12^h 22^m.

1.	2.	3.	4.	5.	6.
Time of observation.	Volume percentage of carbonic acid.	Pressure-gauge, and pressure in lbs. on the square inch equivalent thereto.	Indication of thermometer. Temperature of the bath.	Indication of thermometer. Temperature of the issuing gas.	Cooling effect in Cent. degrees.
h m					
11 5	0	31·6	646·35	479·1	
7	0	31·4	646·3	478·8	
9	0	31·7	646·1	478·05	
11	0	31·6	646·05	478·1	
13	0	31·9	646·05	478·2	
15	0	31·5	646·05	478·35	
	0	31·62 = 91·508 (lbs.)	646·15 = 91·452	478·43 = 90·008	1·444
17	0	31·8	646·2	478·7	
19	0	31·5	646·0	478·6	
21	0	32·0	646·0	478·7	
22	0	32·2	646·1	478·6	
23	0	32·2	646·1	478·1	
24	0	32·0	646·1	478·8	
	0	31·95 = 90·576	646·08 = 91·442	478·58 = 90·043	1·399
25	0	32·0	646·1	477·0	
26	0	32·1	646·4	471·6	
30	95·51	32·2	646·7	469·2	
32	95·51	32·2	646·5	469·5	
33	95·51	32·0	646·45	469·6	
36	95·51	32·6	646·7	469·6	
38	95·51	32·2	646·6	469·9	
40		32·2	646·6	469·98	
	95·51	32·23 = 89·799	646·59 = 91·516	469·63 = 88·044	3·472
43	93·03	32·1	646·6	470·05	
46		32·1	647·0	470·3	
48	90·60	32·1	647·1	470·9	
50		32·1	647·4	471·05	
	91·81	32·1 = 90·162	647·03 = 91·579	470·57 = 88·255	3·324
53	80·82	32·1	647·2	471·2	
55		32·05	647·2	471·75	
58		32·0	647·2	472·05	
12 0	75·65	32·0	647·7	472·6	
4		32·6	647·9	472·9	
6	75·65	32·25	647·8	473·25	
	77·37	32·16 = 90·006	647·5 = 91·647	472·29 = 88·638	3·009
9		32·8	647·95	473·95	
11	65·72	32·4	647·9	474·1	
15	60·83	32·2	647·95	474·8	
20	60·83	32·4	647·95	475·15	
22		32·9	647·95	475·2	
	62·46	32·54 = 88·971	647·94 = 91·711	474·64 = 89·162	2·549
27	0	32·0	647·85	477·0	
29	0	32·0	647·8	480·1	
31	0	31·6	647·5	480·6	
33	0	32·0	647·3	480·6	
35	0	32·1	647·1	480·8	
37	0	32·2	647·0	480·83	
39	0	32·0	647·1	480·9	
41	0	32·1	647·03	481·03	
43	0	32·1	647·1	480·9	
45	0	32·1	647·03	481·02	
47	0	32·4	647·05	481·04	
49	0	32·6	646·98	480·98	
51	0	32·8	646·85	480·9	
	0	32·3 = 89·618	647·02 = 91·578	480·97 = 90·528	1·050

Although hot air had been passed through the plug for half an hour before the readings in the preceding Table were obtained, it is probable that the numbers 1·444 and 1·399, representing the cooling effect of atmospheric air, are not so accurate as the value 1°·050. Taking this latter figure for the effect of an excess of pressure of 89·618—14·68 = 74·938 lbs., we find a considerable decrease of cooling effect owing to elevation of temperature, for that pressure, at the low temperatures previously employed, is able to produce a cooling effect of 1°·309.

In order to obtain the effect of carbonic acid unmixed with atmospheric air, we shall, in accordance with the principle already adhered to, consider the thermal capacities of the gases to be equal for equal volumes. Then the cooling effect of the pure gas

$$= \frac{3 \cdot 472 \times 100 - 1 \cdot 052 \times 4 \cdot 49}{95 \cdot 51} = 3° \cdot 586.$$

Collecting these results, we have,—

Temperature of bath.	Excess of pressure.	Cooling effect.	Cooling effect reduced to 100 lbs. pressure.	Theoretical cooling effect for 100 lbs. pressure.
12·844	60·601	5·049	8·33	8·27
19·077	37·248	2·938	7·89	8·07
91·516	74·938	3·586	4·78	4·96

Note.—The numbers shown in the last column of the Table are calculated by the general expression given in our former paper* for the cooling effect, from an empirical formula for the pressure of carbonic acid, recently communicated by Mr Rankine in a letter, from which the following is extracted.

"GLASGOW, *May* 9, 1854.

"Annexed I send you formulæ for carbonic acid, in which the coefficient *a* has been determined *solely* from Regnault's experiments on the increase of pressure at constant volume between 0° and 100° Cent. It gives most satisfactory results for expansion at constant pressure, compression at constant temperature, and also (I think) for cooling by free expansion" [*i.e.* the cooling effect in our experiments].

* *Philosophical Transactions*, June, 1853.

" Carbonic Acid Gas.

P pressure in pounds per square foot.

V volume of one pound in cubic feet.

P_0 one atmosphere.

V_0 *theoretical* volume, in the state of *perfect gas*, of one lb. at the pressure P_0 and the temperature of melting ice.

P_0V_0 for carbonic acid 17116 feet, $\log P_0V_0 = 4\cdot2334023$.

(P_0V_0 *actually*, at $0°$, 17145.)

K_p dynam. spec. heat at constant pressure $300\cdot7$ feet;

$$\log K_p = 2\cdot4781334.$$

C absolute temperature of melting ice, $274°$ Cent.

" The absolute zeros of gaseous tension and of heat are supposed sensibly to coincide, *i.e.* κ is supposed inappreciably small.

"*Formulæ*:

$$\frac{PV}{P_0V_0} = \frac{T+C}{C} - \frac{a}{T+C}\frac{V_0}{V} \quad\quad\dots\dots\dots\dots(1),$$

$$a = 1\cdot9, \quad \log a = 0\cdot2787536.$$

"Cooling by free expansion, supposing the perfect gas thermometer to give the true scale of absolute temperatures:

$$\delta T = \frac{P_0V_0}{K_p}\cdot\frac{3a}{T+C}\left\{\frac{V_0}{V_1} - \frac{V_0}{V_2}\right\} \quad\dots\dots\dots\dots(2)^*,$$

$$\log \frac{3P_0V_0a}{K_p} = 2\cdot5111438."$$

By substituting for $\dfrac{V_0}{V_1}$ and $\dfrac{V_0}{V_2}$ their approximate values $\dfrac{C}{T+C}\cdot\dfrac{P_1}{P_0}$ and $\dfrac{C}{T+C}\dfrac{P_2}{P_0}$, we reduce it to

$$\delta = \frac{3V_0aC}{K_p(T+C)^2}\cdot\frac{P_1-P_2}{P_0},$$

from which we have calculated the theoretical results for different temperatures shown above, which agree remarkably well with those we have obtained from observation.

The interpretation given above for the experimental results on mixtures of carbonic acid and air depends on the assumption

---

* Obtained by using Mr Rankine's formula (1) in the general expression for the cooling effect given in our former paper, and repeated below as equation (15) of Section V.

(rendered probable as a very close approximation to the truth, by
Dalton's law), that in a mixture each gas retains all its physical
properties unchanged by the presence of the other. This assumption, however, may be only approximately true, perhaps similar
in accuracy to Boyle's and Gay-Lussac's laws of compression and
expansion by heat; and the theory of gases would be very much
advanced by accurate comparative experiments on all the physical
properties of mixtures and of their components separately. Towards
this object we have experimented on the thermal effect of the
mutual interpenetration of carbonic acid and air. In one experiment we found that when 7500 cubic inches of carbonic acid at
the atmospheric pressure were mixed with 1000 cubic inches of
common air and a perfect mutual interpenetration had taken
place, the temperature had fallen by about ·2° Cent. We intend
to try more exact experiments on this subject.

THEORETICAL DEDUCTIONS.

§ 1. *On the Relation between the Heat evolved and the Work
spent in Compressing a Gas kept at constant temperature.*

This relation is not a relation of simple mechanical equivalence,
as was supposed by Mayer* in his 'Bemerkungen über die Kräfte
der unbelebten Natur,' in which he founded on it an attempt to
evaluate numerically the mechanical equivalent of the thermal
unit. The heat evolved may be less than, equal to, or greater
than the equivalent of the work spent, according as the work
produces other effects in the fluid than heat, produces only heat,
or is assisted by molecular forces in generating heat, and according
to the quantity of heat, greater than, equal to, or less than that
held by the fluid in its primitive condition, which it must hold to
keep itself at the same temperature when compressed. The à
priori assumption of equivalence, for the case of air, without some
special reason from theory or experiment, is not less unwarrantable
than for the case of any fluid whatever subjected to compression.
Yet it may be demonstrated† that water below its temperature
of maximum density (39°·1 Fahr.), instead of evolving any heat at

---

* *Annalen* of Wöhler and Liebig, May, 1842.

† Dynamical Theory of Heat, § 63, equation (*b*) [Art. xlviii. above], *Trans. Roy.
Soc. Edinb.*, April, 1851, Vol. xvi. p. 290 ; or *Phil. Mag.*, Dec. 1852, p. 425.

all when compressed, actually absorbs heat, and at higher temperatures evolves heat in greater or less, but probably always very
small, proportion to the equivalent of the work spent; while air,
as will be shown presently, evolves always, at least when kept at
any temperature between 0° and 100° Cent., somewhat more heat
than the work spent in compressing it could alone create.   The
first attempts to determine the relation in question, for the case of
air, established an approximate equivalence without deciding how
close it might be, or the direction of the discrepance, if any.   Thus
experiments "On the Changes of Temperature produced by the
Rarefaction and Condensation of Air*," showed an approximate
agreement between the heat evolved by compressing air into a
strong copper vessel under water, and the heat generated by an
equal expenditure of work in stirring a liquid; and again, conversely, an approximate compensation of the cold of expansion
when air in expanding spends all its work in *stirring* its own mass
by rushing through the narrow passage of a slightly opened stop-
cock.   Again, theory†, without any doubtful hypothesis, showed
from Regnault's observations on the pressure and latent heat of
steam, that unless the density of saturated steam differs very
much from what it would be if following the gaseous laws of
expansion and compression, the heat evolved by the compression
of air must be sensibly less than the equivalent of the work spent
when the temperature is as low as 0° Cent., and very considerably
greater than that equivalent when the temperature is above 40° or
50°.   Mr Rankine is, so far as we know, the only other writer who
independently admitted the necessity of experiment on the subject, and he was probably not aware of the experiments which
had been made in 1844, on the rarefaction and condensation of
air, when he remarked‡, that "the value of κ is unknown; and
as yet no experimental data exist by which it can be determined"
(κ denoting in his expressions a quantity the vanishing of which
for any gas would involve the equivalence in question).   In further

---

* Communicated to the Royal Society, June 20, 1844, and published in the
Philosophical Magazine, May 1845.

† Appendix to "Account of Carnot's Theory" [Art. XLI. above], Roy. Soc.
Edinburgh, April 30, 1849, *Transactions*, Vol. XVI. p. 568; confirmed in the
Dynamical Theory, § 22 [Art. XLVIII. above], *Transactions Roy. Soc. Edinb.*, March
17, 1851; and *Phil. Mag.* July, 1852, p. 20.

‡ Mechanical Action of Heat, Section II. (10), communicated to the Roy. Soc.
Edinb., Feb. 4, 1850, *Transactions*, Vol. XX. p. 166.

observing that probably κ is small in comparison with the reciprocal of the coefficient of expansion, Mr Rankine virtually adopted
the equivalence as probably approximate; but in his article "On
the Thermic Phenomena of Currents of Elastic Fluids*," he took
the first opportunity of testing it closely, afforded by our preliminary experiments on the thermal effects of air escaping through
narrow passages.

We are now able to give much more precise answers to the
question regarding the heat of compression, and to others which
rise from it, than those preliminary experiments enabled us to do.
Thus if K denote the specific heat under constant pressure, of air
or any other gas, issuing from the plug in the experiments described above, the quantity of heat that would have to be supplied,
per pound of the fluid passing, to make the issuing stream have
the temperature of the bath, would be $K\delta$, or $Km(P-P')/\Pi$,
where m is equal to $\cdot26^\circ$ for air and $1\cdot15^\circ$ for carbonic acid, since
we found that the cooling effect was simply proportional to the
difference of pressure in each case, and was $\cdot0176^\circ$ per pound per
square inch, or $\cdot26$ per atmosphere, for air, and about $4\frac{1}{2}$ times as
much for carbonic acid. This shows precisely how much the heat
of friction in the plug falls short of compensating the cold of
expansion. But the heat of friction is the thermal equivalent of
all the work done actually in the narrow passages by the air
expanding as it flows through. Now this, in the cases of air and
carbonic acid, is really not as much as the whole work of expansion, on account of the deviation from Boyle's law to which these
gases are subject; but it exceeds the whole work of expansion in
the case of hydrogen which presents a contrary deviation; since
$P'V'$, the work which a pound of air must do to escape against
the atmospheric pressure, is, for the two former gases, rather
greater, and for hydrogen rather less, than PV, which is the work
done on it in pushing it through the spiral up to the plug. In
any case, w denoting the whole work of expansion, $w-(P'V'-PV)$
will be the work actually spent in friction within the plug; and
$1\{w-(P'V'-PV)\}/J$ will be the quantity of heat into which it is
converted, a quantity which, in the cases of air and carbonic acid,
falls short by $Km(P-P')/\Pi$ of compensating the cold of expansion. If therefore H denote the quantity of heat that would

* Mechanical Action of Heat, Subsection 4, communicated to the Roy. Soc.
Edinb. Jan. 4, 1853, *Transactions*, Vol. xx. p. 580.

exactly compensate the cold of expansion, or which amounts to the same, the quantity of heat that would be evolved by compressing a pound of the gas from the volume V' to the volume V, when kept at a constant temperature, we have

$$\frac{1}{J}\{w - (PV' - PV)\} = H - Km\frac{P - P'}{\Pi},$$

whence

$$H = \frac{w}{J} + \left\{ -\frac{1}{J}(P'V' - PV) + Km\frac{P - P'}{\Pi} \right\}.$$

Now, from the results derived by Regnault from his experiments on the compressibility of air, of carbonic acid, and of hydrogen, at three or four degrees above the freezing-point, we find, approximately,

$$\frac{P'V' - PV}{PV} = f\frac{P - P'}{\Pi},$$

where $f = \cdot00082$ for air,

         $f = \cdot0064$ for carbonic acid,

and $f = - \cdot00043$ for hydrogen.

No doubt the deviations from Boyle's law will be somewhat different at the higher temperature (about 15^0 or 16^0 Cent.) of the bath in our experiments, probably a little smaller for air and carbonic acid, and possibly greater for hydrogen; but the preceding formula may express them accurately enough for the rough estimate which we are now attempting.

We have, therefore, for air or carbonic acid,

$$H = \frac{w}{J} + \left(Km - \frac{PVf}{J}\right)\frac{P - P'}{\Pi} = \frac{w}{J} + \frac{PV}{J}\left(\frac{JKm}{PV} - f\right)\frac{P - P'}{\Pi}.$$

The values of JK and PV for the three gases in the circumstances of the experiments are as follow:—

For atmospheric air $JK = 1390 \times \cdot238 = 331$
For carbonic acid $JK = 1390 \times \cdot217 = 301$
For hydrogen.........$JK = 1390 \times 3\cdot4046 = 4732,$

and for atmospheric air, at 15^0 Cent. $PV = 26224 \ (1 + 15 \times \cdot00366)$
$= 27663,$

for carbonic acid,   at 10^0 Cent. $PV = 17154 \ (1 + 10 \times \cdot00366)$
$= 17782,$

for hydrogen ...... at 10^0 Cent. $PV = 378960 \, (1 + 10 \times \cdot00367)$
$= 393000.$

Hence we have, for air and carbonic acid,

$$H = \frac{w}{J} + \frac{PV}{J} \lambda \frac{P - P'}{\Pi},$$

where λ denotes ·0024 for air, and ·013 for carbonic acid; showing (since these values of λ are positive) that in the case of each of these gases, more heat is evolved in compressing it than the equivalent of the work spent (a conclusion that would hold for hydrogen even if no cooling effect, or a heating effect less than a certain limit, were observed for it in our form of experiment). To find the proportion which this excess bears to the whole heat evolved, or to the thermal equivalent of the work spent in the compression, we may use the expression

$$w = PV \log \frac{P}{P'},$$

as approximately equal to the mechanical value of either of those energies; and we thus find for the proportionate excess,

$$\frac{H - w/J}{w/J} = \lambda \frac{P - P'}{\Pi \log (P/P')} = ·0024 \frac{P - P'}{\Pi \log (P/P')} \text{ for air,}$$

or $$= ·013 \frac{P - P'}{\Pi \log (P/P')} \text{ for carbonic acid.}$$

This equation shows in what proportion the heat evolved exceeds the equivalent of the work spent in any particular case of compression of either gas. Thus for a very small compression from $P' = \Pi$, the atmospheric pressure, we have

$$\log \frac{P}{P'} = \log \left(1 + \frac{P - \Pi}{\Pi} \right) = \frac{P - \Pi}{\Pi} \text{ approximately,}$$

and therefore $$\frac{H - w/J}{w/J} = ·0024 \text{ for air,}$$

or $$= ·013 \text{ for carbonic acid.}$$

Therefore, when slightly compressed from the ordinary atmospheric pressure, and kept at a temperature of about 60° Fahr., common air evolves more heat by $\frac{1}{417}$, and carbonic acid more by $\frac{1}{77}$ than the amount mechanically equivalent to the work of compression. For considerable compressions from the atmospheric pressure, the proportionate excesses of the heat evolved are greater than these values, in the ratio of the Napierian logarithm of the number of times the pressure is increased, to this number diminished by 1. Thus, if either gas be compressed from the standard state to double density, the heat evolved exceeds the thermal equivalent of the work spent, by $\frac{1}{200}$ in the case of air, and $\frac{1}{55}$ in the case of carbonic acid.

As regards these two gases, it appears that the observed cooling effect was chiefly due to an actual preponderance of the mechanical equivalent of the heat required to compensate the cold of expansion over the work of expansion, but that rather more than one-fourth of it in the case of air, and about one-third of it in the case of carbonic acid, depended on a portion of the work of expansion going to do the extra work spent by the gas in issuing against the atmospheric pressure above that gained by it in being sent into the plug. On the other hand, in the case of hydrogen, in such an experiment as we have performed, there would be a heating effect, if the work of expansion were precisely equal to the mechanical equivalent of the cold of expansion, since not only the whole work of expansion, but also the excess of the work done in forcing the gas in above that performed by it in escaping, is spent in friction in the plug. Since we have observed actually a cooling effect, it follows that the heat absorbed in expansion must exceed the equivalent of the work of expansion, enough to over-compensate the whole heat of friction mechanically equivalent, as this is, to the work of expansion together with the extra work of sending the gas into the plug above that which it does in escaping. In the actual experiment* we found a cooling effect of ·076°, with a difference of pressures, $P - P'$, equal to 53·7 lbs. per square inch, or 3·7 atmospheres. Now the mechanical value of the specific heat of a pound of hydrogen is, according to the result stated above, 4732 foot-pounds, and hence the mechanical value of the heat that would compensate the observed cooling effect per pound of hydrogen passing is 360 foot-pounds. But, according to Regnault's experiments on the compression of hydrogen, quoted above, we have

$$PV - P'V' = PV \times ·00043 \frac{P - P'}{\Pi} \text{ approximately};$$

* From the single experiment we have made on hydrogen we cannot conclude that at other pressures a cooling effect proportional to the difference of pressures would be observed, and therefore we confine the comparison of the three gases to the particular pressure used in the hydrogen experiment. It should be remarked too, that we feel little confidence in the value assigned to the thermal effect for the case observed in the experiment on hydrogen, and only consider it established that it is a cooling effect, and very small. [Addition of March 29, 1882. Subsequent experiments showed us, that the effect was in truth very small; and probably not a cooling effect, but a very slight degree of heating. See Part IV. and Abstracts below.]

and as the temperature was about 10° in our experiment, we have, as stated above, $PV = 393000.$

Hence, for the case of the experiment in which the difference of pressures was 3·7 atmospheres, or

$$\frac{P - P'}{\Pi} = 3\text{·}7,$$

we have $PV - P'V' = 625;$

that is, 625 foot-pounds more of work, per pound of hydrogen, is spent in sending the hydrogen into the plug at 4·7 atmospheres of pressure, than would be gained in allowing it to escape at the same temperature against the atmospheric pressure. Hence the heat required to compensate the cold of expansion, is generated by friction from (1) the actual work of expansion, together with (2) the extra work of 625 foot-pounds per pound of gas, and (3) the amount equivalent to 360 foot-pounds which would have to be communicated from without to do away with the residual cooling effect observed. Its mechanical equivalent therefore exceeds the work of expansion by 985 foot-pounds; which is $\frac{1}{630}$ of its own amount, since the work of expansion in the circumstances is approximately $393000 \times \log 4\text{·}7 = 608000$ foot-pounds. Conversely, the heat evolved by the compression of hydrogen at 10° Cent., from 1 to 4·7 atmospheres, exceeds by $\frac{1}{630}$ the work spent. The corresponding excess in the case of atmospheric air, according to the result obtained above, is $\frac{1}{174}$, and in the case of carbonic acid $\frac{1}{32}$.

It is important to observe how much less close is the compensation in carbonic acid than in either of the other gases, and it appears probable that the more a gas deviates from the gaseous laws, or the more it approaches the condition of a vapour at saturation, the wider will be the discrepancy. We hope, with a view to investigating further the physical properties of gases, to extend our method of experimenting to steam (which will probably present a large cooling effect), and perhaps to some other vapours.

In Mr Joule's original experiment[*] to test the relation between heat evolved and work spent in the compression of air, without an

---

[*] The second experiment mentioned in the abstract published in the *Proceedings of the Royal Society*, June 20, 1844, and described in the *Philosophical Magazine*, May 1845, p. 377.

independent determination of the mechanical equivalent of the thermal unit, air was allowed to expand through the aperture of an open stop-cock from one copper vessel into another previously exhausted by an air-pump, and the whole external thermal effect on the metal of the vessels, and a mass of water below which they are kept, was examined. We may now estimate the actual amount of that external thermal effect, which observation only showed to be insensibly small. In the first place it is to be remarked, that, however the equilibrium of pressure and temperature is established between the two air vessels, provided only no appreciable amount of work is emitted in sound, the same quantity of heat must be absorbed from the copper and water to reduce them to their primitive temperature; and that this quantity, as was shown above, is equal to

$$\frac{PV}{J} \times \cdot 0024 \times \frac{P - P'}{\Pi} = \frac{27000 \times \cdot 0024}{1390} \times \frac{P - P'}{\Pi} = \cdot 046 \frac{P - P'}{\Pi}.$$

In the actual experiments the exhausted vessel was equal in capacity to the charged vessel, and the latter contained ·13 of a pound of air under 21 atmospheres of pressure, at the commencement. Hence $P' = \frac{1}{2}P$, and

$$\frac{P - P'}{\Pi} = 10 \cdot 5;$$

and the quantity of heat required from without to compensate the total internal cooling effect must have been

$$\cdot 046 \times 10 \cdot 5 \times \cdot 13 = \cdot 063.$$

This amount of heat, taken from $16\frac{1}{2}$ lbs. of water, 28 lbs. of copper, and 7 lbs. of tinned iron, as in the actual experiment, would produce a lowering of temperature of only ·003° Cent. We need not therefore wonder that no sensible external thermal effect was the result of the experiment when the two copper vessels and the pipe connecting them were kept under water, stirred about through the whole space surrounding them, and that similar experiments, more recently made by M. Regnault, should have led only to the same negative conclusion.

If, on the other hand, the air were neither allowed to take in heat from nor to part with heat to the surrounding matter in any part of the apparatus, it would experience a resultant cooling effect (after arriving at a state of uniformity of temperature as well as

pressure) to be calculated by dividing the preceding expression for
the quantity of heat which would be required to compensate it, by
17, the specific heat of air under constant pressure.   The cooling
effect on the air itself therefore amounts to

$$0^{0.}27 \times \frac{P - P'}{\Pi} \, *,$$

which is equal to $2^{0.}8$, for air expanding, as in Mr Joule's experiment, from 21 atmospheres to half that pressure, and is 900 times
as great as the thermometric effect when spread over the water
and copper of the apparatus.   Hence our present system, in which
the thermometric effect on the air itself is directly observed, affords
a test hundreds of times more sensitive than the method first
adopted by Mr Joule, and no doubt also than that recently
practised by M. Regnault, in which the dimensions of the various
parts of the apparatus (although not yet published) must have
been on a corresponding scale, or in somewhat similar proportions,
to those used formerly by Mr Joule.

§ 2.   On the density of Saturated Steam.

The relation between the heat evolved and the work spent,
approximately established by the air-experiments communicated to
the Royal Society in 1844, was subjected to an independent
indirect test by an application of Carnot's theory, with values of
"Carnot's function" which had been calculated from Regnault's
data as to the pressure and latent heat of steam, and the
assumption (in want of experimental data), that the density varies
according to the gaseous laws.   The verification thus obtained was
very striking, showing an exact agreement with the relation of
equivalence at a temperature a little above that of observation,
and an agreement with the actual experimental results quite
within the limits of the errors of observation; but a very wide
discrepancy from equivalence for other temperatures.   The following Table is extracted from the Appendix to the "Account of

---

* It is worthy of remark that this, the expression for the cooling effect experienced by a mass of atmospheric air expanding from a bulk in which its pressure is
P to a bulk in which, at the same (or very nearly the same) temperature its pressure
is P', and spending all its work of expansion in friction among its own particles,
agrees very closely with the expression, $\cdot26 \, (P - P')/\Pi$, for the cooling effect in the
somewhat different circumstances of our experiments.

Carnot's Theory" [Art. XLI. above] in which the theoretical comparison was first made, to facilitate a comparison with what we now know to be the true circumstances of the case.

" Table of the Values of $\dfrac{\mu(1 + Et)}{E}$" $= [W]$.

" Work requisite to produce a unit of heat by the compression of a gas $\dfrac{[\mu](1 + Et)}{E} = [W]$.	" Temperature of the gas t.	" Work requisite to produce a unit of heat by the compression of a gas $\dfrac{[\mu](1 + Et)}{E} = [W]$.	" Temperature of the gas t.
ft. lbs.	°	ft. lbs.	°
1357·1	0	1446·4	120
1368·7	10	1455·8	130
1379·0	20	1465·3	140
1388·0	30	1475·8	150
1395·7	40	1489·2	160
1401·8	50	1499·0	170
1406·7	60	1511·3	180
1412·0	70	1523·5	190
1417·6	80	1536·5	200
1424·0	90	1550·3	210
1430·6	100	1564·0	220
1438·2	110	1577·8	230 "

We now know, from the experiments described above in the present paper, that the numbers in the first column, and we may conclude with almost equal certainty, that the numbers in the third also, ought to be each very nearly the mechanical equivalent of the thermal unit. This having been ascertained to be 1390 (for the thermal unit Centigrade) by the experiments on the friction of fluids and solids, communicated to the Royal Society in 1849, and the work having been found above to fall short of the equivalent of heat produced, by about $\frac{1}{417}$, at the temperature of the air-experiments at present communicated, and by somewhat less at such a higher temperature as 30°, we may infer that the agreement of the tabulated theoretical result with the fact is perfect at about 30° Cent. Or, neglecting the small discrepance by which the work truly required falls short of the equivalent of heat produced, we may conclude that the true value of $\mu(1 + Et)/E$ for all temperatures is about 1390 ; and hence that if $[W]$ denote the numbers shown for it in the preceding table, μ the true value of Carnot's function, and $[\mu]$ the value tabulated for any temperature in the

" Account of Carnot's Theory " [Art. XLI. above], we must have, to a very close degree of approximation,

$$\mu = [\mu] \times \frac{1390}{[W]} .$$

But if $[\sigma]$ denote the formerly assumed specific gravity of saturated steam, p its pressure, and λ its latent heat per pound of matter, and if ρ be the mass (in pounds) of water in a cubic foot, the expression from which the tabulated values of $[\mu]$ were calculated is

$$[\mu] = \frac{1 - [\sigma]}{\rho [\sigma]} \frac{1}{\lambda} \frac{dp}{dt} ;$$

while the true expression for Carnot's function in terms of properties of steam is

$$\mu = \frac{1 - \sigma}{\rho \sigma} \cdot \frac{1}{\lambda} \frac{dp}{dt} .$$

Hence
$$\frac{\mu}{[\mu]} = \frac{[\sigma]}{\sigma} \cdot \frac{1 - \sigma}{1 - [\sigma]} ;$$

or, approximately, since σ and $[\sigma]$ are small fractions, $\mu/[\mu] = [\sigma]/\sigma$.

We have, therefore, $$\frac{\sigma}{[\sigma]} = \frac{[W]}{1390} ;$$

and we infer that the densities of saturated steam in reality bear the same proportions to the densities assumed, according to the gaseous laws, as the numbers shown for different temperatures in the preceding Table bear to 1390.   Thus we see that the assumed density must have been very nearly correct, about 30° Cent., but that the true density increases much more at the high temperatures and pressures than according to the gaseous laws, and consequently that steam appears to deviate from Boyle's law in the same direction as carbonic acid, but to a much greater amount, which in fact it must do unless its coefficient of expansion is very much less, instead of being, as it probably is, somewhat greater than for air.   Also, we infer that the specific gravity of steam at 100° Cent., instead of being only $\frac{1}{1693\cdot5}$, as was assumed, or about $\frac{1}{1700}$, as it is generally supposed to be, must be as great as $\frac{1}{1645}$.   Without using the preceding Table, we may determine the absolute density of saturated steam by means of a formula obtained as follows.   Since we have seen the true value of W is nearly 1390, we must have, very approximately, $1390E/(1 + Et)$, and hence, according to the preceding expression for μ in terms of the properties of steam,

$$\rho\sigma = \frac{1-\sigma}{1390E}(1+Et)\frac{1}{\lambda}\frac{dp}{dt},$$

or, within the degree of approximation to which we are going (omitting as we do fractions such as $\frac{1}{400}$ of the quantity evaluated),

$$\rho\sigma = \frac{(1+Et)}{1390E.\lambda}\frac{dp}{dt},$$

an equation by which $\rho\sigma$, the mass of a cubic foot of steam in fraction of a pound, or τ, its specific gravity (the value of ρ being 63·887), may be calculated from observations such as those of Regnault on steam. Thus, using Mr Rankine's empirical formula for the pressure which represents M. Regnault's observations correctly at all temperatures, and M. Regnault's own formula for the latent heat; and taking $E = \frac{1}{273}$, we have

$$\rho\sigma = \frac{273+t}{1390}\frac{p\left(\dfrac{\beta}{(274\cdot6+t)^2} + \dfrac{2\gamma}{(274\cdot6+t)^3}\right) \times \cdot4342945}{(606\cdot5+0\cdot305t)-(t+\cdot00002t^2+\cdot0000003t^3)},$$

with the following equations for calculating p and the terms involving β and γ;

$$\log_{10}p = \alpha - \frac{\beta}{t+274\cdot6} - \frac{\gamma}{(274\cdot6+t)^2},$$

$$\alpha = 4\cdot950433 + \log_{10}2114 = 8\cdot275538,$$

$$\log_{10}\beta = 3\cdot1851091,$$

$$\log_{10}\gamma = 5\cdot0827176.$$

The densities of saturated steam calculated for any temperatures, either by means of this formula, or by the expression given above, with the assistance of the Table of values of $[W]$, are the same as those which, in corresponding on the subject in 1848, we found would be required to reconcile Regnault's actual observations on steam with the results of air-experiments which we then contemplated undertaking, should they turn out, as we now find they do, to confirm the relation which the air-experiments of 1844 had approximately established. They should agree with results which Clausius* gave as a consequence of his extension of

---

* Poggendorff's *Annalen*, April and May, 1850.

Carnot's principle to the dynamical theory of heat, and his assumption of Mayer's hypothesis.

§ 3.   *Evaluation of Carnot's Function.*

The importance of this object, not only for calculating the efficiency of steam-engines and air-engines, but for advancing the theory of heat and thermo-electricity, was a principal reason inducing us to undertake the present investigation. Our preliminary experiments, demonstrating that the cooling effect which we discovered in all of them was very slight for a considerable variety of temperatures (from about $0°$ to $77°$ Cent.), were sufficient to show, as we have seen in §§ 1 and 2, that $\mu(1 + Et)/E$ must be very nearly equal to the mechanical equivalent of the thermal unit; and therefore we have

$$\mu = \frac{J}{1/E + t} \text{ approximately,}$$

or, taking for E the standard coefficient of expansion of atmospheric air, ·003665,

$$\mu = \frac{J}{272\cdot85 + t}.$$

At the commencement of our first communication to the Royal Society on the subject, we proposed to deduce more precise values for this function by means of the equation

$$\frac{J}{\mu} = \frac{JK\delta - (P'V' - PV) + w}{\dfrac{dw}{dt}} ;$$

where
$$w = \int_{V}^{V'} p\, dv ;$$

v, V, V' denote, with reference to air at the temperature of the bath, respectively, the volumes occupied by a pound under any pressure p, under a pressure, P, equal to that with which the air enters the plug, and under a pressure, P', with which the air escapes from the plug; and $JK\delta$ is the mechanical equivalent of the amount of heat per pound of air passing that would be required to compensate the observed cooling effect δ. The direct use of this equation for determining J/μ requires, besides our own results, information as to compressibility and expansion which is as yet but very insufficiently afforded by direct experiments, and is consequently very unsatisfactory, so much so that we shall only give an outline, without details, of two plans we have followed, and mention the results. First, it may be remarked that, approximately,

$$w = (1 + Et)H \log \frac{P}{P'}, \text{ and } \frac{dw}{dt} = EH \log \frac{P}{P'},$$

H being the "height of the homogeneous atmosphere," or the product of the pressure into the volume of a pound of air, at 0° Cent.; of which the value is 26224 feet. Hence, if $\mathfrak{E}$ denote a certain mean coefficient of expansion suitable to the circumstances of each individual experiment, it is easily seen that $\dfrac{w}{\dfrac{dw}{dt}}$ may be put under the form $1/\mathfrak{E} + t$, and thus we have

$$\frac{J}{\mu} = \frac{1}{\mathfrak{E}} + t + \frac{JK\delta - (P'V' - PV)}{EH \log \dfrac{P}{P'}},$$

since the numerator of the fraction constituting the last term is so small, that the approximate value may be used for the denominator. The first term of the second member may easily be determined analytically in general terms; but as it has reference to the rate of expansion at the particular temperature of the experiment, and not to the mean expansion from 0° to 100°, which alone has been investigated by Regnault and others who have made sufficiently accurate experiments, we have not data for determining its values for the particular cases of the experiments. We may, however, failing more precise data, consider the expansion of air as uniform from 0° to 100°, for any pressure within the limits of the experiments (four or five atmospheres); because it is so for air at the atmospheric density by the hypothesis of the air-thermometer, and Regnault's comparisons of air-thermometers in different conditions show for all, whether on the constant-volume or constant-pressure principle, with density or pressure from one-half to double the standard density or pressure, a very close agreement with the standard air-thermometer. On this assumption then, when we take into account Regnault's observations regarding the effect of variations of density on the coefficient of increase of pressure, we find that a suitable mean coefficient $\mathfrak{E}$ for the circumstances of the preceding formula for J/μ is expressed, to a sufficient degree of approximation, by the equation

$$\mathfrak{E} = 0036534 + \frac{\cdot 0000441}{3 \cdot 81} \frac{P - P}{\Pi \log P/P}$$

Also, by using Regnault's experimental results on compressibility of air as if they had been made, not at $4°\!\cdot\!75$, but at $16°$ Cent., we have estimated $PV' - PV$ for the numerator of the last term of the preceding expression. We have thus obtained estimates for the value of J/μ, from eight of our experiments (not corresponding exactly to the arrangement in seven series given above), which, with the various items of the correction in the case of each experiment, are shown in the following Table:

No. of experiment.	Pressure of air forced into the plug.	Barometric pressure.	Excess.	Cooling effect.	Correction by cooling effect.	Correction by reciprocal coefficient of expansion.	Correction by compressibility (subtracted).	Value of J divided by Carnot's function for $16°$ Cent.
	P.	P'.	$P-P'$.	δ.	$\dfrac{JK\delta}{EH\log P/P'}$	$\dfrac{1}{E'} - \dfrac{1}{E} \cdot$	$\dfrac{P'V'-PV}{EH\log P/P'}$	$\dfrac{J}{\mu_{16}}\cdot$
I.	20·943	14·777	6·166	0·105	1·031	0·174	0·290	289·4
II.	21·282	14·326	6·956	0·109	0·942	0·168	0·291	289·3
III.	35·822	14·504	21·318	0·375	1·421	0·519	0·412	289·97
IV.	33·310	14·692	18·618	0·364	1·523	0·470	0·372	290·065
V.	55·441	14·610	40·831	0·740	1·892	0·923	0·480	289·705
VI.	53·471	14·571	38·900	0·676	1·814	0·883	0·475	289·59
VII.	79·464	14·955	64·509	1·116	2·272	1·379	0·592	289·69
VIII.	79·967	14·785	65·182	1·142	2·300	1·376	0·586	289·73
							Mean......	289·68

In consequence of the approximate equality of J/μ to $1/E + t$, its value must be, within a very minute fraction, less by 16 at $0°$ than at $16°$; and, from the mean result of the preceding Table, we therefore deduce 273·68 as the value of J/μ at the freezing-point. The correction thus obtained on the approximate estimate $1/E + t = 272\!\cdot\!85 + t$, for J/μ, at temperatures not much above the freezing-point, is an augmentation of ·83.

For calculating the unknown terms in the expression for J/μ, we have also used Mr Rankine's formula for the pressure of air, which is as follows:—

$$pv = H\frac{C+t}{C}\left\{1 - \frac{aC}{(C+t)^2}\left(\frac{1}{\rho v}\right)^{\frac{3}{5}} + \frac{hC}{C+t}\left(\frac{1}{\rho v}\right)^{\frac{1}{2}}\right\},$$

where $C = 274\!\cdot\!6$, $\log_{10}a = \!\cdot\!3176168$, $\log_{10}h = \bar{3}\!\cdot\!8181546$,

$$H = \frac{26224}{1 - a + h};$$

and, v being the volume of a pound of air when at the temperature t and under the pressure p, ρ denotes the mass in pounds of a cubic foot at the standard atmospheric pressure of 29·9218 inches of mercury. The value of p according to this equation, when substituted in the general expression for J/μ, gives

$$\frac{J}{\mu} = C + t$$

$$+ \frac{\frac{JKC}{H}\delta + 3h\,\frac{C^{\frac{3}{8}}}{(C+t)^{\frac{3}{8}}}\left\{\left(\frac{P}{\Pi}\right)^{\frac{1}{2}} - \left(\frac{P'}{\Pi}\right)^{\frac{1}{2}}\right\} - \frac{13}{3}\,a\left(\frac{C}{C+t}\right)^{\frac{3}{5}}\left\{\left(\frac{P}{\Pi}\right)^{\frac{3}{5}} - \left(\frac{P'}{\Pi}\right)^{\frac{3}{5}}\right\}}{\log P/P'}.$$

From this we find, with the data of the eight experiments just quoted, the following values for J/μ at the temperature of 16^{0} Cent.,

289·044, 289·008, 288·849, 289·112, 288·787, 288·722, 288·505, 288·559, the mean of which is 288·82,

giving a correction of only ·03 to be subtracted from the previous approximate estimate $1/E + t$.

It should be observed that Carnot's function varies only with the temperature; and therefore if such an expression as the preceding, derived from Mr Rankine's formula, be correct, the cooling effect, δ, must vary with the pressure and temperature in such a way as to reduce the complex fraction, constituting the second term, to either a constant or a function of t. Now at the temperature of our experiments, δ is very approximately proportional simply to $P - P'$, and therefore all the terms involving the pressure in the numerator ought to be either linear or logarithmic; and the linear terms should balance one another so as to leave only terms which, when divided by $\log P/P'$, become independent of the pressures. This condition is not fulfilled by the actual expression, but the calculated results agree with one another as closely as could be expected from a formula obtained with such insufficient experimental data as Mr Rankine had for investigating the empirical forms which his theory left undetermined. We shall see in § 5 below, that simpler forms represent Regnault's data within their limits of error of observation, and at the same time may be reduced to consistency in the present application.

As yet we have no data regarding the cooling effect, of sufficient accuracy for attempting an independent evaluation of Carnot's function for other temperatures. In the following section, however, we propose a new system of thermometry, the adoption of which will quite alter the form in which such a problem as that of evaluating Carnot's function for any temperature presents itself.

§ 4.   *On an absolute Thermometric Scale founded on the Mechanical Action of Heat.*

In a communication to the Cambridge Philosophical Society* of June, 1848, [Art. xxxix. above] it was pointed out that any system of thermometry, founded either on equal additions of heat, or equal expansions, or equal augmentations of pressure, must depend on the particular thermometric substance chosen, since the specific heats, the expansions, and the elasticities of substances vary, and, so far as we know, not proportionally with absolute rigour for any two substances.   Even the air-thermometer does not afford a *perfect standard,* unless the precise constitution and physical state of the gas used (the density, for a pressure-thermometer, or the pressure, for an expansion-thermometer) be prescribed; but the very close agreement which Regnault found between different air- and gas-thermometers removes, for all practical purposes, the inconvenient narrowness of the restriction to atmospheric air kept permanently at its standard density, imposed on the thermometric substance in laying down a rigorous definition of temperature.   It appears then that the standard of practical thermometry consists essentially in the reference to a certain numerically expressible quality of a particular substance.   In the communication alluded to, the question, "Is there any principle on which an absolute thermometric scale can be founded?" was answered by showing that Carnot's function (derivable from the properties of any substance whatever, but the same for all bodies at the same temperature), or any arbitrary function of Carnot's function, may be defined as temperature, and is therefore the foundation of an absolute system of thermometry.   We may now adopt this suggestion with great advantage, since we have found that Carnot's function varies very nearly in the inverse ratio of what has been called "temperature from the zero of the air-thermometer," that is, Centigrade temperature by the air-thermometer increased by the reciprocal of the coefficient of expansion; and we may define temperature simply

* " On an Absolute Thermometric Scale founded on Carnot's Theory of the Motive Power of Heat, and calculated from Regnault's Observations on Steam," by Prof. W. Thomson, [Art. xxxix. above] *Proceedings Camb. Phil. Soc.*, June 5, 1848, or *Philosophical Magazine*, Oct. 1848.

as the reciprocal of Carnot's function. When we take into account what has been proved regarding the mechanical action of heat*, and consider what is meant by Carnot's function, we see that the following explicit definition may be substituted:—

If any substance whatever, subjected to a perfectly reversible cycle of operations, takes in heat only in a locality kept at a uniform temperature, and emits heat only in another locality kept at a uniform temperature, the temperatures of these localities are proportional to the quantities of heat taken in or emitted at them in a complete cycle of the operations.

To fix on a unit or degree for the numerical measurement of temperature, we may either call some definite temperature, such as that of melting ice, unity, or any number we please; or we may choose two definite temperatures, such as that of melting ice and that of saturated vapour of water under the pressure 29·9218 inches of mercury in the latitude of 45°, and call the difference of these temperatures any number we please, 100 for instance. The latter assumption is the only one that can be made conveniently in the present state of science, on account of the necessity of retaining a connexion with practical thermometry as hitherto practised; but the former is far preferable in the abstract, and must be adopted ultimately. In the mean time it becomes a question, what is the temperature of melting ice, if the difference between it and the standard boiling-point be called 100°? When this question is answered within a tenth of a degree or so, it may be convenient to alter the foundation on which the degree is defined, by assuming the temperature of melting ice to agree with that which has been found in terms of the old degree; and then to make it an object of further experimental research, to determine by what minute fraction the range from freezing to the present standard boiling-point exceeds or falls short of 100. The experimental data at present available do not enable us to assign the temperature of melting ice, according to the new scale, to perfect certainty within less than two- or three-tenths of a degree; but we shall see that its value is probably about 273·7, agreeing with the value of J/μ at 0° found by the first method in § 3. From the very close approximation to equality between J/μ and $1/E + t$, which our experiments have established, we may be sure that

---

* Dynamical Theory of Heat, [Art. XLVIII. above] §§ 42, 43.

temperature from the freezing-point by the new system must agree to a very minute fraction of a degree with Centigrade temperature between the two prescribed points of agreement, 0° and 100°, and we may consider it as highly probable that there will also be a very close agreement through a wide range on each side of these limits. It becomes of course an object of the greatest importance, when the new system is adopted, to compare it with the old standard; and this is in fact what is substituted for the problem, the evaluation of Carnot's function, now that it is proposed to call the reciprocal of Carnot's function, temperature. In the next section we shall see by what kind of examination of the physical properties of air this is to be done, and investigate an empirical formula expressing them consistently with all the experimental data as yet to be had, so far as we know. The following Table, showing the indications of the constant-volume and constant-pressure air-thermometer in comparison for every twenty degrees of the new scale, from the freezing-point to 300° above it, has been calculated from the formulæ (9), (10), and (39) of § 5 below.

Comparison of Air-thermometer with Absolute Scale.

Temperature by absolute scale in Cent. degrees from the freezing-point. $t - 273.7$.	Temperature Centigrade by constant-volume thermometer with air of specific gravity Φ/v. $\theta = 100\,\dfrac{p_t - p_{273\cdot7}}{p_{373\cdot7} - p_{273\cdot7}}$.	Temperature Centigrade by constant-pressure air-thermometer. $\vartheta = 100\,\dfrac{v_t - v_{273\cdot7}}{v_{373\cdot7} - v_{273\cdot7}}$
$\overset{\circ}{0}$	$\overset{\circ}{0}$	$\overset{\circ}{0}$
20	$20 + \cdot0298 \times \dfrac{\Phi}{v}$	$20 + \cdot0404 \times \dfrac{p}{\Pi}$
40	$40 + \cdot0403$,,	$40 + \cdot0477$,,
60	$60 + \cdot0366$,,	$60 + \cdot0467$,,
80	$80 + \cdot0223$,,	$80 + \cdot0277$,,
100	$100 + \cdot0000$,,	$100 + \cdot0000$,,
120	$120 - \cdot0284$,,	$120 - \cdot0339$,,
140	$140 - \cdot0615$,,	$140 - \cdot0721$,,
160	$160 - \cdot0983$,,	$160 - \cdot1134$,,
180	$180 - \cdot1382$,,	$180 - \cdot1571$,,
200	$200 - \cdot1796$,,	$200 - \cdot2018$,,
220	$220 - \cdot2232$,,	$220 - \cdot2478$,,
240	$240 - \cdot2663$,,	$240 - \cdot2932$,,
260	$260 - \cdot3141$,, ·	$260 - \cdot3420$,,
280	$280 - \cdot3610$,,	$280 - \cdot3897$,,
300	$300 - \cdot4085$,,	$300 - \cdot4377$,,

The standard defined by Regnault is that of the constant-volume air-thermometer, with air at the density which it has when at the freezing-point under the pressure of 760 mm. or 22·9218 inches of mercury, and its indications are shown in comparison with the absolute scale by taking $\Phi/v = 1$ in the second column of the preceding Table. The greatest discrepance between $0°$ and $100°$ Cent. amounts to less than $\frac{1}{20}$th of a degree, and the discrepance at $300°$ Cent. is only four-tenths. The discrepancies of the constant-pressure air-thermometer, when the pressure is equal to the standard atmospheric pressure, or $p/\Pi = 1$, are somewhat greater, but still very small.

§ 5. *Physical Properties of Air expressed according to the absolute Thermodynamic scale of Temperature.*

All the physical properties of a fluid of given constitution are completely fixed when its density and temperature are specified; and as it is these qualities which we can most conveniently regard as being immediately adjustable in any arbitrary manner, we shall generally consider them as the independent variables in formulæ expressing the pressure, the specific heats, and other properties of the particular fluid in any physical condition.

Let v be the volume (in cubic feet) of a unit mass (one pound) of the fluid, and t its absolute temperature; and let p be its pressure in the condition defined by these elements.

Let also e be the "mechanical energy*" of the fluid, reckoned from some assumed standard or zero state, that is, the sum of the mechanical value of the heat communicated to it, and of the work spent on it, to raise it from that zero state to the condition defined by (v, t); and let N and K be its specific heats with constant volume, and with constant pressure, respectively. Then denoting, as before, the mechanical equivalent of the thermal unit by J, and the value of Carnot's function for the temperature t by μ, we have†

$$\frac{de}{dv} = \frac{J}{\mu}\frac{dp}{dt} - p \dots\dots\dots\dots\dots\dots(1),$$

---

* Dynamical Theory of Heat, [Art. xlviii. above] Part V.—On the Quantities of Mechanical Energy contained in a Fluid in different States as to Temperature and Density, § 28. *Trans. Roy. Soc. Edin.*, Dec. 15, 1851.

† Ibid. §§ 89, 91.

$$N = \frac{1}{J}\frac{de}{dt} \quad \dots\dots\dots\dots\dots \quad \dots\dots\dots(2),$$

$$K = \frac{1}{J}\frac{de}{dt} + \frac{1}{J}\left(\frac{de}{dv} + p\right)\frac{\dfrac{dp}{dt}}{-\dfrac{dp}{dv}} \dots\dots\dots(3).$$

From these we deduce, by eliminating e,

$$K - N = \frac{1}{\mu}\frac{\left(\dfrac{dp}{dt}\right)^2}{-\dfrac{dp}{dv}} \dots\dots\dots\dots\dots\dots\dots(4),$$

and
$$\frac{dN}{dv} = \frac{d\left(\dfrac{1}{\mu}\dfrac{dp}{dt}\right)}{dt} - \frac{1}{J}\frac{dp}{dt} \dots\dots\dots\dots\dots(5),$$

equations which express two general theorems regarding the specific heats of any fluid whatever, first published* in the *Transactions of the Royal Society of Edinburgh*, March 1851. The former (4) is the extension of a theorem on the specific heats of gases originally given by Carnot†, while the latter (5) is inconsistent with one of his fundamental assumptions, and expresses in fact the opposed axiom of the Dynamical Theory. The use of the absolute thermo-dynamic system of thermometry proposed in Section 4, according to which the definition of temperature is

$$t = \frac{J}{\mu} \quad \dots\dots\dots\dots\dots\dots\dots\dots\dots\dots\dots(6),$$

simplifies these equations, and they become

$$JK - JN = t\,\frac{\left(\dfrac{dp}{dt}\right)^2}{-\dfrac{dp}{dv}} \dots\dots\dots\dots\dots\dots\dots(7),$$

$$\frac{d(JN)}{dv} = t\,\frac{d^2p}{dt^2} \dots\dots\dots\dots\dots\dots\dots(8).$$

To compare with the absolute scale the indications of a thermometer in which the particular fluid (which may be any gas, or even liquid) referred to in the notation p, v, t, is used as the

---

* Ibid. §§ 47, 48. [Art. xlviii. above].

† See "Account of Carnot's Theory," [Art. xli. above] Appendix III., *Trans. Roy. Soc. Edin.*, April 30, 1849, p. 565.

thermometric substance, let p_0 and p_{100} denote the pressures which it has when at the freezing and boiling points respectively, and kept in constant volume, v; and let v_0 and v_{100} denote the volumes which it occupies under the same pressure, p, at those temperatures.   Then if θ and ϑ denote its thermometric indications when used as a constant-volume and as a constant-pressure thermometer respectively, we have

$$\theta = 100\,\frac{p - p_0}{p_{100} - p_0} \quad\ldots\ldots\ldots\ldots\ldots\ldots\ldots(9),$$

$$\vartheta = 100\,\frac{v - v_0}{v_{100} - v_0} \quad\ldots\ldots\ldots\ldots\ldots\ldots\ldots(10).$$

Let also ϵ denote the "coefficient of increase of elasticity with temperature*," and ϵ the coefficient of expansion at constant pressure, when the gas is in the state defined by (v, t); and let E and E denote the mean values of the same coefficients between 0° and 100° Cent.   Then we have

$$\epsilon = \frac{dp}{p_0\,dt} \quad\ldots\ldots\ldots\ldots\ldots\ldots\ldots\ldots\ldots(11),$$

$$\epsilon = \frac{\dfrac{dp}{dt}}{v_0 \times -\dfrac{dp}{dv}} \quad\ldots\ldots\ldots\ldots\ldots\ldots\ldots(12),$$

$$E = \frac{p_{100} - p_0}{100\,p_0} \quad\ldots\ldots\ldots\ldots\ldots\ldots\ldots(13),$$

$$E = \frac{v_{100} - v_0}{100\,v_0} \quad\ldots\ldots\ldots\ldots\ldots\ldots\ldots(14).$$

Lastly, the general expression for J/μ quoted in § 2 from our paper of last year [Part I. of Art. XLIX. above], leads to the following expression for the cooling effect on the fluid when forced through a porous plug as in our air experiments:—

$$\delta = \frac{1}{JK}\left\{\int_V^{V'}\left(t\,\frac{dp}{dt} - p\right)dv + (P'V' - PV)\right\}\ldots\ldots\ldots(15),$$

(p, v) (P', V') (P, V), as explained above, having reference to the fluid in different states of density, but always at the same temperature, t, as that with which it enters the plug.

---

* So called by Mr Rankine.   The same element is called by M. Regnault the coefficient of dilatation of a gas at constant volume.

From these equations, it appears that if p be fully given in terms of v and absolute values of t for any fluid, the various properties denoted by

$$JK - JN, \frac{d(JN)}{dv} , \; \theta, \; \vartheta, \; \epsilon, \; \epsilon, \; E, \; \mathrm{E}, \text{ and } \delta,$$

may all be determined for it in every condition. Conversely, experimental investigations of these properties may be made to contribute, along with direct measurements of the pressure for various particular conditions of the pressure, towards completing the determination of the function which expresses this element in terms of v and t. But it must be remarked, that even complete observations determining the pressure for every given state of the fluid, could give no information as to the values of t on the absolute scale, although they might afford data enough for fully expressing p in terms of the volume and the temperature with reference to some particular substance used thermometrically. On the other hand, observations on the specific heats of the fluid, or on the thermal effects it experiences in escaping through narrow passages, may lead to a knowledge of the absolute temperature, t, of the fluid when in some known condition, or to the expression of p in terms of v, and absolute values of t; and accordingly the formulæ (7), (8), and (15) contain t explicitly, each of them in fact essentially involving Carnot's function. As for actual observations on the specific heats of air, none which have yet been published appear to do more than illustrate the theory, by confirming (as Mr Joule's, and the more precise results more recently published by M. Regnault, do), within the limits of their accuracy, the value for the specific heat of air under constant pressure which we calculated* from the *ratio of the specific heats*, determined according to Laplace's theory by observations on the velocity of sound, and the *difference of the specific heats* determined by Carnot's theorem with the value of Carnot's function estimated from Mr Joule's original experiments on the changes of temperature produced by the rarefaction and condensation of air†, and established to a closer degree of accuracy by our preliminary experiments on expansion through a resisting solid‡. It ought also to be remarked, that the specific heats of air can only be applied to the evaluation of absolute

---

* *Philosophical Transactions*, March 1852, p. 82. [Art. xlviii. Appendix Note iii.]

† *Royal Society Proceedings*, June 10, 1844, or *Phil. Mag.* May, 1845.

‡ *Phil. Mag.* Dec. 1852.

temperature with a knowledge of the mechanical equivalent of the thermal unit: and therefore it is probable that, even when sufficiently accurate direct determinations of the specific heats are obtained, they may be useful rather for a correction or verification of the mechanical equivalent, than for the thermometric object. On the other hand, a comparatively very rough approximation to JK, the mechanical value of the specific heat of a pound of the fluid, will be quite sufficient to render our experiments on the cooling effects available for expressing with much accuracy, by means of the formula (15), a thermo-dynamic relation between absolute temperature and the mechanical properties of the fluid at two different temperatures.

[Note of Jan. 5th, 1882. The remainder of this Article (five quarto pages of the *Transactions*) which was devoted to working out an empirical formula for the thermo-elastic properties of air, and the calculation of specific heats from it, is not reproduced here, because at the conclusion of Part IV. of this series of papers by Dr Joule and myself, a better and simpler empirical formula is derived from more comprehensive experimental data.]

PART III.

[From *Transactions of the Royal Society*, June, 1860.]

On the Changes of Temperature experienced by bodies moving through Air.

This interesting branch of our researches has been prosecuted by us from time to time since 1856. In the spring of that year we commenced our experiments by trying the effect of whirling thermometers in the air. This process has been confidently recommended as a means of obtaining the temperature of the atmosphere, but we were sure that the plan was not absolutely correct, and one of us had*, as early as 1847, explained the phenomena of "shooting-stars" by the heat developed by bodies rushing into our atmosphere. In our early experiments we whirled a thermometer by means of a string, alternately quickly and slowly, and it was found that the thermometer was invariably higher after quick than after slow whirling, in some cases the difference amounting to as much as a degree Fahrenheit. We also suc-

---

* See Joule, "On Shooting Stars," *Philosophical Magazine*, 1848, first half-year, p. 349.

ceeded in exhibiting the same phenomenon by whirling a thermo-electric junction. In 1857 we resumed the subject, using an apparatus consisting of a wheel worked by hand, communicating rapid rotation to an axle, at the extremity of which an arm carrying a thermometer, with its bulb outwards, was fixed. The distance between the centre of the axle and the thermometer bulb was 39 inches. The thermometers made use of were filled with ether or chloroform, and had, the smaller 275, and the larger 330 divisions to the degree C. The lengths of the cylindrical bulbs were $\frac{9}{10}$ and $1\frac{4}{10}$ inch, their diameters ·26 and ·48 of an inch respectively. The method of experimenting was to revolve the thermometer bulb at a certain velocity until we knew by experience that it had obtained the full thermal effect, then to stop it as suddenly as possible and observe the temperature.

Alternately with these observations others were made to ascertain the temperature after a slow velocity, the effect due to which was calculated from the other observations, on the hypothesis that it varied with the square of the velocity. In all cases the results in the Tables are means of several experiments.

SERIES I.—Bulb ·26 inch diameter.

Velocities in the alternate experiments, in feet per second.	Difference of thermal effect.	Estimated effect of low velocity.	Thermal effect of high velocity.	Velocity due to 1° C.
46·9 and 24	0·082	0·018	0·1	148·8
51·5 and 24	0·098	0·018	0·116	151·2
68·1 and 24	0·151	0·018	0·169	165·6
72·7 and 24	0·191	0·018	0·209	159
78·7 and 24	0·228	0·018	0·246	158·6
84·8 and 24	0·251	0·018	0·269	163·5
103·7 and 24	0·333	0·018	0·351	175
130·2 and 24	0·531	0·018	0·549	175·7
133·2 and 24	0·607	0·018	0·625	168·5
145·4 and 24	0·676	0·018	0·694	174·6

SERIES II.—Bulb ·48 inch diameter.

Velocities in the alternate experiments, in feet per second.	Difference of thermal effect.	Estimated effect of low velocity.	Thermal effect of high velocity.	Velocity due to 1° C.
36·3 and 18	0·039	0·015	0·054	156·2
66·6 and 18	0·112	0·015	0·127	186·9
84·8 and 18	0·158	0·015	0·173	203·9
125·6 and 18	0·427	0·015	0·442	189

In the following experiments, made in the spring of 1859, thermo-electric junctions of copper and iron wire were whirled, and the effect measured by a Thomson's reflecting galvanometer. The arrangement will be understood from the adjoining sketch, where

  a is the axle of the whirling apparatus, *b* a block of wood placed on the end of the axle; to this is attached *cc'*, a copper

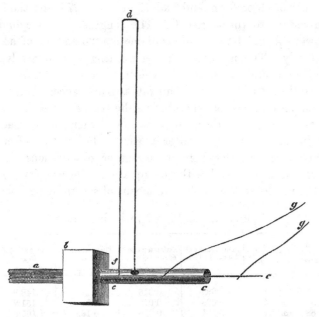

tube, $\frac{3}{16}$th of an inch in diameter, with a hole in its side. *de* is a copper wire, which, entering the hole, passes along the axis of the tube, from which it is insulated by non-conducting material. *df* is an iron wire soldered at *d* to the copper wire. *gg* are thick copper wires, communicating at their remote ends with the galvanometer. They apply to the tube and wire with a springing force, perfect contact being maintained by keeping the touching surfaces clean, and lubricated with oil. A thin piece of wood, not drawn in the sketch, was attached to the block of wood. It was made to extend to within 1, 2, or 3 feet off *d*, according as the velocity was to be slow or quick. The wires being tied to it, were prevented from twisting out of their proper position. The distance of *d* from the axis of revolution was generally 44 inches. The thermal value of the indications of the galvanometer was repeatedly ascertained by direct observations of the effect of heating the junctions.

The following Tables comprise the results of those experiments in which the junction was placed at right angles to the direction of its motion.

SERIES III.—Junction of wires $\frac{1}{100}$th. of an inch in diameter.

Velocities in the alternate experiments, in feet per second.	Difference of thermal effect.	Estimated effect of low velocity.	Thermal effect of high velocity.	Velocity due to 1° C.
53·6 and 21	0̊·21	0̊·037	0̊·247	108
77·8 and 17·5	0·258	0·025	0·283	146·3
105 and 18	0·373	0·026	0·399	166·2
126·4 and 44·8	0·320	0·058	0·378	205·6
146 and 25	0·673	0·030	0·703	174·1
159·3 and 34	0·866	0·041	0·907	167·3
180 and 48	0·671	0·051	0·722	211
186·6 and 48·3	0·967	0·071	1·038	181·3
221 and 47	1·393	0·050	1·443	184
300·5 and 105·8	2·364	0·333	2·697	183
315·6 and 73	3·572	0·202	3·774	162·5
326·5 and 66·2	4·133	0·172	4·305	157·3
372·5 and 66·4	5·21	0·170	5·380	160·6

SERIES IV.—Junction of wires $\frac{1}{40}$th of an inch in diameter.

Velocities in the alternate experiments, in feet per second.	Difference of thermal effect.	Estimated effect of low velocity.	Thermal effect of high velocity.	Velocity due to 1° C.
29 and 17·5	0̊·022	0̊·012	0̊·034	175·2
46·2 and 15·5	0·157	0·026	0·183	108
73·7 and 16·75	0·281	0·024	0·305	133·5
90 and 17	0·363	0·013	0·376	146·8
139·1 and 26	0·61	0·021	0·631	175·1
155·6 and 26	0·878	0·024	0·902	163·9
246·4 and 31	1·482	0·023	1·505	200·8
262·6 and 35	2·087	0·045	2·132	179·8

SERIES V.—Junction of wires $\frac{1}{175}$th of an inch in diameter.

Velocities in the alternate experiments, in feet per second.	Difference of thermal effect.	Estimated effect of low velocity.	Thermal effect of high velocity.	Velocity due to 1° C.
64·03 and 33·96	0̊·127	0̊·049	0̊·176	152·6
91·48 and 41·17	0·204	0·073	0·277	173·8
134·9 and 51	0·67	0·112	0·782	152·5
160·73 and 47·34	0·685	0·097	0·782	181·8
177·9 and 48·25	0·863	0·100	0·963	181·3
208 and 44	1·469	0·071	1·540	168

26—2

SERIES VI.—Junction of wires $\frac{1}{87}$th of an inch in diameter.

Velocities in the alternate experiments, in feet per second.	Difference of thermal effect.	Estimated effect of low velocity.	Thermal effect of high velocity.	Velocity due to 1° C.
93·1 and 39	0·198	0·042	0·240	190
109·6 and 52·4	0·239	0·072	0·311	196·5
133·96 and 58·3	0·432	0·100	0·532	188
163·7 and 55·2	0·654	0·084	0·738	190·5

From the above Tables it is manifest that the thermal effect increases nearly with the square of the velocity; it is, however, a little greater at low velocities than accords with this law. Taking, therefore, the means of the foregoing results, and re-jecting all those obtained from a velocity under 100 feet per second, we obtain the following summary :—

Material of the whirled cylinder.	Diameter.	Velocity due to 1° Cent.
Glass . . .	0·26	173·45
Glass . . .	0·48	189
Copper-iron . .	0·01	177·54
Copper-iron . .	0·025	179·9
Copper-iron . .	0·057	170·9
Copper-iron . .	0·115	190
Mean . . .		180·13

It may be inferred from the above that the thermal effect is independent of the kind of material whirled, provided its surface is smooth ; and that it is likewise independent of the diameter of the cylinder moving in a direction perpendicular to its length.

In the next experiments we whirled the junc-tions parallel to the direction of motion.

SERIES VII.

Diameter of wire.	Velocities in the alternate experiments, in feet per second.	Difference of thermal effect.	Estimated effect of low velocity.	Thermal effect of high velocity.	Velocity due to 1° C.	Means.
·01	123·4 and 45·2	0·415	0·064	0·479	178·3	172·1
	186·7 and 54·2	1·160	0·107	1·267	165·9	
·057	126·5 and 39·6	0·59	0·058	0·648	157·1	160·55
	206·6 and 44	1·518	0·072	1·59	164	
·115	100·5 and 44·8	0·215	0·053	0·268	194·1	182·7
	182·7 and 50	1·053	0·085	1·138	171·3	

The general mean of the velocities due to 1° Cent. is therefore 171·78, which is not notably different from the result obtained when the wire was placed at right angles to the direction of motion. The absence of any considerable effect arising from the shape of the body whirled, was also shown by the following results obtained with a junction of flattened wires a quarter of an inch broad and one-thirtieth of an inch thick.

SERIES VIII.

Position of junction.	Velocities in the alternate experiments, in feet per second.	Difference of thermal effect.	Estimated effect of low velocity.	Thermal effect of high velocity.	Velocity due to 1° C.	Mean.
Flat side against the air............	85·8 and 40	0·165	0·046	0·211	186·6	
	156·5 and 52	0·683	0·085	0·768	178·6	180·6
	164·4 and 48	0·736	0·074	0·810	182·6	
Thin edge against the air	182·4 and 54·3	0·811	0·074	0·885	193·9	

The general mean of all the foregoing results is 179·15 feet per 1° Cent. The phenomena hitherto observed seemed to point to the effect of stopping air as a cause, since 145 feet per second is the velocity of air equivalent to the quantity of heat required to raise its substance, under constant pressure, by 1° Cent. temperature; and it was reasonable to infer that a portion of the effect was lost by radiation. The following experiments, made with a junction of fine wires covered loosely with cotton-wool or tow, enabled us to eliminate all effects but those due to stopped air. Their results will be found to agree closely with theory.

SERIES IX.

Position of junction.	Velocities in the alternate experiments, in feet per second.	Difference of thermal effect.	Estimated effect of low velocity.	Thermal effect of high velocity.	Velocity due to 1° C.	Mean.
Cotton-wool closely tied about the junction of fine wires	91·3 and 34	0·25	0·04	0·29	169·6	
	112 and 26	0·447	0·025	0·472	162·8	
	135·8 and 40·6	1·37	0·14	1·51	110·5	148·76
	140·3 and 24	0·823	0·024	0·847	152·4	
	167·6 and 43·5	1·188	0·085	1·273	148·5	
Junction of fine wires placed in a small wicker basket filled with cotton wool or tow	94·1 and 35	0·494	0·079	0·573	124	
	95·7 and 25·3	0·426	0·032	0·458	141·1	
	105·5 and 27·6	0·33	0·026	0·356	168·3	144·53
	113·5 and 33	0·409	0·038	0·447	169·5	
	116·8 and 33	0·639	0·055	0·694	140·1	
	116·8 and 47·4	0·739	0·145	0·884	124·2	

When the junction was placed in the basket, without any cotton-wool or tow, a velocity of 160·1 ft. per second was required to give 1°. N.B. The basket was so open that its orifices amounted to half the entire area.

In several of our experiments with very slow velocities there appeared to be a greater evolution of heat than could be due to the stopping of air. This circumstance induced us to try various modifications of the surface of the whirled body. In the first instance we covered the bulb of the thermometer used in the second series of experiments with five folds of writing-paper, and then obtained the following results :—

SERIES X.

Velocities in the alternate experiments, in feet per second.	Difference of thermal effect.	Estimated effect of low velocity.	Thermal effect of high velocity.	Velocity due to 1° C., or V_1°, on the hypothesis that $V_{1^\circ} = \dfrac{v}{\sqrt{t}}$.
36·3 and 18	0̊·045	0̊·015	0̊·060	148·2
51·5 and 18	0·115	0·015	0·130	142·8
72·6 and 18	0·146	0·015	0·161	180·9
118 and 18	0·385	0·015	0·400	186·6

It will be seen from the last column that the effect at slow velocities was greater than that which might have been anticipated. We were thus led to try the effect of a further increase of what we may call " fluid friction." In the next series the bulb was wrapped with fine iron wire.

SERIES XI.

Velocities in the alternate experiments, in feet per second.	Difference of thermal effect.	Estimated effect of low velocity.	Thermal effect of high velocity.	Velocity due to 1° C., or V_1°, on the hypothesis that $V_{1^\circ} = \dfrac{v}{\sqrt{t}}$.
15·36 and 7·68	0̊·022	0̊·008	0̊·030	88·8
23·04 and 15·36	0·069	0·03	0·099	73·2
30·71 and 15·36	0·118	0·03	0·148	79·8
46·08 and 15·36	0·177	0·03	0·207	101·3
69·12 and 15·36	0·267	0·03	0·297	126·8
111·34 and 15·36	0·530	0·03	0·560	148·8
126·72 and 15·36	0·598	0·03	0·628	160
153·55 and 15·36	0·850+	0·03	0·880+	163·4 −

In the next experiments the bulb was wrapped with a spiral of fine brass wire.

SERIES XII.

Velocities in the alternate experiments, in feet per second.	Difference of thermal effect.	Estimated effect of low velocity.	Thermal effect of high velocity.	Velocity due to 1° C., or V_{1°, on the hypothesis that $V_{1^\circ} = \dfrac{v}{\sqrt{t}}$.
7·68 and 1·92	0·006	0·002	0·008	86·3
15·36 and 7·68	0·033	0·008	0·041	75·8
23·04 and 15·36	0·070	0·041	0·111	69·1
30·71 and 15·36	0·105	0·041	0·146	80·3
46·08 and 19·2	0·120	0·075	0·195	104·4
76·8 and 23·04	0·203	0·111	0·314	137·1
115·18 and 23·04	0·570	0·111	0·681	139·5
148·78 and 76·8	0·488	0·314	0·802	166·2

The last columns of the above Tables clearly indicate that at slow velocities a source of heat exists besides that from stopped air. It is also evident that, as the velocity increases, this thermal cause decreases; for at a velocity of 150 feet per second the thermal effect is such as would be due to the influence of stopped air alone.

In prosecuting still further this part of our subject we made the following arrangement. A disc of mill-board, 32 inches in diameter, was fixed at the end of the axis of the whirling apparatus. An ether thermometer, whose bulb was one-fourth of an inch in diameter, was tied by its stem to the face of the disc, so that the bulb was 15 inches distant from the axis of revolution, and 1 inch from the margin of the disc. In the following Table the first five experiments were made with the above arrangement, but in the last two a thermo-electric junction of thin copper and iron wires, tied closely to the mill-board, was substituted for the ether thermometer.

Series XIII.

Velocities in the alternate experiments, in feet per second.		Difference of thermal effect.	Estimated effect of low velocity.	Thermal effect of high velocity.	Velocity due to 1° C., or V_1°, on the hypothesis that V_1° $= \dfrac{v}{\sqrt{t}}$.
	3·15 and ·5	$\overset{\circ}{0}$·029	$\overset{\circ}{0}$·005	$\overset{\circ}{0}$·034	17·1
	7·85 and 3·15	0·027	0·034	0·061	31·7
	15·7 and 7·85	0·052	0·061	0·113	46·6
	31·4 and 15·7	0·022	0·113	0·135	85·5
Thermo-electric junction	63·3 and 27·4	0·106	0·120	0·226	133·3
	90·2 and 25	0·286	0·116	0·402	142·3

The surface of the mill-board disc being rather rough, it was judged desirable to make similar experiments with a disc of sheet zinc. This was perfectly smooth, $36\frac{1}{2}$ inches in diameter. The thermometer bulb was fixed at 17·1 inches distance from the axis.

Series XIV.

Velocities in the alternate experiments, in feet per second.	Difference of thermal effect.	Estimated effect of low velocity.	Thermal effect of high velocity.	Velocity due to 1° C., or V_1°, on the hypothesis that V_1° $= \dfrac{v}{\sqrt{t}}$.
1·71 and ·57	$\overset{\circ}{0}$·024	$\overset{\circ}{0}$·010	$\overset{\circ}{0}$·034	9·2
3·42 and 1·71	0·017	0·034	0·051	15·1
8·55 and 3·42	0·027	0·051	0·078	30·7
17·1 and 8·55	0·023	0·078	0·101	53·8
34·2 and 17·1	0·046	0·102	0·148	88·8
57·28 and 17·1	0·070	0·102	0·172	138·3

The last column of the two foregoing Tables clearly show the inapplicability of the law of the increase of temperature with the square of the velocity, at low velocities. The thermal effect appears even to increase at a slower rate than simply with the velocity. This phenomenon may, we think, be ascribed to the internal fluid friction of the particles of air among themselves, which Professor Stokes has proved to exist, by his researches on the motion of pendulums. We may easily apprehend that in such experiments as our last, the entire face of the disc is covered with a film of air, which revolves along with it at very slow velocities. As the velocity increases there will still be a film of air adhering to the disc, but with the difference that it will be constantly replaced by fresh stopped air, the thermal effect of which will ultimately be the only recognizable phenomenon.

A very interesting and important branch of our subject was to inquire into the thermal phenomena which take place at the surface of a sphere passing rapidly through air. Some of our experiments on this subject have been made by blowing air from a large bellows against the ball; others by whirling a ball or sphere in the air by means of the apparatus already described. We shall commence by describing the latter, in some of which a thermo-electric junction was employed, and in others an ether thermometer.

SERIES XV.—Wooden ball 2 inches in diameter, with a thermo-electric junction of fine copper and iron wires made even with the surface.

Position of the junction in respect to the direction of motion.	Velocities in alternate experiments.	Difference of thermal effect.
In front, or anterior ....{	75·6 and 23·1 118·4 and 23·1 141·5 and 39·5	0·269 0·517 0·745
At the side, or equatorial...{	74 and 28·5 115 and 26·3 120 and 40	− 0·146 0·283 0·020
In the rear, or posterior ...{	71·5 and 25 112·4 and 19·3 113·7 and 42	0·093 0·414 0·280

In the above experiments differential results for the several pairs of velocities are alone given, so that, although one of the quantities has a negative sign, there is no proof of actual cooling effect. In the next experiments we whirled a thin glass globe, 3·58 inches in diameter, placed at a distance of 38 inches from the axle of the apparatus. The small bulb of an ether thermometer was kept in contact with the glass.

SERIES XVI.

Position of the bulb of the thermometer in respect to the direction of motion.	Velocities in the alternate experiments, in feet per second.	Difference of thermal effect.	Thermal effect due to low velocity.	Thermal effect due to high velocity.
In front of the globe......	3·84 and 1·92	0̊·002	0̊·001	0̊·003
	7·68 and 3·84	0·007	0·003	0·010
	15·36 and 7·68	0·143	0·010	0·153
	23·04 and 15·36	0·106	0·153	0·259
	38·4 and 15·36	0·133	0·153	0·286
	57·5 and 15·36	0·211	0·153	0·364
At the side of the globe...	3·84 and 1·92	0·029	0·007	0·036
	7·68 and 3·84	0·109	0·036	0·145
	15·36 and 7·68	0·138	0·145	0·283
	23·04 and 7·68	0·181	0·145	0·326
	38·4 and 23·04	0·000	0·326	0·326
	57·5 and 23·04	0·087	0·326	0·413
In the rear of the globe...	3·84 and 1·92	0·011	0·004	0·015
	7·68 and 3·84	0·024	0·015	0·039
	15·36 and 7·68	0·147	0·039	0·186
	23·04 and 15·36	0·076	0·186	0·262
	38·4 and 15·36	0·062	0·186	0·248
	57·5 and 15·36	0·091	0·186	0·277
	70·92 and 15·36	0·204	0·186	0·390

In the next experiments, a 12-inch globe, such as is used in schools, was fixed at a distance of 3 feet from the axis of the revolving apparatus. The ether thermometer was generally employed, as in the last series, but for the highest velocity a thermo-electric junction of thin wires placed close to the globe registered the thermal effect.

SERIES XVII.

Measurer of heat, and its position in respect to the direction of the motion.		Velocities in the alternate experiments, in feet per second.	Difference of thermal effect.	Thermal effect due to low velocity.	Thermal effect due to high velocity.
Anterior.	Ether thermometer	3·72 and 1·24	0̊·019	0̊·009 estimd.	0̊·028
		7·44 and 3·72	0·008	0·028	0·036
		14·88 and 7·44	0·028	0·036	0·064
	Thermo-electric junction	39·68 and 7·44	0·200	0·036	0·236
Equatorial.	Ether thermometer	3·72 and 1·24	0·007	0·003 estimd.	0·010
		7·44 and 3·72	0·013	0·010	0·023
		14·88 and 7·44	0·024	0·023	0·047
	Thermo-electric junction	39·37 and 7·44	0·170	0·023	0·193
Posterior.	Ether thermometer	3·72 and 1·24	0·024	0·012 estimd.	0·036
		7·44 and 3·72	0·022	0·036	0·058
		14·88 and 7·44	0·046	0·058	0·104
	Thermo-electric junction	37·2 and 7·44	0·140	0·058	0·198

In the experiments in which air was blown against a sphere, we made use of a large organ-bellows, from which a constant stream of air could be kept up at velocities dependent upon the weights laid on.   In our first trials, the air issued from a circular aperture 2½ inches in diameter, and the ball was placed, at half an inch distance, in front of the aperture.   We shall, as before, call that point of the ball which was nearest the wind, the Anterior Pole; the most sheltered point, the Posterior Pole; and the intermediate part, the Equator.   The balls were furnished with thermo-electric junctions of thin copper and iron wires, made flat with the surface, the junctions being in each case 90° apart from one another.

SERIES XVIII.—½-inch Wooden Ball.

Velocity of air.
68 ft. per sec. ... Equator 0·114 colder} ...Posterior Pole 0·067 colder than Equator.
            than Anterior Pole }

SERIES XIX.—1-inch Wooden Ball.

Velocity of air.
1·2 ... Equator 0·088 warmer than Anterior Pole.

3·6 ... Equator 0·129 warmer} ... Posterior Pole 0·03 warmer than Equator.
        than Anterior Pole }

7·2 ... Equator 0·160 warmer} ... Posterior Pole 0·022 warmer than Equator.
        than Anterior Pole }

14·4 ... Equator 0·120 warmer} ... Posterior Pole 0·018 colder than Equator.
         than Anterior Pole }

28·8 ... Equator 0·056 warmer} ... Posterior Pole 0·018 colder than Equator.
         than Anterior Pole }

36 ... Equator 0·008 colder than Anterior Pole.

48 ... Equator 0·035 colder than Anterior Pole.

57·6 ... Equator 0·056 colder} ... Posterior Pole 0·090 colder than Equator.
         than Anterior Pole }

73 ... Equator 0·245 colder than Anterior Pole.

105 ... Equator 0·380 colder} ... Posterior Pole 0·232 colder than Equator.
         than Anterior Pole }

In our next series, one junction was placed within the bellows, and the other in contact with the different parts of the 1-inch ball.   All the results will be seen to indicate, as might have been anticipated, that the junction within the bellows was warmer than any part of the ball.

SERIES XX.

Velocity of air.	Pressure of air in the bellows, in inches of water.	Cold of Anterior Pole, in respect to the inner junction.	Cold of Equator, in respect to the inner junction.	Cold of Posterior Pole, in respect to the inner junction.
2·4	0·003 estimated	0°·098	0°·065	0°·028
3·6	0·006 estimated	0·094	0·065	0·028
7·2	0·025 estimated	0·083	0·103	0·060
14·4	0·105 estimated	0·110	0·089	0·109
28·8	0·42 estimated	0·102	......	0·112
73	2·7 estimated	0·188	0·309	0·300
105	5·6 measured	0·195	......	0·360

A further modification of the experiments was made by placing a glass tube 3 feet long and of 1½-inch interior diameter, within the aperture, so that two-thirds of the tube was inside, and one-third outside of the bellows. A ball furnished with junctions 90° distant from each other was placed within the tube.

SERIES XXI.—Wooden Ball, 1-inch diameter.

Velocity of air.

1·8 ... Equator 0°·045 warmer than Anterior Pole ... Posterior Pole 0°·052 warmer than Equator.

2·7 ... Equator 0·056 warmer than Anterior Pole ... Posterior Pole 0·052 warmer than Equator.

5·4 ... Equator 0·074 warmer than Anterior Pole ... Posterior Pole 0·035 warmer than Equator.

10·8 ... Equator 0·052 warmer than Anterior Pole ... Posterior Pole 0·017 warmer than Equator.

21·6 ... Equator 0·037 warmer than Anterior Pole ... Posterior Pole 0·008 colder than Equator.

43·2 ... Equator 0·011 warmer than Anterior Pole ... Posterior Pole 0·013 colder than Equator.

54 ... Equator 0·019 colder than Anterior Pole ... Posterior Pole 0·014 colder than Equator.

62 .... Posterior Pole 0·023 colder than Equator.

83·8.................................... Posterior Pole 0·041 colder than Equator.

108 ... Equator 0·019 colder than Anterior Pole ... Posterior Pole 0·086 colder than Equator.

SERIES XXII.—Wooden Ball, ½-inch diameter.

Velocity of air.

1·8 ... Equator 0°·048 warmer than Anterior Pole ... Posterior Pole 0°·050 warmer than Equator.

5·4 ... Equator 0·030 warmer than Anterior Pole ... Posterior Pole 0·047 warmer than Equator.

10·8 ... Equator 0·023 warmer than Anterior Pole ... Posterior Pole 0·031 warmer than Equator.

21·6 ... Equator 0·008 warmer than Anterior Pole ... Posterior Pole 0·012 warmer than Equator.

Velocity of air.

43·2	... Equator 0·006 colder) than Anterior Pole }	... Posterior Pole 0̊·009 warmer than Equator.
54	... Equator 0·019 colder) than Anterior Pole }	... Posterior Pole 0·006 colder than Equator.
62	... Equator 0·026 colder) than Anterior Pole }	... Posterior Pole 0·014 colder than Equator.
83·8	... Equator 0·040 colder) than Anterior Pole }	... Posterior Pole 0·031 colder than Equator.
108	... Equator 0·068 colder) than Anterior Pole }	... Posterior Pole 0·050 colder than Equator.

The general result is that at slow velocities of air there is a gradual increase of temperature from the anterior to the posterior pole, but the reverse at high velocities. We observed a great effect for slow velocities at the commencement of an experiment, which gradually declined on continued blowing. This phenomenon was apparently owing to circumstances in connexion with the temperature of the orifice and of the bellows.

The causes of the thermal effects on the surface of balls slowly passing through air are very complicated, as they arise from the effects of stopped air, fluid friction, and varied pressures. In order, if possible, to throw some light on these, we made the following observations:—

1st. That when the 12-inch globe passed through the air at the velocity of about 12 feet per second or under, the air at the equatorial part moved in the reverse direction. We did not observe the velocity, if it existed, at which this pheno- menon ceased to take place*.

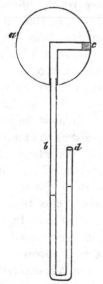

2nd. An ivory ball, 1·7 inch diameter, had holes drilled from points of the surface 90° asunder, which holes met at the centre of the ball. Into the lower one (see adjoining figure) a bent glass tube, partly filled with water, was cemented; in the other, at c, a porous wooden plug was placed. It was then found that when c was made the anterior pole in a blast from the bellows, a pressure was experienced able to produce a difference of level in the tubes b and d equal to 2·5 inches. When c was put in the equatorial position,

* See *Proceedings of the Royal Society*, June 18, 1857, p. 558. [Appendix to present Article below.]

there was, on the contrary, a suction equal to 1·2 inch. When
c was made the posterior pole there was also a suction, equal,
however, to only 0·1 inch. Having tied a thick fold of silk over
the orifice d, we tried the same thing in a strong breeze of
wind, when we found that on making c the anterior pole, we
had a pressure amounting to 0·6 of an inch; on making c equatorial, a suction of 0·3; and on making c the posterior pole, a
suction of 0·05 inch.

We have not hitherto been able to detect any change in the
thermal effect, owing to the whistling sound of wire or other
bodies rapidly whirled. We think it possible that this vibratory
action decreases the resistance and the evolution of heat. Some
of the sounds produced are interesting and worthy of further
investigation. When a small piece of paper was attached to the
revolving wire, we obtained a continuous succession of loud cracks
similar to those of a whip.

But although this and other parts of our subject remain to
be cleared up, we believe that it will be found that at all high
velocities the thermal effect arises entirely from stopped air, and
thus is independent of the shape and mass of the body, and of
the temperature and density of the atmosphere. From some
experiments described in the *Proceedings of the Royal Society* of
June 19, 1856, p. 183 [Appendix to present Article below], we
inferred that a body placed in a stream of air moving with a
velocity of 1780 feet per second, was raised 137° C. above the
temperature of that stream. This gives 152 feet per second as
the velocity due to 1°, while our direct results, given in the present
paper, indicate 179.

It must be obvious that a thermometer placed in the wind
registers the temperature of the air, plus the greater portion, but
not the whole, of the temperature due to the *vis viva* of its
motion. In a place perfectly sheltered from the wind, the temperature of a thermometer immersed in the air will be that of the
wind, plus the whole temperature due to the *vis viva* of the
moving air. In accordance with this we have found that a thermometer placed in a sheltered situation, such as on the top of
a wall opposite the wind, indicates a higher temperature than
when it is exposed to the blast. A minute examination of these
phenomena cannot fail to interest the meteorologist.

PART IV.

[From *Transactions of the Royal Society*, June, 1862.]

In the Second Part of these researches we have given the results of our experiments on the difference between the temperatures of an elastic fluid on the high- and low-pressure sides of a porous plug through which it was transmitted. The gases employed were atmospheric air and carbonic acid. With the former, $0°·0176$ of cooling effect was observed for each pound per square inch of difference of pressure, the temperature on the high-pressure side being $17°·125$. With the latter gas, $0°·0833$ of cooling effect was produced per lb. of difference of pressure, the temperature on the high-pressure side being $12°·844$.

It was also shown that in each of the above gases the difference of the temperatures on the opposite sides of the porous plug is sensibly proportional to the difference of the pressures.

An attempt was also made to ascertain the cooling effect when elastic fluids of high temperature were employed; and it was satisfactorily shown that in this case a considerable diminution of the effect took place. Thus, in air at $91°·58$, the effect was only $0°·014$; and in carbonic acid at $91°·52$ it was $0°·0474$.

In the experiments at high temperatures there appeared to be some grounds for suspecting that the apparent cooling effect was too high; for the quantity of transmitted air was very considerable, and its temperature possibly had not arrived accurately at that of the bath by the time it reached the porous plug.

The obvious way to get rid of all uncertainty on this head was to increase the length of the coil of pipes. Hence in the following experiments the total length of 2-inch copper pipe immersed in the bath was 60 feet instead of 35, as in the former series. The volume of air transmitted in a given time was also considerably less. There could therefore be no doubt that the temperature of the air on its arrival at the plug was sensibly the same as that of the bath.

The nozzle employed in the former series of experiments was of box-wood, — the space occupied by cotton-wool, or other porous material, being 2·72 inches long and an inch and a half in diameter. The box-wood was protected from the water of the bath by being enveloped by a tin can filled with cotton-wool.

This was unquestionably in most respects the best arrangement for obtaining accurate results; but it was found necessary to make each experiment last one hour or more before we could confidently depend on the thermal effect. The oscillations of temperature which took place during the first part of the time were traced to various causes, one of the principal being the length of time which, on account of the large capacity for heat and the small conductivity of the box-wood nozzle, elapsed before the first large thermal effects consequent on the getting up of the pressure were dissipated. No doubt the results we arrived at were very accurate with the elastic fluids employed, viz. atmospheric air and carbonic acid; but we possessed an unlimited supply of the former and a supply of the latter equal to 120 cubic feet, which was sufficient to last for more than half an hour without being exhausted. In extending the inquiry to gases not so readily procured in large quantities, it was therefore desirable to use a porous plug of smaller dimensions enclosed in a nozzle of less capacity for heat, so as to arrive rapidly at the normal effect.

Various alterations of the apparatus were made in order to meet the new requirements of our experiments. A small high-pressure engine of about one-horse power was placed in gear with a double-acting compressing air-pump, which had a cylinder $4\frac{1}{2}$ inches in diameter, with a length of stroke of 9 inches. The engine was able to work the piston of the pump sixty complete strokes in the minute. The quantity of air which it ought to have discharged at low pressure was therefore upwards of 16,000 cubic inches per minute. But much loss, of course, occurred from leakage past the metallic piston, and in consequence of the necessary clearance at the top and bottom of the cylinder when the pressure increased by a few atmospheres; so that in practice we never pumped more than 8000 cubic inches per minute.

The nozzle we employed will be understood by inspecting Plate XXVI. fig. 1, where aa is the upright end of the coil of copper pipes. On a shoulder within the pipe a perforated metallic disk (b) rests. Over this a short piece of india-rubber tube (cc) enclosing a silk plug (d), which is kept in a compressed state by the upper perforated metallic plate (e). This upper plate is pressed down with any required force by the operation of the screw f on the metallic tube gg. A tube of cork (hh) is placed within the metallic tube, in order to protect the bulb of the thermometer from the effects of a too rapid conduction of heat from the bath. Cotton-wool is loosely packed round the bulb, so as to distribute the flowing air as evenly as possible. The glass tube (*ii*) is attached to the nozzle by means of a piece of strong india-rubber tubing, and through it the indications of the thermometer are read. The top of the glass tube is attached to the metallic tube *ll*, for the purpose of conveying the gas to the meter.

The thermometer (*m*) for registering the temperature of the bath is placed with its bulb near the nozzle. The level of the water is shown by *nn*; and *oo* represents the wooden cover of the bath.

When a high temperature was employed, it was maintained by introducing steam into the bath by means of a pipe led from the boiler. The water of the bath was in every case constantly and thoroughly stirred, especially when high temperatures were used.

The general disposition of the apparatus will be understood from fig. 2, in which *A* represents the boiler, *B* the steam-engine geared to the condensing air-pump *C*. From this pump the compressed air passes through a train of pipes 60 feet long and 2 inches in diameter, and then enters the coil of pipes in the bath *D*. Thence, after issuing from the porous plug, it passes through the gasometer *E*, and ultimately arrives again at the pump *C*. This complete circulation is of great importance, inasmuch as it permits the gas which has been collected in the meter to be used for a much longer period than would otherwise have been possible. A glass vessel full of chloride of calcium is placed in the circuit at *F*, and chloride of calcium is also placed in the pipe at *f*. A small tube leading from the coil is carried to the shorter leg of the glass siphon gauge *G*, of which the longer leg is 17 feet, and the shorter 12 feet long.

The thermometers employed were all carefully calibrated, and had about ten divisions to the degree Centigrade. We took the precaution of verifying the air- and bath-thermometers from time to time, especially when high temperatures were used, in which latter case a comparison between the thermometers at high temperature was made immediately after each experiment.

Atmospheric Air.

In the experiments described in the present paper, the air was not deprived of its carbonic acid. It was simply dried by transmitting it in the first place, before it entered the pump, through a cylinder 18 inches long and 12 inches in diameter filled with chloride of calcium, and afterwards, in its compressed state, through a pipe 12 feet long and 2 inches in diameter filled with the same substance. The experiments were principally carried on in the winter season; so that the chloride kept dry for a long time. From its condition after some weeks' use, it was evident that the water was removed, almost as much as chloride of calcium can remove it, after the air had traversed three inches of the chloride contained by the first vessel.

TABLE I.

No. of experiment.	Cubical inches of air transmitted per minute.	Pressure over that of the atmosphere, in inches of mercury.	Temperature of the bath.	Thermal effect.	Correction on account of conduction of heat.	Corrected thermal effect.	Thermal effect reduced to the pressure of 100 inches of mercury.	Time occupied by experiment, in minutes.	Number of observations comprised in each mean.	Extreme range of the temperature of the bath.	Extreme range of the temperature of the air.	Extreme range of the pressure.
1	3000	83·96	4·499	− 0·711	− 0·044	− 0·755	− 0·900	14	5	0·020	0·015	2·25
2	3600	136·19	6·112	− 1·11	− 0·058	− 1·168	− 0·858	24	7	0·017	0·055	1·7
3	2600	156·59	6·082	− 1·307	− 0·094	− 1·401	− 0·895	15	5	0·009	0·065	8·0
4	1750	139·58	7·471	− 1·137	− 0·122	− 1·259	− 0·902	24	15	0·006	0·19	5·3
5	2250	153·9	7·640	− 1·231	− 0·103	− 1·334	− 0·867	12·5	10	0·008	0·028	3·6
6	2300	159·3	8·546	− 1·252	− 0·102	− 1·354	− 0·850	18	20	0·017	0·105	3·0
7	2060	165·73	8·2	− 1·329	− 0·121	− 1·450	− 0·875	14	15	0·034	0·128	4·8
8	1500	129·73	8·72	− 1·019	− 0·127	− 1·146	− 0·883	8	9	0·008	0·135	2·9
9	5000	128·9	24·92	− 0·983	− 0·037	− 1·020	− 0·791	12	8	0·015	0·09	7·0
10	4600	122·8	27·81	− 0·874	− 0·036	− 0·910	− 0·741	26	15	0·029	0·064	0
11	5000	123·5	42·64	− 0·947	− 0·036	− 0·983	− 0·796	8	4	0·127	0·122	7·0
12	4800	137	43·54	− 0·943	− 0·037	− 0·980	− 0·715	6	3	0·02	0·02	0
13	5000	127·5	47·92	− 0·937	− 0·037	− 0·974	− 0·764	35	4	0·058	0·09	2·0
14	3700	147	49·96	− 0·969	− 0·049	− 1·018	− 0·692	28	30	0·14	0·26	0
15	5600	146	53·375	− 0·860	− 0·028	− 0·888	− 0·608	24	30	0·05	0·08	0
16	5700	146	64·9	− 0·870	− 0·029	− 0·899	− 0·616	20	20	0·18	0·23	0
17	2700	112·43	89·901	− 0·469	− 0·033	− 0·502	− 0·446	4	10	0·112	0·23	8·6
18	1700	147	90·353	− 0·821	− 0·091	− 0·912	− 0·620	19	3	0·022	0·085	0
19	1700	153·16	92·486	− 0·756	− 0·083	− 0·839	− 0·547	12	10	0·202	0·273	7·0
20	3150	156·5	92·603	− 0·674	− 0·040	− 0·714	− 0·456	24	10	0·078	0·19	3·5
21	3800	146	93·78	− 0·700	− 0·036	− 0·736	− 0·504	20	20	0·236	0·255	0
22	4600	158·5	97·528	− 0·722	− 0·029	− 0·751	− 0·474		16	0·112	0·115	
13	12	11	10	9	8	7	6	5	4	3	2	1

Oxygen Gas.

This elastic fluid was procured by cautiously heating chlorate of potash mixed with a small quantity of peroxide of manganese. In its way to the meter it passed through a tube containing caustic potash, in order to deprive it of any carbonic acid it might contain. The same drying-apparatus was employed as in the case of atmospheric air.

TABLE II.

1	2	3	4	5	6	7	8	9	10	11	12	13	14	15
No. of experiment.	Cubical inches of elastic fluid transmitted per minute.	Composition of the elastic fluid.	Pressure over that of the atmosphere, in inches of mercury.	Temperature of the bath.	Thermal effect.	Correction on account of conduction of heat.	Corrected thermal effect.	Thermal effect reduced to the pressure of 100 inches of mercury.	Ditto, calculated for pure oxygen.	Time occupied by experiment, in minutes.	Number of observations comprised in each mean.	Extreme range of the temperature of the bath.	Extreme range of the temperature of the elastic fluid.	Extreme range of the pressure.
1	2000	5·095 N / 94·905 O	159·28	8·682	−1·547	−0·145	−1·692	−1·061	−1·075	9	10	0·007	0·35	7·8
2	2000	54·62 N / 45·38 O	161·81	8·75	−1·373	−0·129	−1·502	−0·928	−1·074	11	10	0·017	0·046	0·9
3	1700	3·64 N / 96·36 O	151	89·466	−1·069	−0·118	−1·187	−0·786	−0·800	14	10	0·45	0·43	6·2
4	3150	22·37 N / 77·63 O	159·77	90·8	−0·840	−0·050	−0·890	−0·557	−0·580	12	10	0·326	0·336	8·0
5	3150	51·03 N / 48·97 O	154·1	92·792	−0·734	−0·043	−0·777	−0·504	−0·527	12	10	0·18	0·19	4·0
6	4500	4 N / 96 O	152	95·453	−0·795	−0·033	−0·828	−0·544	−0·570	11	8	0·135	0·158	0

Nitrogen Gas.

In preparing this gas the meter was first filled with air, and then a long shallow tin vessel was floated under it, containing sticks of phosphorus so disposed as to burn in succession. Some hours were allowed to elapse after the combustion had terminated, in order to allow of the deposition of the phosphoric acid formed.

Table III.

1	2	3	4	5	6	7	8	9	10	11	12	13	14	15
No. of experiment.	Cubical inches of elastic fluid transmitted per minute.	Composition of the elastic fluid.	Pressure over that of the atmosphere, in inches of mercury.	Temperature of the bath.	Thermal effect.	Correction on account of conduction of heat.	Corrected thermal effect.	Thermal effect reduced to the pressure of 100 inches of mercury.	Ditto, calculated for pure nitrogen.	Time occupied by experiment, in minutes.	Number of observations comprised in each mean.	Extreme range of the temperature of the bath.	Extreme range of the temperature of the elastic fluid.	Extreme range of the pressure.
1	2050	7·9 O / 92·1 N	163·38	7·204	−1·448	−0·133	−1·581	−0·967	−1·034	7	8	0·008	0·25	6·2
2	2500	2·2 O / 97·8 N	162·65	91·415	−0·857	−0·064	−0·921	−0·567	−0·576	13	10	0·036	0·48	4·5
3	2500	12·5 O / 87·5 N	164·61	91·965	−0·869	−0·065	−0·934	−0·567	−0·691	12	9	0·337	0·378	3·0

Carbonic Acid.

This gas was formed by adding sulphuric acid to a solution of carbonate of soda. It was dried in the same manner as all the other gases.

TABLE IV.

No. of experiment.	Cubical inches of elastic fluid transmitted per minute.	Composition of the elastic fluid.	Pressure over that of the atmosphere, in inches of mercury.	Temperature of the bath.	Thermal effect.	Correction on account of conduction of heat.	Corrected thermal effect.	Thermal effect reduced to the pressure of 100 inches of mercury.	Ditto, calculated for pure carbonic acid, calling its sp. heat for equal vol. 1·39.	Time occupied by experiment, in minutes.	Number of observations comprised in each mean.	Extreme range of the temperature of the bath.	Extreme range of the temperature of the elastic fluid.	Extreme range of the pressure.
1	2450	Air 68·42, CO$_2$ 31·58	163·7	7·362	− 2·699	− 0·190	− 2·889	− 1·765	− 3·166	12	10	0	0·16	3·2
2	2350	CO$_2$ 89·16, Air 10·84	148·82	7·360	− 1·621	− 0·125	− 1·746	− 1·173	− 2·990	14	10	0·004	0·282	9·2
3	3100	Air 3·52, CO$_2$ 96·48	164·07	7·384	− 6·719	− 0·299	− 7·018	− 4·277	− 4·367	6·5	6	0·008	0·021	1·4
4	2500	CO$_2$ 62·5, Air 37·5	162·925	7·407	− 2·839	− 0·191	− 3·030	− 1·860	− 3·052	8	8	0·007	0·11	5·8
5	2300	Air 88·13, CO$_2$ 11·87	158·08	7·433	− 1·682	− 0·132	− 1·814	− 1·147	− 2·648	10	10	0·005	0·107	2
6	2260	CO$_2$ 97·46, Air 2·54	163·52	7·608	− 1·407	− 0·116	− 1·523	− 0·931	− 7·253	8	8	0·007	0·064	2·0
7	3300	Air 4·0, H 5·286, CO$_2$ 90·714	161·97	7·960	− 6·131	− 0·262	− 6·393	− 3·947	− 4·215	6	8	0	0·18	4·8
8	3000	CO$_2$ 4·23, Air 46·47, H 49·3	153·72	8·020	− 2·189	− 0·117	− 2·306	− 1·500	− 2·631	5	5	0	0·19	1·6
9	1500	Air 7·09, H 67·05, CO$_2$ 25·86	97·56	8·296	− 0·543	− 0·063	− 0·606	− 0·622	− 1·940	15	15	0·012	0·146	5·4
10	2925	Air 2·11, CO$_2$ 97·89	167·25	93·523	− 3·418	− 0·160	− 3·578	− 2·139	− 2·164	10	10	0·382	0·49	4·0
11	2925	Air 56·78, CO$_2$ 43·22	167·4	91·26	− 1·746	− 0·099	− 1·845	− 1·102	− 1·674	30	20	0·292	0·49	11·0
12	2925	Air 77·77, CO$_2$ 22·23	146·83	91·642	− 1·292	− 0·077	− 1·369	− 0·938	− 2·053	9	6	0·045	0·245	3·5
13	5500	Air 0·83, CO$_2$ 99·17	146	54·0	− 4·184	− 0·104	− 4·288	− 2·937	− 2·951	24	16	0·24	0·46	0
14	5300	Air 67·7, CO$_2$ 32·3	147	49·703	− 1·832	− 0·059	− 1·891	− 1·286	− 2·225	24	16	0·025	0·17	0
15	5600	Air 87·77, CO$_2$ 12·23	145	49·764	− 1·250	− 0·032	− 1·282	− 0·884	− 2·025	20	16	0·01	0·11	0
16	5100	Air 1·83, CO$_2$ 98·17	127·5	35·604	− 4·186	− 0·112	− 4·298	− 3·371	− 3·407	18	15	0·03	0·095	0
17	5000	Air 1·66, CO$_2$ 98·34	151	97·553	− 3·11	− 0·084	− 3·194	− 2·115	− 2·135	20	16	0·292	0·272	0

Hydrogen.

Our method of procuring this elastic fluid was to pour sulphuric acid, prepared from sulphur, into a carboy nearly filled with water and containing fragments of sheet zinc. The gas was passed through a tube filled with rags steeped in a solution of sulphate of copper, and then through a tube filled with sticks of caustic potash. The rags became speedily browned, and we therefore adopted the plan of pouring a small quantity of solution of sulphate of copper from time to time into the carboy itself. This succeeded perfectly; the rags retained their blue colour, and the gas was rendered perfectly inodorous, whilst at the same time its evolution became much more free and regular.

TABLE V.

1	2	3	4	5	6	7	8	9	10	11	12	13	14	15
No. of experiment.	Cubical inches of elastic fluid transmitted per minute.	Composition of the elastic fluid.	Pressure over the atmosphere, in inches of mercury.	Temperature of the bath.	Thermal effect.	Correction on account of conduction of heat.	Corrected thermal effect.	Thermal effect reduced to the pressure of 100 inches of mercury.	Ditto, calculated for pure hydrogen.	Time occupied by experiment, in minutes.	Number of observations comprised in each mean.	Extreme range of the temperature of the bath.	Extreme range of the temperature of the elastic fluid.	Extreme range of the pressure.
1	3000	17·635 Air / 82·365 H	64·1	6·34	−0·114	−0·009	0·153	−0·239	−0·104	3	3	0	0	0
2	3000	75·16 H / 24·84 Air	99·86	6·355	−0·564	−0·035	−0·599	−0·600	+0·226	10	4	0	0·15	1·5
3	3900	4·866 Air / 95·134 H	49·91	6·132	+0·033	+0·002	+0·035	+0·070	+0·118	12·	6	0·002	0·06	1·2
4	2900	78·295 Air / 21·705 H	99·657	5·808	−0·535	−0·034	−0·569	−0·571	+0·525	34	12	0·03	0·11	1·85
5	2800	9·2 / 90·8 Air	86·885	7·244	+0·041	+0·003	+0·044	+0·05	+0·143	27	10	0·034	0·033	1·75
6	3300	1·798 Air / 98·202 H	79·84	7·572	+0·043	+0·003	+0·046	+0·058	+0·075	23	8	0·008	0·023	2·85
7	2950	4·795 Air / 95·205 H	74·08	6·654	+0·054	+0·004	+0·058	+0·078	+0·126	17	10	0·016	0·11	6·6
8	2650	67·75 Air / 32·25 H	130·97	6·717	−0·571	−0·040	−0·611	−0·466	+0·383	12	6	0·01	0·07	2·6
9	3800	4·07 / 95·93 H	100·72	6·781	+0·039	+0·002	+0·041	+0·041	+0·08	10	10	0·012	0·078	2·6
10	2700	58·29 Air / 41·71 H	144·02	6·846	−0·504	−0·035	−0·539	−0·375	+0·317	8·5	8	0·011	0·07	3·6
11	1900	91·81 / 8·19 H	152·67	7·406	−1·002	−0·099	−1·101	−0·721	+0·904	9	8	0	0·225	9·0
12	1760	97·56 Air / 2·44 H	138·55	7·474	−1·032	−0·11	−1·142	−0·825	+0·814	13	8	0·001	0·053	8·2
13	3100	4·375 Air / 95·625 H	87·74	88·66	+0·178	+0·011	+0·189	+0·215	+0·248	14	6	0·08	0·17	4·6
14	3300	6·08 Air / 93·92 H	91·52	92·951	+0·081	+0·005	+0·086	+0·094	+0·132	18	8	0·157	0·07	3·2
15	3000	5·043 Air / 94·957 H	73·99	90·353	+0·072	+0·005	+0·077	+0·104	+0·136	20	10	0·18	0·11	1·65
16	3000	2·99 / 97·01 H	85·15	89·242	+0·111	+0·007	+0·118	+0·139	+0·159	42	15	0·472	0·44	3·2
17	2900	4·13 Air / 95·87 H	104·72	89·858	+0·073	+0·004	+0·077	+0·073	+0·098	15·5	10	0·09	0·035	6·2

Remarks on the Tables.

The correction for conduction of heat through the plug, inserted in column 6 of Table I., and in column 7 of the rest of the Tables, was obtained from data furnished by experiments in which the difference between the temperature of the bath and the air was purposely made very great. It was considered as directly proportional to the difference of temperature, and inversely to the quantity of elastic fluid transmitted in a given time.

The 10th column of Tables II., III., IV., and V. is calculated on the hypothesis that, in mixtures with other gases, atmospheric air retains its thermal qualities without change. This hypothesis is almost certainly incorrect, since it is reasonable to expect that the effect of mixture on the physical character is experienced by each of the constituent gases. The column is given as one method of showing the effect of mixture.

Effect of Mixture on the Constituent Gases.—Although the experiments on nitrogen given in Table III. are not so numerous as might be desired, we may infer from them, and the results in Table II., that common air and all other mixtures of oxygen and nitrogen behave more like a perfect gas, *i.e.* give less cooling effect than either one or the other gas alone. We might expect the mixture to be something intermediate between the two. But this does not appear to be the case. The two are very nearly equal in their deviations from the condition of a perfect gas. Nitrogen deviates less than oxygen, but oxygen mixed with nitrogen differs less than nitrogen!

In the case of carbonic acid, which at low temperatures (7°) deviates five times as much as atmospheric air, we might expect that a mixture of CO_2 and air would deviate more than air and less than CO_2. This is the case (see Table IV.). Further, we might expect the two to contribute each its proportion of cooling effect according to its own amount, and its specific heat volume for volume. But do the mixtures exhibit such a result? No! See column 10, Table IV., in which also note, under experiments 8 and 9, the great diminution produced by the admixture of hydrogen.

If, instead of attributing to air and carbonic acid moments in proportion to their specific heats, or 1 : 1·39, as we have done in column 10, we use 1 : ·7, we obtain more consistent results.

Let δ denote the cooling effect experienced by air per 100 inches of mercury, δ' that by carbonic acid, and Δ that by a mixture of volume V of air, and V' of carbonic acid; then we may take

$$\Delta = \frac{m V \delta + m' V' \delta'}{m V + m' V'}$$

to represent the cooling effect for the mixture, where m and m' are numbers which we may call the moments (or importances) of the two in determining the cooling effect for the mixture. The ratio of m to m' is the proper result of each experiment on a mixture, if we knew with perfect accuracy the cooling effect for each gas with none of the other mixed. Now for common air we have direct experiments (Table I.), and know the cooling effect for it better than from any inferences from mixtures. But for pure CO_2 we know the effect, for the most part, only inferentially. Hence, having tried making $m : m' :: 1 : 1\cdot39$ without obtaining consistent results, we tried other proportions; and, after various attempts, found that $m : m' :: 1 : \cdot7$, for all temperatures and pressures within the limits of our experiments, gives results as consistent with one another as the probable errors of the experiments justify us in expecting. Thus, using the formula

$$\Delta = \frac{V \delta + V' \delta' \times \cdot7}{V + V' \times \cdot7},$$

we have, for calculating the effect for CO_2 from any experiment on a mixture, the following formula,

$$\delta' = \frac{(V + V' \times \cdot7) \Delta - V \delta}{V' \times \cdot7}.$$

Hence, using the numbers in columns 3 and 9 of Table IV. which relate to mixtures of air and carbonic acid alone, we find

TABLE VI.

No of experiment.	Proportions of mixtures.		Temperature of bath.	Thermal effect for air.	Deduced thermal effect for pure CO_2.
	Air	CO^2			
1	68·42	31·58	7·36	− ·88	− 4·51
2	89·16	10·84	7·36	− ·88	− 4·61
3	3·52	96·48	7·38	− ·88	·· 4·46
4	62·5	37·5	7·41	− ·88	− 4·19
5	88·13	11·87	7·43	− ·88	− 3·98
6	97·46	2·54	7·61	− ·88	− 3·89
16	1·83	98·17	35·6	− ·75	− 3·44
14	67·7	32·3	49·7	− ·7	− 3·04
15	87·77	12·23	49·76	− ·7	− 2·77
13	0·83	99·17	54	− ·66	− 2·96
10	2·11	97·89	93·52	− ·51	− 2·19
11	56·78	43·22	91·26	− ·51	− 2·21
12	77·77	22·23	91·64	− ·51	− 3·08
17	1·66	98·34	97·55	− ·49	− 2·16
1	2		3	4	5

The agreement for each set of results at temperatures nearly agreeing (with one exception, No. 12), shows that the assumption $m : m' :: 1 : ·7$ cannot be far wrong within our limits of temperature.

[Received subsequently to the reading of the Paper.]

Application of the preceding results to deduce approximately the Equation of Elasticity for the gases experimented on.

The "Equation of Elasticity" for any fluid is the most appropriate name for the equation expressing the relation between the pressure and the volume of any portion of the fluid. As this relation depends on the temperature, the equation expressing it involves essentially three variables, which, as in our previous communications on this subject, we shall denote by p, v, t. Of these, p is the pressure in units of force per unit of area, v the volume of a unit mass of the fluid, and t the temperature according to the absolute thermodynamic system of thermometry* which we have proposed. As before, we shall still adopt a degree, or thermometric unit, agreeing approximately with the degree Centigrade of the air-thermometer; according to which, as we have

* *Philosophical Transactions*, 1854, p. 350. [Present Article, Part II. sec. IV. above.]

demonstrated by experiment*, the value of t for the freezing-point is within a few tenths of a degree of 273·7 (its value at the standard boiling-point being, by definition of the Centigrade scale, 100° more than at the freezing-point).

Instead of, as in our previous communications, taking v and t as independent variables, we shall now take p and t; and we shall accordingly consider the object of the equation of elasticity as being to express v explicitly as a function of p and t. Whatever may be the relation between these elements, the thermal effect, $d\vartheta$ (reckoned as positive when it is a rise in temperature), produced by forcing the fluid in a continuous stream through a narrow passage or porous plug by an infinitely small difference of pressures, dp, will be given by the formula

$$\frac{d\vartheta}{dp} = -\frac{1}{JK}\left(t\frac{dv}{dt} - v\right),$$

where K denotes the thermal capacity, under constant pressure, of unit of mass of the fluid. This formula may be derived from equation (15) of our previous communication already referred to, by substituting p, v, and $-\vartheta$ for P, V, and δ in that equation, changing to p and t, instead of v and t, as independent variables, and differentiating with reference to p. It is scarcely necessary to remark that a direct demonstration of our present formula, founded on elementary thermodynamic principles, may be readily obtained.

Each experiment, of the several series recorded above, gives a value for $d\vartheta/dp$, which is found by multiplying the "corrected thermal effect" by $\frac{299218}{2114}$, to reduce from the amounts per 100 inches of mercury to the amounts per pound per square foot. Now by examining carefully the series of results for different temperatures, in the cases of atmospheric air and of carbonic acid, we find that they follow very closely the law of varying inversely as the square of the absolute temperature (or temperature Centigrade with 273·7 added). Thus for air the formula

$$0·92 \times \left(\frac{273·7}{t}\right)^2$$

and for carbonic acid $$4°·64 \times \left(\frac{273·7}{t}\right)^2$$

---

* *Philosophical Transactions*, 1854, p. 352. [Present Article, Part II. sec. IV. above.]

express, the former almost accurately, the latter with a deviation
which we shall hereafter investigate, the results through the whole
range of temperature for which the investigation has been
carried out.

Air.

Temperature.	Actual cooling effect.	Theoretical cooling effect.
$\overset{0}{0}$	$\overset{0}{\cdot}92$	$\overset{0}{\cdot}92$
7·1	·88	·87
39·5	·75	·70
92·8	·51	·51

Carbonic acid.

Temperature.	Actual cooling effect.	Theoretical cooling effect.
$\overset{0}{0}$	$\overset{0}{4}\cdot64$	$\overset{0}{4}\cdot64$
7·4	4·37	4·4
35·6	3·41	3·63
54·0	2·95	3·23
93·5	2·16	2·57
97·5	2·14	2·52

We have not experiments enough to establish the law of
variation with temperature of the thermal effect for the pure
gases oxygen and nitrogen, or for any stated mixture of them
other than common air; but there can be no doubt, from the
general character of the results, that the same law will be about
as approximately followed by them as it is by air.

Hence we may presume that in all these cases the cooling
effect is very well represented by the formula

$$-\frac{d\vartheta}{dp} = A\left(\frac{273\cdot7}{t}\right)^2.$$

Comparing this with the general formula given above, we find

$$t\frac{dv}{dt} - v = AJK\left(\frac{273\cdot7}{t}\right)^2.$$

The general integral of this differential equation, for v in terms
of t, is

$$v = Pt - \tfrac{1}{3}AJK\left(\frac{273\cdot7}{t}\right)^2,$$

P denoting an arbitrary constant with reference to t, which, so far as this integration is concerned, may be an arbitrary function of p. To determine its form, we remark in the first place, in consequence of Boyle's law, that it must be approximately C/p, C being independent of both pressure and temperature; and thus, if we omit the second term, we have two gaseous laws expressed by the approximate equation $v = Ct/p$.

Now it is generally believed [but is certainly false]* that at higher and higher temperatures the gases approximate more and more nearly to the rigorous fulfilment of Boyle's law. If this is true, the complete expression for P must be of the form C/p, since any other would simply show a deviation from Boyle's law at very high temperatures, when the second term of our general integral disappears. Assuming then that no such deviation exists, we have, as the complete solution

$$v = \frac{Ct}{p} - \tfrac{1}{3}AJK \left(\frac{273 \cdot 7}{t}\right)^{2}.$$

This is an expression of exactly the same form as that which Professor Rankine found applicable to carbonic acid, in the first place to express its deviations from the laws of Boyle and Gay-Lussac, as shown by Regnault's experiments, and which he afterwards proved to give correctly the law and the absolute amount of the cooling effect demonstrated by our first experiments on that gas†.

That more complicated formulæ were found for the law of elasticity for common air both by Mr Rankine and by ourselves, now seems to be owing to an irreconcileability among the data we had from observation. The whole amounts of the deviations from the gaseous laws are so small, for common air, that very small absolute errors in observations of so heterogeneous a character as those of Regnault on the law of compression and on the changes produced by pressure in the coefficients of expansion, and our own on the thermo-dynamic property on which we have experimented, may readily present us with results either absolutely inconsistent with one another, or only reconcileable by very strained assumptions. It is satisfactory now to find, when we have succeeded in extending our observations through a con-

* Note of Aug. 1879.

† *Philosophical Transactions*, 1854, Part II. p. 336.

siderable range of temperature, that they lead to so simple a law ; and it is probable that the formula we have been led to by these observations alone, will give the deviations from Boyle's law, and the changes produced by pressure in the coefficients of expansion, with more accuracy than has hitherto been attained in attempts to determine these deviations by direct observation. We must, however, reserve for a future communication the comparison between such results of our theory and experiments, and Regnault's direct observations. In the mean time we conclude by putting the integral equation of elasticity into a more convenient form, by taking $C = \mathfrak{H}/t_0$, where $\mathfrak{H}$ denotes the "height of the homogeneous atmosphere" for the gas under any excessively small pressure, at any temperature t_0, and taking t_0 to denote the absolute temperature of freezing water, in which case we shall have, as nearly as observations hitherto made allow us to determine, $t_0 = 273°.7$. Then, in terms of this notation, and of that above explained, in which t, p, v denote absolute temperature, pressure in pounds weight per square foot, and volume in cubic feet of one pound of air, the equation of elasticity investigated above becomes

$$v = \frac{\mathfrak{H}t}{pt_0} - \tfrac{1}{3}AJK\left(\frac{t_0}{t}\right)^2,$$

where A denotes the amount of the thermal effect per pound per square foot for temperature t_0, determined by our observations, reckoned positive when it is a depression of temperature.

APPENDIX.

ABSTRACT OF PART I. ABOVE.

[From the *Proceedings of the Royal Society*, Vol. VI., June, 1853.]

ON THE THERMAL EFFECTS OF ELASTIC FLUIDS.

The authors had already proved by experiments conducted on a small scale, that when dry atmospheric air, exposed to pressure, is made to percolate a plug of non-conducting porous material, a depression of temperature takes place increasing in some proportion with the pressure of the air in the receiver. The numerous sources of error which were to be apprehended in experiments of this kind conducted on a small scale, induced the authors to apply for the means of executing them on a larger scale ; and the present paper contains the introductory part of their researches with apparatus furnished by the Royal Society, comprising a force pump worked by a steam-engine and capable of propelling 250 cubic inches of air per second, and a series of tubes by which the elastic fluid is conveyed through a bath of water, by which its temperature is regulated, a flange at the terminal permitting the attachment of any nozzle which is desired.

Preliminary experiments were made in order to illustrate the thermal phenomena which result from the rush of air through a single aperture. Two effects were anticipated, one of heat arising from the *vis viva* of air in rapid motion, the other of cold arising from dilatation of the gas and the consequent conversion of heat into mechanical effect. The latter was exhibited by placing the bulb of a very small thermometer close to a small orifice through which dry atmospheric air, confined under a pressure of 8 atmospheres, was permitted to escape. In this case the thermometer was depressed 13° Cent. below the temperature of the bath. The former effect was exhibited by causing the stream of air as it issued from the orifice to pass in a very narrow stream between the bulb of the thermometer and a piece of gutta percha tube in which the latter was enclosed. In this experiment, with a pressure of 8 atmospheres, an elevation of temperature equal to 23 Cent. was observed. The same phenomenon was even more strikingly exhibited by pinching the rushing stream with the finger and thumb, the heat resulting therefrom being insupportable.

The varied effects thus exhibited in the "rapids" neutralize one another at a short distance from the orifice, leaving however a small cooling effect, to ascertain the law of which and its amount for various gases, the present researches have principally been instituted. A plug of cotton wool was employed, for the purpose at once of preventing the escape of thermal effect in the rapids, and of mechanical effect in the shape of sound. With this arrangement a depression of $0°·31$ Cent. was observed, the temperature of the dry atmospheric air in the receiver being $14°·5$ Cent., and its pressure $34·4$ lbs. on the square inch, and the pressure of the atmosphere being $14·7$ lbs. per square inch.

ABSTRACT OF PART II. ABOVE.

[From the *Proceedings of the Royal Society*, Vol. VII., June, 1854.]

ON THE THERMAL EFFECTS OF FLUIDS IN MOTION.

The first experiments described in this paper show that the anomalies exhibited in the last table of experiments, in the paper preceding it[*], are due to fluctuations of temperature in the issuing stream consequent on a change of the pressure with which the entering air is forced into the plug. It appears from these experiments, that when a considerable alteration is suddenly made in the pressure of the entering stream, the issuing stream experiences remarkable successions of augmentations and diminutions of temperature, which are sometimes perceptible for half-an-hour after the pressure of the entering stream has ceased to vary.

Several series of experiments are next described in which air is forced (by means of the large pump and other apparatus described in the first paper) through a plug of cotton wool, or unspun silk pressed together, at pressures varying in their excess above the atmospheric pressure, from five or six up to fifty or sixty pounds on the square inch. By these it appears that the cooling effect which the air, as found in the authors' previous experiments, always experiences in passing through the porous plug, varies proportionally to the excess of the pressure of the air on entering the plug above

---

[*] Communicated to the Royal Society, June 1853, and published in the *Transactions*. [Present Article, Part I. above.]

that with which it is allowed to escape. Seven series of experiments, in each of which the air entered the plug at a temperature of about 16° Cent., gave a mean cooling effect of about ·0175° Cent., per pound on the square inch, or ·27° Cent. per atmosphere, of difference of pressure. Experiments made at lower and at higher temperatures showed that the cooling effect is very sensibly less for high than for low temperatures, but have not yet led to sufficiently exact results at other temperatures than that stated (16° Cent.), to indicate the law according to which it varies with the temperature.

Experiments on carbonic acid at different temperatures are also described, which show that at about 16° Cent., this gas experiences $4\frac{1}{2}$ times as great a cooling effect as air. They agree well at all the different temperatures with a theoretical result derived according to the general dynamical theory from an empirical formula for the pressure of carbonic acid in terms of its temperature and density, which was kindly communicated by Mr Rankine to the authors, having been investigated by him upon no other experimental data than those of Regnault on the expansion of the gas by heat and its compressibility.

Experiments were also made on hydrogen gas, which, although not such as to lead to accurate determinations, appeared to indicate very decidedly a cooling effect* amounting to a small fraction, perhaps about $\frac{1}{16}$, of that which air would experience in the same circumstances.

The following theoretical deductions from these experiments are made:—

I. The relations between the heat generated and the work spent in compressing carbonic acid, air and hydrogen, are investigated from the experimental results. In each case the relation is nearly that of equivalence, but the heat developed exceeds the equivalent of the work spent, by a very small amount for hydrogen, considerably more for air, and still more for carbonic acid. For slight compressions with the gases kept about the temperature 16°, this excess amounts to about $\frac{1}{77}$ of the whole heat emitted in the case of carbonic acid, and $\frac{1}{420}$ in the case of air.

II. It is shown by the general dynamical theory, that the air experiments, taken in connexion with Regnault's experimental

---

* Subsequent experiments showed this to be a mistake. See Note to this passage in Part II. above.

results on the latent heat and pressure of saturated steam, make it certain that the density of saturated steam increases very much more with the pressure than according to Boyle's and Gay-Lussac's gaseous laws, and numbers are given expressing the theoretical densities of saturated steam at different temperatures, which it is desired should be verified by direct experiments.

III. Carnot's function in the "Theory of the Motive Power of Heat" is shown to be very nearly equal to the mechanical equivalent of the thermal unit divided by the temperature from the zero of the air-thermometer (that is, temperature Centigrade with a number equal to the reciprocal of the coefficient of expansion added), and corrections, depending on the amount of the observed cooling effects in the new air experiments, and the deviations from the gaseous laws of expansion and compression determined by Regnault, are applied to give a more precise evaluation.

IV. An absolute scale of temperature, that is, a scale not founded on reference to any particular thermometric substance or to any special qualities of any class of bodies, is founded on the following definition:—

If a physical system be subjected to cycles of perfectly reversible operations and be not allowed to take in or to emit heat except in localities, at two fixed temperatures, these temperatures are proportional to the whole quantities of heat taken in or emitted at them respectively during a complete cycle of the operations.

The principles upon which the unit or degree of temperature is to be chosen, so as to make the difference of temperatures on the absolute scale, agree with that on any other scale for a particular range of temperatures. If the difference of temperatures between the freezing and the boiling-points of water be made 100° on the new scale, the absolute temperature of the freezing-point is shown to be about $273\cdot7$; and it is demonstrated that the temperatures from the freezing-point on the new scale will agree very closely with Centigrade temperature by the standard air-thermometer; quite within the limits of the most accurate practical thermometry when the temperature is between 0° and 100° Cent., and very nearly if not quite within these limits for temperatures up to 300° Cent.

V. An empirical formula for the pressure of air in terms of its density, and its temperature on the absolute scale, is investigated,

28—2

by using forms such as those first proposed and used by Mr Rankine, and determining the constants so as to fulfil the conditions (1) of giving the observed cooling effects, (2) of agreeing with Regnault's observations on expansion by heat, and (3) of agreeing with Regnault's experimental results on compressibility at a particular temperature.

A table of comparison of temperature by the air-thermometer under varied conditions of temperature and pressure with the absolute scale, is deduced from this formula.

Expressions for the specific heats of any fluid in terms of the absolute temperature, the density, and the pressure, derived from the general dynamical theory, are worked out for the case of air according to the empirical formula; and tables of numerical results derived exclusively from these expressions and the ratio of the specific heats as determined by the theory of sound, are given. These tables show the mechanical values of the specific heats of air at different constant pressures, and at different constant densities. Taking 1390 as the mechanical equivalent of the thermal unit as determined by Mr Joule's experiment on the friction of fluids, the authors find, as the mean specific heat of air under constant pressure,

·2390, from 0° to 100° Cent.
·2384, from 0° to 300° Cent.

[From the *Proceedings of the Royal Society*, Vol. VIII., February, 1856.]

A VERY great depression of temperature has been remarked by some observers when steam of high pressure issues from a small orifice into the open air. After the experiments we have made on the rush of air in similar circumstances, it could not be doubted that a great elevation of temperature of the issuing steam might be observed as well as the great depression usually supposed to be the only result. The method to obtain the entire thermal effect is obviously that which we have already employed in our experiments on permanently elastic fluids, viz. to transmit the steam through a porous material and to ascertain its temperature as it enters into and issues from the resisting medium. We have made a preliminary experiment of this kind which may be sufficiently interesting to place on record before proceeding to obtain more exact numerical results.

A short pipe an inch and a half diameter was screwed into an elbow pipe inserted into the top of a high pressure steam-boiler. A cotton plug placed in the short pipe had a fine wire of platina passed through it, the ends of which were connected with iron wires passing away to a sensitive galvanometer. The deflection due to a given difference of temperature of the same metallic junctions having been previously ascertained, we were able to estimate the difference of temperature of the steam at the opposite ends of the plug. The result of several experiments showed that for each lb. of pressure by which the steam on the pressure side exceeded that of the atmosphere on the exit side there was a cooling effect of 0·2 Cent. The steam, therefore, issued at a temperature above 100° Cent., and, consequently, *dry;* showing the correctness of the view which we brought forward some years ago * as to the non-scalding property of steam issuing from a high-pressure boiler.

[From the *Proceedings of the Royal Society*, Vol. VIII., June, 1856.]

On the Temperature of Solids exposed to Currents of Air.

IN examining the thermal effects experienced by air rushing through narrow passages, we have found, in various parts of the stream, very decided indications of a lowering of temperature (see *Phil. Trans.* June, 1853) [present Article, Part I., above], but never nearly so great as theoretical considerations at first led us to expect, in air forced by its own pressure into so rapid motion as it was in our experiments. The theoretical investigation is simply as follows:—Let P and V denote the pressure and the volume of a pound of the air moving very slowly up a wide pipe towards the narrow passage. Let p and v denote the pressure and the volume per pound in any part of the narrow passage, where the velocity is q. Let also $e - E$ denote the difference of intrinsic energies of

---

* See a letter from Mr Thomson to Mr Joule, published in the *Philosophical Magazine*, Nov. 1850. [Art. XLVII. above.]

the air per pound in the two situations. Then the equation of mechanical effect is

$$\frac{q^2}{2g} = (PV - pv) + (E - e),$$

since the first member is the mechanical value of the motion, per pound of air; the first bracketed term of the second member is the excess of work done in pushing it forward, above the work spent by it in pushing forward the fluid immediately in advance of it in the narrow passage; and the second bracketed term is the amount of intrinsic energy given up by the fluid in passing from one situation to the other.

Now, to the degree of accuracy to which air follows Boyle's and Gay-Lussac's laws, we have

$$pv = \frac{t}{T} PV,$$

if t and T denote the temperatures of the air in the two positions reckoned from the absolute zero of the air-thermometer. Also, to about the same degree of accuracy, our experiments on the temperature of air escaping from a state of high pressure through a porous plug, establish Mayer's hypothesis as the thermo-dynamic law of expansion; and to this degree of accuracy we may assume the intrinsic energy of a mass of air to be independent of its density when its temperature remains unaltered. Lastly, Carnot's principle, as modified in the dynamical theory, shows that a fluid which fulfils those three laws must have its capacity for heat in constant volume constant for all temperatures and pressures,—a result confirmed by Regnault's direct experiments to a corresponding degree of accuracy. Hence the variation of intrinsic energy in a mass of air is, according to those laws, simply the difference of temperatures multiplied by a constant, irrespectively of any expansion or condensation that may have been experienced. Hence, if N denote the capacity for heat of a pound of air in constant volume, and J the mechanical value of the thermal unit, we have

$$E - e = JN(T - t).$$

Thus the preceding equation of mechanical effect becomes

$$\frac{q^2}{2g} = PV\left(1 - \frac{t}{T}\right) + JN(T - t).$$

Now (see "Notes on the Air-Engine," *Phil. Trans.* March, 1852, p. 81 [Art. XLVIII., Appendix III. above], or "Thermal Effects of Fluid in Motion" [present Article], Part 2, *Phil. Trans.* June, 1854, p. 361) we have

$$JN = \frac{1}{k-1}\frac{H}{t_0} = \frac{1}{k-1}\frac{PV}{T},$$

where k denotes the ratio of the specific heat of air under constant pressure to the specific heat of air in constant volume; H, the product of the pressure into the volume of a pound, or the "height of the homogeneous atmosphere" for air at the freezing-point (26,215 feet, according to Regnault's observations on the density of air), and t_0 the absolute temperature of freezing (about 274^0 Cent.). Hence we have

$$\frac{q^2}{2g} = PV\left(1 + \frac{1}{k-1}\right)\left(1 - \frac{t}{T}\right) = \frac{kPV}{k-1}\left(1 - \frac{t}{T}\right).$$

Now the velocity of sound in air at any temperature is equal to the product of $\sqrt{k}$ into the velocity a body would acquire in falling under the action of a constant force of gravity through half the height of the homogeneous atmosphere; and therefore if we denote by α the velocity of sound in air at the temperature T, we have

$$\alpha^2 = kgPV.$$

Hence we derive from the preceding equation,

$$\frac{T - t}{T} = \frac{k-1}{2}\left(\frac{q}{\alpha}\right)^2,$$

which expresses the lowering of temperature, in any part of the narrow channel, in terms of the ratio of the actual velocity of the air in that place to the velocity of sound in air at the temperature of the stream where it moves slowly up towards the rapids. It is to be observed, that the only hypothesis which has been made is, that in all the states of temperature and pressure through which it passes the air fulfils the three gaseous laws mentioned above; hence whatever frictional resistance, or irregular action from irregularities in the channel, the air may have experienced before coming to the part considered, provided only it has not been allowed either to give out heat or to take in heat from the matter round it, nor to lose any mechanical energy in sound, or in other motions not among its own particles, the preceding formulæ will give the lowering of temperature it experiences in acquiring the velocity q. It is to be observed that this is not the velocity the air would have in issuing in the same quantity at the density which it has in the slow stream approaching the narrow passage. Were no fluid friction operative in the circumstances, the density and pressure would be the same in the slow stream flowing away from, and in the slow stream approaching towards the narrow passage; and each would be got by considering the lowering of temperature from T to t as simply due to expansion, so that we should have

$$\frac{t}{T} = \left(\frac{V}{v}\right)^{k-1}$$

by Poisson's formula. Hence if Q denote what we may call the "reduced velocity" in any part of the narrow channel, as distinguished from q, the actual or true velocity in the same locality, we have

$$Q = \frac{V}{v}\, q = \left(\frac{t}{T}\right)^{\frac{1}{k-1}} q,$$

and the rate of flow of the air will be, in pounds per second, wQA, if w denote the weight of the unit of volume, under pressure P, and A the area of the section in the narrow part of the channel considered. The preceding equation, expressed in terms of the "reduced velocity," then becomes

$$1 - \frac{t}{T} = \frac{k-1}{2} \left(\frac{T}{t}\right)^{\frac{2}{k-1}} \left(\frac{Q}{\alpha}\right)^{2},$$

and therefore we have

$$\frac{Q}{\alpha} = \sqrt{\left\{ \frac{2}{k-1} \left(\frac{t}{T}\right)^{\frac{2}{k-1}} \left(1 - \frac{t}{T}\right) \right\}}.$$

The second member, which vanishes when $t = 0$, and when $t = T$, attains a maximum when

$$t = \cdot 83T,$$

the maximum value being

$$\frac{Q}{\alpha} = \cdot 578.$$

Hence, if there were no fluid friction, the "reduced velocity" could never, in any part of a narrow channel, exceed ·578 of the velocity

of sound, in air of the temperature which the air has in the wide
parts of the channel, where it is moving slowly. If this temperature be 13° Cent. above the freezing-point, or 287° absolute
temperature (being 55° Fahr., an ordinary atmospheric condition),
the velocity of sound would be 1115 feet per second, and the
maximum reduced velocity of the stream would be 644 feet per
second. The cooling effect that air must, in such circumstances,
experience in acquiring such a velocity would be from 287° to 238°·2
absolute temperature, or 48°·8 Cent.

The effects of fluid friction in different parts of the stream
would require to be known in order to estimate the reduced
velocity in any narrow part, according to either the density on the
high-pressure side or the density on the low-pressure side. We
have not as yet made any sufficient investigation to allow us to
give even a conjectural estimate of what these effects may be in
any case. But it appears improbable that the "reduced velocity,"
according to the density on the high-pressure side, could ever with
friction exceed the greatest amount it could possibly have without
friction. It therefore seems improbable that the "reduced velocity" in terms of the density on the high-pressure side can ever,
in the narrowest part of the channel, exceed 644 feet per second,
if the temperature of the high-pressure air moving slowly be about
the atmospheric temperature of 13° Cent. used in the preceding
estimate.

Experiments in which we have forced air through apertures of
$\frac{29}{1000}$, $\frac{53}{1000}$, and $\frac{84}{1000}$ths of an inch in diameter drilled in thin
plates of copper, have given us a maximum velocity, reduced to the
density of the high-pressure side, equal to 550 feet per second.
But there can be little doubt that the stream of air, after issuing
from an orifice in a thin plate, contracts as that of water does
under similar circumstances. If the velocity were calculated from
the area of this contracted part of the stream, it is highly probable
that the maximum velocity reduced to the density on the high-
pressure side would be found as near 644 feet as the degree of
accuracy of the experiments warrants us to expect.

As an example of the results we have obtained on examining
the temperature of the rushing stream by a thermo-electric junction
placed $\frac{1}{8}$th of an inch above the orifice, we cite an experiment, in
which the total pressure of the air in the receiver being 98 inches
of mercury, we found the velocity in the orifice equal to 535 and

1780 feet respectively as reduced to the density on the high-pressure and that on the atmospheric side. The actual velocity in the small aperture must have been greater than either of these, perhaps not much greater than 1780, the velocity reduced to atmospheric density. If it had been only this, the cooling effect would have been exactly $T\dfrac{k-1}{2}\left(\dfrac{1780}{1115}\right)^{2}$, that is, a lowering of temperature amounting to 150° Cent. But the amount of cooling effect observed in the experiment was only 13° Cent.; nor have we ever succeeded in observing (whether with thermometers held in various positions in the stream, or with a thermo-electric arrangement constituted by a narrow tube through which the air flows, or by a straight wire of two different metals in the axis of the stream, with the junction in the place of most rapid motion, and in other positions on each side of it,) a greater cooling effect than 20° Cent.; we therefore infer *that a body round which air is flowing rapidly acquires a higher temperature than the average temperature of the air close to it all round.* The explanation of this conclusion probably is, that the surface of contact between the air and the solid is the locality of the most intense frictional generation of heat that takes place, and that consequently a stratum of air round the body has a higher average temperature than the air further off; but whatever the explanation may be, it appears certainly demonstrated that the air does not give its own temperature even to a tube through which it flows, or to a wire or thermometer-bulb completely surrounded by it.

Having been convinced of this conclusion by experiments on rapid motion of air through small passages, we inferred of course that the same phenomenon must take place universally whenever air flows against a solid or a solid is carried through air. If a velocity of 1780 feet per second in the foregoing experiment gave 137° Cent. difference of temperature between the air and the solid, how probable is it that meteors moving at from six to thirty miles per second even through a rarefied atmosphere, really acquire, in accordance with the same law, all the heat which they manifest! On the other hand, it seemed worth while to look for the same kind of effect on a much smaller scale in bodies moving at moderate velocities through the ordinary atmosphere. Accordingly, although it has been a practice in general undoubtingly followed, to whirl a thermometer through the air for the purpose of finding the atmospheric temperature, we have tried and found, with thermometers of different sizes and variously shaped bulbs, whirled through the air at the end of a string, with velocities of from 80 to 120 feet per second, temperatures always higher than when the same thermometers are whirled in exactly the same circumstances at smaller velocities. By alternately whirling the same thermometers for half a minute or so fast, and then for a similar time slow, we have found differences of temperature sometimes little if at all short of a Fahrenheit degree. By whirling a thermo-electric junction alternately fast and slow, the same phenomenon is most satisfactorily and strikingly exhibited by a galvanometer. This last experiment we have performed at night, under a cloudy sky, with the galvanometer within doors, and the testing thermo-electric apparatus whirled in the middle of a field; and thus, with as little as can be conceived of disturbing circumstances, we confirmed the result we had previously found by whirling thermometers.

Velocity of Air escaping through Narrow Apertures*.

In the foregoing part of this communication, referring to the circumstances of certain experiments, we have stated our opinion that the velocity of atmospheric air impelled through narrow orifices was, in the narrowest part of the stream, greater than the reduced velocity corresponding to the atmospheric pressure; in other words, that the density of the air, kept at a constant temperature, was, in the narrowest part, less than the atmospheric density. In order to avoid misconception, we now add, that this holds true only when the difference of pressures on the two sides is small, and friction plays but a small part in bringing down the velocity of the exit stream. If there is a great difference between the pressures on the two sides, the reduced velocity will, on the contrary, be *less* than that corresponding with the atmospheric pressure; and even if the pressure in the most rapid part falls short of the atmospheric pressure, the density may, on account of the cooling experienced, exceed the atmospheric density.

We stated that, at 57° Fahr., the greatest velocity of air passing through a small orifice is 550 feet per second, if reduced to the

* Received June 19, 1856.

density on the high-pressure side. The experiments from which we obtained this result enable us also to say that this maximum occurs, with the above temperature and a barometric pressure of 30·14 inches, when the pressure of the air is equal to about 50 inches of mercury above the atmospheric pressure. At a higher or lower pressure, a smaller volume of the compressed air escapes in a given time.

Surface Condenser.—A three-horse power high-pressure steam-engine was procured for our experiments. Wishing to give it equal power with a lower pressure, we caused the steam from the eduction port to pass downwards through a perpendicular iron gas-pipe, ten feet long and an inch and a half in diameter, placed within a larger pipe through which water was made to ascend. The lower end of the gas-pipe was connected with the feed-pump of the boiler, a small orifice being contrived in the pump cover in order to allow the escape of air before it could pass, along with the condensed water, into the boiler. This simple arrangement constituted a "surface condenser" of a very efficient kind, giving a vacuum of 23 inches, although considerable leakage of air took place, and the apparatus generally was not so perfect as subsequent experience would have enabled us to make it. Besides the ordinary well-known advantages of the "surface condenser," such as the prevention of incrustation of the boiler, there is one which may be especially remarked as appertaining to the system we have adopted, of causing the current of steam to move in an opposite direction to that of the water employed to condense it. The refrigerating water may thus be made to pass out of the condenser at a high temperature, while the vacuum is that due to a low temperature; and hence the quantity of water used for the purpose of condensation may be materially reduced. We find that our system does not require an amount of surface so great as to involve a cumbrousness or cost which would prevent its general adoption, and have no doubt that it will shortly supersede that at the present time almost universally used.

[From the *Proceedings of the Royal Society*, Vol. VIII., June, 1857.]

Temperature of a Body moving slowly through Air.

THE motion of air in the neighbourhood of a body moving very slowly through it, may be approximately determined by treating the problem as if air were an incompressible fluid. The ordinary hydro-dynamical equations, so applied, give the velocity and the pressure of the fluid at any point; and the variations of density and temperature actually experienced by the air are approximately determined by using the approximate evaluation of the pressure thus obtained. Now, if a solid of any shape be carried uniformly through a perfect liquid*, it experiences fluid-pressure at different parts of its surface, expressed by the following formula,—

$$p = \Pi + \tfrac{1}{2}\rho \left(V^2 - q^2 \right),$$

where Π denotes the fluid-pressure at considerable distances from the solid, ρ the mass of unity of volume of the fluid, V the velocity of translation of the solid, and q the velocity of the fluid relatively to the solid, at the point of its surface in question. The effect of this pressure on the whole is, no resultant force, and only a resultant couple which vanishes in certain cases, including all in which the solid is symmetrical with reference to the direction of motion. If the surface of the body be everywhere convex, there will be an augmentation of pressure in the fore and after parts of it, and a diminution of pressure round a medium zone. There are clearly in every such case just two points of the surface of the solid, one in the fore part, and the other in the after part, at which the velocity of the fluid relatively to it is zero, and which we may call the fore and after pole respectively. The middle region round the body in which the relative velocity exceeds V, and where consequently the fluid pressure is diminished by the motion, may be called the equatorial zone; and where there is a definite middle line, or line of maximum relative velocity, this line will be called the equator.

If the fluid be air instead of the ideal " perfect liquid," and if the motion be slow enough to admit of the approximation referred to above, there will be a heating effect on the fore and after parts

---

* That is, as we shall call it for brevity, an ideal fluid, perfectly incompressible and perfectly free from mutual friction among its parts.

of the body, and a cooling effect on the equatorial zone. If the dimensions and the thermal conductivity of the body be such that there is no sensible loss on these heating and cooling effects by conduction, the temperature maintained at any point of the surface by the air flowing against it, will be given by the equation

$$t = \Theta \left(\frac{p}{\Pi}\right)^{\frac{\cdot41}{1\cdot41}},$$

where Θ denotes the temperature of the air as uninfluenced by the motion, and p and Π denote the same as before*. Hence, using for p its value by the preceding equation, we have

$$t = \Theta \left\{1 + \frac{\rho}{2\Pi}(V^2 - q^2)\right\}^{\frac{\cdot41}{1\cdot41}}.$$

But if H denote the length of a column of homogeneous atmosphere of which the weight is equal to the pressure on its perpendicular section, and if g denote the dynamical measure of the force of gravity (32·2 in feet per second of velocity generated per second), we have

$$g\rho H = \Pi \, ;$$

and if we denote by α the velocity of sound in the air, which is equal to $\sqrt{1\cdot41 \times gH}$, the expression for the temperature becomes

$$t = \Theta \left\{1 + \frac{1\cdot41}{2} \cdot \frac{V^2 - q^2}{\alpha^2}\right\}^{\frac{\cdot41}{1\cdot41}}.$$

According to the supposition on which our approximation depends, that the velocity of the motion is small, that is, as we now see, a small fraction of the velocity of sound, this expression becomes

$$t = \Theta \left\{1 + \cdot41 \times \frac{V^2 - q^2}{2\alpha^2}\right\}.$$

At either the fore or after pole, or generally at every point where the velocity of the air relatively to the solid vanishes (at a

---

* The temperatures are reckoned according to the absolute thermodynamic scale which we have proposed, and may, to a degree of accuracy correspondent with that of the ordinary "gaseous laws," be taken as temperature Centigrade by the air-thermometer, with 273°·7 added in each case. See the authors' previous paper "On the Thermal Effects of Fluids in Motion," [present Article] Part II., *Philosophical Transactions*, 1854, Part 2, p. 353.

re-entrant angle for instance, if there is such), we have $q = o$, and therefore an elevation of temperature amounting to

$$\cdot 41 \times \frac{V^2}{2a^2}\,\Theta.$$

If, for instance, the absolute temperature, Θ, of the air at a distance from the solid be 287° (that is 55° on the Fahr. scale), for which the velocity of sound is 1115 per second, the elevation of temperature at a pole, or at any point of no relative motion, will be, in degrees Centigrade,

$$58^{0\cdot}8 \times \left(\frac{V}{a}\right)^2, \text{ or } 58^{0\cdot}8 \times \left(\frac{V}{1115}\right)^2,$$

the velocity V being reckoned in feet per second. If, for instance, the velocity of the body through the air be 88 feet per second (60 miles an hour). the elevation of temperature at the points of no relative motion is $\cdot 36°$, or rather more than $\frac{1}{3}$ of a degree Centigrade.

To find the greatest depression of temperature in any case, it is necessary to take the form of the body into account. If this be spherical, the absolute velocity of the fluid backwards across the equator will be half the velocity of the ball forwards; or the relative velocity (q) of the fluid across the equator will be $\frac{3}{2}$ of the velocity of the solid. Hence the depression of temperature at the equator of a sphere moving slowly through the air will be just $\frac{9}{4}$ of the elevation of temperature at each pole. It is obvious from this that a spheroid of revolution, moving in the direction of its axis, would experience at its equator a depression of temperature, greater if it be an oblate spheroid, or less if it be a prolate spheroid, than $\frac{9}{4}$ of the elevation of temperature at each pole.

It must be borne in mind, that, besides the limitation to velocities of the body, small in comparison with the velocity of sound, these conclusions involve the supposition that the relative motions of the different parts of the air are unresisted by mutual friction, a supposition which is not even approximately true in most cases that can come under observation. Even in the case of a ball pendulum vibrating in air, Professor Stokes* finds that the motion is seriously influenced by fluid friction. Hence with velo-

* " On the Effect of the Internal Friction of Fluids on the Motion of Pendulums," read to the Cambridge Philosophical Society, Dec. 9, 1850, and published in Vol. IX. Part 2, of their *Transactions*, [or *Mathematical and Physical Papers*, Stokes, Vol. II.].

cities which could give any effect sensible on even the most delicate of the ether thermometers yet made (330 divisions to a degree), it is not to be expected that anything like a complete verification or even illustration of the preceding theory, involving the assumption of no friction, can be had. It is probable that the forward polar region of heating effect will, in consequence of fluid friction, become gradually larger as the velocity is increased, until it spreads over the whole equatorial region, and does away with all cooling effects.

Our experimental inquiry has hitherto been chiefly directed to ascertain the law of the thermal effect upon a thermometer rapidly whirled in the air. We have also made some experiments on the modifying effects of resisting envelopes, and on the temperatures at different parts of the surface of a whirled globe. The whirling apparatus consisted of a wheel worked by hand, communicating rapid rotation to an axle, at the extremity of which an arm carrying the thermometer with its bulb outwards was fixed. The distance between the centre of the axle and the thermometer bulb was in all the experiments 39 inches. The thermometers made use of were filled with ether or chloroform, and had, the smaller 275, and the larger 330 divisions to the degree Centigrade. The lengths of the cylindrical bulbs were $\frac{9}{10}$ and $1\frac{4}{10}$ inch, their diameters ·26 and ·48 of an inch respectively.

TABLE I.—*Small bulb Thermometer.*

Velocity in feet per second.	Rise of temperature in divisions of the scale.	Rise divided by square of velocity.
46·9	27½	·0125
51·5	32	·0121
68·1	46½	·0100
72·7	57½	·0109
78·7	67½	·0109
84·8	74	·0103
104·5	91	·0083
130·2	151	·0089
133·2	172	·0097
145·4	191	·0090
	Mean...	·01026

The above Table shows an increase of temperature nearly proportional to the square of the velocity.

$$V = \sqrt{\frac{275}{\cdot01026}} = 163\cdot7 = \text{the velocity in feet per second, which,}$$

in air of the same density, would have raised the temperature 1° Centigrade.

TABLE II.—*Larger bulb Thermometer.*

Velocity in feet per second.	Rise of temperature in divisions of scale.	Rise divided by square of velocity.
36·3	18	·0125
66·6	42	·0095
84·8	57	·0079
125·6	146	·0093
	Mean...	·0098

In this instance $V = \sqrt{\dfrac{330}{\cdot0098}} = 183\cdot5$ feet per second for 1° Centigrade. It is however possible that the full thermal effect was not so completely attained in three minutes (the time occupied by each whirling) as with the smaller bulb. On the whole it did not appear to us that the experiments justified the conclusion, that an increase of the dimensions of the bulb was accompanied by an alteration of the thermal effect.

TABLE III.—*Larger bulb Thermometer covered with five folds of writing-paper.*

Velocity in feet per second.	Rise of temperature in divisions of scale.	Rise divided by square of velocity.
36·3	20	·0152
51·5	43	·0162
72·6	53	·0101
118.0	132	·0095

The increased thermal effect at comparatively slow velocities, exhibited in the above Table, appeared to be owing to the friction of the air against the paper surface being greater than against the polished glass surface.

One quarter of the enveloping paper was now removed, and the bulb whirled with its bared part in the rear. The results were as follow :—

T.   29

TABLE IV.—*Paper removed from posterior side.*

Velocity in feet per second.	Rise of temperature in divisions of scale.	Rise divided by square of velocity.
75·6	60	·0105
96·8	87	·0093

On whirling in the contrary direction, so that the naked part of the bulb went first, we got,—

TABLE V.—*Paper removed from anterior side.*

Velocity in feet per second.	Rise of temperature in divisions of scale.	Rise divided by square of velocity.
81·7	56	·0084
93·8	72	0082

On rotating with the bare part, posterior and anterior in turns, at the constant velocity of 90 feet per second, the mean result did not appear to indicate any decided difference of thermal effect.

Another quarter of paper was now removed from the opposite side. Then on whirling so that the bared parts were anterior and posterior, we obtained a rise of 83 divisions with a velocity of 93·8. But on turning the thermometer on its axis one quarter round, so that the bared parts were on each side, we found the somewhat smaller rise of 62 divisions for a velocity of 90·8 feet per second.

The effect of surface friction having been exhibited at slow velocities with the papered bulb, we were induced to try the effect of increasing it by wrapping iron wire round the bulb.

TABLE VI.—*Larger bulb Thermometer wrapped with iron wire.*

Velocity in feet per second.	Rise in divisions of scale.	Rise divided by square of velocity.
15·36	10·25	·0434
23·04	33	·0623
30·71*	49·25	·0522
46·08	68·75	·0324
69·12	98	·0206
111·34	185	·0149
126·72	207	·0129
153·55	above 280	above ·0118

* The whirring sound began at this velocity. According to its intensity the thermal effect must necessarily suffer diminution; unless indeed it gives rise to increased resistance.

On inspecting the above Table, it will be seen that the thermal effect produced at slow velocities was five times as great as with the bare bulb. This increase is evidently due to friction. In fact, as one layer of wire was employed, and the coils were not so close as to prevent the access of air between them, the surface must have been about four times as great as that of the uncovered bulb. At high velocities, it is probable that a cushion of air which has not time to escape past resisting obstacles makes the actual friction almost independent of variations of surface, which leave the magnitude of the body unaltered. In conformity with this observation, it will be seen that at high velocities the thermal effect was nearly reduced to the quantity observed with the uncovered bulb. Similar remarks apply to the following results obtained after wrapping round the bulb a fine spiral of thin brass wire.

TABLE VII.—*Bulb wrapped with a spiral of fine brass wire.*

Velocity in feet per second.	Rise in divisions of scale.	Rise divided by square of velocity.
7·08	2·5	·0424
15·36	13·5	·0572
23·04	36·5	·0687
30·71	48	·0509
46·08	64·5	·0304
76·8	103·5	·0175
115·18	224·5	·0169
148·78	264	·0119

The thermal effects on different sides of a sphere moving through air, have been investigated by us experimentally by whirling a thin glass globe of 3·58 inches diameter along with the smaller thermometer, the bulb of which was placed successively in three positions, viz. in front, at one side, and in the rear. In each situation it was placed as near the glass globe as possible without actually touching it.

Table VIII.—*Smaller Thermometer whirled along with glass globe.*

Rise in divisions of scale.

Velocity in feet per second.	Therm. in front.	Therm. at side.	Therm. in rear.
3·84	·66	10	4
7·68	2·66	40	10·5
15·36	41·9	78	51
23·04	71·2	90	71·7
38·4	78·4	90	68
57·5	99·9	112	76
70·92			107

The effects of fluid friction are strikingly evident in the above results, particularly at the slow velocities of 3 and 7 feet per second. It is clear from these, that the air, after coming in contact with the front of the globe, traverses with friction the equatorial parts, giving out an accumulating thermal effect, a part of which is carried round to the after pole. At higher velocities the effects of friction seem rapidly to diminish, so that at velocities between 23 and 38 feet per second, the mean indication of thermometers placed all round the globe would be nearly constant. Our anticipation (written before these latter experiments were made), that a complete verification of the theory propounded at the commencement was impossible with our present means, is thus completely justified.

It may be proper to observe, that in the form of experiment hitherto adopted by us, the results are probably, to a trifling extent, influenced by the vortex of air occasioned by the circular motion.

We have on several occasions noticed the effect of sudden changes in the force of wind on the temperature of a thermometer held in it. Sometimes the thermometer was observed to rise, at other times to fall, when a gust came suddenly on. When a rise occurred, it was seldom equivalent to the effect, as ascertained by the foregoing experiments, due to the increased velocity of the air. Hence we draw the conclusion, that the actual temperature of a gust of wind is lower than that of the subsequent lull. This is probably owing to the air in the latter case having had its *vis viva* converted into heat by collision with material objects. In fact we find that in sheltered situations, such for instance as one or two inches above a wall opposite to the wind, the thermometer indicates a higher temperature than it does when exposed to the blast. The question, which is one of great interest for meteorological science, has hitherto been only partially discussed by us, and for its complete solution will require a careful estimate of the temperature of the earth's surface, of the effects of radiation, &c., and also a knowledge of the causes of gusts in different winds.

[From the *Proceedings of the Royal Society*, Vol. x., June, 1860.]

In our paper published in the *Philosophical Transactions* for 1854 [Present Art. Part II. above], we explained the object of our experiments to ascertain the difference of temperature between the high- and low-pressure sides of a porous plug through which elastic fluids were forced. Our experiments were then limited to air and carbonic acid. With new apparatus, obtained by an allotment from the Government grant, we have been able to determine the thermal effect with various other elastic fluids. The following is a brief summary of our principal results at a low temperature (about $7°$ Cent.):

Elastic Fluid.				Thermal effect per 100 lbs. pressure on the square inch, in degrees Centigrade.	
		Air.		$1·6$	Cold.
$3·9$	Air	$+96·1$	Hydrogen	$0·116$	Heat.
$7·9$	Air	$+92·1$	Nitrogen	$1·772$	Cold.
$5·1$	Air	$+94·9$	Oxygen	$1·936$	Cold.
$3·5$	Air	$+96·5$	Carbonic acid	$8·19$	Cold.
$58·3$	Air	$+41·7$	Hydrogen	$0·7$	Cold.
$62·5$	Air	$+37·5$	Carbonic acid	$3·486$	Cold.
$54·6$	Nitrogen	$+45·4$	Oxygen	$1·696$	Cold.
$4·23$	Air	$\begin{cases} +46·47 \\ +49·3 \end{cases}$	Hydrogen........... Carbonic acid	$2·848$	Cold.

Further experiments are being made at high temperatures, which show, in the gases in which a cooling effect is found, a decrease of this effect, and an increase of the heating effect in hydrogen. The results at present arrived at, indicate invariably that a mixture of gases gives a smaller cooling effect than that deduced from the average of the effects of the pure gases.

ABSTRACT OF PART III. ABOVE.

[From the *Proceedings of the Royal Society*, Vol. x., June, 1860.]

An abstract of a great part of the present paper has appeared in the *Proceedings*, Vol. VIII. p. 556 [see also present Appendix above]   To the experiments then adduced a large number have since been added, which have been made by whirling thermometers and thermo-electric junctions in the air.   The result shows that at high velocities the thermal effect is proportional to the square of the velocity, the rise of temperature of the whirled body being evidently that due to the communication of the velocity to a constantly renewed film of air.   With very small velocities of bodies of large surface, the thermal effect was very greatly increased by that kind of fluid friction the effect of which on the motion of pendulums has been investigated by Prof. Stokes*.

ABSTRACT OF PART IV. ABOVE.

[From the *Proceedings of the Royal Society*, Vol. XII., June, 1862.]

A brief notice of some of the experiments contained in this paper has already appeared in the *Proceedings* [see also present Appendix above]   Their object was to ascertain with accuracy the lowering of temperature, in atmospheric air and other gases, which takes place on passing them through a porous plug from a state of high to one of low pressure.   Various pressures were employed, with the result (indicated by the authors in their Part II.) that the thermal effect is approximately proportional to the difference of pressure on the two sides of the plug.   The experiments were also tried at various temperatures, ranging from 5° to 98° Cent, and have shown that the thermal effect, if one of cooling, is approximately proportional to the inverse square of the absolute temperature.   Thus, for example, the refrigeration at the freezing temperature is about twice that at

---

* " On the effect of the Internal Friction of Fluids on the Motion of Pendulums," *Transactions of the Cambridge Philosophical Society*, Dec. 1850, Vol. IX., Part II.; or *Mathematical and Physical Papers*, Stokes, Vol. II.

100° Cent. In the case of hydrogen, the reverse phenomenon of a rise of temperature on passing through the plug was observed, the rise being doubled in quantity when the temperature of the gas was raised to 100°. This result is conformable with the experiments of Regnault, who found that hydrogen, unlike other gases, has its elasticity increased more rapidly than in the inverse ratio of the volume. The authors have also made numerous experiments with mixtures of gases, the remarkable result being, that the thermal effect (cooling) of the compound gas is less than it would be, if the gases, after mixture, retained in integrity the physical characters they possessed while in a pure state.

[From the *Philosophical Magazine*, May, 1879.]

ART. L.   ON THERMODYNAMIC MOTIVITY.

AFTER having for some years felt with Professor Tait the want of a word "to express the Availability for work of the heat in a given magazine, a term for that possession the waste of which is called Dissipation*," I suggested three years ago the word *Motivity* to supply this want, and made a verbal communication to the Royal Society of Edinburgh defining and illustrating the application of the word; but as the communication was not given in writing, only the title of the paper, "Thermodynamic Motivity," was published. In consequence of Professor Tait's letter to me, published in the present number of the *Philosophical Magazine*, I now offer, for publication in the *Proceedings of the Royal Society of Edinburgh* and in the *Philosophical Magazine*, the following short abstract of the substance of that communication.

In my paper "On the Restoration of Energy from Unequally Heated Space" [Art. LXIII. below], published in the *Philosophical Magazine* in January, 1853, I gave the following expression for the amount of "mechanical energy" derivable from a body, *B*, given with its different parts at different temperatures, by the equalization of the temperature throughout to one common temperature†

---

* Tait's *Thermodynamics*, first edition (1868), § 178: quoted also in Professor Tait's letter in the present number of the *Philosophical Magazine*.

† In the present article I suppose this temperature to be the given temperature of the medium in which *B* is placed; and thermodynamic engines to work with their recipient and rejectant organs respectively in connexion with some part of *B* at temperature *t*, and the endless surrounding matter at temperature *T*. In the original paper this supposition is introduced subordinately at the conclusion. The chief purpose of the paper was the solution of a more difficult problem, that of finding the value of *T*,—a kind of average temperature of *B* to fulfil the condition that the quantities of heat rejected and taken in by organs of the thermodynamic engines at temperature *T* are equal. The burden of the problem was the evaluation of this thermodynamic average; and I failed to remark that when the value which the solution gave for *T* is substituted in the formula of the text it reduces it to

$$J \iiint dx\, dy\, dz \int_T^t c\, dt,$$ which was not instantly obvious from the analytical form

T, by means of perfect thermodynamic engines,

$$W = J \iiint dx\, dy\, dz \int_T^t c\, dt \left(1 - \epsilon^{-\frac{1}{J} \int_T^t \mu\, dt}\right) \quad \ldots\ldots\ldots\ldots(1),$$

where t denotes the temperature of any point x, y, z of the body, c the thermal capacity of the body's substance at that point and that temperature, J Joule's equivalent, and μ Carnot's function of the temperature t.

Further on in the same paper a simplification is introduced thus:—

"Let the temperature of the body be measured according to an absolute scale, founded on the values of Carnot's function, and expressed by the following equation,

$$t = \frac{J}{\mu} - \alpha,$$

where α is a constant which may have any value; but ought to have for its value the reciprocal of the expansibility of air, in order that the system of measuring temperature here adopted may agree approximately with that of the air-thermometer. Then we have

$$\epsilon^{-\frac{1}{J} \int_0^t \mu\, dt} = \frac{\alpha}{t + \alpha} \text{"} \quad \ldots\ldots\ldots\ldots\ldots(2).$$

It was only to obtain agreement with the zero of the ordinary Centigrade scale of the air-thermometer that the α was needed; and in the joint paper by Joule and myself [Art. XLIX., Part II. above], published in the *Transactions of the Royal Society* (London) for June, 1854, we agreed to drop it, and to define temperature simply as the reciprocal of Carnot's function, with a constant coefficient proper to the unit or degree of temperature adopted. Thus definitively, in equation (6) of § 5 of that paper, we took $t = J/\mu$, and have used this expression ever since as the expression for temperature on the arbitrarily assumed thermodynamic scale. With it we have

$$\epsilon^{-\frac{1}{J} \int_T^t \mu\, dt} = \frac{T}{t} \quad \ldots\ldots\ldots\ldots\ldots(3);$$

of my solution, but which we immediately see must be the case by thinking of the physical meaning of the result; for the sum of the excesses of the heats taken in above those rejected by all the engines must, by the first law of thermodynamics, be equal to the work gained by the supposed process. This important simplification was first given by Professor Tait in his *Thermodynamics* (first and second editions). It does not, however, affect the subordinate problem of the original paper, which is the main problem of this one.

and by substitution (1) becomes

$$W = J \iiint dx\,dy\,dz \int_T^t c\,dt \left(1 - \frac{T}{t}\right) \dots\dots\dots\dots(4).$$

Suppose now B to be surrounded by other matter all at a common temperature T. The work obtainable from the given distribution of temperature in B by means of perfect thermodynamic engines is expressed by the formula (4). If, then, there be no circumstances connected with the gravity, or elasticity, or capillary attraction, or electricity, or magnetism, of B in virtue of which work can be obtained, that expressed by (4) is what I propose to call the whole Motivity of B in its actual circumstances. If, on the other hand, work is obtainable from B in virtue of some of these other causes, and if V denote its whole amount, then

$$\mathfrak{M} = V + W\dots\dots\dots\dots\dots\dots(5)$$

is what I call the whole Motivity of B in its actual circumstances according to this more comprehensive supposition.

We may imagine the whole Motivity of B developed in an infinite variety of ways. The one which is obvious from the formula (5) is first to keep every part of B unmoved, and to take all the work producible by perfect thermodynamic engines equalizing its temperature to T; and then keeping it rigorously at this temperature, to take all the work that can be got from it elastically, cohesively, electrically, magnetically, and gravitationally, by letting it come to rest unstressed, diselectrified, demagnetized, and in the lowest position to which it can descend. But instead of proceeding in this one definite way, any other order of procedure whatever leading to the same final condition may be followed; and, provided nothing is done which cannot be undone (that is to say, in the technical language of thermodynamics, provided all the operations are reversible), the same whole quantity of work will be obtained in passing from the same initial condition to the same final condition, whatever may have been the order of procedure. Hence the Motivity is a function of the temperature, volume, figure, and proper independent variables for expressing the cohesive, the electric, and the magnetic condition of B, with the gravitational potential of B simply added (which, when the force of gravity is sensibly constant and in parallel lines, will be simply the product of the gravity of B into the height of its centre of

gravity above its lowest position). So also is the *Energy* of
a body *B* (as I first pointed out, for the case of *B* a fluid, in
Part V. of "Dynamical Theory of Heat" [Art. XLVIII. above;
Transactions of the Royal Society of Edinburgh for December 15,
1851], entitled, "On the Quantities of Mechanical Energy contained in a Fluid in Different States as to Temperature and
Density.") Consideration of the *Energy* and the *Motivity*, as two
functions of all the independent variables specifying the condition
of *B* completely in respect to temperature, elasticity, capillary
attraction, electricity, and magnetism, leads in the simplest and
most direct way to demonstrations of the theorems regarding the
thermodynamic properties of matter which I gave in Part III.
of the Dynamical Theory of Heat (March 1851) [Art. XLVIII.
above]; in Part VI. of Dynamical Theory of Heat, Thermo-electric
Currents (May 1, 1854) [Art. XLVIII. above]; in a paper in the
Proceedings for 1858 of the Royal Society of London, entitled,
"On the Thermal Effect of Drawing out a Film of Liquid;" and
in a communication to the Royal Society of Edinburgh (*Proc.
R. S. E.* 1869—70), "On the Equilibrium of Vapour at the Curved
Surface of a Liquid;" and in my article on the Thermo-elastic and
Thermomagnetic Properties of Matter, in the first number of the
Quarterly Journal of Mathematics (April 1855) [Art. XLVIII.
Part. VII. above]; and in short articles in Nichol's *Cyclopædia*
under the titles "Thermomagnetism," "Thermo-electricity," and
"Pyro-electricity," put together and republished with additions in
the *Philosophical Magazine* for January, 1878, under the title "On
the Thermo-elastic, Thermomagnetic, and Pyro-electric Properties
of Matter" [Art. XLVIII. Part VII. above].

It would be beyond the scope of the present article to enter
in detail into these applications, which were merely indicated in
my communication to the Royal Society of Edinburgh of three
years ago, as a very short and simple analytical method of setting
forth the whole non-molecular theory of Thermodynamics.

UNIVERSITY OF GLASGOW,
April 11, 1879.

[From the *Proceedings of the Royal Society*, Vol. VII. May, 1854.]

ART. LI. EXPERIMENTAL RESEARCHES IN THERMO-
ELECTRICITY.

§ I. *On the Thermal Effects of Electric Currents in Unequally
Heated Conductors.*

THEORETICAL considerations (communicated in December 1851
to the Royal Society of Edinburgh), [Art. XLVIII. App. I. above,]
founded on observations which had been made regarding the law of
thermo-electric force in an unequally heated circuit of two metals,
led me to the conclusion that an electric current must exercise a
convective effect on heat in a homogeneous metallic conductor of
which different parts are kept at different temperatures. A special
application of the reasoning to the case of a compound circuit of
copper and iron was made, and it is repeated here because of the
illustration it affords of the mechanical principles on which the
general reasoning is founded.

Becquerel discovered that if one junction of copper and iron,
in a circuit of the two metals, be kept at an ordinary atmospheric
temperature, while the other is raised gradually to a red or white
heat, a current first sets from copper to iron through the hot
junction, increasing in strength only as long as the temperature
is below about 300° Cent.; and becoming feebler with farther
elevations of temperature until it ceases, and a current actually
sets in the contrary direction when a high red heat is attained.
Many experimenters have professed themselves unable to verify
this extraordinary discovery, but the description which M. Becquerel
gives of his experiments leaves no room for the doubts which some
have thrown upon his conclusion, and establishes the thermo-
electric inversion between iron and copper, not as a singular case
(extraordinary and unexpected as it appeared), but as a phenomenon to be looked for between any two metals, when tried
through a sufficient range of temperature, especially any two which
lie near one another in the thermo-electric series for ordinary
temperatures. M. Regnault has verified M. Becquerel's conclusion
so far, in finding that the strength of the current in a circuit of

copper and iron wire did not increase sensibly for elevations of
temperature above 240° Cent., and began to diminish when the
temperature considerably exceeded this limit; but the actual
inversion observed by M. Becquerel is required to show that the
diminution of strength in the current is due to a real falling off in
the electromotive force, and not to the increased resistance known
to be produced by an elevation of temperature.

From Becquerel's discovery it follows that, for temperatures
below a certain limit, which, for particular specimens of copper and
iron wire, I have ascertained, by a mode of experimenting described
below, to be 280° Cent., copper is on the negative side of iron in
the thermo-electric series, and on the positive side for higher
temperatures; and at the limiting temperature copper and iron
are thermo-electrically neutral to one another.   It follows, according to the general mechanical theory of thermo-electric currents
referred to above, that electricity passing from copper to iron
causes the absorption or the evolution of heat according as the
temperature of the metals is below or above the neutral point;
but neither evolution nor absorption of heat, if the temperature be
precisely that of neutrality (a conclusion which I have already
partially verified by experiment).   Hence, if in a circuit of copper
and iron, one junction be kept about 280°, that is, at the neutral
temperature, and the other at any lower temperature, a thermo-
electric current will set from copper to iron through the hot,
and from iron to copper through the cold junction; causing the
evolution of heat at the latter, and the raising of weights too if it
be employed to work an electro-magnetic engine, but not causing
the absorption of any heat at the hot junction.   Hence there must
be an absorption of heat at some part or parts of the circuit
consisting solely of one metal or of the other, to an amount
equivalent to the heat evolved at the cold junction, together with
the thermal value of any mechanical effects produced in other
parts of the circuit.   The locality of this absorption can only be
where the temperatures of the single metals are non-uniform,
since the thermal effect of a current in any homogeneous uniformly
heated conductor is always an evolution of heat.   Hence there
must be on the whole an absorption of heat, caused by the current
in passing from cold to hot in copper, and from hot to cold in iron.
When a current is forced through the circuit against the thermo-
electric force, the same reasoning establishes an evolution of heat

to an amount equivalent to the sum of the heat that would be then taken in at the cold junction, and the value in heat of the energy spent by the agency (chemical or of any other kind) by which the electromotive force is applied. The aggregate reversible thermal effect, thus demonstrated to exist in the unequally heated portions of the two metals, might be produced in one of the metals alone, or (as appears more natural to suppose) it may be the sum or difference of effects experienced by the two. Adopting as a matter of form the latter supposition, without excluding the former possibility, we may assert that either there is absorption of heat by the current passing from hot to cold in the copper, and evolution, to a less extent, in the iron of the same circuit; or there is absorption of heat produced by the current from hot to cold in the iron, and evolution of heat to a less amount in the copper; or there must be absorption of heat in each metal, with the reverse effect in each case when the current is reversed. The reversible effect in a single metal of non-uniform temperature may be called a convection of heat; and to avoid circumlocution, I shall express it, that the vitreous electricity carries heat with it, or that the specific heat of vitreous electricity is positive, when this convection is in the nominal "direction of the current," and I shall apply the same expressions to "resinous electricity" when the convection is against the nominal direction of the current. It is established then that one or other of the following three hypotheses must be true :—

Vitreous electricity carries heat with it in an unequally heated conductor whether of copper or iron; but more in copper than in iron.

Or Resinous electricity carries heat with it in an unequally heated conductor whether of copper or iron; but more in iron than in copper.

Or Vitreous electricity carries heat with it in an unequally heated conductor of copper, and Resinous electricity carries heat with it in an unequally heated conductor of iron.

Immediately after communicating this theory to the Royal Society of Edinburgh, I commenced trying to ascertain by experiment which of the three hypotheses is the truth, as Theory with only thermo-electric data could not decide between them. I had a slight bias in favour of the first rather than the second, in consequence of the positiveness which, after Franklin, we habitually attribute to the vitreous electricity, and a very strong feeling of the improbability of the third.   With the able and persevering exertions of my assistant, Mr M‘Farlane, applied to the construction of various forms of apparatus and to assist me in conducting experiments, the research has been carried on, with little intermission, for more than two years.   Mr Robert Davidson, Mr Charles A. Smith, and other friends have also given much valuable assistance during the greater part of this time, in the different experimental investigations of which results are now laid before the Royal Society.   Only nugatory results were obtained until recently from multiplied and varied experiments both on copper and iron conductors; but the theoretical anticipation was of such a nature that no want of experimental evidence could influence my conviction of its truth.   About four months ago, by means of a new form of apparatus, I ascertained that *resinous electricity carries heat with it in an unequally heated iron conductor.* A similar equally sensitive arrangement showed no result for copper.   The second hypothesis might then have been expected to hold; but to ascertain the truth with certainty I have continued ever since, getting an experiment on copper nearly every week with more and more sensitive arrangements, and at last, in two experiments, I have made out with certainty, that *vitreous electricity carries heat with it in an unequally heated copper conductor.*

The third hypothesis is thus established: a most unexpected conclusion I am willing to confess.

I intend to continue the research, and I hope not only to ascertain the nature of the thermal effects in other metals, but to determine its amount in absolute measure in the most important cases, and to find how it varies, if at all, with the temperature; that is, to determine the character (positive or negative) and the value of the specific heat, varying or not with the temperature, of the unit of current electricity in various metals.

§ II.   *On the Law of Thermo-electric Force in an unequally heated circuit of two Metals.*

A general relation between the specific heats of electricity in two different metals, and the law of thermo-electric force, in a

circuit composed of them according to the temperatures of their
junctions, was established in the communication to the Royal
Society of Edinburgh referred to above, and was expressed by
an equation* which may now be simplified by the thermometric
assumption

$$t = \frac{J}{\mu} \; ;$$

(μ denoting Carnot's function, J Joule's equivalent, and t the
temperature measured from an absolute zero, about $273\frac{1}{2}°$ Cent.
below the freezing-point,) since this assumption defines a system of
thermometry in absolute measure, which the experimental researches
recently made by Mr Joule and myself establish as not differing
sensibly from the scale of the air-thermometer between ordinary
limits.   The equation, when so modified, takes the following
form :—

$$F = J \left\{ \frac{\Theta_S}{S} (S - T) + \int_T^S \vartheta \left(1 - \frac{T}{t} \right) dt \right\},$$

where ϑ denotes the excess of the specific heat of electricity in the
metal through which the current goes from cold to hot above the
specific heat of the same electricity in the other metal, at the temperature t; F the thermo-electric force in the circuit when the
two junctions are kept at the temperatures S and T respectively,
of which the former is the higher; and Θ_S the amount of heat
absorbed per unit of electricity crossing the hot junction.   The
following relation (similarly simplified in form) was also established :—

$$\vartheta = \frac{\Theta}{t} - \frac{d\Theta}{dt} .$$

These relations show how important it is towards the special
object of determining the specific heats of electricity in metals, to
investigate the law of electromotive force in various cases, and to
determine the thermal effect of electricity in passing from one
metal to another at various temperatures.   Both of these objects of
research are therefore included in the general investigation of the
subject.

The only progress I have as yet made in the last-mentioned

* See *Proceedings R.S.E.* Dec. 1851, or *Philosophical Magazine*, 1852. [Art. xlviii.
App. I. above.]

branch of the inquiry, has been to demonstrate experimentally that there is a cooling or heating effect produced by a current between copper and iron at an ordinary atmospheric temperature according as it passes from copper to iron or from iron to copper, in verification of a theoretical conclusion mentioned above : but I intend shortly to extend the verification of theory to a demonstration that reverse effects take place between those metals at a temperature above their neutral point of about 280° Cent.; and I hope also to be able to make determinations in absolute measure of the amount of the Peltier effect for a given strength of current between various pairs of metals.

With reference to laws of electromotive force in various cases, I have commenced by determining the order of several specimens of metals in the thermo-electric series, and have ascertained some very curious facts regarding varieties in this series which exist at different temperatures. In this I have only followed Becquerel's remarkable discovery, from which I had been led to the reasoning and experimental investigation regarding copper and iron described above. My way of experimenting has been to raise the temperature first of one junction as far as the circumstances admit, keeping the other cold, and then to raise the temperature of the other gradually, and watch the indications of a galvanometer during the whole process. When an inversion of the current is noticed, the changing temperature is brought back till the galvanometer shows no current; and then (by a process quite analogous to that followed by Mr Joule and Dr Lyon Playfair in ascertaining the temperature at which water is of maximum density) the temperatures of the two junctions are approximated, the galvanometer always being kept as near zero as possible When the difference between any two temperatures on each side of the neutral point which give no current is not very great, their arithmetical mean will be the neutral temperature. A regular deviation of the mean temperature from the true neutral temperature is to be looked for with wide ranges, and a determination of it would show the law according to which the difference of the specific heat of electricity in the two metals varies with the temperatures; but I have not even as yet ascertained with certainty the existence of such a deviation in any particular case. The following is a summary of the principal results I have already obtained in this department of the subject.

T.   30

The metals tried being,—three platinum wires (P_1 the thickest, P_2 the thinnest, and P_3 one of intermediate thickness), brass wires (B), a lead wire (L'), slips of sheet lead (L), copper wires (C), and iron wire (I), I find that the specimens experimented on stand thermo-electrically at different temperatures in the order shown in the following Table, and explained in the heading by reference to bismuth and antimony, or to the terms "negative" and "positive" as often used :—

Temp. Cent.	Bismuth "Negative."	Antimony "Positive."
—20	... P_3 ... c ...... P_2 .... P_1	I .. ..
0	... P_3 ... l' ...... P_2 ......C ...... P_1	I ....
37	... P_3 ......b...{$L'P_2$}...... C ... P_1	I ....
64	... P_3 .... P_2 .. b ... l' {CP_1}...............	I ....
130	... P_3 .. .. P_2 .... {BP_1} ...L...C...... ...	I ....
140	... P_3 .... P_2 .... P_1 ...{BL} ......C ...	I ....
280	... P_3 .... P_2 .... P_1 .... b......	{CI} ......
300	... P_3 .... P_2 .... P_1 .... b	I C

It must be added, by way of explanation, that the bracket enclosing the symbols of any two of the metallic specimens indicates that they are neutral to one another at the corresponding temperature, and the arrow-head below one of them shows the direction in which it is changing its place with reference to the other, in the series, as the temperature is raised. When there is any doubt as to a position as shown in the Table, the symbol of the metal is a small letter instead of a capital.

The rapidity with which copper changes its place among some of the other metals (the platinums and iron) is very remarkable. Brass also changes its place in the same direction possibly no less rapidly than copper; and lead changes its place also in the same direction but certainly less rapidly than brass, which after passing the thick platinum wire (P_1) at 130° Cent. passes the lead at 140°, the lead itself having probably passed the thick platinum at some temperature a little below 130°*.

* I have since found that it does pass the thick platinum, at the temperature 118°. [May 16, 1854.]

The conclusion as regards specific heats of electricity in the different metals, from the equation expressing thermo-electric force given above, is that the specific heat of vitreous electricity is greater in each metal passing another from left to right in the series as the temperature rises than in the metal it passes: thus in particular,—

The specific heat of vitreous electricity is greater in copper than in platinum or in iron; greater in brass than in platinum or in lead; and greater in lead than in platinum.

It is probable enough from the results regarding iron and copper mentioned above, that the specific heat of vitreous electricity is positive in brass; very small positive, or else negative, in platinum, perhaps of about the same value as in iron. It will not be difficult to test these speculations either by direct experiment on the convective effects of electric currents in the different metals, or by comparative measurements of thermo-electric forces for various temperatures in circuits of the metals, and I trust to be able to do so before long.

§ III. *On Thermo-electricity in crystalline metals, and in metals in a state of mechanical strain.*

Having recently been occupied with an extension of the mechanical theory to the phenomena of thermo-electricity in crystalline metals, I have been led to experimental investigation on this branch of the subject. The difficulty of obtaining actual metallic crystals of considerable dimensions made it desirable to imitate crystalline structure in various ways. The analogies of the crystalline optical properties which have been observed in transparent solids, in a state of strain, and of the crystalline structure as regards magnetic induction which Dr Tyndall's remarkable experiments show to be produced not only in bismuth but in wax, thick paste of flour, and "the pith of fresh rolls" by pressure, made it almost certain that pressure or tension on a mass of metal would give it the thermo-electric properties of a crystal. The only case which I have as yet had time to try, verifies this anticipation. I have found that copper wire stretched by a weight bears to

similar copper wire unstretched, exactly the thermo-electric relation which Svanberg discovered in a bar cut equatorially from a
crystal of bismuth or antimony compared with a bar cut axially
from a crystal of the same metal.   Thus I found that:—

   If part of a circuit of copper wire be stretched by a considerable
force and the remainder left in its natural condition, or stretched
by a less force, and if either extremity of the stretched part be
heated, *a current sets from the stretched to the unstretched part
through the hot junction:* and if the wire be stretched and unstretched on the two sides of the heated part alternately, the
current is reversed (as far as I have been able yet to test, instantaneously) with each change of the tension.

   I intend to make similar experiments on other metallic wires;
also to try the effect of transverse as well as of longitudinal tension
on slips of sheet metal *with their ends at different temperatures,*
when placed longitudinally in an electric circuit; and the effects of
oblique tension on slips of metal similarly placed in a circuit,
but kept with their ends at the same temperature and their
lateral edges unequally heated.   I have no doubt of being able so to
verify every thermo-electric characteristic of crystalline structure,
in metals in a state of strain.

GLASGOW COLLEGE,
      March 30, 1854.

   P.S.   April 19, 1854.—I have to-day found by experiment that
iron wire when stretched by a considerable force bears a thermo-
electric relation to unstretched iron wire, the opposite of that
which I had previously discovered in the case of copper wire; and I
have ascertained that when the wire is alternately stretched and
unstretched on the two sides of a heated part the current is
reversed along with the change of tension, always passing from
the unstretched to the stretched part, through the hot locality.

   I hope before the end of the present Session to have a complete
account of all the experiments of which the results are stated above,
ready to communicate to the Royal Society.

§ IV. *Account of Experimental Researches in Thermoelectricity.*

[*British Association Report.* Part II. 1854.]

The experiments described in this paper have been made for the purpose of continuing the investigation of various branches of the subject commenced in researches, of which an account was published in the *Proceedings of the Royal Society* for May last. [Preceding portion of present Article.] In one class of experiments, thermo-electric inversions, of the kind first discovered by Prof. Cumming, were sought for between various pairs of metals; and many remarkable variations of order in the thermoelectric series were found. The following table exhibits the results of observations to determine the neutral points in various cases in which thermo-electric inversions had been ascertained. The first column is the temperature, Centigrade, at which the two metals opposite are thermo-electrically neutral to one another, and the latter metal is that which *passes the other from bismuth towards antimony* as the temperature rises:—

-14 Cent.	......	P_3—Brass.
$-12 \cdot 2$,,	......	P_1—Cadmium.
$-1 \cdot 5$,,	......	P_1—Silver.
$8 \cdot 2$,,	......	P_1—Zinc.
36 ,,	......	P_2—Lead.
38 ,,	......	P_2—Brass.
44 ,,	......	P_2—Tin.
44 ,,	......	Lead—Brass.
64 Cent.	......	P_1—Copper.
99 ,,	......	P_1—Brass.
121 ,,	......	P_1—Lead.
130 ,,	......	P_1—Tin.
$162 \cdot 5$,,	......	Iron—Cadmium.
237 ,,	......	Iron—Silver.
280 ,,	......	Iron—Copper.

*** P_1, P_2, P_3 denote three different specimens of platinum wire, which have been found to differ very markedly and constantly in their thermo-electric qualities.

It was also found that brass becomes neutral to copper, and copper becomes neutral to silver, at some high temperature, estimated at from 800° to 1400° Cent., in the former case, and from 700° to 1000° in the latter. A diagram, showing the results of these observations, and the orders of the different metals in the thermoelectric series at different temperatures, was exhibited to the Section. In other experiments the effects of magnetization on

the thermo-electric qualities of iron were tested. It was found
that both soft iron wire longitudinally magnetized, when actually
under the influence of a galvanic helix, and steel magnetized
permanently in an ordinary way and removed from the magnetic
influence by which the magnetic state is induced, exhibit very
decided thermo-electric effects of the magnetization: in each case
a thermo-electric current in metal homogeneous except as regards
magnetization, could be produced with great facility ; the current
being always from unmagnetized to magnetized through the hot
junctions. It is intended to continue the experiments to ascertain
the effects of transverse magnetization on portions of a circuit of
soft iron or of steel wire; and to test differences of thermo-electric
properties in different directions in a magnetized mass, which the
author anticipated from certain considerations more fully explained
in a theoretical paper on thermo-electricity in crystals recently communicated to the Royal Society of Edinburgh [Art. XLVIII. Part VI.
above]. Experiments on the effects of temporary and permanent
strains, by tension, and by either lateral or longitudinal compression, on various metals were described. The thermo-electric effects
of temporary tension in the cases of copper and iron, which had
been communicated by the author to the Royal Society in May
[beginning portion of present Article] (namely, current from
stretched to unstretched through hot, in copper, and from unstretched to stretched through hot, in iron), were verified by means
of a new form of apparatus so simple that he hoped to be able with
great ease to test the corresponding property for a great variety of
metals. In those two cases the permanent thermo-electric effects
of permanent extension by drawing the wires through a draw-
plate, which had been discovered by Magnus, were the reverse of
temporary effects of temporary extension discovered by the author
of the present paper. The thermo-electric effects produced by
permanent lateral compression, always found to agree with those
of permanent longitudinal extension, were tested in the cases of
zinc, tin, cadmium, and lead, and were found to be the same as in
the case of copper. The current was from unstrained to strained
through hot in each case. Some of these results agreed, while others
appeared at variance, with what might have been expected from
Magnus's careful experiments ; but as the author had not completed a cycle of experiments which he proposed to make on the
thermo-electric effects, either temporary or permanent, of longi-

tudinal extensions, lateral compressions, longitudinal compressions and lateral extensions, he refrained from any further remarks on the present occasion.

ART. LII. ON THE THEORY OF MAGNETIC INDUCTION IN CRYSTALLINE AND NON-CRYSTALLINE SUBSTANCES.

[*Phil. Mag.* March, 1851. ELECTROSTATICS AND MAGNETISM, Art. xxx.]

ART. LIII. On the Mechanical Theory of Electrolysis.

[*Phil. Mag.* Dec. 1851.]

1. Certain principles discovered by Mr Joule, and published
for the first time in his various papers in this Magazine, must
ultimately become an important part of the foundation of a
mechanical theory of chemistry. The object of the present communication is to investigate, according to those principles, the
relation, in any case of electrolysis, between the electro-motive
intensity, the electro-chemical equivalents of the substances
operated on, and the mechanical equivalent of the chemical effect
produced in the consumption of a given amount of the materials;
and by means of it to determine in absolute measure the electro-
motive intensity of a single cell of Daniell's battery, and the
electro-motive intensity required for the electrolysis of water,
from experimental data which Mr Joule has kindly communicated
to me.

2. If a galvanic current, produced by means of a magneto-
electric machine, be employed in electrolysis, it will generate,
in any time, less heat throughout its entire circuit than the equivalent of the work spent, by an amount which may be called the
thermal equivalent of the chemical action which has been effected,
being the quantity of heat which would be obtained by recom-
bining the elements of the decomposed substance, and reducing
the compound to its primitive condition in every respect; or
generally, by undoing all the action which has been done in the
electro-chemical apparatus. Now the quantity of heat which is
equivalent to the work done is obtained by dividing the number
which measures the work by the number which measures by the
same unit the mechanical equivalent of the unit of heat. Hence
if the mechanical equivalent of the thermal unit be denoted by
J, the work done in any time by W, the total quantity of heat
evolved in the same time throughout the circuit by H, and the
thermal equivalent of the chemical effect produced by Θ, we have

$$H = \frac{W}{J} - \Theta \dots\dots\dots\dots(1);$$

an equation which may also be written in the form

$$W = JH + M \dots\dots\dots\dots\dots\dots(2),$$

if M be used to denote the value of $J\Theta$, or, as it may be called, the mechanical equivalent of the chemical effect produced in the stated period of time.

3.   To avoid the necessity of considering variable or discontinuous currents, let us suppose the "machine" to consist of a metallic disc, touched at its centre and at its circumference by fixed wires, and made to revolve in its own plane about an axis through its centre, held in any position not at right angles to the direction of the earth's magnetic force*.   [Note of April 16, 1882.   Better have said *an axis coincident with the direction of the earth's magnetic force*.]   If these wires be connected by contact between their ends, there will, as is known, be a current produced in them of a strength proportional directly to the angular velocity of the disc, and inversely to the resistance through the whole circuit.   Hence there will be between the ends of the wires, if separated by an insulating medium, an electromotive force the intensity of which will be constant and proportional to the angular velocity of the disc.

4.   Let us now suppose the wires to be connected with the electrodes of an electro-chemical apparatus, for instance a galvanic battery of any kind, or an apparatus for the decomposition of water; and let us conceive the electro-motive intensity between them to be sufficient to produce a current in its own direction. The preceding equations, when applied to this case, will have each of their terms proportional to the time, since the action is continuous and uniform, and therefore it will be convenient to consider the unit of time as the period during which the amounts of work and heat denoted by W and H, and the amount of chemical action of which the thermal and the mechanical equivalents are denoted respectively by Θ and M, are produced.   If r denote the radius of the disc, ω the angular velocity with which it is moved, F the component of the earth's magnetic force perpendicular to its plane, and γ the strength of the current which is induced; the work done in a unit of time in moving the disc against the resistance which it experiences in virtue of the earth's

---

* This is in fact the "new electrical machine" suggested by Faraday in the Bakerian Lecture of 1832. (*Experimental Researches*, § 154.)

magnetic action on the current through it, will be expressed by the integral $\int_0^r \omega z . F . \gamma dz$; as is easily proved, whether the current be supposed to pass directly between the centre of the disc and the point of its circumference touched by the fixed wire, or to be, as it in reality must be, more or less diffused from the direct line, on account of the lateral extension of the revolving conductor. Hence we have

$$W = \frac{1}{2} r^2 F \gamma \omega \quad \ldots\ldots\ldots\ldots\ldots\ldots\ldots(3).$$

5.   Let E denote the quantity (in units of matter, as grains for instance) of one of the elements concerned in the chemical action, which is electrolysed or combined in the unit of time, and let θ denote the quantity of heat absorbed in the chemical action during the electrolysis or combination of a unit quantity of that element.   Then we have

$$\Theta = \theta . E \ldots\ldots\ldots\ldots\ldots\ldots\ldots\ldots\ldots\ldots(4),$$
$$M = J . \theta E \ldots\ldots\ldots\ldots\ldots\ldots\ldots\ldots\ldots(5).$$

Now it has been shown by Faraday, that in electro-chemical action of any known kind, produced by means of a continuous current, the amount of the action in a given time is approximately if not rigorously proportional to the strength of the current; and all subsequent researches on the subject have tended to confirm this conclusion.   The only exception to it which, so far as I am aware, has yet been discovered, is the fact established by Faraday, that various electrolytes can conduct a continuous current, when the electro-motive intensity is below certain limits, without experiencing any continued decomposition *; but from it we may infer as probable, that in general the quantity decomposed with high or low electro-motive intensities is not quite rigorously proportional to the strength of the current.

* It is probable that when an electromotor of an intensity below a certain limit is put in connexion with two platinum electrodes immersed in water, there is at the first instant no electrolytic resistance; and a decomposing current passes which gradually falls off in strength, until the electrodes are, by the separated oxygen and hydrogen, put into a certain state, such that with the water between them, they exert a resisting electric force very nearly equal to that of the electromotor; after which a uniform current of excessively reduced strength passes without producing further decomposition.   I hope before long to be able to communicate to the Magazine an account of some experiments I have made to illustrate these circumstances.

This non-electrolytic conducting power is, however, at least
in the case of water, found to be excessively feeble; and it is
not probable that when electrolysis is actually going on in any
ordinary case, the quantity of electricity conducted by means of
it is ever considerable compared with that which is electrically
conducted; and the normal law of true electrolytic conduction
will therefore be assumed as applicable to the conduction through
the electro-chemical apparatus, subject to modification in any
case in which the deviations from it can be determined.   If,
then, we denote by ϵ the electro-chemical equivalent of the
particular element referred to for measuring the chemical action,
that is, the quantity of it which is electrolysed or combined in a
unit of time by the operation of a current of unit strength, since
the actual strength of the current is γ, we have

$$E = \epsilon\gamma \ldots\ldots\ldots\ldots\ldots\ldots\ldots\ldots\ldots\ldots\ldots(6).$$

The deviations from the normal law which may exist in any
particular case may be represented by giving ϵ a variable value.
For instance, if it were true that when the electro-motive intensity in an apparatus for the decomposition of water exceeds a
certain limit, there is decomposition at a rate precisely proportional to the strength of the current; and when the intensity is
below that limit, a slight current passes without any decomposition; ϵ would be a discontinuous function of the intensity,
having a constant value when the intensity is above, and being
zero when the intensity is below, the limit for decomposition.

6.   According to Joule's law of the generation of heat in the
galvanic circuit, the quantity of heat developed in a unit of time
would be rigorously proportional to the square of the strength
of the current, if the total resistance were constant in all the
circumstances considered; and therefore we may conveniently
assume

$$H = R\gamma^2 \ldots\ldots\ldots\ldots\ldots\ldots\ldots\ldots(7);$$

but as we are not sure that the whole resistance is independent
of the strength of the current when an electrolysed fluid forms
part of the circuit, we must not assume that R is constant.   In
what follows, all that is assumed regarding the value of R is,

that it is neither infinitely great nor infinitely small in any of the circumstances considered*.

7. If we substitute the expressions (3), (4) and (6), (7) for the three terms of the original equation (1), we have

$$R\gamma^2 = \frac{\frac{1}{2}r^2F\gamma\omega}{J} - \theta\epsilon\gamma \dots\dots\dots\dots(8),$$

from which we deduce

$$\gamma = \frac{\frac{1}{2}r^2F\omega - J\theta\epsilon}{JR} \dots\dots\dots\dots(9).$$

8. It appears from this result that the value of γ will be positive or negative according as the angular velocity of the disc exceeds or falls short of a certain value Ω, given by the equation

$$\Omega = \frac{J\theta\epsilon}{\frac{1}{2}r^2F} \dots\dots\dots\dots(10);$$

and therefore we conclude that, when the angular velocity has exactly this value, the electro-motive intensity of the disc is just equal to the intensity of the reverse electro-motive force exerted on the fixed wires, by the electro-chemical apparatus with which they are connected.

9. If we adopt as the unit of electro-motive intensity that which is produced by a conductor of unit length, carried, in a

* Since the present article was put into the Editor's hands, I have become acquainted with a paper by Mr Joule "On the Heat evolved during the Electrolysis of Water," published by the Literary and Philosophical Society of Manchester in 1843 (Vol. VII. part 3, second series), in which it is shown, that in some cases of electro-chemical action (for instance, when hydrogen is evolved at an electrode or battery-plate of a metal possessing a considerable affinity for oxygen) there is a "resistance to electrolysis without chemical change," producing "a reaction on the intensity of the battery," and causing the evolution of heat to an amount exactly equivalent to the loss of heating power, or of external electro-motive force, which the battery thus suffers. [Note of April 17, 1882. For further development of this subject see Art. LV. below.] In any electro-chemical apparatus in which this kind of resistance occurs, the quantity of heat developed by a current of strength γ will be expressible in the form $A\gamma + B\gamma^2$, where A and B are finite when γ is infinitely small. Consequently what is denoted in the text by R will be equal to $\frac{A}{\gamma} + B$, and will therefore be infinitely great when γ is infinitely small. The modification required for such cases will be simply to use B in place of R, and to diminish the value of I found in the text (12) by JA; but the assumption that R does not become infinite in any of the circumstances considered is, I believe, quite justifiable in the two special cases which form the subject of the present communication.—W. T., Nov. 1, 1851.

magnetic field of unit force, with unit velocity, in a direction
which is both perpendicular to its own length and to the lines of
force in the magnetic field, it is easily shown that the electro-
motive force of the disc, in the circumstances specified above, is
given by the equation

$$i = \frac{1}{2} r^2 F \omega \dots\dots\dots\dots\dots\dots(11).$$

Hence if I denote the electro-motive force of the disc when it
just balances that of the chemical apparatus, we have by (10)

$$I = J\theta\epsilon\dots\dots\dots\dots\dots\dots\dots(12).$$

This equation comprehends a general expression of the conclusion long since arrived at by Mr Joule, that the quantities of heat
developed by different chemical combinations are, for quantities
of the chemical action electrically equivalent, proportional to the
intensities of galvanic arrangements adapted to allow the combinations to take place without any evolution of heat in their own
localities; and it may be stated in general terms thus :—

*The intensity of an electro-chemical apparatus is, in absolute
measure, equal to the mechanical equivalent of as much of the
chemical action as goes on with a current of unit strength during
a unit of time.*

10.   When ω is less than Ω, γ is (§ 8) negative; and hence
equations (3), (5) and (6), show W and M to be negative also.
In this case the direction of the current is contrary to the electro-
motive force of the disc; the chemical action is the source of
the current instead of being an effect of it; and the disc by its
rotation produces mechanical effect as an electro-magnetic engine,
instead of requiring work to be spent upon it to keep it moving
as a magneto-electric machine.   If we assume

$$\gamma = -\gamma', \quad M = -M', \quad W = -W',$$

so that when γ, M, and W are negative their absolute values
may be represented by γ', M' and W', we find by (9), (10), (5),
(6), (2), (3) the following expressions for these quantities :

$$\gamma' = \frac{\frac{1}{2} r^2 F}{JR} (\Omega - \omega)\dots\dots\dots\dots\dots(13),$$

$$M' = J\theta\epsilon\gamma' = \frac{\frac{1}{2} r^2 F . \theta\epsilon}{R} (\Omega - \omega)\dots\dots\dots\dots(14),$$

$$W' = M' - JH = \frac{1}{2} r^2 F \omega . \gamma' = \frac{\omega}{\Omega} M'\dots\dots(15).$$

The first of the three expressions (15) for W' merely shows that the mechanical effect produced by the disc in any period of time is less than M', the full mechanical equivalent of the consumption of materials in the electro-chemical apparatus, by the mechanical equivalent of the heat generated in the whole circuit during that period. From the third we infer, that the fraction of the entire duty of the consumption which is actually performed by the engine is equal to $\frac{\omega}{\Omega}$. If ω were precisely equal to Ω, the electro-motive force of the battery would be precisely balanced, and there could be no current, and hence the performance of the engine cannot be perfect; but if ω be less than Ω by an infinitely small amount, the battery will be allowed to act very slowly; a very slight current, with a very small consumption of materials, will be generated; and the mechanical effect produced from it will be infinitely nearly equal to the whole duty, and infinitely greater than the portion of the effect wasted in the creation of heat throughout the circuit.

11.   A condition precisely analogous to that of *reversibility*, established by Carnot and Clausius as the criterion of perfection for a thermo-dynamic engine*, is applicable to this electro-magnetic engine; and is satisfied by it when the disc revolves with an angular velocity infinitely nearly equal to Ω, since then γ', M', and W' are each of them proportional to $\Omega - \omega$, whether this quantity be positive or negative; and consequently if the motion of the disc relatively to a state of rotation with the angular velocity Ω be reversed, all the physical and mechanical agencies are reversed.

12.   From experiments made at Manchester in the year 1845 by Mr Joule, on the quantity of zinc electrolysed from a solution of sulphate of zinc by means of a galvanic current measured by his tangent galvanometer, I have found the electro-chemical equivalent of zinc to be ·07284† [= ·003311 c. g. s.]; and I am

---

* "If an engine be such that, when it is worked backwards, the physical and mechanical agencies in every part of its motions are all reversed, it produces as much mechanical effect as can be produced by any thermo-dynamic engine, with the same temperatures of source and refrigerator, from a given quantity of heat." (From § 9 of "Dynamical Theory of Heat." [Art. XLVIII. above.] *Transactions of the Royal Society of Edinburgh*, March 17, 1851, Vol. xx. part 2.)

† See note on "Electro-chemical Equivalents" published at the end of this paper.

informed by him, that from other experiments which he has made, he finds that the entire heat developed by the consumption of a grain of zinc in a Daniell's battery is as much as would raise the temperature of 769 grains of water from $0°$ to $1°$ Cent.* Hence, if we wish to apply the preceding investigations to the case in which the electro-chemical apparatus (§ 4) is a single cell of Daniell's battery, we may consider the consumption of a grain of zinc as the unit of the chemical action which takes place, and therefore we have

$$\epsilon = ·07284, \quad \theta = 769.$$

Again, according to Mr Joule's last researches on the mechanical equivalent of heat, the work done by a grain of matter in descending through 1390 feet is capable of raising the temperature of a grain of water from $0°$ to $1°$. Hence, since the unit of force adopted in the measurement of galvanic strength on which the preceding value of ϵ is founded, is that force which, operating during one second of time upon one grain of matter, would generate a velocity of one foot per second, and is consequently $\dfrac{1}{32\cdot2}$ of the weight of a grain at Manchester, we have

$$J = 1390 \times 32\cdot2 = 44758.$$

Substituting these values for ϵ, θ, and J in (12), we have

$$I = 2507100 \ [= 1\cdot074 \text{ Volts}]$$

for the "intensity" or "electro-motive force" of a cell of Daniell's battery in absolute measure. To compare this with the electro-motive intensity of a revolving disc such as we have considered (§ 3), let the axis of rotation of the disc be vertical or nearly vertical, and, the vertical component of the terrestrial magnetic force at Manchester being about 9·94, let us suppose that we have

---

* By experiments on the friction of fluids, Mr Joule has found that the quantity of work necessary to raise the temperature of a pound, or 7000 grains, of water from $0°$ to $1°$ Cent. is 1390 foot-pounds. Hence the mechanical equivalent of the consumption of a grain of zinc in Daniell's battery is 152·7, or nearly 153 foot-pounds. Messrs Scoresby and Joule, in their paper "On the Powers of Electro-magnetism, Steam and Horses" (*Phil. Mag.*, Vol. xxvi. 1846, p. 451), use 158 as the number expressing this equivalent according to earlier experiments made by Mr Joule. The experiments from which he deduced the thermal equivalents of chemical action, communicated to me for this paper, are described in a paper communicated to the French Institute, and acknowledged in the *Comptes Rendus* for Feb. 9, 1846, but not yet published.

$F = 10$ exactly, which would be the case with a disc exactly horizontal in localities a little north of Manchester, and might be
made the case in any part of Great Britain by a suitable adjust-
ment of the axis of the disc.   Then we have by (11),

$$i = 5\omega r^2;$$

or if n be the number of turns per second,

$$i = 5 \times 2\pi n r^2 = 31\cdot416 \times n r^2.$$

Hence

$$\frac{i}{I} = \frac{31\cdot416 \times n r^2}{2507100} = \frac{n r^2}{79803}.$$

It appears, therefore, that if the radius of the disc be one foot,
it would, when revolving at the rate of one turn per second, produce an intensity $\dfrac{1}{79803}$ of that of a single cell of Daniell's
and it would consequently have to make more than 79803
turns per second to reverse the action of such a cell in the
arrangement described in § 4.*   We conclude also, that a disc of
one foot radius, touched at its centre and circumference by the
electrodes of a single cell of Daniell's, and allowed to turn about
a vertical axis by the action of the earth upon the current passing
through it, would revolve with a continually accelerated motion
approaching to the limiting rate of 79803 turns per second, if it
were subject to no frictional or other resistance; and that if, by resisting forces, it were kept steadily revolving at the rate of n turns
per second, it would, in overcoming those forces, be performing
$\dfrac{n}{79803}$ of the whole work due to the consumption of zinc and
deposition of copper in the battery.

13.   If the electro-chemical apparatus mentioned in § 4 be a
vessel of pure water with two plates of platinum immersed in it,
we may consider a grain of hydrogen electrolysed as the unit for
measuring the chemical action which takes place.   Now Mr Joule
finds that, in the electrolysis of one grain of hydrogen from

---

* Hence in the multiple form of "the new electrical machine" suggested by
Faraday, about 800 discs, each one foot in radius, would be required, so that with a
rotation at the rate of 100 turns per second about a vertical axis in any part
of Great Britain, it might give an intensity equal to that of a single cell of
Daniell's.

water acidulated with sulphuric acid, as much heat is absorbed as would raise the temperature of 33553 grains of water from 0° to 1°. Hence θ must be less than 33553 by the quantity of heat evolved when as much pure water as contains one grain of hydrogen is mixed with acidulated water, such as that used by Mr Joule; but, without appreciable error on this account, we may take

$$\theta = 33553.$$

I have found also, from results of experiments on the electro-lysis of water made by Mr Joule at Manchester in 1845, that the electro-chemical equivalent of hydrogen is ·002201. Using this value for ϵ, and the values indicated above for θ and J, we have by (12),

$$I = 3305400 \ [= 1\cdot416 \text{ Volts}]$$

for the electro-motive force, in absolute measure, required for the decomposition of water. This exceeds the electro-motive force of a single cell of Daniell's battery, found above, in the ratio of 1·318 to 1. Hence at least two cells of Daniell's battery are required for the electrolysis of water; but fourteen cells of Daniell's battery connected in one circuit with ten electrolytic vessels of water with platinum electrodes would be sufficient to effect gaseous decomposition in each vessel.

14.   In the Bakerian Lecture of 1832, "On Terrestrial Mag-neto-electric Induction," Faraday, after describing some experi-ments he had made at Waterloo Bridge, without however arriving at any positive results, to test the existence of an inductive effect of the terrestrial magnetic force upon the flowing water of the Thames, brought forward some very remarkable speculations regarding the possible effects of magneto-electric induction upon large masses in motion relatively to the earth, or upon the earth itself in motion with reference to surrounding space. The pre-ceding investigations enable us to compare the electro-motive forces in such cases with electro-motive forces the effects of which are familiarly known to us, and so to form some estimate, it may be very vague, of the anticipated effects. Thus let us conceive a mass of air or water, or any other substance moving relatively to the earth with a velocity V, and let A and B be two fixed points in it or at its two sides, at a distance a apart, in a line perpendicular to the direction of motion. Then if F

be the component of the terrestrial magnetic force perpendicular
to the plane of AB and the lines of motion across it, there will
be between A and B, or between any fixed conductors connected
with them, and insulated in all other places from the moving
mass, an electro-motive force, the intensity of which is given by
the equation

$$i = F V a.$$

15. If, for instance, the velocity be one mile per hour, we
should have $V = 1\cdot4667$; and if we take $F = 10$, which will be
nearly the case for a mass moving horizontally in any part of
Great Britain*, we have

$$i = 14\cdot667 \times a.$$

If we take $a = 960$, we find $i = 14080$ for the electro-motive
force between two platinum plates immersed, as in Faraday's
experiment, below the surface of the Thames, at a distance of
960 feet apart across the stream, when the tide is in such a state
that the current is at the rate of a mile an hour. The electro-
motive force varying directly as the rate of the current, must
therefore, when there is a current of two miles and a half an hour,
be 35200, which is very little more than $\dfrac{1}{100}$ of that which
was found in § 13 for the intensity required to decompose water;
and as there is probably in no state of the tide a current of more
than three or four miles an hour, it is not to be wondered at
that no galvanic current was discovered in a wire connecting the
platinum plates.

16. An experiment on a much larger scale might be performed
by means of the telegraph wires which have recently been laid
between England and France, across the straits of Dover, by
simply connecting the ends of one of these wires with platinum
plates immersed in the sea on the two sides of the channel. If
the distance between the plates be twenty miles, in a direction
on the whole at right angles to the direction of the motion of the

---

* In June, 1846, the horizontal magnetic force was found to be 3·7284, and the
dip 68° 58', at Woolwich (*Philosophical Transactions*, 1846, p. 246). Hence the
vertical force was 3·7284 × tan 68° 58', or 9·696. At the same period it was 9·94 at
Manchester, and it must have been 10 exactly at localities in England or Scotland
not far north of Manchester.

water through the channel, and if, in a particular state of the tide, there be an average velocity of a mile an hour, there would, as we find from the preceding expression, by substituting 20×5280 for a, be an electro-motive force of 1,549,000, or very nearly half of that which is required for the decomposition of water. It is not probable that the current produced by the action of this force alone through the wire connecting the platinum plates would be found to be sensible; since a sensible continuous current with water and platinum electrodes in the circuit can scarcely be obtained by any electro-motive force less than that which is required for the continued gaseous decomposition of water. The existence of the inductive action might, however, I think, be tested by using a galvanic battery of very low intensity, to assist the electro-motive force arising from induction, and by adding a little nitric acid to the liquid till it is found that a sensible current is produced. It might then be observed whether or not, when the tide turns and the water flows in the other direction through the channel, the electrical current becomes insensible, or becomes less than it was; and whether it goes on again as before when the tide turns again, and the water flows as it did at first. There would probably be some difficulty experienced in keeping the electro-motive force of the battery sufficiently constant during twelve hours to make the experiment perfectly satisfactory, and many difficulties that could not be foreseen might occur. If, however, in any state of the tide the mean rate per hour of the stream in the channel exceeds two miles or two miles and a half, it is probable that the inductive action might produce a sensible electric current in the telegraph wire without such assistance. It is very much to be desired that the experiment should be tried, as it would afford probably the best test that could at present be applied, to find whether electrolysable liquids possess the property of magneto-electric induction discovered by Faraday in metallic conductors.

17. The possible magneto-electric effects of the earth's rotation were also considered by Faraday, and it was conjectured that electricity may, in virtue of them, rise to considerable intensity. The general nature of the effect was shown to be a tendency for electricity to flow through the earth from the equator towards the poles, from whence it would endeavour to return externally to the equatorial regions. If the distribution of terrestrial mag-

31—2

netism were perfectly symmetrical about the axis of rotation, there could be no other kind of effect than this produced by the rotatory motion; and, neglecting at present the currents in complete external circuits, which may exist in virtue of the want of this symmetry, we may endeavour to form a rough estimate of the electro-motive force that would exist between the equatorial regions of the revolving mass and a quadrantal conductor fixed relatively to the earth's centre, with one end near the surface at the equator and the other touching the surface at one pole. The electrical circumstances would be the same if the earth were at rest, and the conductor were made to revolve once round in $23^h\ 56^m\ 4^s$, with one end always touching at the pole, and the other close to the surface at the equator. In such circumstances there would be an electro-motive force equal to $f v\, ds$ on any infinitely small element ds of the moving conductor, if v denote the velocity of its motion, and f the vertical magnetic force at the part of the earth over which it is passing. Now if θ be the latitude of the element ds, and V the velocity of the surface at the equator, we have

$$v = V\cos\theta\,;$$

if the distribution of magnetic force at the surface be, as in making this rough estimate we may assume it to be, of the simplest type, we have

$$f = F\sin\theta,$$

where F denotes the vertical magnetic force at the pole; and if r denote the earth's radius, we have

$$ds = rd\theta.$$

The intensity of the total electro-motive force between the equatorial end of the moving conductor and the earth, being the sum of the electro-motive forces on all its elements, will consequently be equal to

$$\int_0^{\frac{1}{2}\pi} FVr\sin\theta\cos\theta\, d\theta\,;$$

and hence, denoting it by i, we have

$$i = \frac{1}{2}\,FVr.$$

Now the earth's diameter being about 7912 miles, we have $r = 3956 \times 5280$; and, by dividing the number of feet in the earth's circumference by 86164, the number of seconds in the sidereal day, we find $V = 1523$. If we take $F = 14$, we find, by substituting these values for the factors of the preceding expression,

$$i = 222,700,000,000.$$

This is about 88800 times the intensity of a single cell of Daniell's battery (§ 12), and may therefore be about 50 times that of the battery of two thousand pairs of copper and zinc plates, charged with nitro-sulphuric acid, by which Sir Humphry Davy only obtained sparks half an inch long in the exhausted receiver of an air-pump. Now the electro-motive force we have been considering could in reality only produce galvanic currents by forcing a passage through the whole thickness of the atmosphere, upwards from the surface about the poles, and downwards to the earth in the equatorial regions, and we may conclude *that it does not produce galvanic currents.*

18.   From the smallness of the electro-motive intensity in this extreme case, we may infer that no part of the phenomena of atmospheric electricity can be attributed to the inductive action of the terrestrial magnetism on masses of air or water in motion near the surface of the earth.

19.   If the space surrounding the earth, beyond the limits of the atmosphere, were capable of conducting electricity, and were affected as a fixed conductor by the motion of a magnet in the neighbourhood of it, there would be electrical currents in complete external circuits, induced both by the earth's rotatory motion, on account of the distribution of magnetism not being symmetrical about the axis of rotation, and by its motion through space; and it is I think far from improbable that the phenomena of aurora borealis and australis are so produced. It is quite impossible, in the present state of science regarding the relative motion of the earth or of the solar system, and the medium filling all space, which by its undulations transmits light and radiant heat, to form any estimate on satisfactory principles of the inductive electro-motive forces which may arise from the motion of translation of the terrestrial magnet through this medium; but we may form some idea of those which its rotatory motion may produce by calculating the total electro-motive force on a closed

conductor held externally in a fixed position with reference to the
earth's centre. Thus let us conceive a circular conductor, of
radius R, to be held with a diameter coincident with the earth's
axis of rotation [and its centre coincident with the earth's]; and
let i be the intensity of the total electromotive force which it
would experience if it were made to revolve round the earth once
in $23^h\ 56^m\ 4^s$, and the earth held at rest. Denoting by P the
radial component of the terrestrial magnetic force at any element
of this conductor, and in other respects using the same notation
as before, we have

$$i = \int_{-\pi}^{\pi} P \frac{R}{r}\, V \cos \theta . R d\theta = V \frac{R^2}{r} \int_{-\pi}^{\pi} P \cos \theta d\theta.$$

If we assume the distribution of magnetic force at the earth's
surface to be of the simplest type, the force at either magnetic
pole to be 15, and the magnetic axis to be inclined at an angle
of 20^0 to the axis of rotation, we have, at the time when the
moving conductor is passing over the earth's magnetic poles,

$$P = 15 \frac{r^3}{R^3} \sin (\theta + 20^0) ;$$

and in these circumstances we have consequently

$$i = 15 V \frac{r^2}{R} \int_{-\pi}^{\pi} \sin (\theta + 20^0) \cos \theta d\theta = 15 \times 1523 . \frac{r^2}{R} \pi \sin 20^0.$$

If we take $r = R$, we find, from this,

$$i = 512{,}700{,}000{,}000,$$

which is about 204000 times the intensity of a single cell of
Daniell's. One-half or one-third of this amount would be the
electro-motive force experienced by a fixed circular conductor of
twice or three times the earth's diameter, at the time when the
earth's magnetic poles are passing under it.

GREYPOINT, COUNTY DOWN,
October 6, 1851.

Note on Electro-chemical Equivalents.

The electro-chemical equivalents of zinc and hydrogen used in the preceding paper were deduced from experiments made by Mr Joule on the electrolysis of sulphate of zinc and of water acidulated with sulphuric acid, in which the galvanic currents used were measured by means of a tangent galvanometer consisting of a needle half an inch in length, suspended in the centre of a circular conductor one foot in diameter, fixed in the plane of the magnetic meridian.   The electro-chemical equivalent of a substance, being defined as the mass (in grains) electrolysed from any combination in a second of time by the action of a current of unit strength, will be found by dividing the mass of the substance electrolysed per second in any experiment by the strength of the current.   One way of combining several experiments so as to obtain a mean result, will be to take the arithmetical mean of the quantities of the substance found to be electrolysed per second in the different experiments, and divide it by the mean of the observed strengths of the currents.   In the tangent galvanometer, the tangents of the angles of deflection are proportional to the strengths of the currents, and consequently the arithmetical mean of the tangents of the angles of deflection in different experiments will be the tangent of the angle of deflection corresponding to a current of mean strength.   The mean results, taken in this way, of four experiments on the electrolysis of sulphate of zinc, and of four experiments on the electrolysis of acidulated water, made at Manchester on the 8th, 9th, 15th and 16th of September, 1845, are as follows :—

Electrolysis of Sulphate of Zinc.

Mean corrected tangent of deflection.	Zinc deposited per second.
·7345	·01508 grain.

Electrolysis of Acidulated Water.

Mean corrected tangent of deflection.	Hydrogen liberated per second.
1·7600	·001092 grain.

To determine the strength of the current (γ) in absolute measure, which produces a deflection (δ) of the needle in the tangent galvanometer, we have the equation

$$\gamma = \frac{rH}{2\pi} \tan \delta,$$

where r denotes the radius of the circular conductor, and H the horizontal component of the earth's magnetic force, in absolute measure; since the magnetic axis of the needle will be drawn from the magnetic meridian into a vertical plane containing the resultant of the horizontal force H in the magnetic meridian, and the force $\frac{2\pi\gamma}{r}$ perpendicular to the plane of the conductor, and consequently to the magnetic meridian. It is impossible at present to assign with accuracy the values of the horizontal magnetic force at Manchester at the times when the experiments were made; but according to data which Colonel Sabine has kindly communicated to me, it must have been nearly 3·542 in 1846, and cannot probably at any time of observation during that or the preceding year have differed by as much as $\frac{1}{100}$ of its value from that amount. Taking, therefore, 3·542 for H, and taking $\frac{1}{2}$ for r (the diameter of the conductor being one foot), we have $\frac{rH}{2\pi} = ·28186$; and consequently, for observations made with Mr Joule's tangent galvanometer at Manchester in 1846,

$$\gamma = ·28183 \times \tan \delta.$$

Hence from the preceding experimental results, we find for the electro-chemical equivalent of zinc,

$$\frac{·01508}{·28183 \times ·7345}, \text{ or } ·07284; \ [= ·00331 \text{ c. g. s.}]$$

and for the electro-chemical equivalent of hydrogen,

$$\frac{·001092}{·28183 \times 1·7600}, \text{ or } ·002201. \ [= ·000102 \text{ c. g. s.}]$$

From the mean results of a series of four experiments on the electrolysis of sulphate of copper, communicated to me by Mr Joule, I have found for the electro-chemical equivalent of copper,

$$07052.$$

In Dove's *Repertorium* (Vol. VIII. p. 273), values of the electro-chemical equivalents of water and zinc, determined by Weber, who was the first to give an electro-chemical equivalent in absolute electro-magnetic measure, and by other experimenters, are given in absolute measure according to the French units. To reduce these to British measure, we must multiply by (2·1692), the square root of the fraction obtained by dividing the number (15·432) of grains in a gramme, by the number (3·2809) of feet in a metre. The electro-chemical equivalent of water is obtained by multiplying that of hydrogen by 9; and according to the theory of equivalence in electro-chemistry, it might also be obtained by multiplying the electro-chemical equivalent of zinc by $\frac{9}{32\cdot3}$, and that of copper by $\frac{9}{31\cdot7}$. The following table shows the values of the electro-chemical equivalent of water in British absolute measure obtained in these different ways.

Observers.	Galvanometer used.	Electro-chemical action observed.	Deduced electro-chemical equivalent of water.*
Weber..........	The "electro-dynamometer"	Decomposition of water	·02034
Bunsen.........	Tangent galvanometer	Decomposition of water	·02011
Bunsen.........	Ditto.	Dissolution of zinc	·01995
Casselmann...	Ditto.	Decomposition of water in acid and saline solutions	·02033
Casselmann...	Ditto.	Zinc [deposited or dissolved?]	·02021
Joule	Ditto.	Decomposition of water	·01981
Joule	Ditto.	Deposition of zinc from solution of sulphate of zinc	·02030
Joule	Ditto.	Deposition of copper from solution of sulphate of copper	·02002

* [Note of June 1st, 1882. These values may be compared with others on the c. g. s. system, by observing that an electro-chemical equivalent of ·02 in the old "British absolute measure" (foot-grain-second), is equal to ·000922 on the c. g. s. system of units.]

ART. LIV. APPLICATIONS OF THE PRINCIPLE OF MECHANICAL
EFFECT TO THE MEASUREMENT OF ELECTRO-MOTIVE FORCES,
AND OF GALVANIC RESISTANCES, IN ABSOLUTE UNITS.

[*Phil. Mag.* Dec. 1851.]

1. IN a short paper "On the Theory of Electro-magnetic
Induction," [Art. XXXV. above], communicated to the British
Association in 1848*, I demonstrated that "the amount of mechanical effect continually *lost* or spent in some physical agency
(according to Joule, the generation of heat) during the existence
of a galvanic current in a given closed wire, is, for a given time,
proportional to the square of the strength of the current;" and
I showed that Neumann's beautiful analytical expression for the
electro-motive force experienced by a linear conductor moving
relatively to a magnet of any kind, is, in virtue of this proposition,
an immediate consequence of the general principle of mechanical
effect. At that time I did not see clearly how the reasoning could
be extended to inductive effects produced by a magnet (either of
magnetized matter or an electro-magnet) of varying power upon a
fixed conductor in its neighbourhood, or to "the induction of a varying current on itself;" but I have recently succeeded in making this
extension, and found that the same general principle of mechanical
effect is sufficient to enable us to found on a few elementary
facts, a complete theory of electro-magnetic or electro-dynamic
induction. The present communication, which is necessarily very
brief, contains some propositions belonging to that part of the
theory which was communicated to the British Association; but
it is principally devoted to practical applications with reference
to the measurement of electro-motive forces arising from chemical
action, and to the system of measurement of "galvanic resistance
in absolute units," recently introduced by Weber†.

2. PROP. I.—*If a current of uniform strength be sustained
in a linear conductor, and if an electro-motive force act in this*

---

* *Report*, 1848; *Transactions of Sections*, p. 9.

† "Messungen galvanischer Leitungswiderstände nach einem absoluten Maasse;"
von Wilhelm Weber.—Poggendorff's *Annalen*, March 1851, No. 3.

conductor in the same direction as the current, it will produce work at a rate equal to the number measuring the force multiplied by the number measuring the strength of the current.

3.   Let the electro-motive force considered be produced by the motion of a straight conductor of unit-length, held at right angles to the lines of force of a magnetic field of unit-intensity, and carried in a direction perpendicular to its own length and to those lines of force. The velocity of the motion will be numerically equal to the electro-motive force, which will be denoted by F, thus inductively produced, since the unit of electro-motive force adopted by those who have introduced or used absolute units in electro-dynamics is that which would be produced in the same circumstances if the velocity of the motion were unity. If the ends of the moveable conductor be pressed on two fixed conductors, connected with one another either simply by a wire, or through any circuit excited by electro-motive forces, so that a current of strength γ is sustained through it, it will experience an electro-magnetic force in a direction perpendicular to its own length and to the lines of magnetic force in the field across which it is moving, of which the amount will be the product of γ into the intensity of the magnetic force, or, since this is unity, simply to γ *. The motion of the conductor being in that line, the force will be directly opposed to it when the current is in the direction in which it would be if it were produced solely by the electro-motive force we are considering; and therefore, if we regard γ as positive when this is the case, the work done in moving the conductor during any time will be equal to the product of γ into the space through which it is moved, and will therefore in the unit of time be $F\gamma$, since F is numerically equal to the velocity of the motion. But this work produces no other effect than making the electro-motive force act, and therefore the electro-motive force must produce some kind of effect mechanically equivalent to it. Now if an equal electro-motive force were produced in any other way (whether chemically, thermally, or by a common frictional electrical machine) between the same two conductors, connected in the same way, it would produce the same effects. Hence, universally, the mechanical value of the work done in a unit of time by an electro-motive

---

* This statement virtually expresses the definition of the "strength" of a current, according to the electro-magnetic unit now generally adopted.

force F, on a circuit through which a current of strength γ is passing, is $F\gamma$.

4. If the algebraic signs of F and γ be different, that is if the electro-motive force act against the direction of the current, the amount of work done by it is negative, or effect is gained by allowing it to act. This is the case with the inductive re-action, by which an electro-magnetic engine at work resists the current by which it is excited, or with the electrolytic resistance experienced in the decomposition of water.

5. The application of the proposition which has just been proved, to chemical and thermal electro-motive forces is of much importance. I hope to make a communication to the Royal Society of Edinburgh before the end of this year, in which, by the application to thermal electro-motive forces, the principles explained in a previous communication* "On the Dynamical Theory of Heat," will be extended so as to include a mechanical theory of thermo-electric currents. The application to chemical electro-motive forces leads immediately to the expression for the electro-motive force of a galvanic battery, which was obtained by virtually the same reasoning, in another paper published in this Volume of the Magazine† (p. 429) [Art. LIII. above]: for if ϵ be the electro-chemical equivalent of one of the substances concerned in the chemical action; if θ be the quantity of heat evolved by as much of the chemical action concerned in producing the current as takes place during the consumption of a unit of mass of this substance; and if J be the mechanical equivalent of the thermal unit, the mechanical value of the chemical action which goes on in a unit of time will be $J\theta\epsilon\gamma$, and this must therefore be equal to $F\gamma$, the work done by the electro-motive force which results. Hence we have

$$F = J\theta\epsilon,$$

which is the expression given in the paper referred to above, for the electro-motive force of a galvanic battery in absolute measure.

6. In applying this formula to the case of Daniell's battery, I used a value for θ derived from experiments made by Mr Joule,

---

* March, 1851. Published in the *Transactions*, Vol. XX. Part II. [Art. XLVIII. Parts I. II. III. above.]

† "On the Mechanical Theory of Electrolysis." [Art. LIII. above.]

the details of which have not yet been published, but which I
believe to have consisted of observations of phenomena depending
on the actual working electro-motive forces of the battery. I am
now enabled to compare that value of the thermal equivalent,
with the results of observations made directly on the heat of
combination, by Dr Andrews *, who has kindly communicated to
me the following *data* :—

(1) The heat evolved by the combination of one grain of zinc with gaseous oxygen amounts to

1301 units.

(2) The heat evolved by the combination of the 1·246 grains of oxide thus formed with dilute sulphuric acid amounts to

369 units.

(3) The heat evolved by the combination of the equivalent quantity, ·9727 of a grain of copper, with oxygen, amounts to

588·6 units.

(4) The heat evolved by the combination of the 1·221 grains of oxide thus formed, with dilute sulphuric acid, amounts to

293 units.

Hence the thermal equivalent of the whole chemical action which
goes on in a Daniell's battery during the consumption of a grain
of zinc is

$$1301 + 369 - (588 \cdot 6 + 293), \text{ or } 788 \cdot 4 \dots\dots\dots\dots(I):$$

the thermal equivalent of the part of it which consists of oxidation and deoxidation alone is

$$1301 - 588 \cdot 6, \text{ or } \qquad 712 \cdot 4 \dots\dots\dots\dots(II).$$

The thermal equivalent which I used formerly is

$$769 \dots\dots\dots\dots\dots\dots\dots\dots(III).$$

If the opinion expressed by Faraday, in April, 1834 (*Exper.
Researches*, 919), with reference to the galvanic batteries then
known, that the oxidation alone is concerned in producing the
current, and the dissolution of the oxide in acid is electrically inoperative, be true for Daniell's battery, the number (II) is the
thermal equivalent of the electrically effective chemical action.
Joule's number (III) is considerably greater than this, and falls
but little short of (I), the thermal equivalent of the *whole* chemi-

---

* Published in his papers " On the Heat disengaged during the Combination of
Bodies with Oxygen and Chlorine" (*Phil. Mag.* May and June, 1848), "On the
Heat disengaged during Metallic Substitutions" (*Phil. Transactions*, Part I. for
1848), "On the Heat developed during the Combination of Acids and Bases"
(*Trans. Royal Irish Academy*, Vol. XIX. Part II.), &c.

cal action that goes on during the consumption of a grain of zinc. If we take successively (I), (II), (III) as the value of θ, and take for ϵ and J the values ·07284 and 44758, which were used in my former paper, we find the following values for the product $J\theta\epsilon$:—

(I)   2570300, which would be the electro-motive force (in [= 1·101 Volts]   British absolute units) of a single cell of Daniell's battery if the whole chemical action were electrically efficient.

(II)   2322550, which would be the electro-motive force of a [= ·995 Volts]   single cell of Daniell's battery if only the oxidation and deoxidation of the metals were electrically efficient.

(III) 2507100, which is the electro-motive force of a single cell [= 1·074 Volts]   of Daniell's battery, according to Joule's experiments.

7.   The thermal equivalent of the whole chemical action in a cell of Smee's battery (zinc and platinized silver in dilute sulphuric acid), or of any battery consisting of zinc and a less oxidizable metal immersed in dilute sulphuric acid, is found by subtracting the quantity of heat that might be obtained by burning in gaseous oxygen the hydrogen that escapes, from the quantity of heat that would be obtained in the formation of the sulphate if the zinc were oxidized in gaseous oxygen instead of by combination with oxygen derived from the decomposition of water. Now the quantity of hydrogen that escapes during the consumption of a grain of zinc is $\frac{1}{32\cdot53}$ of a grain (if 32·53, which corresponds to the equivalents used by Dr Andrews, be taken as the equivalent of zinc, instead of 32·3 which I used in my former paper). According to Dr Andrews' experiments, the combination of this with gaseous oxygen would evolve

$$\frac{1}{32\cdot53} \times 33808, \text{ or } 1039\cdot3 \text{ units of heat.}$$

Hence the thermal equivalent of the whole chemical action corresponding to the consumption of a grain of zinc in Smee's battery is

$$1301 + 369 - 1039\cdot3, \text{ or } 630\cdot7 \ldots\ldots\ldots\ldots\ldots(I).$$

The equivalent of that part which consists of the oxidation of zinc and the deoxidation of hydrogen is

$$1301 - 1039\cdot3, \text{ or } \qquad 261\cdot7\ldots\ldots\ldots\ldots\ldots\text{(II).}$$

Hence (I) if the whole chemical action be efficient in producing the current, the electro-motive force is 2056200.

(II) If only the oxidation and deoxidation be efficient, the electro-motive force is 853190.

The *external* electro-motive force (or the electro-motive force with which the battery operates on a very long thin wire connecting its plates), according to either hypothesis, would be found by subtracting the "chemical resistance*" due to the evolution of hydrogen at the platinized silver, from the whole electro-motive force : but, on account of the feeble affinity of the platinized surface for oxygen, it is probable that this opposing electro-motive force, if it exist at all, is but very slight.

(III) The external electro-motive force of a single cell of Smee's battery is, according to Joule's experiments †, ·65 of that of a single cell of Daniell's; and therefore if we take the preceding number (III), derived from his own experiments, as the true external electro-motive force of a single cell of Daniell's, that of a single cell of Smee's is

$$1,629,600.$$

This number is nearly double that which was found for the electro-motive force on the supposition that the oxidation and deoxidation alone are electrically efficient; but it falls considerably short of what was found on the suppositions that the whole chemical action is efficient, and that there is no "chemical resistance."

8. It is to be remarked that the external electro-motive force determined for a single cell of Smee's, according to the preceding principles, by subtracting the "chemical resistance" from the value of $J\theta\epsilon$, is the *permanent working* external electro-motive force. The electro-statical tension, which will determine the initial working external electro-motive force, depends on the

---

* See foot note on § 6 of my paper "On the Mechanical Theory of Electrolysis." [Art. LIII. above. See also Art. LV. below.]

† *Phil. Mag.*, Jan.—June, 1844, XXIV. p. 115, and Dove's *Rep.*, Vol. VIII. p. 341.

primitive state of the platinized silver plate. It could never be greater than to make the initial working force be

$$J \times 1670 \times \epsilon, \text{ or } 5444500,$$

corresponding to the combination of zinc with gaseous oxygen and of the oxide with sulphuric acid. It might possibly reach this limit if the platinized surface had been carefully cleaned, and kept in oxygen gas until the instant of immersion, or if it had been used as the positive electrode of an apparatus for decomposing water, immediately before being connected with the zinc plate; and then it could only reach it if the whole chemical action were electrically efficient, and if there were no "chemical resistance" due to the affinity of the platinized surface for oxygen.

9. It is also to be remarked, that the permanent working electro-motive force of a galvanic element, consisting of zinc and a less oxidizable metal immersed in sulphuric acid, can never exceed the number 2056200, derived above from the *full* thermal equivalent for the single cell of Smee's, since the chemical action is identical in all such cases, and the mechanical value of the external effect can never exceed that of the chemical action. In a pair consisting of zinc and tin, the electro-motive force has been found by Poggendorf* to be only about half that of a pair consisting of zinc and copper, and consequently less than half that of a single cell of Smee's. There is therefore an immense loss of mechanical effect in the external working of a galvanic battery composed of such elements; which *must* be compensated by heat produced within the cells. I believe with Joule, that this compensating heat is produced at the surface of the tin in consequence of hydrogen being forced to bubble up from it, instead of the metal itself being allowed to combine with the oxygen of the water in contact with it. A most curious result of this theory of "chemical resistance" is, that in experiments (such as those of Faraday, *Exp. Researches*, 1027, 1028) in which an electrical current passing through a trough containing dilute sulphuric acid, is made to traverse a diaphragm of an oxidizable metal (zinc or tin), dissolving it on one side and evolving bubbles of hydrogen

---

* "Berl. Acb. 46, 242," Pogg. *Ann.*, LXX. 60. Dove's *Repertorium*, Vol. VII. p. 341.

on the other; part (if not all) of the heat of combination will be
evolved, not on the side on which the metal is eaten away, but on
the side at which the bubbles of hydrogen appear. It will be very
interesting to verify this conclusion, by comparing the quantities
of heat evolved in two equal and similar electrolytic cells, in the
same circuit, each with zinc for the positive electrode, and one
with zinc, the other with platinum or platinized silver for the
positive electrode. The electro-motive force of the latter cell
would be sufficient to excite a current through the circuit, but it
might be found convenient to add electro-motive force from some
other source *.

10. PROP. II. *The resistance of a metallic conductor, in terms
of Weber's absolute unit, is equal to the product of the quantity of
heat developed in it in a unit of time by a current of unit strength,
into the mechanical equivalent of a thermal unit.*

11. If H denote the quantity of heat developed in the conductor in a unit of time, by a current of strength γ, the mechanical
value of the whole effect produced in it will, according to the
principles established by Joule, be JH. But this effect is produced by the electro-motive force, F, and therefore, by Prop. I.,
we have

$$JH = F\gamma.$$

Now, according to Ohm's original definition of galvanic resistance,
if k denote the resistance of the given conductor, we have

$$\gamma = \frac{F}{k}.$$

If the electro-motive force and the strength of the current be
measured in absolute units of the kind explained above, the

* An examination of the thermal effects of a current through four equal and
similar vessels containing dilute sulphuric acid, and connected by means of electrodes [immersed plates] of zinc and platinum, varied according to the four per-
mutations of double zinc, double platinum, zinc-platinum, platinum-zinc, in one
circuit, excited by an independent galvanic battery or other electromotor, would
throw great light on the theory of chemical electro-motive forces and resistances.
Vessels containing electrodes of other metals, such as tin, variously combined, and
direct and reverse cells of Daniell's battery, might all be introduced into the same
circuit. If the exteriors of all the cells were equal and similar, the excesses
of their permanent temperatures above that of an equal and similar cell in the
neighbourhood, containing no source of heat within it, would be very nearly
proportional to the rates at which heat is developed in them. [Compare Article LV.
below.]

number k, expressing the resistance in this formula, will express it in terms of the absolute unit introduced by Weber. Using the value $k\gamma$ derived from this, for F, in the preceding equation, we have

$$JH = k\gamma^2.$$

This equation expresses the law of the excitation of heat in the galvanic circuit discovered by Joule; and if we take $\gamma = 1$, it expresses the proposition to be proved.

12. In Mr Joule's original paper on the heat evolved by metallic conductors of electricity*, experiments are described, in which the strengths of the currents used are determined in absolute measure, the unit employed being the strength which a current must have to decompose 9 grains of water in an hour of time. But the electro-chemical equivalent of water, according to the system of absolute measurement introduced by Weber, is, in British units, very nearly ·02, and therefore a current of unit strength would decompose 72 grains of water in an hour. Hence Joule's original unit is very exactly ⅛th of the British electro-magnetic unit for measuring current electricity. By using the formula $k = JH/\gamma^2$, and taking for γ one-eighth the number of Mr Joule's "degrees of current;" for H the quantity of heat (measured by grains of water raised 1° Cent.) evolved by the current through the conductor experimented on; and for J the value 44758; I have found

$$k = 13{,}240{,}000$$

as the absolute resistance of a certain wire used by Mr Joule for an absolute standard of resistance in the experiments on the heat evolved in electrolysis, described in the second part of the same paper †.

13. The "specific resistance" of a metal referred to unity of volume, may be defined as the absolute resistance of a unit length

---

* *Proceedings of the Royal Society*, Dec. 17, 1840; *Philosophical Magazine*, Vol. XIX. (Oct. 1841), p. 260.

† The three experiments from which the number in the text was deduced as a mean result (described in §§ 25, 26, 27 of the paper, *Phil. Mag.* Vol. XIX. Oct. 1841, p. 266), lead separately to the following values for the resistance:—

13260000
13360000
13090000;

none of which differs by as much as $\frac{1}{50}$ th from the mean given in the text.

of a conductor of unit section; and the specific resistance of a metal referred to unity of mass, or simply "the specific resistance of a metal" (since the term, which was introduced by Weber, is, when unqualified, so used by him), is defined as the absolute resistance of a conductor of uniform section, and of unit length and unit weight. Hence, since the resistance of conductors of similar substance are inversely proportional to their sections, and directly proportional to their lengths, we have

$$\sigma_v = k\omega/l^*, \quad \sigma = km/l^2 ;$$

if l be the length, ω the area of the section, and m the mass (or weight) of a conductor, k its absolute resistance, and σ_v and σ the specific resistances of its substance referred respectively to unity of volume and to unity of mass.

14.   The absolute resistance of a certain silver wire, and of a column of mercury contained in a spiral glass tube, may be determined from experimental *data* extracted from a paper of Mr Joule's laid before the French Institute (*Comptes Rendus*, Feb. 9, 1846), and communicated to me by the author.   In four experiments on the silver wire, and in four similar experiments on the mercury tube, a current measured by a tangent galvanometer was passed through the conductor, and, in each experiment, the quantity of heat evolved during ten minutes was determined by the elevation of temperature produced in a measured mass of water, the temperature of the conductors during all the experi-

---

* By means of this I have found 4·1 for the specific resistance of copper, according to the statement made in § 24 of Mr Joule's paper, that his standard conductor was "10 feet long and ·024 of an inch thick;" but there must be some mistake here, as it will be seen below that this is about double what we might expect it to be. [Note of July 27, 1882.   It is more probable that the wire was of less than half the proper conductivity for copper wire than that Joule had made any mistake in stating its dimensions.   Such deficiencies of conductivity in ordinary copper wire as, nine years later, I found even in wires supplied for submarine telegraph cables, some of which I found to have less than 40 per cent. of the conductivity of others, were not imagined possible at the date (1851) of this paper.]   I have found 2·17 for the specific resistance of copper referred to unity of volume, according to the experiment described in § 9, on a wire stated to be 2 yards long and $\frac{1}{28}$th of an inch thick; and 1·78 and 1·98, according to the experiments described in §§ 9 and 11, on a wire stated to be 2 yards long and $\frac{1}{40}$th of an inch thick; also 7·7 for that of iron, in a wire stated (§ 11) to be 2 yards long and $\frac{1}{37}$th of an inch thick.   It is to be remarked, however, that no attempt was made by Mr Joule to determine the sections of his wires with accuracy, and that the "thicknesses" are merely mentioned in round numbers, as descriptive of the kinds of wire used in his different experiments.

ments having been nearly 50° Fahr. The mean result of the four experiments on each conductor is expressed in terms of the square root of the sum of the squares of the tangents of the galvanometer-deflections, and the mean quantity of heat evolved in ten minutes. The weight of the silver wire in air and in water, the weight of the mercury contained in the glass tube, and the exact length of each conductor, were determined a short time ago, at my request, by Mr Joule, and the areas of the sections of the conductors have been deduced. The same galvanometer having been used as was employed in the experiments on electrolysis, referred to in the "Note on Electro-chemical Equivalents," contained in this Volume of the Magazine, [Art. LIII. above], and the experiments at present referred to having also been made at Manchester in 1845, the strength of the current in absolute measure is found by multiplying the tangent of deflection by ·28186. The various experimental data thus obtained are as follows:

Conductor.	Length in feet.	Mass in grains.	Sectional area in square feet.	Mean corrected tangent of deflection.	Mean strength of current in absolute units.	Mean quantity of heat produced in 10 minutes.
Silver wire..	27⅔	434·51	·0000034462	1·4526	·40943	{ 19375 grs. of water raised 1°·7718 C. in temperature.
Mercury in glass tube	} 5⅛	1511·5	·000048119			
				The resistance of the mercury conductor was found to be ·74964 of that of the silver wire.		

Taking as the thermal unit the quantity of heat required to raise the temperature of a grain of water by 1° Cent., we find 57·213 as the heat generated in the silver wire in one second, of which the mechanical equivalent is $44758 \times 57·213$. Dividing this by the square of the strength of the current, we find 15276000 for the absolute resistance of the silver wire; and by multiplying by ·74964, we deduce 11451000 for the absolute resistance of the mercury conductor. Multiplying each absolute resistance by the sectional area of the conductor to which it corresponds, and dividing by the length; and again, multiplying each resistance by the mass, and dividing by the square of the length, we obtain the following results with reference to the specific resistances of silver and mercury at about 10° Cent. of temperature.

Metal.	Specific resistance referred to unity of volume.		"Specific resistance."
	British system.	[C. G. S.]	
Silver	1·9028	[1768]	8671500
Mercury	106·65	[99081]	648410000

15. The "conducting powers" of metals, as ordinarily defined, are inversely proportional to their specific resistances referred to unity of volume. Hence, according to the preceding results, the conducting powers of silver and mercury at about 10° Cent. of temperature are in the proportion of 1 to ·01784. According to the experiments of M. E. Becquerel (Dove's *Repertorium*, Vol. VIII. p. 193), the conducting powers of silver and mercury at 0° Cent. are in the proportion of 1 to ·017387; and at 100° Cent., of 1 to ·022083: at 10° Cent. they must therefore be nearly in the proportion of 1 to ·01786, which agrees very closely with the preceding comparative result. Again, according to M. Becquerel's experiments, the conducting powers of silver and copper are,—

at 0° in the proportion of 1 to ·91517,
at 100° ... 1 to ·91030,
and therefore at 10° ... 1 to ·915.

Hence the specific resistance of copper at about 10° Cent. referred to unity of volume, may be found by dividing that of silver by ·915; and from the preceding result, it is thus found to be 2·080. Multiplying this by 3810500, the weight in grains of a cubic foot of copper (found by taking 8·72 as the specific gravity of copper), we obtain for the "specific resistance" of copper the value 7925800.

16. Weber, in first introducing the measurement of resistances in absolute units, gave two experimental methods, both founded virtually on a comparison of the electromotive forces with the strengths of the currents produced by them, in the conductors examined; and he actually applied them to various conductors, and obtained results which, reduced to British units, are shown in the following table. The first four numbers in the second column are deduced from M. Weber's results, on the hypothesis that the specific gravity of each specimen of copper is 8·72. The only numbers given on the authority of M. Weber are the first four of the column headed "Specific resistance." Some of the specific resistances derived above from Mr Joule's experiments are shown in the same table for the sake of comparison.

Quality of metal, &c.	Specific resistance referred to unity of volume.		"Specific resistance."
	British system.	[C. G. S.]	
No. 1. Jacobi's copper wire......................	2·851	[2649]	10870000
No. 2. Kirchhoff's copper wire..................	2·365	[2197]	9225000
No. 3. Weber's copper wire......................	2·303	[2139]	8778000
No. 4. Wire of electrolytically precipi- } tated copper........................ }	2·079	[1931]	7924000
No. 5. Copper at about 10° Cent., accord- } ing to Joule and Becquerel........ }	2·080	[1932]	7926000
[One of Joule's copper wires]	[4·1]	[3809]	
[Another of Joule's copper wires] ...	[1·78]	[1653]	
[Pure copper at 0° Cent.]		[1640]	
No. 6. Joule's silver wire at about 10° C.....	1·903	[1768]	8671000
No. 7. Mercury at about 10° Cent.	106·65	[99081]	648400000

The great discrepancies among the first four numbers of the
third column, each of which is probably correct in three of its
significant figures, show how very much the specific resistances
of the substance of different specimens of copper wire may differ
from one another. The specific resistance of copper (No. 5),
deduced indirectly from Joule's absolute by means of Becquerel's
relative determinations, agrees very closely with that of the electrolytically precipitated copper (No. 4) experimented on by Weber.

17. It is very much to be desired that Weber's direct process,
and the indirect method founded on estimating, according to
Joule's principles, the mechanical value of the thermal effects of a
galvanic current, should be both put in practice to determine the
absolute resistance of the same conductor, or that the resistance of
two conductors to which the two methods have been separately
applied, should be accurately compared. Such an investigation
could scarcely be expected to give a more approximate value of the
mechanical equivalent of a thermal unit than has been already
found by means of experiments on the friction of fluids; but it
would afford a most interesting illustration of those principles by
which Mr Joule has shown how to trace an equivalence between
work spent and mechanical effect produced, in all physical agencies
in which heat is concerned.

GLASGOW COLLEGE, *Nov.* 19, 1851.

[Note of June 1, 1882. In the preceding paper as originally published, values
stated in absolute measure were given on the British system, with the foot, grain
and second as the fundamental units. Where it seemed advisable there has now
been added, in square brackets, beside the original figures, the corresponding
values on the C. G. S. system of units.]

ART. LV.   ON THE SOURCES OF HEAT GENERATED BY THE GALVANIC BATTERY.

[*Brit. Ass. Rep.* 1852.   (Part 2.)]

IT has been stated as an objection to the chemical theory of the galvanic battery, that the chemical action being the same in all elements consisting of zinc and any less oxidizable metal, their electromotive force ought according to that theory to be the same ; which is contrary to experience, the electromotive force of a zinc and tin element in dilute sulphuric acid, for instance, being found by Poggendorff to be only about half that of a zinc and platinum element in the same liquid.   Mr Joule in 1841 gave (in his paper on the heat of electrolysis) the key to the explanation of all such difficulties, by pointing out that the heat must be generated in *different* quantities by the electrical evolution of *equal* quantities of hydrogen at equal surfaces of *different* metals.   The author of the present communication, reasoning on elementary mechanical and physical principles, from Faraday's experiments, which show that a zinc diaphragm in a trough of dilute sulphuric acid exercises no sensible resistance to the continued passage of a feeble electric current, demonstrated *that a feeble continued current, passing out of an electrolytic cell by a zinc electrode, must generate exactly as much more heat at the zinc surface than the same amount of current would develope in passing out of an electrolytic cell by a platinum electrode, as a zinc-platinum pair working against great external resistance would develope in the resistance wire by the same amount of current.*   A series of experiments, commenced for illustrating this conclusion, were described and a few of the conclusions stated.   It was found that in two equal and similar electrolytic cells in the same circuit, which differed from one another in one of them having its exit electrodes of zinc, and the other of platinum, very sensibly more heat was developed in the former than in the latter, verifying so far the conclusion stated.   By separating the two electrodes by means of porous diaphragms, it was found that, at least with low strengths of current, more heat was developed at the negative than at the positive electrode, when both electrodes were of zinc; while when both were of platinum, much more heat was found at the positive electrode than was found at the negative, for all strengths of current which gave sufficient thermal effects to be tested in this respect. The last-mentioned result, which had not been anticipated by the author, appears to be in accordance with experimental conclusions announced by De la Rive.

Many other results of a remarkable nature were obtained in a series of experiments on the heat evolved in different parts of various electrolytic and chemical-electromotive arrangements, but much difficulty had been found in interpreting them correctly on account of initial irregularities depending on "polarization," which often appeared to last as long as the experiments could be continued without introducing other sources of disturbance, and which produced marked effects on the observed thermal phenomena.

This communication was brought forward principally for the purpose of calling attention to what may be done if experimenters can be induced to undertake researches on the evolution of heat in all parts of a galvanic battery or of any electro-thermal apparatus, but partly also on account of the novelty of some of the results which have been already obtained by the author.

ART. LVI. ON CERTAIN MAGNETIC CURVES; WITH APPLICATIONS TO PROBLEMS IN THE THEORIES OF HEAT, ELECTRICITY, AND FLUID MOTION.

[*Brit. Ass. Rep.* 1852. (Part 2.) "ELECTROSTATICS AND MAGNETISM," Art. xxxv.]

ART. LVII. ON THE EQUILIBRIUM OF ELONGATED MASSES OF FERRO-MAGNETIC SUBSTANCE IN UNIFORM AND VARIED FIELDS OF FORCE.

[*Brit. Ass. Rep.*, Sept. 1852. (Part 2.) "ELECTROSTATICS AND MAGNETISM," Art. xxxv.]

Art. LVIII. On the Mechanical Action of Radiant Heat
  or Light: On the Power of Animated Creatures over
  Matter: On the Sources available to Man for the
  production of Mechanical Effect*.

On the Mechanical Action of Radiant Heat or Light.

It is assumed in this communication that the undulatory theory
of radiant heat and light, according to which light is merely radiant
heat, of which the vibrations are performed in periods between
certain limits of duration, is true. "The chemical rays," beyond
the violet end of the spectrum, consist of undulations of which
the full vibrations are executed in periods shorter than those of
the extreme visible violet light, or than about the eight hundred
million millionth of a second. The periods of the vibrations of
visible light lie between this limit and another, about double as
great, corresponding to the extreme visible red light. The vibrations of the obscure radiant heat beyond the red end are executed
in longer periods than this; the longest which has yet been experimentally tested being about the eighty million millionth of a
second.

The elevation of temperature produced in a body by the inci-
dence of radiant heat upon it is a mechanical effect of the dynamical
kind, since the communication of heat to a body is merely the
excitation or the augmentation of certain motions among its particles. According to Pouillet's estimate of heat radiated from the
sun in any time, and Joule's mechanical equivalent of a thermal
unit, it appears that the mechanical value of the solar heat incident
perpendicularly on a square foot above the earth's atmosphere is
about eighty-four foot-pounds per second.

Mechanical effect of the statical kind might be produced from
the solar radiant heat, by using it as the source of heat in a
thermo-dynamic engine. It is estimated that about 556 foot-

* From the Proceedings of the Royal Society of Edinburgh, February, 1852.

pounds per second of ordinary mechanical effect, or about the work of "one horse power," might possibly be produced by such an engine exposing 1800 square feet to receive solar heat, during a warm summer day in this country; but the dimensions of the moveable parts of the engine would necessarily be so great as to occasion practical difficulties in the way of using it with economical advantage that might be insurmountable.

The *chemical* effects of light belong to the class of mechanical effects of the statical kind; and reasoning analogous to that introduced and experimentally verified in the case of electrolysis by Joule, leads to the conclusion that when such effects are produced there will be a loss of heating effect in the radiant heat or light which is absorbed by the body acted on, to an extent thermally equivalent to the mechanical value of the work done against forces of chemical affinity.

The deoxidation of carbon and hydrogen from carbonic acid and water, effected by the action of solar light on the green parts of plants, is (as the author recently found was pointed out by Helmholtz* in 1847) a mechanical effect of radiant heat. In virtue of this action combustible substances are produced by plants; and its mechanical value is to be estimated by determining the heat evolved by burning them, and multiplying by the mechanical equivalent of the thermal unit. Taking, from Liebig's Agricultural Chemistry, the estimate 2600 pounds of dry fir-wood for the annual produce of one Hessian acre, or 26,910 square feet of forest land (which in mechanical value appears not to differ much from estimates given in the same treatise for produce of various kinds obtained from *cultivated* land), and assuming, as a very rough estimate, 4000 thermal units Centigrade as the heat of combustion of unity of mass of dry fir-wood, the author finds 550,000 foot-pounds (or the work of a horse power, for 1000 seconds) as the mechanical value of the mean annual produce of a square foot of the land. Taking 50° 34' (that of Giessen) as the latitude of the locality, the author estimates the mechanical value of the solar heat, which, were none of it absorbed by the atmosphere, would fall annually on each square foot of the land, at 530,000,000 foot pounds; and infers that probably a good deal

---

* *Ueber die Erhaltung der Kraft*, von Dr H. Helmholtz.   Berlin, 1847.

more, $\frac{1}{1000}$ of the solar heat, which actually falls on growing plants, is converted into mechanical effect.

When the vibrations of light thus act during the growth of plants, to separate, against forces of chemical affinity, combustible materials from oxygen, they must lose *vis viva* to an extent equivalent to the statical mechanical effect thus produced; and therefore quantities of solar heat are actually put out of existence by the growth of plants, but an equivalent of statical mechanical effect is stored up in the organic products, and may be reproduced as heat, by burning them. All the heat of fires, obtained by burning wood grown from year to year, is in fact solar heat reproduced.

The actual convertibility of radiant heat into statical mechanical effect, by inanimate material agency, is considered in this paper as subject to Carnot's principle; and a possible connexion of this principle with the circumstances regarding the quality of the radiant heat (or the colour of the light), required to produce the growth of plants, is suggested.

On the Power of Animated Creatures over Matter.

The question, "Can animated creatures set matter in motion in virtue of an inherent power of producing a mechanical effect?" must be answered in the negative, according to the well established theory of animal heat and motion, which ascribes them to the chemical action (principally *oxidation*, or a combustion at low temperatures) experienced by the food. A principal object of the present communication is to point out the relation of this theory to the dynamical theory of heat. It is remarked, in the first place, that both animal heat and weights raised or resistance overcome, are *mechanical* effects of the chemical forces which act during the combination of food with oxygen. The former is a dynamical mechanical effect, being thermal motions excited; the latter is a mechanical effect of the statical kind. The whole mechanical value of these effects, which are produced by means of the animal mechanism in any time, must be equal to the mechanical value of the work done by the chemical forces. Hence, when an animal is going up-hill or working against resisting force, there is less heat generated than the amount due to the oxidation of the food, by the thermal equivalent of the mechanical effect pro-

duced.   From an estimate made by Mr Joule, it appears that from
$\frac{1}{4}$ to $\frac{1}{8}$ of the mechanical equivalent of the complete oxidation of
all the food consumed by a horse may be produced, from day to
day, as weights raised.   The oxidation of the whole food consumed
being, in reality, far from complete, it follows that a less proportion
than $\frac{5}{8}$, perhaps even less than $\frac{3}{4}$, of the heat due to the whole
chemical action that actually goes on in the body of the animal, is
given out as heat.   An estimate, according to the same principle,
upon very imperfect data, however, is made by the author, regarding the relation between the thermal and the non-thermal mechanical effects produced by a man at work; by which it appears
that probably as much as $\frac{1}{8}$ of the whole work of the chemical
forces arising from the oxidation of his food during the twenty-four
hours, may be directed to raising his own weight, by a man walking
up-hill for eight hours a day; and perhaps even as much as $\frac{1}{4}$ of
the work of the chemical forces may be directed to the overcoming
of external resistances by a man exerting himself for six hours a
day in such operations as pumping.   In the former case there
would not be more than $\frac{5}{8}$, and in the latter not more than $\frac{3}{4}$ of
the thermal equivalent of the chemical action emitted as animal
heat, on the whole, during the twenty-four hours, and the quantities of heat emitted during the times of working would bear much
smaller proportions respectively than these, to the thermal equivalents of the chemical forces actually operating during those
times.

A curious inference is pointed out, that an animal would be
sensibly less warm in going up-hill than in going down-hill, were
the breathing not greater in the former case than in the latter.

The application of Carnot's principle, and of Joule's discoveries
regarding the heat of electrolysis and the calorific effects of magneto-electricity, is pointed out; according to which it appears
nearly certain that, when an animal works against resisting force,
there is not a *conversion of heat into external mechanical effect*,
but the full thermal equivalent of the chemical forces is *never
produced;* in other words, that the animal body does not act as a
thermo-dynamic engine; and very probable that the chemical forces
produce the external mechanical effects through electrical means.

Certainty regarding the means in the animal body by which
external mechanical effects are produced from chemical forces

acting internally, cannot be arrived at without more experiment and observation than has yet been applied; but the relation of mechanical equivalence, between the work done by the chemical forces, and the final mechanical effects produced, whether solely heat, or partly heat and partly resistance overcome, may be asserted with confidence. Whatever be the nature of these means, consciousness teaches every individual that they are, to some extent, subject to the direction of his will. It appears, therefore, that animated creatures have the power of immediately applying, to certain moving particles of matter within their bodies, forces by which the motions of these particles are directed to produce desired mechanical effects.

On the Sources available to Man for the production of Mechanical Effect.

Men can obtain mechanical effect for their own purposes either by working mechanically themselves, and directing other animals to work for them, or by using natural heat, the gravitation of descending solid masses, the natural motions of water and air, and the heat, or galvanic currents, or other mechanical effects produced by chemical combination, but in no other way at present known. Hence the stores from which mechanical effect may be drawn by man belong to one or other of the following classes:

I.   The food of animals.

II.   Natural heat.

III.   Solid matter found in elevated positions.

IV.   The natural motions of water and air.

V.   Natural combustibles (as wood, coal, coal-gas, oils, marsh gas, diamond, native sulphur, native metals, meteoric iron).

VI.   Artificial combustibles (as smelted or electrolytically deposited metals, hydrogen, phosphorus).

In the present communication, known facts in natural history and physical science, with reference to the sources from which these stores have derived their mechanical energies, are adduced to establish the following general conclusions:—

1. *Heat radiated from the sun* (sunlight being included in this term) *is the principal source of mechanical effect available to man**. From it is derived the whole mechanical effect obtained by means of animals working, water-wheels worked by rivers, steam-engines, and galvanic engines, and part at least of the mechanical effect obtained by means of windmills, and the sails of ships not driven by the trade winds †.

2. The motions of the earth, moon, and sun, and their mutual attractions, constitute an important source of available mechanical effect. From them all, but chiefly, no doubt, from the earth's motion of rotation, is derived the mechanical effect of water-wheels driven by the tides. The mechanical effect so largely used in the sailing of ships by the trade winds is derived partly perhaps principally, from the earth's motion of rotation, and partly from solar heat †.

3. The other known sources of mechanical effect available to man are either terrestrial—that is, belonging to the earth, and available without the influence of any external body,—or meteoric,—that is, belonging to bodies deposited on the earth from external space. Terrestrial sources, including mountain quarries and mines, the heat of hot springs, and the combustion of native sulphur, perhaps also the combustion of all inorganic native combustibles, are actually used, but the mechanical effect obtained from them is very inconsiderable, compared with that which is obtained from sources belonging to the two classes mentioned above. Meteoric sources, including only the heat of newly-fallen meteoric bodies, and the combustion of meteoric iron, need not be reckoned among those available to man for practical purposes.

* A general conclusion equivalent to this was published by Sir John Herschel in 1833. See his Astronomy, edit. 1849, § (399).

† [Note of June 1, 1882. These conclusions (1 and 2) as originally printed contained a serious dynamical error, which I first noticed and corrected in my opening address to section A of the British Association at York last year. The paragraphs are printed as originally published, but with deleting marks to shew the necessary corrections.]

ART. LIX. On a Universal Tendency in Nature to the Dissipation of Mechanical Energy*

THE object of the present communication is to call attention to the remarkable consequences which follow from Carnot's proposition, that there is an absolute waste of mechanical energy available to man when heat is allowed to pass from one body to another at a lower temperature, by any means not fulfilling his criterion of a "perfect thermo-dynamic engine," established, on a new foundation, in the dynamical theory of heat. As it is most certain that Creative Power alone can either call into existence or annihilate mechanical energy, the "waste" referred to cannot be annihilation, but must be some transformation of energy†. To explain the nature of this transformation, it is convenient, in the first place, to divide *stores* of mechanical energy into two classes—*statical* and *dynamical*. A quantity of weights at a height, ready to descend and do work when wanted, an electrified body, a quantity of fuel, contain stores of mechanical energy of the statical kind. Masses of matter in motion, a volume of space through which undulations of light or radiant heat are passing, a body having thermal motions among its particles (that is, not infinitely cold), contain stores of mechanical energy of the dynamical kind.

The following propositions are laid down regarding the *dissipation* of mechanical energy from a given store, and the *restoration* of it to its primitive condition. They are necessary consequences of the axiom, "*It is impossible, by means of inanimate material agency, to derive mechanical effect from any portion of matter by cooling it below the temperature of the coldest of the surrounding objects.*" (Dynamical Theory of Heat [Art. XLVIII. above], § 12.)

* From the Proceedings of the Royal Society of Edinburgh for April 19, 1852, also *Philosophical Magazine*, Oct. 1852.

† See the Author's previous paper on the Dynamical Theory of Heat, § 22 [Art. XLVIII. above].

I. When heat is created by a reversible process (so that the mechanical energy thus spent may be *restored* to its primitive condition), there is also a transference from a cold body to a hot body of a quantity of heat bearing to the quantity created a definite proportion depending on the temperatures of the two bodies.

II. When heat is created by any unreversible process (such as friction), there is a *dissipation* of mechanical energy, and a full *restoration* of it to its primitive condition is impossible.

III. When heat is diffused by *conduction*, there is a *dissipation* of mechanical energy, and perfect *restoration* is impossible.

IV When radiant heat or light is absorbed, otherwise than in vegetation, or in chemical action, there is a *dissipation* of mechanical energy, and perfect *restoration* is impossible.

In connexion with the second proposition, the question, *How far is the loss of power experienced by steam in rushing through narrow steam-pipes compensated, as regards the economy of the engine, by the heat* (containing an exact equivalent of mechanical energy) *created by the friction?* is considered, and the following conclusion is arrived at:—

Let S denote the temperature of the steam (which is nearly the same in the boiler and steam-pipe, and in the cylinder till the expansion within it commences); T the temperature of the condenser; μ the value of Carnot's function, for any temperature t; and R the value of

$$e^{-\frac{1}{J}\int_T^S \mu dt}.$$

Then $(1-R)w$ expresses the greatest amount of mechanical effect that can be economized in the circumstances from a quantity w/J of heat produced by the expenditure of a quantity w of work in friction, whether of the steam in the pipes and entrance ports, or of any solids or fluids in motion in any part of the engine; and the remainder, Rw, is absolutely and irrecoverably wasted, unless some use is made of the heat discharged from the condenser. The value of $1-R$ has been shown to be not more than about $\frac{1}{4}$ for the best steam-engines, and we may infer that in them at least three-fourths of the work spent in any kind of friction is utterly wasted.

In connexion with the third proposition, the quantity of work that could be got by equalizing the temperature of all parts of a solid body possessing initially a given non-uniform distribution of heat, if this could be done by means of perfect thermo-dynamic engines without any conduction of heat, is investigated. If t be the initial temperature (estimated according to any arbitrary system) at any point xyz of the solid, T the final uniform temperature, and c the thermal capacity of unity of volume of the solid the required mechanical effect is of course equal to

$$J \iiint c(t-T)dx\,dy\,dz,$$

being simply the mechanical equivalent of the amount of heat put out of existence. Hence the problem becomes reduced to that of the determination of T. The following solution is obtained:—

$$T = \frac{\iiint \epsilon^{-\frac{1}{J}\int_0^t \mu dt} ct\,dx\,dy\,dz}{\iiint \epsilon^{-\frac{1}{J}\int_0^t \mu dt} c\,dx\,dy\,dz}.$$

If the system of thermometry adopted* be such that $\mu = \dfrac{J}{t+a}$, that is, if we agree to call $J/\mu - a$ the *temperature* of a body, for which μ is the *value of Carnot's function* (a and J being constants), the preceding expression becomes

$$T = \frac{\iiint c\,dx\,dy\,dz}{\iiint \dfrac{c}{t+a}\,dx\,dy\,dz} - a.$$

The following general conclusions are drawn from the propositions stated above, and known facts with reference to the mechanics of animal and vegetable bodies:—

---

* According to "Mayer's hypothesis," this system coincides with that in which equal differences of temperature are defined as those with which the same mass of air under constant pressure has equal differences of volume, provided J be the mechanical equivalent of the thermal unit, and a^{-1} the coefficient of expansion of air.

1.   There is at present in the material world a universal tendency to the dissipation of mechanical energy.

2.   Any *restoration* of mechanical energy, without more than an equivalent of dissipation, is impossible in inanimate material processes, and is probably never effected by means of organized matter, either endowed with vegetable life or subjected to the will of an animated creature.

3.   Within a finite period of time past, the earth must have been, and within a finite period of time to come the earth must again be, unfit for the habitation of man as at present constituted, unless operations have been, or are to be performed, which are impossible under the laws to which the known operations going on at present in the material world are subject.

[From the *Glasgow Phil. Soc. Proc.* Vol. III. Dec. 1852.]

ART. LX.   ON THE ECONOMY OF THE HEATING OR COOLING OF
BUILDINGS BY MEANS OF CURRENTS OF AIR*.

IF it be required to introduce a certain quantity of air at a
stated temperature higher than that of the atmosphere into a
building, it might at first sight appear that the utmost economy
would be attained if all the heat produced by the combustion
of the coals used were communicated to the air; and in fact the
greatest economy that has yet been aimed at in heating air or
any other substance, for any purpose whatever, has had this for
its limit.   If an engine be employed to pump in air for heating
and ventilating a building (as is done in Queen's College, Belfast), all the waste heat of the engine, along with the heat of
the fire not used in the engine, may be applied by suitable
arrangements to warm the entering current of air; and even the
heat actually converted into mechanical effect by the engine,
will be reconverted into heat by the friction of the air in the
passages, since the overcoming of resistance depending on this
friction is the sole work done by the engine.   It appears therefore that whether the engine be economical as a converter of
heat into mechanical work, or not, there would be perfect economy
of the heat of the fire if all the heat escaping in any way from
the engine, as well as all the residue from the fire, were applied
to heating the air pumped in, and if none of this heat were

---

* Mathematical demonstrations of the results stated in this paper have since
been published in the *Camb. and Dub. Math. Journal*, Nov. 1853. [Art. XLVIII.
note III. above.]

allowed to escape by conduction through the air passages. It is not my present object to determine how nearly in practice this degree of economy may be approximated to; but to point out how the limit which has hitherto appeared absolute, may be surpassed, and a current of warm air at such a temperature as is convenient for heating and ventilating a building may be obtained mechanically, either by water power without any consumption of coals, or, by means of a steam engine, driven by a fire burning actually less coals than are capable of generating by their combustion the required heat; and secondly, to show how, with similar mechanical means, currents of cold air, such as might undoubtedly be used with great advantage to health and comfort for cooling houses in tropical countries*, may be produced by motive power requiring (if derived from heat by means of steam engines) the consumption of less coals perhaps than are used constantly for warming houses in this country.

In the mathematical investigation communicated with this paper, it is shown in the first place, according to the general principles of the dynamical theory of heat, that any substance may be heated thirty degrees above the atmospheric temperature by means of a properly contrived machine, driven by an agent spending not more than about $\frac{1}{35}$ of energy of the heat thus communicated; and that a corresponding machine, or the same machine worked backwards, may be employed to produce cooling effects, requiring about the same expenditure of energy in working it to cool the same substance through a similar range of temperature. When a body is heated by such means, about $\frac{34}{35}$

---

* The mode of action and apparatus proposed for this purpose differs from that proposed originally by Professor Piazzi Smyth for the same purpose, only in the use of an egress cylinder, by which the air is made to do work by its extra pressure and by expansion in passing from the reservoir to the locality where it is wanted, which not only saves a great proportion of the motive power that would be required were the air allowed simply to escape through a passage, regulated by a stop-cock or otherwise, but is absolutely essential to the success of the project, as it has been demonstrated by Mr Joule and the author of this communication, that the cold of expansion would be so nearly compensated by the heat generated by friction, when the air is allowed to rush out without doing work, as to give not two-tenths of a degree of cooling effect in apparatus planned for 30 degrees. The use of an egress cylinder has (as the meeting was informed by Mr Macquorn Rankine), recently been introduced into plans adopted by a committee of the British Association appointed to consider the practicability of Professor Piazzi Smyth's suggestion, with a view to recommending it to government for public buildings in India.

of the heat is drawn from surrounding objects, and $\frac{1}{35}$ is created by the action of the agent; and when a body is cooled by the corresponding process, the whole heat abstracted from it, together with a quantity created by the agent, equal to about $\frac{1}{35}$ of this amount, is given out to the surrounding objects.

A very good steam engine converts about $\frac{1}{10}$ of the heat generated in its furnace into mechanical effect; and consequently, if employed to work a machine of the kind described, might raise a substance thirty degrees above the atmospheric temperature by the expenditure of only $\frac{10}{35}$, or $\frac{2}{7}$, that is, less than one-third of the coal that would be required to produce the same elevation of temperature with perfect economy in a direct process. If a water-wheel were employed, it would produce by means of the proposed machine the stated elevation of temperature, with the expenditure of $\frac{1}{35}$ of the work, which it would have to spend to produce the same heating effect by friction.

The machine by which such effects are to be produced must have the properties of a "perfect thermo-dynamic engine," and in practice would be either like a steam engine, founded on the evaporation and re-condensation of a liquid (perhaps some liquid of which the boiling point is lower than that of water), or an air engine of some kind. If the substance to be heated or cooled be air, it will be convenient to choose this itself as the medium operated on in the machine. For carrying out the proposed object, including the discharge of the air into the locality where it is wanted, the following general plan was given as likely to be found practicable. Two cylinders, each provided with a piston, ports, valves, and expansion gearing, like a high-pressure double-acting steam engine, are used, one of them to pass air from the atmosphere into a large receiver, and the other to remove air from this receiver and discharge into the locality where it is wanted. The first, or ingress cylinder and the receiver, should be kept with their contents as nearly as possible at the atmospheric temperature, and for this purpose ought to be of good conducting material, as thin as is consistent with the requisite strength, and formed so as to expose as much external surface as possible to the atmosphere, or still better, to a stream of water. The egress cylinder ought to be protected as much as possible from thermal communication with the atmosphere or surrounding objects. According as the air is to be heated, or cooled, the pistons and valve gearing must

be worked so as to keep the pressure in the receiver below, or
above, that of the atmosphere.   If the cylinders be of equal dimensions, the arrangement when the air is to be heated, would
be as follows:—The two pistons working at the same rate, air
is to be admitted freely from the atmosphere into the ingress
cylinder, until a certain fraction of the stroke, depending on the
heating effect required, is performed, then the entrance port is to
be shut, so that during the remainder of the stroke the air
may expand down to the pressure of the receiver, into which,
by the opening of another valve, it is to be admitted in the
reverse stroke; while the egress cylinder* is to draw air freely
from the receiver through the whole of each stroke on one side
or the other of its piston, and in the reverse stroke first to
compress this air to the atmospheric pressure (and so heat it as
required), and then discharge it into a pipe leading to the locality
where it is to be used.   If it be required to heat the air from
50° to 80° Fahr., the ratio of expansion to the whole stroke in
the ingress cylinder would be $\frac{18}{100}$, the pressure of the air in the
receiver would be $\frac{82}{100}$ of that of the atmosphere (about 2·7 lbs.
on the square inch below the atmospheric pressure), and the
ratio of compression to the whole stroke in the egress cylinder
would be $\frac{18}{100}$.   If 1 lb. of air (or about 13½ cubic feet, at the
stated temperature of 80°, and the mean atmospheric pressure,)
be to be delivered per second, the motive power required for
working the machine would be ·283 of a horse power, were the
action perfect, with no loss of effect, by friction, by loss of expansive power due to cooling in the ingress cylinder, or otherwise.   If each cylinder be four feet in stroke, and 26·3 inches
diameter, the pistons would have to be worked at 26·1 double
strokes per minute.

On the other hand, if it be desired to cool air, either the
ingress piston must be worked faster than the other, or the stroke

* In this case the egress cylinder acts merely as an air pump, to draw air from
the receiver and discharge it into the locality where it is wanted, and the valves
required for this purpose might be ordinary self-acting pump-valves. A similar
remark applies to the action of the ingress cylinder in the use of the apparatus for
producing a cooling effect on the air transmitted, which will then be that of a
compressing air-pump to force air from the atmosphere into the receiver. But
in order that the same apparatus may be used for the double purpose of heating or
cooling, as may be required at different seasons, it will be convenient to have the
valves of each cylinder worked mechanically, like those of a steam engine.

of the other must be diminished, or the ingress cylinder must be larger, or an auxiliary ingress cylinder must be added. The last plan appears to be undoubtedly the best, as it will allow the two principal pistons to be worked stroke for stroke together, and consequently to be carried by one piston rod, or by a simple lever, without the necessity of any variable connecting gearing, whether the machine be used for heating or for cooling air; all that is necessary to adapt it to the latter purpose, besides altering the valve gearing, being to connect a small auxiliary piston to work beside the principal ingress cylinder, with which it is to have free communication at each end. If it were required to cool air from 80^{0} to 50^{0} Fahr., the auxiliary cylinder would be required to have its volume $\frac{1}{17}$ of that of each of the principal cylinders; and, if its stroke be the same, its diameter would therefore be a little less than a quarter of theirs. The valves would have to be altered to give compression in the ingress cylinder during the same fraction of the stroke as is required for expansion when the air is heated through the same range of temperature, and the valves of the egress cylinder would have to give the same proportion of expansion as is given of compression in the other case; and the pressure kept up in the receiver, by the action of the pistons thus arranged, would be $1\frac{18}{100}$ atmos., or about 3·2 lbs. on the square inch above the atmospheric pressure. The principal cylinders being of the same dimensions as those assumed above, and the quantity of air required being the same (1 lb. per second), the pistons would have to be worked at only 21·4 double strokes per minute instead of 26·1, and the horse power required would be ·288, instead of as formerly ·283, when the same machine was used for giving a supply of heated air.

[Note added June 26, 1881. The method of cooling air in unlimited quantities described in this article has been realized by Mr Coleman, first in refrigerators used for the distillation of paraffin, and after that in the Bell-Coleman refrigerator, for carrying supplies of fresh meat from North America to Europe; in a great refrigerator recently sent out for the Abattoir at Brisbane, Queensland; and other large practical applications of a similar kind. The Bell-Coleman machine sends large quantities of air cooled 10^{0} or 20^{0} C. below freezing point into the chamber to be kept cool; and the general temperature of this chamber is thus maintained at the desired point, which, for the case of carrying fresh meat from America to this country is about 35° F.

The method of heating air described in the article remains unrealized to this day. When Niagara is set to work for the benefit of North America through electric conductors, it will no doubt be largely employed for the warming of houses over a considerable part of Canada and the United States. But it is probable that it will also have applications though less large in other cold countries, to multiply the heat of coal and other fuel, and to utilize wind and water power (with aid of electric accumulators) for warming houses.]

ART. LXI. ON THE MECHANICAL VALUES OF DISTRIBUTIONS OF ELECTRICITY, MAGNETISM, AND GALVANISM.

Glasgow Phil. Soc. Proc. Vol. III., *Jan.* 1853, with additions, enclosed in brackets [ ], from Nichol's *Cyclopædia*, 2nd edition, 1860, Article "Dynamical Relations of Magnetism"; and other additions, enclosed in double brackets [[ ]], of date July, 1882, not hitherto published.

I. ELECTRICITY AT REST.

To electrify an insulated conductor (a Leyden phial, for instance, or any mass of metal resting on supports of glass), in the ordinary way by means of an electrical machine, requires the expenditure of work in turning the machine. But inasmuch as part, obviously by far the greater part, of the work done in this operation goes to generate heat by means of friction, and of the small residue some, it may be a considerable proportion, is wasted in generating heat (electrical light being included in the term) by the flashes, illuminated points, and sparks, which accompany the transmission of the electricity from the glass of the machine where it is first excited, to the conductor which receives it, the mechanical value of the electrification thus effected would be enormously overestimated if it were regarded as equivalent to the work that has been spent. Notwithstanding, the mechanical value of any electrification of a conductor has a perfectly definite character, and may be calculated with ease in any particular case, by means of formulæ demonstrated in this communication. The simplest case is that of a single conductor insulated at a distance from other conductors, or with only uninsulated conducting matter in its neighbourhood. In this case the mechanical value of the electrification of the conductor, is equal to *half the square of the quantity of electricity, multiplied by the* capacity of the conductor*.

---

* A term introduced by the author to signify the proportion of the quantity of electricity that the conductor would retain to that which it would communicate to a conducting ball of unit radius, insulated at a great distance from other conducting matter, if connected with it by means of a fine wire.

In any case whatever, the total mechanical value of all the distributions of electricity on any number of [[non-conductors, or]] separate insulated conductors, electrified with any quantities of electricity, is demonstrated by the author to be equal to half the sum of the products obtained by multiplying the "potential*" in each conductor [[or infinitesimal part of a non-conductor,]] by the quantity of electricity with which it is charged. Each term of this expression does not represent the independent value of the actual distribution on the conductor to which it corresponds, inasmuch as the "potential" in each depends on the presence of the others, when they are near enough to exert any sensible mutual influence; but independent expressions of these independent values are readily obtained, although not in a form convenient for statement here; and the author proves that their sum is equal to the total value, as calculated by the preceding expression. When a conductor is discharged without other mechanically valuable effects being developed, the heat generally, as for instance in the sparks produced when the knob of a Leyden phial is put in communication with the outside coating, or when a flash of lightning takes place, is equal in mechanical value to the distribution of electricity lost. Hence, by what precedes, the amount of heat is proportional to the *square* of the quantities discharged, as was first demonstrated by Joule, in a communication to the Royal Society in 1840, although it had been announced by Sir W. Snow Harris, as an experimental result, to be simply proportional to the quantity. Mr Joule's result has been verified by independent experimenters in France, Italy, and Germany. The author pointed out other applications of his investigation, some of a practical kind, and others in the Mathematical Theory of Electricity †. He mentioned, that although he had first arrived at the results in 1845, and used

* A term first introduced by Green, which may be defined as the quantity of mechanical work that would have to be spent to bring a unit of electricity from a great distance up to the surface of the conductor, supposed to retain its distribution unaltered.

† [Among the latter, an analytical investigation of the mutual attraction and repulsion between two electrified spherical conductors may be referred to. Results of this investigation were published as early as 1845 [[*Camb. and Dub. Math. Jour.* Nov., 1845; *Electrostatics and Magnetism*, §§ 30, 31]], although it was not till some years later that the author succeeded in finding a synthetical verification, which, along with the original analytical solution is published in the [[*Electrostatics and Magnetism*, Art. vi.]] *Philosophical Magazine* for April, 1853.]

them in papers published in that year, the first explicit publication
of the theorem regarding the mechanical value of the electrification of a conductor appears to be in 1847, in a paper entitled
"Ueber die Erhaltung der Kraft," by Helmholtz, [who had independently arrived at the same theorem.

The excellent terms—potential energy and actual* energy—
which have been introduced by Prof. Rankine to designate the
statical and dynamical forms of mechanical energy, are well illustrated by their application to this subject. Thus, what has been
defined above as the mechanical value of an electric charge, is its
"potential energy." When two electrified bodies repel one
another, and experience no sensible resistance to their motions,
a portion of the potential energy of the electrical system is
converted into actual energy of motion. If a body is repelled
or attracted upward by electric force, there is a conversion of the
potential energy of electricity into potential energy of gravitation.
If an electrified conductor is discharged without being allowed
to produce electrolytic, or ordinary mechanical effects, its potential
energy is, as has been remarked above, wholly converted into
actual energy of heat and light in the flash. If in its discharge it
breaks a body into fragments thrown violently asunder, and some
of them raised against gravity by the electric force, and at the
same time decomposes water; its potential energy is converted
partly into actual energy of motion, partly into potential energy
of chemical affinity and partly into potential energy of gravitation;
and the remainder into actual energy of heat.]

II. Magnetism.

If a piece of soft iron be allowed to approach a magnet very
slowly from a distant position, and be afterwards drawn away so
rapidly that at the instant when it reaches its primitive position,
where it is left at rest, it retains as yet sensibly unimpaired † the

---

[[* The name kinetic energy, which I subsequently gave as seeming preferable
to "actual energy," has been generally adopted; but Rankine's name "potential
energy" remains to this day, and is universally used to designate energy of the
static kind.—W. T., July 18, 1882.]]

† [[Note of July 17, 1882. It is not certain, I think I may say not probable,
that this condition could be fulfilled without so great a velocity as to thoroughly

magnetization it had acquired at the nearest position, a certain amount of work must have been finally expended on the motion of the iron. For during the approach, the iron has only the magnetization due to the action of the magnet on it in its actual position at each instant, but at each instant of the time in which the iron is being drawn away, it has the whole magnetization due to the action of the magnet on it when it was at the nearest. Hence it is drawn away against more powerful forces of attraction by the magnet, than those with which the magnet attracts it during its approach; from which it follows that more work is spent in drawing the iron away than had been gained in letting it approach the magnet. The sole effect due to this excess of work is the magnetization which the iron carries away with it; and consequently, the mechanical value of this magnetization must be precisely equal to the mechanical value of the balance of work spent in producing it.

After a very short time has elapsed with the piece of soft iron at a great distance from the magnet, it will have lost, as is well known, all or nearly all the magnetization which it had acquired temporarily in the neighbourhood of the magnet; and in this short time some energy, equivalent to that of the magnetization lost, must have been produced. Mr Joule's experiments show that this energy consists of heat, which is generated during the demagnetization of the iron; and we infer the remarkable conclusion, that at the end of the process, which has been described, or of any motion of a piece of soft iron in the neighbourhood of a magnet, from a certain position and back to the same, the iron will be as much the warmer than it was at the beginning, as it would have been without any magnetic action, if it had received the heat that would be generated by the expenditure of the same amount of work on mere friction. [[The heat generated by the electro-magnetically induced electric currents in the iron, ignored in the preceding statements, is included in the reckoning of heat and work to which Joule's conclusion applies.]]

The same considerations are applicable to the magnetization of a piece of steel, with this difference, that according to the hardness of the steel, the magnetization which it receives in the

disturb all the supposed magnetic conditions, at least by heating to incandescence, if not by more serious disturbance of the luminiferous ether. See Joule & Thomson, "Thermal Effects of Fluids in Motion," Part III., p. 400 above.]]

nearest position will be more or less permanent, and if there be any demagnetization after removal from the magnet, it will be much less complete than in the case of soft iron, and that heat will be necessarily generated both during the magnetization which takes place during the gradual approach, and in the subsequent demagnetization. Further, by putting together a number of pieces of steel, each separately magnetized, a complete magnet will be formed, of which the mechanical value will be equal to the sum of the mechanical values of its parts, increased or diminished by the amount of work spent or gained in bringing them together.

[A curious experiment illustrating these principles is easily made by subjecting a piece of soft iron at rest to alternate magnetizations and demagnetizations, or reverse magnetizations, through the action of an electro-magnet, and observing the changes of temperature which it experiences, care being taken to prevent any sensible effect of conduction of heat from the coils of the electro-magnet. In a variety of ways it is easy to show by this action an elevation of temperature amounting to several degrees centigrade, in a piece of unmoved soft iron. Surely when this subtle magnetic influence, producing heat at a distance through bodies seemingly insensible to its effects, is known, even as we may now imagine " Knowing " it, we shall know that it is a dynamical quality—a modification of the constitutional motions—of all the matter occupying space in which it rests.

For theoretical considerations regarding reverse thermal effects which it is anticipated must be experienced by a body according as it is moved to or from a magnet, see "Thermo-Magnetism."] [[§§ 207, 208 of Art. XLVIII. above.]]

Upon the principles which have been explained, the author has investigated the mechanical value of any conceivable distribution of magnetism, in any kind of substance. [[(A) Let the whole substance become, for a time, endowed with perfect coercivity; so that, whatever we do to it, every part of it shall retain absolutely unchanged the magnetization it had at the beginning of our operations. Now divide the whole of the magnetized substance into infinitely fine filaments along the lines of magnetization; and divide each of these filaments into infinitely short parts, but still infinitely long in comparison with transverse diameters. Each of these parts will be infinitely nearly straight, and infinitely nearly uniform from end to end. Now, very slowly to avoid heating by induced electric currents, separate these parts to infinitely great distances from one another. To do this requires, on the part of the agent performing the operations, the absorption of an amount of work, essentially positive, which is easily proved (and is in fact proved by (10) below) to be equal to

$$\frac{1}{8\pi} \int_{-\infty}^{\infty} \int_{-\infty}^{\infty} \int_{-\infty}^{\infty} dx\,dy\,dz\,R^2 \quad \dotsc\dotsc\dotsc\dotsc(a),$$

where R denotes the resultant magnetic force at any point in an infinitely fine crevasse in the substance, tangential to its lines of magnetization in its given condition. Now let us have access to an auxiliary magnet of perfect coercivity in some distant place, giving a field F, in which we can find any needed intensity of magnetizing force. Take each of our infinitesimal bars to such a position of F, that the force it experiences there suffices to maintain its magnetism unchanged without aid from any coercivity of its own. Let now the ideal temporary "perfect coercivity" be done away with, and the original quality of the bar given back to it. This quality is necessarily, for some parts of the given substance, a condition of partial if not of perfect coercivity; else there would be no magnetism to deal with. For some parts of the substance it may be a condition of zero coercivity, that is to say, of perfectly free susceptibility. Lastly, move the bar very slowly through such a series of positions of F, to an infinitely great distance from F, that we have it finally unmagnetized in a place of zero force. Let the whole heat generated during this last journey, through and away from F, be measured. It will necessarily amount to a positive quantity for every part of the substance which has partial coercivity: but to zero for every part if any there is which has either perfect coercivity or no coercivity at all. Whatever be the condition of the given substance as to more or less of coercivity, and as to degree of susceptibility, a *positive* amount of work, including the heat generated, must have been absorbed by the agent in these journeys to and from F: this positive amount, for an infinitesimal bar of volume b, is, in the notation explained below, represented by

$$\lambda q^2 \quad \dotsc\dotsc\dotsc\dotsc\dotsc\dotsc(b).]]$$

The result, which cannot be well expressed, except in mathematical language, is as follows:

$$\iiint \lambda q^2 dx\,dy\,dz + \frac{1}{8\pi} \iiint R^2 dx\,dy\,dz,$$

where R denotes the resultant magnetic force* at any internal or external point (x, y, z), q the intensity of magnetization at a point (x, y, z), of the magnet, and λ a quantity depending on the nature of the substance at this point.

The integral constituting the first term of this expression, includes the whole of the magnetized substance, and expresses the sum of the separate mechanical values of the distributions in all the parts obtained by infinitely minute division along the lines of magnetization. The second term [[, an integral extending through all space,]] expresses the amount of work that would have to be spent to put these parts together, were they given separately, each with the exact magnetization that it is to have when in its place in the whole. [[In the first term, λq^2 is, for every part of the space occupied by intrinsically magnetic matter, as magnetized steel, an arbitrary formula for the experimentally determinable "mechanical value" of this magnetism, supposed given in infinitely thin bars.]] If [[any part of]] the substance be perfectly free in its susceptibility for magnetization or demagnetization, [[or perfectly devoid of "coercivity"; and if the intensity of magnetization of this part of the substance be assumed to be in simple proportion to the magnetizing force; in other words, if its "susceptibility" be independent of the magnitude of the force,]] λ will [[, for that part of the substance,]] express such a function of the inductive capacity that if a ball of similar substance be placed in a magnetic field where the force is F, the intensity of the magnetization induced in it will be

$$F/(2\lambda + 4\pi/3).$$

[[(B) The quantity here denoted by λ is half the reciprocal of what was afterwards, in *Electrostatics and Magnetism*, § 611, 3, Def. 2, called the "magnetic susceptibility" of the substance, as we see by § 626 of the same. Thus if we denote the susceptibility by μ, and neglect the part $\iiint \lambda q^2 dx dy dz$, consisting of the mechanical value of the intrinsic magnetization supposed given in infinitely thin bars and kept in virtue of coercivity, and take into our reckoning only the part of $\iiint \lambda q^2 dx dy dz$ which depends on

---

* [[Reckoned according to the "polar" definition of *Electrostatics and Magnetism*, § 480. Compare § 517, foot-note, and Postscript of Nov. 1881.]]

induced magnetization in substance perfectly devoid of coercivity, the preceding expression becomes

$$\iiint dx\,dy\,dz\,\frac{q^2}{2\mu} + \frac{1}{8\pi}\iiint dx\,dy\,dz\,R^2 \ldots\ldots\ldots\ldots(1) ;$$

which may be verified, for isotropic substance, as follows* : Let ξ, η, ζ be the components of q, the induced magnetism, and X, Y, Z those of R, the magnetizing force. By the definition of "susceptibility" we have, for isotropic substance,

$$\xi = \mu X, \quad \eta = \mu Y, \quad \zeta = \mu Z \ldots\ldots\ldots\ldots\ldots(2),$$

whence

$$q^2/\mu = \xi X + \eta Y + \zeta Z \ldots\ldots\ldots\ldots\ldots(3);$$

and the first term of (1) may be written

$$\frac{1}{2}\int_{-\infty}^{\infty}\int_{-\infty}^{\infty}\int_{-\infty}^{\infty} dx\,dy\,dz\,(X\xi + Y\eta + Z\zeta) \ldots\ldots\ldots(4),$$

the integral being taken through all space, but its elements vanishing in virtue of the vanishing of ξ, η, ζ, wherever the susceptibility is zero. Now because X, Y, Z are the components of a force due to a distribution of the ideal "magnetic matter," $X\,dx + Y\,dy + Z\,dz$ is a complete differential, and, V denoting the corresponding potential, we have

$$X = -\frac{dV}{dx}, \quad Y = -\frac{dV}{dy}, \quad Z = -\frac{dV}{dz} \ldots\ldots\ldots\ldots(5).$$

Also, $$\frac{dX}{dx} + \frac{dY}{dy} + \frac{dZ}{dz} = -\nabla^2 V = 4\pi\,(\rho + \sigma) \ldots\ldots\ldots(6);$$

and (*Electrostatics and Magnetism*, § 473)

$$\frac{d\xi}{dx} + \frac{d\eta}{dy} + \frac{d\zeta}{dz} = -\sigma \ldots\ldots\ldots\ldots\ldots\ldots(7);$$

if ρ denote the density of the ideal magnetic matter representing the given intrinsic magnetism, kept by coercivity, and σ that of the magnetism (ξ, η, ζ) induced by it.

Using (5) in (4), integrating by parts in the well-known manner, and remarking that the integrated parts $-\frac{1}{2}V\xi$, $-\frac{1}{2}V\eta$,

---

* For aeolotropic substance, with Cartesian formulas necessarily more complicated and notation more elaborate, substantially the same proof is applicable. See *Electrostatics and Magnetism*, §§ 700—730.

$-\frac{1}{2} V\zeta$ vanish at the respective limits $x = \pm \infty$, $y = \pm \infty$, $z = \pm \infty$, we find the following equivalent

$$\frac{1}{2} \int_{-\infty}^{\infty} \int_{-\infty}^{\infty} \int_{-\infty}^{\infty} dx\, dy\, dz\, V \left(\frac{d\xi}{dx} + \frac{d\eta}{dy} + \frac{d\zeta}{dz} \right) \dots\dots\dots\dots(8),$$

Hence, and by (7) we find, for the first term of (1),

$$\iiint dx\, dy\, dz\, \frac{q^2}{2\mu} = -\frac{1}{2} \iiint dx\, dy\, dz\, V\sigma \dots\dots\dots\dots(9).$$

For the second term of (1) we similarly find, by (5), and integration by parts, and (6),

$$\frac{1}{8\pi} \iiint dx\, dy\, dz\, R^2 = \frac{1}{2} \iiint dx\, dy\, dz\, V (\rho + \sigma) \dots\dots\dots(10).$$

Thus finally (1) is reduced to

$$\tfrac{1}{2} \iiint dx\, dy\, dz\, V\rho \dots\dots\dots\dots\dots\dots(11).$$

Now imagine that instead of the given distribution of rigid magnetism we have a similar distribution less strong in the ratio of 1 to θ; θ denoting any numeric less than unity: so that instead of ρ we now have $\theta\rho$. If the susceptibility were independent of the intensity of the magnetizing force we should now have θV instead of V: but this supposition we have known to be far from the truth ever since Joule made the discovery forty years ago of the limit to the amount of the magnetization of which soft iron is capable; and to avoid erroneous assumption we shall therefore, for the potential at (x, y, z) due to $\theta\rho$ and the magnetism induced by it, take $\backsim V$, where $\backsim$ denotes a function of (x, y, z, θ); vanishing when $\theta = 0$; and equal to unity for all values of x, y, z, when θ is unity. Denoting now by dw the work required to bring ideal magnetic matter from infinite distances so as to increase $\theta\rho$ to $(\theta + d\theta)\rho$, we have

$$dw = \iiint dx\, dy\, dz\, \backsim V \cdot \rho\, d\theta = d\theta \iiint dx\, dy\, dz\, \backsim V\rho \dots\dots\dots(12).$$

Hence, if W denote the work required to bring ideal magnetic matter from infinite distances to constitute the given distribution ρ in the presence of the given susceptible substance, we have

$$W = \int_0^1 d\theta \iiint dx\, dy\, dz\, \backsim V\rho \dots\dots\dots\dots\dots(13):$$

and, on the assumption of constant susceptibility, which makes $\backsim = \theta$, we therefore have

T.   34

$$W = \int_0^1 d\theta \, \theta \iiint dx \, dy \, dz \, V\rho = \tfrac{1}{2} \iiint dx \, dy \, dz \, V\rho \ \ldots\ldots(14).$$

Returning hence to (11) and (1), we verify the theorem.

By (3) and (2) we may put (1) into the following form:

$$\frac{1}{8\pi} \iiint dx \, dy \, dz \, (4\pi\mu + 1) \, R^2 \ldots\ldots\ldots\ldots\ldots(15).$$

Here $(4\pi\mu + 1)$ is what is called the "magnetic permeability" in *Electrostatics and Magnetism*, § 628, and denoted by ϖ. See also § 717 (55), which is the equivalent for aeolotropic substance to our present formula (15) for isotropic substance. The α^2 of Art. XXXVI. above corresponds to the $(4\pi\mu + 1)$ of our present notation; which is the "permeability" in the magnetic and hydrokinetic applications (*El. and Mag.* §§ 751—763); the thermal conductivity in the application to Fourier's *Theory of Heat* (*Ibid.* §§ 1—10); and Faraday's "specific inductive capacity" in the electrostatic application (*Ibid.* §§ 44—49.)]]

III. ELECTRICITY IN MOTION.

If an electric current be excited in a conductor, and then left without electro-motive force, it retains energy to produce heat, light, and other kinds of mechanical effect, and it lasts with diminishing, or it may be with alternately diminishing and increasing strength, before it finally ceases and electrical equilibrium is established: as is amply demonstrated by the experiments of Faraday and Henry, on the spark which takes place when a galvanic circuit is opened at any point, and by those of Weber, Helmholtz, and others on the electro-magnetic effects of varying currents. The object of the present communication is to show how the mechanical value of all the effects that a current in a close circuit can produce after the electro-motive force ceases, may be obtained by a determination, (founded on the known laws of electro-dynamic induction,) of the mechanical value of the energy of a current of given strength, circulating in a linear conductor (a bent wire, for instance) of any form. To do this;—in the first place it may be remarked, that although a current, once instituted in a conductor, will circulate in it with diminishing strength after the electro-motive force ceases, just as if the electricity had inertia, it will diminish in strength according to the same, or nearly the same, laws as a current of water or other fluid, once set in motion and left without moving force, in a pipe forming a closed circuit. But according to Faraday, who found that an electric circuit consisting of a wire doubled on itself, with the two parts close together, gives no sensible spark when suddenly opened, compared to that given by an equal length of wire bent into a coil, it appears that the effects of ordinary *inertia* either do not exist for electricity in motion, or are but small compared with those which, in a suitable arrangement, are produced by the "induction of the current upon itself." In the present state of science it is only these effects that can be determined by a mathematical investigation; but the effects of electrical inertia, should it be found to exist, will be taken into account by adding a term of determinate form to the fully determined result of the present investigation which expresses the mechanical value of a current in a linear conductor, as far as it depends on the induction of the current on itself.

The general principle of the investigation is this;—if two conductors, with a current sustained in each by a constant electromotive force, be slowly moved towards one another, and there be a certain *gain of work* on the whole, by electro-dynamic force, operating during the motion, there will be twice as much as this of work spent by the electro-motive forces (for instance, twice the equivalent of chemical action in the batteries, should the electromotive forces be chemical,) over and above that which they would have had to spend in the same time if the conductors had been at rest merely to keep up the currents: because the electro-dynamic induction produced by the motion will augment the currents; while on the other hand, if the motion be such as to require the *expenditure* of work against electro-dynamic forces to produce it, there will be twice as much work saved off the action of the electromotive forces by the currents being diminished during the motion. Hence the aggregate mechanical value of the currents in the two conductors, when brought to rest will be increased in the one case by an amount equal to the work done by mutual electro-dynamic forces in the motion, and will be diminished by the corresponding amount in the other case. The same considerations are applicable to relative motions of two portions of the same linear conductor (supposed perfectly flexible). Hence it is concluded that the mechanical value of a current of given strength in a linear conductor of any form, is determined by calculating the amount of work against electro-dynamic forces, required to double it upon itself, while a current of constant strength is sustained in it. The mathematical problem thus presented leads to an expression for the required mechanical value consisting of two factors, of which one is determined according to the form and dimensions of the line of the conductor in any case, irrespectively of its section, and the other is the square of the strength of the current. If it be found necessary to take inertia into account, it will be necessary to add to this expression a term consisting of two factors, of which one is directly proportional to the length of the conductor, and inversely proportional to the area of its section, and the other is the square of the strength of the current, to obtain the complete mechanical value of the electrical motion.

[The mechanical value of a current in a closed circuit, determined on these principles, may be calculated by means of the following simple formula not hitherto published :—

$$\frac{1}{8\pi} \iiint R^2 dx\,dy\,dz$$

where R denotes the resultant electro-magnetic force at any point (x, y, z).

This expression may be useful in the dynamical theory of electro-magnetic engines, and of magneto-electric machines. As an example let the circuit consist of a length l of wire, wrapped in a helix approximating to a succession of circles on a cylinder of length a. The mechanical value of a current of strength γ flowing through it is approximately

$$\tfrac{1}{2}\frac{l^2}{a}\gamma^2,$$

whatever be the diameter of the cylinder provided it be small in proportion to its length. For instance let 100 feet of wire be wrapped on a cylinder an inch or two in diameter, and one foot long. The mechanical value of a current of unit strength (or a current which would decompose about $\frac{1}{50}$ of a grain of water per second) flowing through it, is 50,000 dynamical units.

We must divide this by 32·2 to reduce to "foot-grains," and we find therefore that the "vis viva," or "actual energy," or "mechanical value," of the current in that case (which is just such as a very ordinary experimental illustration might be) is 1550 times as much as is produced by gravity upon a grain of water descending one foot. If we divide again this number by 1390, to reduce to thermal units, we find 1·12; and conclude that about a grain and a tenth of water would be raised in temperature one degree Cent. by the spark on breaking circuit or that 1600 such sparks would give the same elevation of temperature to a quarter of a pound of water. It is well known that these effects are immensely increased by inserting soft iron into the cylindrical space surrounded by the coil. The same theory gives complete indications for finding how much, when experimental determinations of the law and amount of magnetization are complete enough to supply the requisite numerical data.

The following definition is of importance in applications of the theory of the energy of electricity in motion :—

The electro-dynamic capacity of a linear conductor of any form is double the mechanical value of a current of unit strength circulating in it.]

ART. LXII. ON TRANSIENT ELECTRIC CURRENTS.

Abstract.

[From *Glasgow Phil. Soc. Proc.* Jan. 1853.]

Followed by the full Mathematical Paper.

[From *Phil. Mag.* June, 1853.]

THE object of this communication is to determine the motion of electricity at any instant after an electrified conductor of given capacity, is put in connexion with the earth by means of a wire or other linear conductor of given form and given resisting power. The solution is founded on the *equation of energy* (corresponding precisely to "the equation of vis-viva" in ordinary dynamics) which is sufficient for the solution of every mechanical problem, involving only one variable element to be determined in terms of the time. That there is only one such variable in the present case follows from two assumptions which are made regarding the data, namely,

(1) That the electrical capacity of the first mentioned, or principal conductor, as it will be called, is so great in comparison with that of the second or discharger, as to allow no appreciable proportion of its original charge to be contained in the discharger at any instant of the discharge, which will imply that the strength of the current at each instant must be sensibly uniform through the whole length of the discharger.

(2) That there is no sensible resistance to conduction over the principal conductor, so that the amount of charge left in it at any instant of the discharge will be distributed on it in sensibly the same way as if there was complete electrical equilibrium.

The theorems demonstrated in the first and third parts of the previous communication [Art. LXI. above] give expressions for the mechanical values of the charge left in the principal conductor, and the electrical motion in the discharger, at any instant, in terms of the amount of that charge, and the rate at which it is diminishing. The sum of these two quantities, constitutes the whole electro-statical and electro-dynamical energy in the apparatus, and the diminution which it experiences in any time, must be mechanically compensated by heat generated in the same time. We have thus an equation between the diminution of the electrical energy in any infinitely small time, and the expression according to Joule's law for the heat generated in the same time in the discharger multiplied by the mechanical equivalent of the thermal unit. The equation so obtained is in the form of a well-known differential equation, of which the integral gives the quantity of electricity left at any instant in the principal conductor, and consequently expresses the complete solution of the problem. Precisely the same equation and solution are applicable to the circumstances of a pendulum, drawn through a small angle from the vertical, and let go in a viscous fluid, which exercises a resistance simply proportional to the velocity of the body moving through it.

The interpretation of the solution indicates two kinds of discharge, presenting very remarkable distinguishing characteristics; a continued discharge, and an oscillatory discharge; one or other of which will take place in any particular case. In the continued discharge the quantity of electricity on the principal conductor diminishes continuously, and the discharging current first increases to a maximum, and then diminishes continuously until after an infinite time equilibrium is established. In the oscillatory discharge, the principal conductor first loses its charge, becomes charged with a less amount of the contrary kind of electricity, becomes again discharged, and again charged with a still smaller amount of electricity, but of the same kind as the initial charge, and so on for an infinite number of times, until equilibrium is established; the strength of the current and its direction, in the discharger, has corresponding variations; and the instants when the charge of either kind of electricity on the principal conductor is at the greatest, being also those where the current in the discharger is on the turn, follow one another at equal intervals of time. The continued or the oscillatory discharge takes place in any particular case, according to the electrical capacity of the principal conductor, the electro-dynamical capacity of the discharger, and the resistance of the discharger to the conduction of electricity. Thus, if the discharger be given, it will effect a continued or an oscillatory discharge, according as the capacity of the principal conductor exceeds or falls short of a certain limit. If the principal conductor, and the length and substance of the discharger, be given, the discharge will be continued or oscillatory according as the electro-dynamic capacity of the latter, depending as it does on the form into which it is bent, falls short of, or exceeds a certain limit. Lastly, if the principal conductor, and the length and form of the discharger be given, the discharge will be continued or oscillatory, according as the resistance of the discharger to conduction exceeds or falls short of a certain limit.

It ought to be remarked that, although the electrical equilibrium is not rigorously attained, whatever kind of discharge it may be, in any finite time; yet practically, in all ordinary experimental cases the discharge is finished almost instantaneously as regards all appreciable effects; and the great obstacle in the way of experimenting at all on the subject arises from the difficulty of arranging the circumstances, so that the periods of time indicated by the theory for the succession of various phenomena, (as for instance, the alternations of the charges of the contrary electricity on the principal conductor), may not be inappreciably small.

It is not improbable that double, triple, and quadruple flashes of lightning which are frequently seen on the continent of Europe, and sometimes, though not so frequently, in this country, lasting generally long enough to allow an observer, after his attention is drawn by the first light of the flash, to turn his head round and see distinctly the course of the lightning in the sky, result from the discharge possessing the oscillatory character. A corresponding phenomenon might probably be produced artificially on a small scale, by discharging a Leyden phial or other conductor across a very small space of air, and through a linear conductor of large electro-dynamic capacity and small resistance. Should it be impossible, on account of the too great rapidity of the successive flashes, for the unaided eye to distinguish them, Wheatstone's method of a revolving mirror might be employed, and might show the spark as several points or short lines of light separated by dark intervals, instead of a single point of light, or of an unbroken line of light, as it would be if the discharge were instantaneous, or were continuous and of appreciable duration.

The experiments by Riess and others on the magnetization of fine steel needles by the discharge of electrified conductors, illustrate in a very remarkable manner the oscillatory character of the discharge in certain circumstances; not only when, as in the case with which we are at present occupied, the whole mechanical effect of the discharge is produced within a single linear conductor, but when induced currents in secondary conductors generate a portion of the final thermal equivalent.

The decomposition of water by electricity from an ordinary electrical machine, in which, as has been shown by Faraday, more than the electro-chemical equivalent of the whole electricity that passes appears in oxygen and hydrogen rising mixed from each pole, is probably due to electrical oscillations in the discharger consequent on the successive sparks*. Thus, if the general law of electro-chemical decomposition be applicable to currents of such very short duration as that of each alternation in such an oscillatory discharge as may take place in these circumstances, there will be decomposed altogether as much water as is electro-chemically equivalent to the sum of the quantities of electricity that pass in all the successive currents in the two directions, while the quantities of oxygen and hydrogen which appear at the two electrodes will differ by the quantities arising from the decomposition of a quantity of water electro-chemically equivalent to only the quantity of electricity initially contained by the principal conductor. The mathematical results of the present communication lead to an expression for the quantity of water decomposed by an oscillatory discharge in any case to which they are applicable, and show that the greater the electro-dynamic capacity of the charger, the less its resistance, and the less electro-statical capacity of the principal conductor, the greater will be the quantity of water decomposed. Probably the best arrangement in practice would be one in which merely a small ball or knob is substituted for a principal conductor fulfilling the conditions prescribed above; but those conditions not being fulfilled, the circumstances would not be exactly expressed by the formulæ of the present communication; the resistance would be much diminished, and consequently the whole quantity of water decomposed much increased, by substituting large platinum electrodes for the mere points used by Wollaston;

---

* This conjecture was first, so far as I am aware, given by Helmholtz, the existence of electrical oscillations in many cases of discharge having been indicated by him as a probable conclusion from the experiments of Riess, alluded to in the text.

but then the oxygen and hydrogen separated during the first
direct current would adhere to the platinum plates and would be
in part neutralized by combination with the hydrogen and oxygen
brought to the same plates respectively by the succeeding reverse
current; and so on through all the alternations of the discharge.
In fact, if the electrodes be too large, all the equivalent quantities
of the two gases brought successively to the same electrode will
recombine, and at the end of the discharge there will be only
oxygen at the one electrode and only hydrogen at the other, in
quantities electro-chemically equivalent to the initial charge of
the principal conductor.   Hence we see the necessity of using very
minute electrodes, and of making a considerable quantity of electricity pass in each discharge, so that each successive alternation
of the current may actually liberate from the electrodes some of
the gases which it draws from the water.   [The above results may
be applied to determine the laws, according to which a current
varies at the commencement and end of any period, during which
a constant electro-motive force, such as that of a galvanic battery,
acts in a conductor of given electro-dynamic capacity and resistance, and to show how the relation between the electro-statical
and electro-dynamical units of electrical quantity and electro-
motive force may be experimentally determined.

Induction Coils.—In the single coil apparatus, the recipient arc
completes a circuit with the induction coil.   A battery sends its
current divided through the "coil" and the "recipient arc," in
quantities inversely proportional to the resistances of these two
channels.   At the instant when the battery circuit is broken, the
current in the coil, with its great *momentum*, overbalances the
comparatively small *momentum* of the current previously excited
in the recipient arc by the direct action of the battery and gives
rise to the induced current.

In the double apparatus—with primary and secondary coils—
the impulse induced in the secondary coil is equal in absolute
measure to the strength of the current stopped in the primary,
multiplied by a coefficient of induction.   The whole quantity of
the current which it produces is equal to its own measure divided
by the resistance in the whole circuit of secondary coil and
recipient arc.   The more sudden the stoppage of the primary
current, the more intense and the shorter in duration is the shock
in the secondary, and hence also the improved effect produced

by Fizeau's addition of the condenser *.] Probably the most effective arrangement would be one in which a Leyden phial or other body of considerable capacity is put in connection with the machine and discharged in sparks through a powerful discharger, not only of great electro-dynamic capacity, and of as little resistance as possible except where the metallic communication is broken in the electrolytic vessel, but of considerable electro-statical capacity, so that all, or as great a portion as possible, of the oscillating electricity may remain in it and not give rise to successive sparks across the space of air separating the discharger from the source of the electricity.

The paper is concluded with applications of the results to determine the laws, according to which a current varies at the commencement and end of any period, during which a constant electro-motive force, such as that of a galvanic battery, acts in a conductor of given electro-dynamic capacity and resistance, and to show how the relation between the electro-statical and electro-dynamic units of electrical quantity and electro-motive force may be experimentally determined.

* [Note of June 12, 1882. This addition is made from the article on Dynamical Relations of Magnetism, contributed by the Author to Nichol's Cyclopedia, 2nd Edition, 1860.]

Art. LXII. (COMPLETED). On Transient Electric Currents.

[From *Phil. Mag.* June, 1853.]

THE object of this communication is to determine the motion
of electricity at any instant after an electrified conductor, of given
capacity, charged initially with a given quantity of electricity, is
put in connexion with the earth by means of a wire or other linear
conductor of given form and resisting power. This linear conductor, which, to distinguish it from the other or principal conductor, will be called the discharger, is supposed to be of such
small electrical capacity that the whole quantity of free electricity
in it at any instant during the discharge is excessively small
compared with the original charge of the principal conductor.
Now any difference that can exist in the strength of the current
at any instant in different parts of the discharger must produce
accumulations of free electricity in the discharger itself, and therefore must be very small* compared with the actual strength of the
current depending on the discharge of the principal conductor.
The strength of the current throughout the discharger will therefore be considered as the same at each instant, and, being
measured by the quantity of electricity discharged per second, will
be denoted by γ. Again, the conducting property and extent
of surface of the principal conductor, and the resistance of the
discharger, will be considered as so related that the potential
throughout the principal conductor is uniform at each instant.
Hence if q denote the quantity of electricity which the principal
conductor possesses at any time t, we have

$$\gamma = -\frac{dq}{dt} \quad\dots\dots\dots\dots\dots\dots\dots(1).$$

Now, if C denote the electrical capacity of the principal conductor, that is, the quantity of electricity which it takes to make
the potential within it unity, the mechanical value or the "potential
energy" of the distribution of a quantity q upon it is $\frac{1}{2}\frac{q^2}{C}$. As
this diminishes from the commencement of the discharge, and

* [Contrast this with the discharge of a submarine telegraph cable. W. T.
Aug. 11, 1882.]

varies during the whole period of the discharge, corresponding mechanical effects must be produced in the discharger according to the general law of *"vis viva,"* or of the preservation of mechanical energy. The mechanical effects in the discharger are of two kinds,—first, the excitation or alteration of electrical motion; secondly, the generation of heat. To estimate the first of these, it is necessary to know the mechanical value or the "actual energy" [kinetic energy] of an electrical current of given strength established and left without electromotive force in the discharger. In investigations which I have made towards a mechanical theory of electro-magnetic induction [Art. LXI. above], I have found that the mechanical value of a current in a closed linear conductor is equal to the quantity of work that would have to be spent against the mutual electro-magnetic forces between its parts in bending it from its actual shape into any other shape, while a current of constant strength is sustained in it by an external electromotive force, together with the mechanical value of the current in the conductor thus altered. According to Faraday's experiments (*Experimental Researches*, § 1090, &c.), it appears that the actual energy of a current in a linear conductor doubled upon itself throughout its whole extent, is either nothing, or such as to produce no sensible spark when the circuit is suddenly opened at any point; that is, that what can be obviously interpreted as inertia of electricity either does not exist, or produces but insensible effects compared to those which have been attributed to the "induction of a current upon itself." According to these views, the actual energy of an electric current of given strength in a given closed linear conductor would be determined analytically by calculating the amount of work against mutual electro-magnetic actions required to double it upon itself throughout its whole extent; but it may be that a more complete knowledge of the circumstances will show a term depending on electrical inertia which must be added to the quantity determined in that way to give the entire mechanical value of the current. However this may be, and whether the linear conductor be open or closed, it is obvious that the actual energy of a current established in it and left without electromotive force must be proportional to the square of the strength of the current, and this is all that is required to be known for the present investigation. Let then $\frac{1}{2} A \gamma^2$ denote the actual energy of a current of strength, γ, in the linear conductor which serves for discharger in the arrangement which forms the subject of the present investigation, A being a constant which may be called the electrodynamic capacity [see end of Art. LXI. above] of the discharger. The work spent in exciting electrical motion during the time dt will be

$$d\left(\frac{1}{2}A\gamma^2\right).$$

Again, the work done in generating heat in the same time is, according to Joule's law,

$$k\gamma^2 dt,$$

if k denote the "galvanic resistance" of the discharger, or the mechanical equivalent of the heat generated in it, in the unit of time by a current of unit strength*. Now the loss of potential energy from the principal conductor, in the time dt, being

$$-d\left(\frac{1}{2}\frac{q^2}{C}\right),$$

is entirely spent in producing these effects; and therefore

$$-d\left(\frac{1}{2}\frac{q^2}{C}\right)=d\left(\frac{1}{2}A\gamma^2\right)+k\gamma^2 dt\dots\dots\dots\dots(2).$$

This equation and (1), with the conditions

$$q=Q,\text{ and }\gamma=0,\text{ when }t=0\dots\dots\dots\dots(3),$$

are sufficient for the determination of q and γ for any value of t, that is, for the complete solution of the problem.

By (1) we have

$$-d\left(\frac{1}{2}\frac{q^2}{C}\right)=\frac{q}{C}\gamma dt,$$

and (2) becomes

$$\frac{q}{C}\gamma dt=A\gamma d\gamma+k\gamma^2 dt,$$

from which we find

$$q=C\left(A\frac{d\gamma}{dt}+k\gamma\right)\dots\dots\dots\dots\dots\dots(4).$$

* See a paper entitled "Application of the Principle of Mechanical Effect to the Measurement of Electromotive Forces and Galvanic Resistances in absolute units," *Phil. Mag.*, Dec., 1851. [Art. LIV., above.]

Substituting for γ its value by (1), we obtain

$$\frac{d^2q}{dt^2} + \frac{k}{A}\frac{dq}{dt} + \frac{1}{CA}q = 0 \quad \ldots\ldots\ldots\ldots\ldots(5).$$

The general solution of this equation is

$$q = K\epsilon^{\rho t} + K'\epsilon^{\rho' t},$$

where ρ and ρ' are the roots of the equation

$$x^2 + \frac{k}{A}x + \frac{1}{CA} = 0.$$

Using equations (3) and (1) to determine the arbitrary constants K and K', and to derive an expression for γ, we obtain a complete solution of the problem which is expressed most conveniently by one or the other of the following sets of formulæ, according as ρ and ρ' are real or imaginary:—

$$q = \frac{Q}{2\alpha A}\epsilon^{-\frac{k}{2A}t}\left\{\left(\alpha A + \frac{k}{2}\right)\epsilon^{\alpha t} + \left(\alpha A - \frac{k}{2}\right)\epsilon^{-\alpha t}\right\}$$

$$\gamma = \frac{Q}{2\alpha AC}\epsilon^{-\frac{k}{2A}t}\left\{\epsilon^{\alpha t} - \epsilon^{-\alpha t}\right\}$$

$$\left.\vphantom{\begin{array}{c}a\\a\\a\\a\end{array}}\right\} \ldots\ldots(6),$$

where

$$\alpha = \left(\frac{k^2}{4A^2} - \frac{1}{CA}\right)^{\frac{1}{2}},$$

$$q = \frac{Q}{\alpha' A}\epsilon^{-\frac{k}{2A}t}\left\{\alpha' A \cos(\alpha' t) + \frac{k}{2}\sin(\alpha' t)\right\}$$

$$\gamma = \frac{Q}{\alpha' AC}\epsilon^{-\frac{k}{2A}t}.\sin(\alpha' t)$$

$$\left.\vphantom{\begin{array}{c}a\\a\\a\\a\end{array}}\right\} \ldots\ldots\ldots\ldots(7).$$

where

$$\alpha' = \left(\frac{1}{CA} - \frac{k^2}{4A^2}\right)^{\frac{1}{2}}.$$

Among numerous other beautiful applications of his "electrodynamometer," Weber has shown a method of determining what he calls the "duration"* of a transient electric current. In accordance with the terms he uses, the duration, and the *mean*

---

* "Bestimmung der Dauer momentaner Ströme mit dem Dynamometer nebst Anwendung auf physiologische Versuche," § 13 of Weber's *Electro-dynamische Maasbestimmungen*, Leipsic, 1846.

strength of a transient current may be defined respectively as the duration and the strength that a uniform current must have to produce the same effects on the electro-dynamometer and on an ordinary galvanometer; so that if T and Γ denote the duration and the mean strength of a current, of which the actual strength at any instant is γ, we have

$$\left. \begin{aligned} T &= \frac{\left\{ \int_0^\infty \gamma\, dt \right\}^2}{\int_0^\infty \gamma^2\, dt} \\[2em] \Gamma &= \frac{\int_0^\infty \gamma^2\, dt}{\int_0^\infty \gamma\, dt} \end{aligned} \right\} \quad \dots\dots\dots\dots\dots\dots(8);$$

since the electro-dynamometer indicates the value of $\int_0^\infty \gamma^2 dt$, and the ordinary galvanometer that of $\int_0^\infty \gamma\, dt$. If for γ we use the expression in either (6) or (7), we find

$$\int_0^\infty \gamma^2 dt = \frac{1}{2}\frac{Q^2}{kC} \dots\dots\dots\dots\dots\dots(9),$$

as might have been foreseen, independently of the complete solution, by considering that, as the heat generated in the discharger is the sole final effect produced by the discharge, the mechanical value of the whole heat generated, or $\int_0^\infty k\gamma^2 dt$, must be equal $\frac{1}{2}\frac{Q^2}{C}$, the mechanical value of the primitive charge, and that k has been assumed to have a constant value during the discharge. Again, we derive from (1) and (3),

$$\int_0^\infty \gamma\, dt = Q \dots\dots\dots\dots\dots\dots(10).$$

Hence in the present case the expressions for the duration and mean strength of the current becomes

$$\left. \begin{aligned} T &= 2kC \\[1em] \Gamma &= \frac{Q}{2kC} \end{aligned} \right\} \quad \dots\dots\dots\dots\dots\dots(11).$$

We conclude that the "duration" of the discharge is proportional to the capacity of the principal conductor, and to the resistance of the discharger; and that it is independent of the quantity of electricity in the primitive charge, and of the electrodynamical capacity (denoted above by A) of the discharger. The only doubtful assumption involved in the preceding investigation is that of the constancy of k during the discharge. Joule's experiments show that the value of k remains unchanged for the same metallic conductor kept at the same temperature, whatever be the strength of the current passing through it; but that it would be increased by any elevation of temperature, whether produced by the current itself or by any other source of heat, since an elevation of temperature always increases the galvanic resistance of a metal. When large quantities of electricity are discharged, or when the discharger is a very fine wire, great augmentations and diminutions may therefore take place in the value of k, and therefore the solution obtained above is not applicable to such cases. If, however, k denote the mean resistance of the discharger during the discharge, that is, a quantity such that

$$ k \int_0^\infty \gamma^2 dt = \int_0^\infty \kappa \gamma^2 dt \dots\dots\dots\dots\dots\dots(12), $$

where κ denotes the actual resistance at any instant of the discharge, the last equations (11) become merely the expressions for the elements determined by Weber from observations by means of the two instruments, and they are therefore applicable to all cases.

In the experiments described by Weber, the discharger consisted of a wet cord of various lengths, and all the wire of the electrodynamometer and the ordinary galvanometer. The "durations" of the discharge in different cases were found to be nearly proportional to the length of the wet cord, and equal to ·0851 sec., or about $\frac{1}{12}$ of a second, when the length of the cord was 2 metres. As the principal resistance must undoubtedly have been in the wet cord, we may infer from equations (11) and (12) that the mean resistances in all the different discharges must have been nearly proportional to its lengths. In some of the experiments the length was only a $\frac{1}{4}$ of a metre, and the value of T was about 0095 of a second. Hence the current in the string must have been about eight times as intense as when the duration was $\frac{1}{12}$, since the quantities of electricity discharged were nearly equal in the different cases. We conclude that the intensity of the current cannot have materially affected the resisting power of the cord; probably not nearly so much as inevitable differences arising from accidental circumstances in the different experiments. Hence, although nothing is known with certainty regarding the non-electrolytic resistance of liquid conductors in general, it is probable that the whole resistance of the discharger in Weber's experiments must have been nearly independent of the strength of the current at each instant; and we may therefore consider the general solution expressed above by (6) or (7) as at least approximately applicable to these cases.

The two forms (6) and (7) of the solution of the general problem indicate two kinds of discharge presenting very remarkable distinguishing characteristics. Thus in all cases in which $\dfrac{k^2}{4A^2}$ exceeds $\dfrac{1}{CA}$, the exponentials in (6) are all real; and the solution expressed by these equations shows that the quantity of electricity on the principal conductor diminishes continuously, and that the discharging current commences and gradually increases in strength up to a time given by the equation

$$-\left(\frac{k}{2A}-\alpha\right)\epsilon^{-\left(\frac{k}{2A}-\alpha\right)t}+\left(\frac{k}{2A}+\alpha\right)\epsilon^{-\left(\frac{k}{2A}+\alpha\right)t}=0,$$

or

$$t=\frac{1}{2\left(\dfrac{k^2}{4A^2}-\dfrac{1}{CA}\right)^{\frac{1}{2}}}\log\frac{\dfrac{k}{2A}+\left(\dfrac{k^2}{4A^2}-\dfrac{1}{CA}\right)^{\frac{1}{2}}}{\dfrac{k}{2A}-\left(\dfrac{k^2}{4A^2}-\dfrac{1}{CA}\right)^{\frac{1}{2}}}\dots\dots\dots(13);$$

after which it diminishes gradually, and, as well as the quantity of electricity on the principal conductor, becomes nothing when $t=\infty$. On the other hand, when $\dfrac{1}{CA}$ exceeds $\dfrac{k^2}{4A^2}$, the exponentials and trigonometrical functions in (7) are all real; and the solution expressed by these equations shows that the principal conductor loses its charge, becomes charged with a less quantity of the contrary kind of electricity, becomes again discharged, and after that charged with a still less quantity of the same kind of electricity as at first, and so on for an infinite number of times before equilibrium is established. The times at which the charge of either kind of electricity on the principal conductor is a maximum, being those at which γ vanishes, are the roots of the equation $\sin(\alpha't) = 0$, and therefore follow successively from the commencement at equal intervals $\frac{\pi}{\alpha'}$. The quantities constituting the successive maximum charges are

$$Q, \quad -Qe^{-\frac{k\pi}{2A\alpha'}}, \quad Qe^{-\frac{2k\pi}{2A\alpha'}}, \quad -Qe^{-\frac{3k\pi}{2A\alpha'}}, \&c.\ldots\ldots(14);$$

each being less in absolute magnitude than that which precedes it in the ratio of $1 : e^{\frac{k\pi}{2A\alpha'}}$, and of the opposite kind. The strength of current will be a maximum in either direction when $\frac{d\gamma}{dt} = 0$, or when

$$\frac{k}{2A} \sin(\alpha't) = \alpha' \cos(\alpha't);$$

and therefore if T_1, T_2, &c. denote the successive times when this is the case, measured from the commencement of the discharge, and γ_1, γ_2, &c. the corresponding maximum values of the strength of the current, and if θ denote the acute angle satisfying the equation $\tan\theta = \frac{2A\alpha'}{k}$, we have

$$T_1 = \frac{\theta}{\left(\frac{1}{CA} - \frac{k^2}{4A^2}\right)^{\frac{1}{2}}}, \quad T_2 = \frac{\theta + \pi}{\left(\frac{1}{CA} - \frac{k^2}{4A^2}\right)^{\frac{1}{2}}}, \quad T_3 = \frac{\theta + 2\pi}{\left(\frac{1}{CA} - \frac{k^2}{4A^2}\right)^{\frac{1}{2}}}, \&c.,$$

$$\gamma_1 = \frac{Q}{A\left(\frac{1}{CA} - \frac{k^2}{4A^2}\right)^{\frac{1}{2}}} e^{-\frac{k}{2A}T_1}, \quad \gamma_i = \left(\frac{-1}{e^{\frac{k\pi}{2A\alpha'}}}\right)^i \gamma_1.$$

It is probable that many remarkable phenomena which have been observed in connexion with electrical discharges are due to the oscillatory character which we have thus found to be possessed when the condition

$$\frac{1}{CA} > \frac{k^2}{4A^2} \text{ or } C < \frac{4A}{k^2} \ldots\ldots\ldots\ldots(15)$$

is fulfilled.   Thus if the interval of time

$$\frac{\pi}{\left(\dfrac{1}{CA} - \dfrac{k^2}{4A^2}\right)^{\frac{1}{2}}},$$

at which the successive instants when the strength of the current is a maximum follow one another, be sufficiently great, and if the evolution of heat in any part of the circuit by the current during several of its alternations in directions be sufficiently intense to produce visible light, a succession of flashes diminishing in intensity and following one another rapidly at equal intervals will be seen.   It appears to me not improbable that double, triple, and quadruple flashes of lightning which I have frequently seen on the continent of Europe, and sometimes, though not so frequently in this country, lasting generally long enough to allow an observer, after his attention is drawn by the first light of the flash, to turn his head round and see distinctly the course of the lightning in the sky, result from the discharge possessing this oscillatory character.   A corresponding phenomenon might probably be produced artificially on a small scale by discharging a Leyden phial or other conductor across a very small space of air, and through a linear conductor of large electro-dynamic capacity and small resistance.   Should it be impossible on account of the too great rapidity of the successive flashes for the unaided eye to distinguish them, Wheatstone's method of a revolving mirror might be employed, and might show the spark as several points or short lines of light separated by dark intervals, instead of a single point of light, or of an unbroken line of light, as it would be if the discharge were instantaneous, or were continuous and of appreciable duration.

The experiments of Riess, Feddersen and others on the magnetization of fine steel needles by the discharge of electrified conductors, illustrate in a very remarkable manner the oscillatory character of the discharge in certain circumstances; not only when, as in the case with which we are at present occupied, the whole mechanical effect of the discharge is produced within a single linear conductor, but when induced currents in secondary conductors generate a portion of the final thermal equivalent.

The decomposition of water by electricity from an ordinary electrical machine, in which, as has been shown by Faraday, more than the electro-chemical equivalent of the whole electricity that passes appears in oxygen and hydrogen rising mixed from each pole, is probably due to electrical oscillations in the discharger consequent on the successive sparks*. Thus, if the general law of electro-chemical decomposition be applicable to currents of such very short duration as that of each alternation in such an oscillatory discharge as may take place in these circumstances, there will be decomposed altogether as much water as is electrochemically equivalent to the sum of the quantities of electricity that pass in all the successive currents in the two directions, while the quantities of oxygen and hydrogen which appear at the two electrodes will differ by the quantities arising from the decomposition of a quantity of water electro-chemically equivalent to only the quantity of electricity initially contained by the principal conductor. The formulæ investigated above will be applicable to this case if the end of the discharging train next the machine be placed in metallic communication with an insulated conductor, satisfying the conditions laid down with reference to the "principal conductor" at the commencement of this paper, and if this conductor be successively electrified by sparks from the machine. The whole quantity of water decomposed will therefore be the electro-chemical equivalent of the sum of the absolute values of the quantities of electricity flowing out of and into the principal conductor during the successive alternations of the current, that is, according to the preceding formulæ, the electro-chemical equivalent of the quantity,

$$Q\left(1 + 2\epsilon^{-\frac{k\pi}{2Aa'}} + 2\epsilon^{-\frac{2k\pi}{2Aa'}} + \&c.\right), \quad \text{or} \quad Q\frac{1 + \epsilon^{-\frac{k\pi}{2Aa'}}}{1 - \epsilon^{-\frac{k\pi}{2Aa'}}},$$

* This explanation occurred to me about a year and a half ago, in consequence of the conclusions regarding the oscillatory nature of the discharge in certain circumstances drawn from the mathematical investigation. I afterwards found that it had been suggested, as a conjecture by Helmholtz, in his *Erhaltung der Kraft* (Berlin, 1847), in the following terms:—

" ** It is easy to explain this law if we assume that the discharge of a battery is not a simple motion of the electricity in one direction, but a backward and forward motion between the coatings, in oscillations which become continually smaller until the entire *vis viva* is destroyed by the sum of the resistances. The notion that the current of discharge consists of alternately opposed currents is favoured by the alternately opposed magnetic actions of the same; and secondly, by the phenomena observed by Wollaston while attempting to decompose water by of electricity. This quantity will be the greater the more nearly $\epsilon^{-\frac{k\pi}{2A\alpha'}}$ approaches to unity, that is the greater is $\dfrac{2A\alpha'}{k}$ or $\left(\dfrac{4A}{kC}-1\right)^{\frac{1}{2}}$, or the greater is $\dfrac{4A}{kC}$. Hence the greater the electro-dynamic capacity of the discharger, the less its resistance, and the less the electro-statical capacity of the principal conductor, the greater will be the whole quantity of water decomposed. Probably the best arrangement in practice would be one in which merely a small ball or knob is substituted for a principal conductor fulfilling the conditions prescribed above; but those conditions not being fulfilled, the circumstances would not be exactly expressed by the formulæ of the present communication. The resistance would be much diminished, and consequently the whole quantity of water decomposed much increased, by substituting large platinum electrodes for the mere points used by Wollaston; but then the oxygen and hydrogen separated during the first direct current would adhere to the platinum plates, and would be in part neutralized by combination with the hydrogen and oxygen brought to the same plates respectively by the succeeding reverse current; and so on through all the alternations of the discharge. In fact, if the electrodes be too large, all the equivalent quantities of the two gases brought successively to the same electrode will recombine, and at the end of the discharge there will be only oxygen at the one electrode and only hydrogen at the other, in quantities electrochemically equivalent to the initial charge of the principal conductor. Hence we see the necessity of using very minute electrodes, and of making a considerable quantity of electricity pass in each discharge, so that each successive alternation of the current may actually liberate from the electrodes some of the gases which it draws from the water. Probably the most effective arrangement would be one in which a Leyden phial or other body of considerable capacity is put in connexion with the machine and discharged in sparks through a powerful discharger, not only of great electro-dynamic capacity and of as little resistance as possible except where the metallic communication is broken in the electrolytic vessel, but of great electro-statical capacity also, so that

electric shocks, that both descriptions of gases are exhibited at both electrodes."
[Quoted from the translation in Taylor's *New Scientific Memoirs*, Part II.]

all, or as great a portion as possible, of the oscillating electricity may remain in it and not give rise to successive sparks across the space of air separating the discharger from the source of the electricity.

The initial effect of a uniform electromotive force in establishing a current in a linear conductor may be determined by giving C and Q infinite values in the preceding formulæ, and $\frac{Q}{C}$ a finite value V, which will amount to supposing the potential at one end of the discharger to be kept constantly at the value V, while the potential at the other end is kept at zero. The formulæ suitable to this case, which is obviously a case of non-oscillatory discharge, are (6); and from them we deduce

$$\gamma = \frac{V}{k}(1 - \epsilon^{-\frac{k}{A}t}) \quad\quad\quad\quad\text{.....................(16),}$$

which agrees with conclusions arrived at by Helmholtz and others.

This result shows how, when a linear conductor, initially in a state of electrical equilibrium, becomes subjected to a constant electromotive force V between its extremities, a current commences in it and rises gradually in strength towards the limit $\frac{V}{k}$. This limit cannot be perfectly reached in any finite time, although in reality only a very minute time elapses from the commencement in ordinary cases, until the current acquires so nearly the full strength $\frac{V}{k}$ that no further augmentation is perceptible*.

The equations (6), expressing generally a continuous discharge, assume the following forms when A is infinitely small,

$$\left.\begin{aligned}\gamma &= \frac{Q}{Ck}\epsilon^{-\frac{t}{kC}}\\[2mm]q &= Q\epsilon^{-\frac{t}{kC}}\end{aligned}\right\} \quad\quad\quad\text{.....................(17),}$$

which show how, when anything like electrical inertia is insensible, the current commences instantly with its maximum strength, and then gradually sinks as the charge gradually and permanently leaves the principal conductor.

---

* See a paper by Helmholtz in Poggendorff's *Annalen*, 1852, which contains valuable researches, both theoretical and experimental, on this subject.

One of the results of the preceding investigations shows a very important application that may be made of Weber's experimental determination of the "duration" of a transient current, to enable us to determine the numerical relation between electro-statical and electro-magnetic units. For if σ denote the quantity of electricity in electro-statical measure which passes in the unit of time to constitute a current of unit strength in electro-magnetic measure, the strength of a current expressed above by γ will be $\frac{1}{\sigma}\gamma$, in terms of the electro-magnetic unit; and if K denote the resistance, in absolute electro-magnetic measure, of a linear conductor of which the resistance measured as above in terms of the electro-statical unit is k, we have

$$K\left(\frac{\gamma}{\sigma}\right)^2 = k\gamma^2,$$

which gives

$$k = \frac{K}{\sigma^2} \dots\dots\dots\dots\dots\dots(18).$$

Hence the first of equations (11) gives

$$\sigma = \left(\frac{2KC}{T}\right)^{\frac{1}{2}} \dots\dots\dots\dots\dots\dots(19).$$

Now Weber has not only determined T, in certain cases alluded to above, but has shown how K may be determined for any linear conductor. Again, the value of C for a Leyden phial is, according to Green and Faraday, expressed by the equation

$$C = I\frac{S}{4\pi\tau} \dots\dots\dots\dots\dots\dots\dots(20),$$

where S denotes the area of one side of the coated glass, τ the thickness of the glass, and I the specific inductive capacity of its substance* Thus, either by using a Leyden phial or some other conductor of which the electro-statical capacity can be found, and by determining the "duration" of a discharge from it through a linear conductor, of which the resistance in absolute electro-magnetic measure has been determined, we have everything that

* The value of I for flint-glass is, according to Faraday (*Experimental Researches*, Series XI.), greater than 1·76, for shell-lac about 2, for sulphur rather more than 2·2. See a paper "On the Elementary Laws of Statical Electricity," § 8. (*Cambridge and Dublin Mathematical Journal*, Vol. I., Nov. 1845.) [Art. xxi. above.]

is required for calculating the value of σ by means of equation (19). The determination of this quantity enables us to compare the electro-statical and electro-magnetic measures of electromotive force. For if V denote a constant difference of potentials kept up between the two extremities of a linear conductor, and if γ denote the strength of the uniform current that results, we have, according to the conclusions drawn above from (16), $\gamma = V/k$. But if F denote the electromotive force between the ends of the linear conductor in electro-magnetic measure, we have

$$\frac{1}{\sigma}\gamma = \frac{F}{K} = \frac{F}{\sigma^2 k},$$

and therefore $\qquad F = \sigma V$ ....(21).

Many different ways of determining the value of this important element, σ, besides that suggested above, might probably be put in practice. Perhaps the most accurate would be to take a multiple galvanic battery of constant and known electromotive force (consisting, for instance, of a hundred or more cells of Daniell's), and measure the force of attraction between two plane conductors parallel to one another at a very small measured distance asunder, with only air between them. If X be the force of attraction thus measured, a the distance between the conductors, S the area of each or of that portion of each which is directly opposed to the other, and V the difference of the electrical potential kept up between them by the battery, we should have $V = a\,(8\pi X/S)^{\frac{1}{2}}$; and therefore, if F be the electromotive force of the battery in electro-magnetic measure, $\sigma = F/(8\pi X a^2/S)^{\frac{1}{2}}$.

[Note added Aug. 11, 1882. The theory of oscillatory electric discharge investigated in the preceding paper, of 1853, soon received an interesting illustration in Feddersen's beautiful photographic investigation of the electric spark (Pogg. *Ann.* Vol. cviii. p. 497, 1859; and Vols. cxii. and cxiii., 1861). More recently (see Pogg. *Ann.* clii., 1874, p. 535) it has been made the subject of a very important and ably performed experimental research ("Einige experimentelle Untersuchungen über elektrische Schwingungen; von N. Schiller") in Helmholtz's laboratory in Berlin; and among other valuable results the specific inductive capacities of several solid insulating substances have been determined from the measurements of periods of the observed oscillations.]

[From *Phil. Mag.* Vol. v. Feb. 1853.]

ART. LXIII.   ON THE RESTORATION OF MECHANICAL ENERGY FROM AN UNEQUALLY HEATED SPACE.

WHEN heat is diffused by conduction from one part to another of an unequally heated body, the body is put into such a state that it is *impossible* to derive as much mechanical effect of a non-thermal kind* from it as could have been derived from the body in its given state†.   Hence, if a body be given in an envelope impermeable to heat, with its different parts at different temperatures, a *dissipation* of mechanical energy within it, going on until the temperatures of all its parts become the same, can only be avoided by immediately restoring a portion of its mechanical energy from the state of heat, and equalizing the temperature of all its parts, wholly by the operation of perfect thermo-dynamic engines.   Let T be the uniform temperature to which the body is brought by this process of restoration; let t be the temperature of the body in its given condition at any point xyz; and let $cdt.dx\,dy\,dz$ be the quantity of heat that an infinitely small element, $dx\,dy\,dz$, of the body at this point must part with to go down

---

* [Note added Jan. 14, 1853.]   Instead of "mechanical effect of a non-thermal kind," I should have said simply *potential energy*, had I at the time of writing this paper learned the use of the admirable terms "potential" and "actual" introduced by Mr Rankine in his paper "On the Transformation of Energy" (communicated to the Glasgow Philosophical Society at its last meeting, Jan. 5, and published in the present number of the *Philosophical Magazine*), to designate the two kinds of energy which I had previously distinguished by the inconvenient adjectives of "statical" and "dynamical." (See *Proc. Roy. Soc., Edinb.*, April 19, 1852; or *Phil. Mag.*, Oct., 1852, p. 304.)   [Art. LIX. above.]

† *Proceedings of the Royal Society of Edinburgh* for April 19, 1852 (p. 139), or p. 305 of the last Volume of this Journal [Art. LIX. above].   The formulæ given in that paper which have reference to the subject of the present communication, require corrections, which are indicated in "Errata" published in the last Number.

in temperature from t to $t - dt$. Let us suppose that this quantity of heat enters a perfect thermo-dynamic engine, of which the hot part is at the temperature t, and the cold part at the temperature T. The quantity of work that will be derived from it will be

$$dx\,dy\,dz\,J\,.\,cdt\,.\,(1 - \epsilon^{\frac{-1}{J}\int_T^t \mu dt})\dots\dots\dots\dots(a),$$

where μ denotes Carnot's function, and J the mechanical equivalent of the thermal unit (see "Dynamical Theory of Heat," § 25, *Trans. Royal Soc., Edinb.*, 1851, or *Phil. Mag.*, Aug. 1852) [Art. XLVIII. above], and the part of it rejected as waste into the refrigerator at the temperature T will be

$$dx\,dy\,dz\,.\,cdt\,.\,\epsilon^{\frac{-1}{J}\int_T^t \mu dt}\dots\dots\dots\dots\dots(b),$$

or

$$\frac{dx\,dy\,dz.cdt.\epsilon^{\frac{-1}{J}\int_0^t \mu dt}}{\epsilon^{\frac{-1}{J}\int_0^T \mu dt}}\dots\dots\dots\dots\dots(c).$$

Hence the whole work obtained by lowering, by means of a perfect thermo-dynamic engine, the temperture of the element $dx\,dy\,dz$ from t to T is

$$dx\,dy\,dz\int_T^t J.cdt.(1 - \epsilon^{\frac{-1}{J}\int_T^t \mu dt})\dots\dots\dots\dots(d);$$

and the part of the heat taken from this element, which is rejected into the refrigerator, is

$$\frac{dx\,dy\,dz\int_T^t cdt.\,\epsilon^{\frac{-1}{J}\int_0^t \mu dt}}{\epsilon^{-\frac{1}{J}\int_0^T \mu dt}},$$

or

$$\frac{dx\,dy\,dz\left\{\int_0^t cdt\,\epsilon^{\frac{-1}{J}\int_0^t \mu dt} - \int_0^T cdt\,\epsilon^{\frac{-1}{J}\int_0^t \mu dt}\right\}}{\epsilon^{-\frac{1}{J}\int_0^T \mu dt}}\dots\dots\dots\dots(e).$$

These expressions are of course equally applicable to parts of the body of which the temperatures are lower than T; and, without change of form, will express respectively (in virtue of the algebraic signs corresponding to any case in which t is $< T$), (d) the quantity of work obtained by *raising* the temperature of an element $dx\,dy\,dz$

from t to T by heat drawn from a source at the temperature T by means of a perfect thermo-dynamic engine, and (e) the *negative quantity* of heat *given to* the part of the engine at the temperature T in this process. Now if such a process be completed for every part of the body without either taking or giving heat by communication with other bodies, the whole quantity of heat given to the refrigerators at T by the engines worked from all the parts of the body for which $t > T$, must be equal to the whole quantity given by the sources at T to the engines working to raise the temperature of those parts of the body for which $t < T$; or, algebraically, the sum of the quantities of heat given by all the engines to the parts of them at T must be 0, that is, according to (e),

$$\frac{\iiint dx\,dy\,dz \left\{ \int_0^t cdt \cdot \epsilon^{\frac{-1}{J}\int_0^t \mu dt} - \int_0^T cdt \cdot \epsilon^{\frac{-1}{J}\int_0^t \mu dt} \right\}}{\epsilon^{\frac{-1}{J}\int_0^T \mu dt}} = 0 \ldots\ldots\ldots(1).$$

From this we deduce

$$\iiint dx\,dy\,dz \cdot \int_0^T cdt \cdot \epsilon^{-\frac{1}{J}\int_0^t \mu dt} = \iiint dx\,dy\,dz \int_0^t cdt \cdot \epsilon^{\frac{-1}{J}\int_0^t \mu dt} \ldots..(2),$$

by which the value of T may be found. When T is determined, the whole work obtained in the process may be calculated by adding all the terms given by the expression (d); and thus, if W denote its value, we find

$$W = J \iiint dx\,dy\,dz \int_T^t cdt \cdot (1 - \epsilon^{\frac{-1}{J}\int_T^t \mu dt}) \ldots\ldots\ldots\ldots(3).$$

The first member of the former of these equations may be put into a simpler form if we take Θ to denote the thermal capacity of the whole body at the temperature θ, a quantity which must generally be considered as a function of θ. Thus, if $c_\theta\,dx\,dy\,dz$ denote the thermal capacity of the portion $dx\,dy\,dz$ when at the temperature θ, we have

$$\iiint c_\theta dx\,dy\,dz = \Theta \ldots\ldots\ldots\ldots\ldots\ldots(4);$$

and the equation becomes

$$\int_0^T \Theta d\theta \cdot \epsilon^{\frac{-1}{J}\int_0^\theta \mu dt} = \iiint dx\,dy\,dz \int_0^t cdt \cdot \epsilon^{\frac{-1}{J}\int_0^t \mu dt} \ldots\ldots\ldots\ldots(5).$$

In order that there may be data enough for solving the problem, the nature of the given body must be specified so that the value of c for each point xyz is known, not only for the given initial temperature t of that point, but for all other temperatures through which we have to suppose it to vary in this investigation; and Θ is therefore to be regarded as a known function of θ. Hence the value of the integral $\displaystyle\int_0^\theta \Theta d\theta \, \epsilon^{\frac{-1}{J}\int \mu dt}$ may be regarded as a known function of θ, and may be tabulated for different values of this variable. The value of θ, for which this function is equal to the second member of equation (5), is the required quantity T.

The solution of the problem may be put under a very simple form, if the thermal capacity of each part of the body be independent of the temperature, in the following manner. Let the temperature of the body be measured according to an absolute scale, founded on the values of Carnot's function, and expressed by the following equation,

$$ t = \frac{J}{\mu} - \alpha \quad\dots\dots\dots\dots\dots\dots\dots\dots\dots(6), $$

where α is a constant which might have any value, but ought to have for its value the reciprocal of the coefficient of expansion of air, in order that the system of measuring temperature here adopted may agree approximately with that of the air-thermometer. Then we have

$$ \epsilon^{-\frac{1}{J}\int_0^t \mu dt} = \frac{\alpha}{t+\alpha}, \text{ and } \epsilon^{-\frac{1}{J}\int_0^\theta \mu dt} = \frac{\alpha}{\theta+\alpha}; $$

and since, according to the hypothesis that is now made, Θ is con-stant, the first member of equation (5) becomes simply

$$ \Theta\alpha \log \frac{T+\alpha}{\alpha}; $$

and we have explicitly, for the value of T, the equation

$$ \left. \begin{aligned} T &= \alpha \left\{ \epsilon^{\frac{1}{\Theta}\iiint \log \frac{t+\alpha}{\alpha} \cdot c\,dx\,dy\,dz} - 1 \right\} \\[2mm] T &= \epsilon^{\dfrac{\iiint \log(t+\alpha)\cdot c\,dx\,dy\,dz}{\iiint c\,dx\,dy\,dz}} - \alpha \end{aligned} \right\} \quad\dots\dots\dots\dots\dots(7); $$

or

and equation (3) takes the simpler form,

$$W = J \iiint c \, dx \, dy \, dz \left\{ t - T - (T + \alpha) \log \frac{t + \alpha}{T + \alpha} \right\} \dots\dots\dots (8).$$

If the given body be of infinite extent, and if the temperature of all parts of it have a uniform value, T, with the exception of a certain limited space of finite extent through which there is a given varied distribution of temperature, any of the equations (2), (5), or (7) leads to the result

$$T = \mathrm{T},$$

which might have been foreseen without analysis. In this case, then, equation (3) gives (in terms of a definite integral of which the elements vanish for all points at which t has the value T) the work that may be obtained by bringing the temperature of all the matter to T; and the same result is expressed more simply by (8) when the specific heat of all the matter in the space through which the initial distribution of temperature is non-uniform does not vary with the temperature.

END OF VOL. I.

CAMBRIDGE: PRINTED BY C. J. CLAY, M.A. AND SON, AT THE UNIVERSITY PRESS.

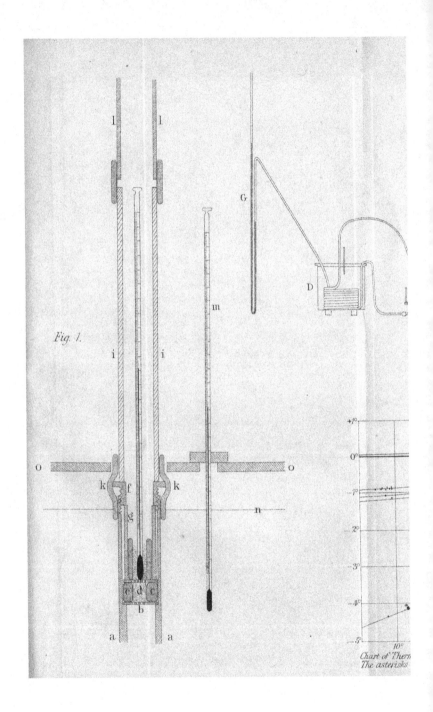

Fig. 1.

Chart of Therm
The asterisks

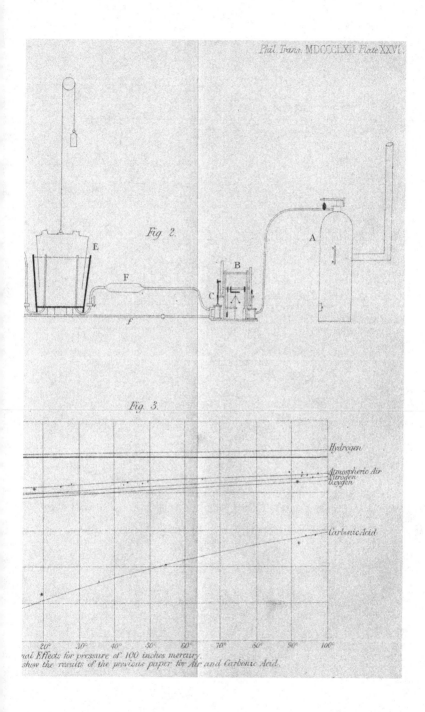

Fig. 2.

E

F

f

B

C

A

Fig. 3.

Hydrogen

Atmospheric Air
Nitrogen
Oxygen

Carbonic Acid

20°   30°   40°   50°   60°   70°   80°   90°   100°

nal Effects for pressure of 100 inches mercury.
show the results of the previous paper for Air and Carbonic Acid.

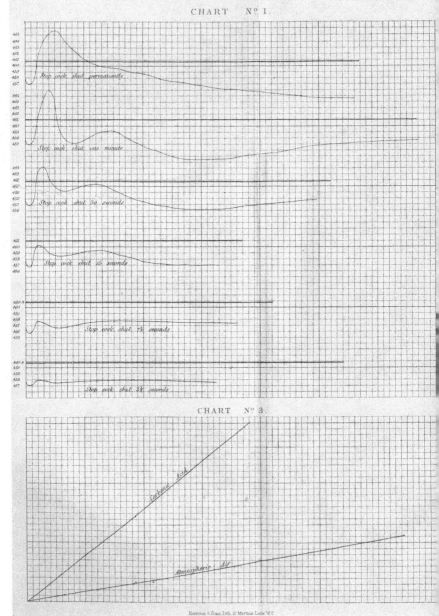

CHART Nº 1.

Stop cock shut permanently

Stop cock shut one minute

Stop cock shut 30 seconds

Stop cock shut 15 seconds

Stop cock shut 7½ seconds

Stop cock shut 3¾ seconds

CHART Nº 3.

Carbonic Acid

Atmospheric Air

Harrison & Sons, Lith. St Martins Lane W.C.

J.Basire

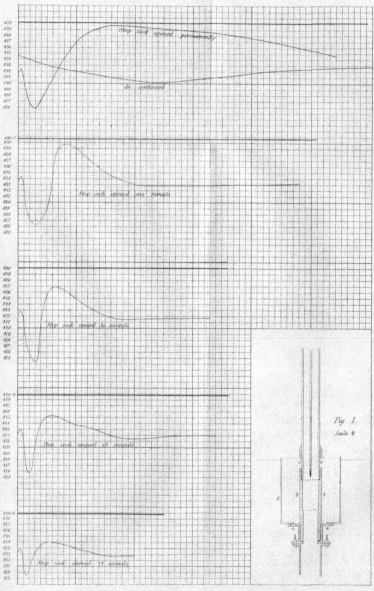

CHART Nº 2.

Stop cock opened permanently.

do. continued.

Stop cocks opened one minute.

Stop cock opened 30 seconds.

Stop cock opened 15 seconds.

Stop cock opened 7½ seconds.

Fig. 1.
Scale ½

J. Basire sc.

Printed in the United States
By Bookmasters